Processing and Preservation of Aquatic Products

Processing and Preservation of Aquatic Products

Editors

Tao Yin
Liu Shi

MDPI • Basel • Beijing • Wuhan • Barcelona • Belgrade • Manchester • Tokyo • Cluj • Tianjin

Editors
Tao Yin
Huazhong Agricultural University
China

Liu Shi
Hubei Academy of Agricultural Science
China

Editorial Office
MDPI
St. Alban-Anlage 66
4052 Basel, Switzerland

This is a reprint of articles from the Special Issue published online in the open access journal *Foods* (ISSN 2304-8158) (available at: https://www.mdpi.com/journal/foods/special_issues/processing_aquatic).

For citation purposes, cite each article independently as indicated on the article page online and as indicated below:

LastName, A.A.; LastName, B.B.; LastName, C.C. Article Title. *Journal Name* **Year**, *Volume Number*, Page Range.

ISBN 978-3-0365-7894-1 (Hbk)
ISBN 978-3-0365-7895-8 (PDF)

Contents

About the Editors

Tao Yin

Tao Yin, PH.D., is an associate professor at the College of Food Science and Technology of Huazhong Agricultural University. He graduated from the College of Food Science and Technology of Huazhong Agricultural University and studied at Oregon State University for one year. Currently, he serves as the Receiving Editor of the *Journal of Aquatic Food Product Technology* and the Guest Editor of *Foods*. He is a member of the Excellent Young and Middle-aged Scientific and Technological Innovation Team in Hubei Province's higher education institutions. He is mainly engaged in (1) freshwater fish transportation and theoretical and technological innovation in the processing of freshwater fish surimi and surimi products; (2) the efficient utilization of by-products, such as fish bones; (3) and research on the impact of nano-processing on the structure, physicochemical properties, and nutritional quality of aquatic raw materials. He has received funding for several research projects, including from the National Natural Science Foundation, the National Key R&D Plan sub project, the Hubei Provincial Major Science and Technology Innovation Plan sub project, the International Cooperation Project, etc. He received second prize for Scientific and Technological Progress from the Ministry of Education, has published over 100 papers, and obtained 8 patents.

Liu Shi

Liu Shi is an associate researcher at the Institute of Agricultural Products Processing and Nuclear-agricultural Technology at Hubei Academy of Agricultural Sciences (since 2016). Her research mainly focusses on the preservation and quality control technology of freshwater resources during processing, storage, and cold-chain transportation, in particular for fish and red swamp crayfish. Additionally, she is the owner of five national patents, two achievements, and one science award at provincial level. She obtained a Ph.D. degree in the Processing and Storage Engineering of Agricultural Products (2016), and a bachelor's degree in Food Quality and Safety (2006) from Huazhong Agricultural University.

foods

MDPI

Editorial

Processing and Preservation of Aquatic Products

Tao Yin [1,*] and Liu Shi [2]

1 College of Food Science and Technology, Huazhong Agricultural University, Wuhan 430070, China
2 Institute for Agricultural Products Processing and Nuclear—Agricultural Technology, Hubei Academy of Agricultural Science, Wuhan 430064, China; shiliu@hbaas.com
* Correspondence: yintao@mail.hzau.edu.cn; Tel.: +86-130-7277-4989

Citation: Yin, T.; Shi, L. Processing and Preservation of Aquatic Products. *Foods* **2023**, *12*, 2061. https://doi.org/10.3390/foods12102061

Received: 8 May 2023
Revised: 15 May 2023
Accepted: 16 May 2023
Published: 19 May 2023

1. Introduction

Aquatic products represent an important food source; they include products such as fish, shrimp, shellfish, crab, and seaweed, and provide high-quality proteins, fatty acids, minerals, and other nutritional elements. Generally, raw aquatic materials must go through a production chain, including growth, harvesting, processing, storage, and circulation—from the waters to the dining table. Among these stages, processing is one of the most important. Through processing, product quality improvement, diversification, and comprehensive economic benefits can be achieved. Like processing, storage is also an important production stage. Extending the storage shelf life of aquatic products through preservation techniques can reduce loss, and also increases the convenience of product consumption. Using a cross-sectional consumer survey covering the main cities of the seven regions of Turkey, Güney et al. [1] applied an ordered probit model and descriptive statistics to analyze the effects of attitudes and socio-demographic characteristics (as independent variables) on fish consumption and purchase intention (as the dependent variables) in 421 participants. They found that taste, physical appearance, convenience, etc., have a significant positive relationship with the frequency of fish purchase and consumption. Therefore, selecting appropriate processing and preservation techniques to improve the quality of aquatic products and meet consumer needs represents the key to achieving enhanced economic value.

2. Processing of Aquatic Products

2.1. Raw Materials

Compared to raw food materials such as grains, wheat, and cattle, there is a much richer variety of aquatic products available for processing, including more than 2000 species. The qualities of a single species of fish, such as its nutritional composition, efficacy, and sensory properties, are influenced by environmental factors, size, season, gender, and other factors [2,3]. Snakehead fish (*Channa argus*), commonly known as black fish, are widely distributed in rivers, lakes, ponds, and swamps in China, Korea, Japan, Southeast Asia, India, and the Russian Far East. Snakehead fish are mainly farmed at high density or live in wild fish ponds. Ren et al. [2] prepared soups from farmed and wild snakehead fish and analyzed the nutritional composition of the soups. According to their results, there were no obvious differences in the nutritional composition or antioxidant activity between the soups. The protein content (1.90%) of the soup made from wild snakehead fish was relatively lower, but the total fatty acid (16.22 g/100 g), polyunsaturated fatty acid (7.17 g/100 g), and Zn (12.57 mg/kg) contents were significantly higher, suggesting that it may have better efficacy in promoting wound healing. Catfish are a significant commercial food commodity. Haque et al. [3] compared the proximate composition, retained water, and bacterial loads of two sizes of hybrid catfish (*Ictalurus furcatus* × *Ictalurus punctatus*) fillets. Additionally, they analyzed the impact of season on the differences in the raw materials. It was found that the water content (78.0 vs. 76.0%) was higher, and fat content (6.0% vs. 8.0%) was lower for smaller (50–150 g) fillets compared to larger fillets (150–450 g). Furthermore, higher levels of psychrotrophics (~4.2 vs. ~3.0) and total coliform (~3.4 vs. ~1.7) were observed in fillets in the warmer seasons (April–July) compared to the cold seasons (Feb–April). Therefore, understanding the differences in the raw materials of aquatic products is beneficial for ensuring quality control during processing, and ultimately, maintaining high end product quality.

The quality of crab and fish cultured in outdoor earthen ponds can be affected by numerous factors, such as the culture environment [4] and feeding [5]. For instance, compared with fresh water, fattening male Chinese mitten crabs (*eriocheir sinensis*) in saline and alkaline water led to numerous advantages, including improvements in carotenoid accumulation in freeze-dried carapaces, DHA, EPA, total free essential amino acids, total free amino acids, and total umami value in the hepatopancreas and muscle [4]. During transportation, fish experience a decrease in muscle quality due to the stress induced by various factors (e.g., ammonia nitrogen, collision, noise, and temperature). Peng et al. [5] temporarily raised Wuchang bream fish (*Megalobrama amblycephala*) for 48 h before transportation, and found that it significantly improved the muscle quality of the fish, including the water holding capacity and shear force. The reason for this may be that the intestinal emptying of the fish during the temporary respite process reduces ammonia nitrogen accumulation and fish stress during transportation. Improving the quality of raw materials before processing has positive significance.

2.2. *Pretreatment*

The initial stage of aquatic product processing, also known as pretreatment, generally includes grading, bleeding, cooling, scaling, head removal, visceral removal, etc. The main purpose of pretreatment is to remove non-edible portions such as the scales, viscera, head, and tail, which comprise up to 70% of the total body weight. In addition, this stage also includes maintaining the clean, white appearance of the fish meat and inhibiting microbial growth. Pretreatment is a crucial step in processing, and determines the quality, yield, and profits. Cutting comprises a set of crucial operations in fish pretreatment, whereby the whole fish is divided into smaller pieces or certain sections are cut off to produce fish products (e.g., fish fillets, steaks, surimi, slices, etc.). Traditionally, the cutting process was performed by humans using metal knives, which is labor-intensive. This process has two drawbacks. During cutting, a considerable amount of debris is created, which leads to a large amount of raw fish waste. Additionally, heavy metal food contamination is expected under an aqueous medium when a metal knife is used. In recent years, novel techniques (such as waterjet cutting, machine vision, and artificial intelligence) have been invented to overcome the drawbacks of metal blade cutting, and to achieve higher cutting efficiency and quality [6].

2.3. *Aquatic Product Processing*

Aquatic processing products include surimi and its products, frozen headed and gutted fish and fish fillets, fermented fish meat, dried squid and seaweed, canned fish meat, etc. Surimi is a concentrated myofibril protein, and is an intermediate raw material for processing surimi products. Washing is a key process in surimi processing. Water-soluble cathepsin, blood and fishy substances, as well as most fats can be removed through washing. The number of washes directly affects the quality and economic benefits of surimi. By comparing different numbers of washes (0, 1, 2, 3), Zhang et al. [7] found that the yield, types of fatty acids, redness (a*), total volatile basic nitrogen, and thiobarbituric acid-reactive substances of the surimi decreased, and the whiteness, pH, gel strength, and water retention increased, with increasing numbers of washes. They suggested that two washes was optimal, taking into account yield, water consumption, and surimi quality. The production of surimi using the traditional washing method has the disadvantages of bony structures, high levels of cathepsins, and a muddy off-odor occurring in the final product, the latter of which is mainly caused by geosmin (GEO) and 2-methylisoborneol (MIB).

The use of a pH-shifting process not only significantly improves yield, but also effectively removes GEO and MIB. Guo et al. [8] studied the effects of acid and alkali extractions on the quality of surimi from silver carp (*Hypophthalmichthys molitrix*), and claimed that the alkali extraction method was a better alternative for making water-washed surimi from silver carp. However, currently, it is still difficult to achieve commercialization using the method of pH-shifting in surimi processing, which may be due to the difficulty in scaling up production and the high level of environmental pollution.

Surimi products are a type of deeply processed aquatic product with characteristics such as high protein, low fat, and an elastic texture. The processing of surimi products generally includes sequential processes such as thawing, cutting, chopping, the addition of ingredients, shaping, heating, and packaging. Adding functional ingredients such as starch, calcium ions, and transglutaminase can significantly improve the quality of surimi products. Sang et al. [9] added different proportions (0 to 4.5%) of calcium lactate to large yellow croaker (*Pseudosciaena crocea*) surimi and found that calcium lactate could significantly improve its gelation properties, including its gel strength, whiteness, cooking loss and water holding capacity. The mechanism of calcium lactate enhancing surimi gelation was found to be related to the promotion of myosin stretching by divalent calcium ions and the formation of salt bridges between adjacent negatively charged proteins. The authors claimed that 1.5% was the optimal amount of calcium lactate.

Fermented aquatic products are also an important type of aquatic processing products. In the processing of aquatic products, it is generally necessary to use methods such as ozone water washing and to operate at low temperatures to reduce microorganisms, in order to prevent raw materials from spoiling and deteriorating. Contrary to other aquatic products, fermented aquatic products utilize the activity of microorganisms to partially decompose aquatic proteins and partially oxidize fats, producing unique flavors. Due to differences in raw materials and fermentation processes, various fermented aquatic products with local characteristics are formed, such as "Zhayu", which is a popular fish product in central provinces of China. Compared with natural fermentation, inoculated microbial fermentation has the significant advantages of shortening production cycles and improving product quality stability. An et al. [10] found that inoculating lactic acid bacteria could increase the crude fat and protein content of "Zhayu", reduce the fishy smell, and make the product tender and soft. According to their research results, the samples prepared with 10^9 cfu/100 g lactic acid bacteria presented the best overall qualities. In addition, they also reported significant differences in the quality of "Zhayu" produced via inoculation with three different types of lactic acid bacteria. "Zhayu" fermented with *L. plantarum* and *P. acidilactici* showed the strongest sourness, while the samples prepared with *P. pentosaceus* showed the strongest umami taste, as evidenced by the fact that they had the highest Asp (25.1 mg/100 g) and Glu (67.8 mg/100 g) content.

Besides fish, seaweed is also widely available and of great commercial importance. At present, seaweed is mainly used for extracting polysaccharides, or dried for consumption in dishes and casual snacks. In a recent study, it was stated that powdered seaweed (*Ulva intestinalis*) from coastal areas of Bangladesh is incorporated into cookies to enrich them in unsaturated fatty acids [11].

3. Preservation of Aquatic Products

3.1. Preserving Agents

The muscle tissue of fish and other aquatic products is soft and tender, with high water content. The enzymes in a fish's body exhibit strong activity at room temperature. After the fish dies, these enzymes quickly decompose the fish protein into a large number of low-molecular metabolites and free amino acids, which become nutrients for bacteria. Therefore, aquatic products are generally prone to spoilage. Thanks to the advancement of preservation and rapid detection technologies (such as employing the peroxide activity of platinum cubes as an indicator of fish freshness [12]), the shelf life of aquatic products can be extended and safety hazards can be reduced. Adding a preserving agent is an effective means of extending the shelf life of aquatic products. In recent years, the use of weak acids [13–15], essential oils [16,17], and epigallocatechin gallate (EGCG) [18] as aquatic preserving agents has become a research hotspot.

Zhan et al. [13] applied sous vide (SV) cooking treatments to half-shell scallop (*Chlamys farreri*), and compared its physicochemical qualities and the volatility of its flavor with a control (cooked at 100 °C for 10 min) during 30 d of chilled storage. Upon monitoring the volatile basic nitrogen (TVBN), pH, texture, malondialdehyde (MDA) content, and myofibrillar protein (MP) extraction rate of the scallop samples, they reported that the SV cooking treatments effectively maintained acceptable and stable physicochemical qualities during storage. The effects of different acids on the quality, nutrition, and sensory quality of aquatic products vary. For example, Baltic herring (*Clupee harengus membras*) pickled in acetic acid and malic acid showed lower fat content during storage compared with those pickled in citric, lactic, and tartaric acids [14]. Replacing acetic acid with other weak acids resulted in pickled and marinated Baltic herring with novel and milder sensory profiles. There is a synergistic effect between weak acids and other physical preservation technologies. Lactic acid in combination with high hydrostatic pressure effectively inhibited microbiological growth and physicochemical changes (pH, sensory evaluation, flush, and texture) of large-mouth bass fillets [15].

Essential oils (EO) are important natural plant extracts and have attracted much interest from scientists for their antibacterial and antioxidant properties. By comparing four essential oils (oregano essential oil (OEO), tea tree essential oil (TTEO), and wild orange essential oil (WOEO)), Qian et al. [16] reported that OEO displayed the highest antimicrobial effect in Pacific white shrimp (*Litopenaeus vannamei*) during cold storage. Additionally, they found that the combination of OEO and CLEO had a synergistic effect, and displayed the highest efficacy in preventing melanosis, bacterial growth, and protein hydrolysis in shrimp. In order to overcome the volatile nature of essential oils, Qiu et al. [17] processed microcapsules with porous starch as an adsorbed substrate to store SOEO (PS/SOEO), sodium alginate (SA), and chitosans (CMCSs) as shell materials, in order to delay the volatilization of SOEO. They found that CMCS and SA improved the slow reducing ability of SOEO microcapsules, and the shelf life of crawfish (*Procambarus clarkii*) could be extended to 6 days by SOEO microcapsules (1/10 g, SOEO microcapsules/crowfish) at room temperature.

Like essential oils, tea polyphenols are an emerging food preserving agent for aquatic products due to their vital antioxidant and bacteriostatic properties. According to a study by Tian et al. [18], besides maintaining texture and gel strength, the addition of EGCG can inhibit microbial growth and the formation of off-odor compounds such as total volatile basic nitrogen (TVB-N) and malondialdehyde (MDA). They recommended that EGCG at a concentration of 0.01–0.02% could effectively preserve surimi gels during freeze–thaw cycles.

3.2. Non-Thermal Processing

In recent years, non-thermal processing has been increasingly applied to the preservation of aquatic products. Compared with thermal sterilization such as pasteurization, it can better preserve the nutrition, texture, and flavor of aquatic products. Common non-thermal processing methods for aquatic products include ultrasound [19], high-voltage atmospheric cold plasma [20], electron-beam irradiation [21], high hydrostatic pressure [15], etc.

Ling et al. [19] investigated ozone water and ultrasound cleaning in the storage of crayfish through microbial viable counts and 16S rRNA gene sequencing. They reported that the ozone water in combination with ultrasound cleaning could significantly reduce the total viable count, psychrophilic viable count, mesophilic viable count, pseudomonas, hydrogen sulfide-producing bacteria, molds, and yeasts. With the extension of high-voltage atmospheric cold plasma treatment (0 to 300 s), the number of colonies on the surface of tilapia (*Oreochromis mossambicus*) fillets inoculated with *Salmonella enterica serovar enteritid* is (*S. enteritis*) and/or *Listeria monocytogenes* (*L. monocytogenes*) gradually decreased [20].

However, the pH, b* value, elasticity, chewiness, thiol value, and TVB-N value were not significantly different. Electron-beam irradiation treatment with a dose of 4 kGy was able to retain the muscle water content and preserve their quality of silver carp (*Hypophthalmichthys molitrix*) chunks stored at 4 °C [21]. Additionally, high hydrostatic pressure treatment at 400 MPa effectively inhibited microbiological growth and physiochemical changes in large-mouth bass fillets [15]. The mechanism by which non-thermal processing exerts its preservation effect is generally related to its damage to microbial cell membranes and enzyme/protein structures. Briefly, non-thermal processing can be used as a cleaning strategy to control the microbiological quality of aquatic products, and has much lighter effects on their quality.

3.3. *Packaging*

Recently, in addition to air packaging and vacuum packaging, modified atmosphere packaging (MAP) has also been utilized in aquatic preservation. Regarding MAP, O_2, CO_2, and N_2 gas are mixed into high-barrier-performance packaging material to inhibit microbial growth, enzymatic reactions, and lipid oxidation. In a recent study, Zhang et al. reported that MAP with a 70% CO_2/30% N_2 gas ratio was optimized to inhibit the quality deterioration of golden pompano (*Trachinotus ovatus*) fillets [22].

Aquatic products are an important source of food for humans. The processing and preservation techniques for aquatic products are constantly improving, with the aim of improving nutrition, taste, and convenience. In the production process of aquatic products, enterprises are also increasingly emphasizing energy conservation, loss reduction, and environmental protection. It can be foreseen that these techniques, which are still in their experimental and pilot stages, will play an increasingly important role in future industrial production. New processing and preservation techniques will also continue to emerge.

In this editorial, with regard to aquatic product processing and preservation, we introduce methods of raw material quality improvement (using fattening techniques in saline and alkaline water and respite as examples), pretreatments (using cutting as examples), and new techniques for processing aquatic products (using surimi and its products, fermented products, and seaweed products as examples). In addition, we introduce new technological advancements in the preservation of aquatic products from the perspectives of preserving agents (using weak acids, essential oils, and EGCG as examples), non-thermal processing (using ultrasound, high-voltage atmospheric cold plasma, electron-beam irradiation, and high hydrostatic pressure as examples), and packaging (using MAP as an example). This editorial is suitable for undergraduate and graduate students majoring in food science and technology.

We are grateful to the contributing authors of this volume for their efforts and excellent work. It is hoped that readers of this text will find it useful and direct constructive comments regarding its content (as well as printing errors) to our attention.

Author Contributions: Writing—review and editing, L.S.; supervision, T.Y. All authors have read and agreed to the published version of the manuscript.

Data Availability Statement: Not applicable.

Conflicts of Interest: The authors declare no conflict of interest.

References

1. Güney, O.I.; Özşahinoğlu, I.; Erçen, Z.; Yeldan, H.; Dikel, Ç.; Sangün, L. The Mediator Role of Attitudes in Fish Choice Behavior: A Turkish Market Survey. *Foods* **2022**, *11*, 3180. [CrossRef]
2. Ren, M.; Yin, T.; You, J.; Liu, R.; Huang, Q.; Xiong, S. Comparative Study of the Nutritional Composition and Antioxidant Ability of Soups Made from Wild and Farmed Snakehead Fish (*Channa Argus*). *Foods* **2022**, *11*, 3294. [CrossRef]
3. Haque, M.; Silva, J.L. Proximate Composition, Retained Water, and Bacterial Load for Two Sizes of Hybrid Catfish (*Ictalurus furcatus* × *Ictalurus punctatus*) Fillets at Different Process Steps. *Foods* **2023**, *12*, 1112. [CrossRef]
4. Wang, S.; Guo, K.; Luo, L.; Zhang, R.; Xu, W.; Song, Y.; Zhao, Z. Fattening in Saline and Alkaline Water Improves the Color, Nutritional and Taste Quality of Adult Chinese Mitten Crab *Eriocheir sinensis*. *Foods* **2022**, *11*, 2573. [CrossRef] [PubMed]

5. Peng, L.; You, J.; Wang, L.; Xiong, S.; Huang, Q.; Yin, T. Effect of Respite Time before Live Transportation on Muscle Quality of Blunt Snout (Wuchang) Bream. *Foods* **2022**, *11*, 2254. [CrossRef]
6. Liu, W.; Lyu, J.; Wu, D.; Cao, Y.; Ma, Q.; Lu, Y.; Zhang, X. Cutting Techniques in the Fish Industry: A Critical Review. *Foods* **2022**, *11*, 3206. [CrossRef]
7. Zhang, X.; Zhang, Y.; Ding, H.; Zhang, W.; Dai, Z. Effect of Washing Times on the Quality Characteristics and Protein Oxidation of Silver Carp Surimi. *Foods* **2022**, *11*, 2397. [CrossRef]
8. Guo, W.; Zhan, M.; Liu, H.; Fu, X.; Wu, W. Effect of pH-Shifting Process on the Cathepsin Activity, Muddy Off-Odor Compounds' Content and Gelling Properties of Isolated Protein from Silver Carp. *Foods* **2023**, *12*, 939. [CrossRef]
9. Sang, S.; Chen, X.; Qin, Y.; Tong, L.; Ou, C. A Study on the Effects of Calcium Lactate on the Gelling Properties of Large Yellow Croaker (*Pseudosciaena crocea*) Surimi by Low-Field Nuclear Magnetic Resonance and Raman Spectroscopy. *Foods* **2022**, *11*, 3197. [CrossRef]
10. An, Y.; Cai, X.; Cong, L.; Hu, Y.; Liu, R.; Xiong, S.; Hu, X. Quality Improvement of Zhayu, a Fermented Fish Product in China: Effects of Inoculated Fermentation with Three Kinds of Lactic Acid Bacteria. *Foods* **2022**, *11*, 2756. [CrossRef]
11. Mohibbullah, M.; Amin, A.; Talha, M.A.; Baten, M.A.; Rana, M.M.; Sabuz, A.A.; Newaz, A.W.; Choi, J.-S. Physicochemical and Nutritional Characteristics of Cookies Prepared with Untapped Seaweed *Ulva intestinalis*: An Approach to Value Addition as a Functional Food. *Foods* **2023**, *12*, 205. [CrossRef] [PubMed]
12. Ma, X.; Feng, T.; Zhang, P.; Zhang, H.; Hu, X.; Yang, Y.; Wang, Z.; Zhang, H.; Peng, D.; Li, X.; et al. Downregulation of Peroxidase Activity of Platinum Cube Enables Minute–Time Scale Colorimetric Signaling of Hypoxanthine for Fish Freshness Monitoring. *Foods* **2023**, *12*, 291. [CrossRef] [PubMed]
13. Zhan, Y.; Tu, C.; Jiang, H.; Benjakul, S.; Ni, J.; Dong, K.; Zhang, B. Effects of Sous Vide Cooking on the Physicochemical and Volatile Flavor Properties of Half-Shell Scallop (*Chlamys farreri*) during Chilled Storage. *Foods* **2022**, *11*, 3928. [CrossRef] [PubMed]
14. Logrén, N.; Hiidenhovi, J.; Kakko, T.; Välimaa, A.-L.; Mäkinen, S.; Rintala, N.; Mattila, P.; Yang, B.; Hopia, A. Effects of Weak Acids on the Microbiological, Nutritional and Sensory Quality of Baltic Herring (*Clupea harengus membras*). *Foods* **2022**, *11*, 1717. [CrossRef]
15. Zhou, M.; Ling, Y.; Chen, F.; Wang, C.; Qiao, Y.; Xiong, G.; Wang, L.; Wu, W.; Shi, L.; Ding, A. Effect of High Hydrostatic Pressure Combined with Sous-Vide Treatment on the Quality of Largemouth Bass during Storage. *Foods* **2022**, *11*, 1931. [CrossRef]
16. Qian, Y.-F.; Lin, T.; Liu, X.; Pan, J.; Xie, J.; Yang, S.-P. In-Vitro Study on the Antibacterial and Antioxidant Activity of Four Commercial Essential Oils and In-Situ Evaluation of Their Effect on Quality Deterioration of Pacific White Shrimp (*Litopenaeus vannamei*) during Cold Storage. *Foods* **2022**, *11*, 2475. [CrossRef]
17. Qiu, L.; Ma, H.; Luo, Q.; Bai, C.; Xiong, G.; Jin, S.; Wang, J.; Zu, X.; Li, H.; Liao, T. Preparation, Characterization, and Application of Modified Starch/Chitosan/Sweet Orange Oil Microcapsules. *Foods* **2022**, *11*, 2306. [CrossRef]
18. Tian, Z.; Jiang, X.; Xiao, N.; Zhang, Q.; Shi, W.; Guo, Q. Assessing the Gel Quality and Storage Properties of *Hypophythalmalmichthys molitrix* Surimi Gel Prepared with Epigallocatechin Gallate Subject to Multiple Freeze-Thaw Cycles. *Foods* **2022**, *11*, 1612. [CrossRef]
19. Ling, Y.; Tan, H.; Shen, L.; Wei, L.; Xiong, G.; Wang, L.; Wu, W.; Qiao, Y. Microbial Evaluation of Ozone Water Combined with Ultrasound Cleaning on Crayfish (*Procambarus clarkii*). *Foods* **2022**, *11*, 2314. [CrossRef]
20. Wang, J.; Fu, T.; Wang, Y.; Zhang, J. Effects of High-Voltage Atmospheric Cold Plasma Treatment on Microbiological and Quality Characters of Tilapia Fillets. *Foods* **2022**, *11*, 2398. [CrossRef]
21. Li, H.-L.; Li, M.-J.; Zhao, Q.; Huang, J.-J.; Zu, X.-Y. Analysis of Water Distribution and Muscle Quality of Silver Carp (*Hypophthalmichthys molitrix*) Chunks Based on Electron-Beam Irradiation. *Foods* **2022**, *11*, 2963. [CrossRef] [PubMed]
22. Zhang, X.; Pan, C.; Chen, S.; Xue, Y.; Wang, Y.; Wu, Y. Effects of Modified Atmosphere Packaging with Different Gas Ratios on the Quality Changes of Golden Pompano (*Trachinotus ovatus)* Fillets during Superchilling Storage. *Foods* **2022**, *11*, 1943. [CrossRef] [PubMed]

foods

MDPI

Article

The Mediator Role of Attitudes in Fish Choice Behavior: A Turkish Market Survey

Osman Inanç Güney [1,*], Ilgın Özşahinoğlu [1], Zeynep Erçen [1], Hacer Yeldan [2], Çiğdem Dikel [3] and Levent Sangün [1]

1 Vocational School of Adana, Cukurova Universitesi, Adana 01330, Turkey
2 Marine Biology, Basic Fisheries Science, Cukurova University, Adana 01330, Turkey
3 Department of Biotechnology, Cukurova University, Adana 01330, Turkey
* Correspondence: iguney@cu.edu.tr; Tel.: +90-5336582730

Abstract: Due to the dynamic nature of demand, it is becoming increasingly important for the fish industry to investigate the changing choice behaviors of consumers in the face of increasing demand. This research investigated the role of attitudes and socio-demographic characteristics, which are the main factors in the fish choice behavior of consumers and in fish consumption behavior. In this context, an ordered probit model was constructed to analyze the effect of attitudes and socio-demographic characteristics as independent variables on fish consumption and purchase intention as the dependent variables. In addition, descriptive statistics were also used to reveal the current preferences related to fish. The data required for the model and descriptive statistics were obtained from 421 participants using a cross-sectional consumer survey covering the main cities of the seven regions of Turkey. The results show that while consumers prefer fish more than red meat and less than poultry, they mostly buy fresh fish from fish markets. Moreover, taste, physical appearance, convenience, wild fish, and seller trust attitudes have a significant and positive relationship with the dependent variable (the frequency of fish purchase and consumption) and price has a negative and significant relationship. Moreover, an increase in education level has a positive and significant relationship with the frequency of fish consumption. The research results provide important suggestions and information for decision-makers in the fish industry to implement effective policies and meet the consumer expectations of producers and distributors in the fish industry. In addition, the current study provides guidance for future research.

Keywords: fish choice parameters; attitudes; socio-economic features; consumer-seller determinants

Citation: Güney, O.I.; Özşahinoğlu, I.; Erçen, Z.; Yeldan, H.; Dikel, Ç.; Sangün, L. The Mediator Role of Attitudes in Fish Choice Behavior: A Turkish Market Survey. *Foods* **2022**, *11*, 3180. https://doi.org/10.3390/foods11203180

Academic Editors: José Antonio Beltrán Gracia, Tao Yin and Liu Shi

Received: 9 August 2022
Accepted: 9 October 2022
Published: 12 October 2022

1. Introduction

1.1. The Food Choice Behavior Phenomenon

Food choice is a dynamic, multidimensional, and complex phenomenon influenced by many interconnected factors; so, it has been the focus of food market research analysis over the last 50 years [1–5]. The studies of [6], who calculated that consumer make more than 220 food-related decisions per day, demonstrate the great frequency and significance of this choice. The authors of [7] suggested that individuals' food choices are influenced by three categories: (i) the nutritional content and physical properties of the food, (ii) sensory attributes (taste, odor, texture, etc.) and psychological factors of the individuals, and (iii) the economic and social environment. Factors that influence food choice can also be summarized as sensory factors (taste, odor, and texture) and non-sensory factors (price, health claims, availability, sustainability, convenience, personal values, etc.) [8].

Food choice behavior has also been affected by economic and social drivers such as increases in per capita income, changes in agri-food sector practices, diversification of production methods, fluctuating food prices, trade liberalization, supply chain developments, and the increased role of the media and marketing over the past several decades [9–12].

The importance of the influencing categories is likely to vary between consumers and products, making it crucial to understand which factors are important among specific segments for evaluating food choice and the drivers behind it [13].

Revealing the factors underlying consumers' food preferences is one of the main issues that agri-food marketing experts have focused on for many years. Globalization and developments in communication opportunities have affected the speed and dynamics of changes in food preferences. Therefore, the scope of research in this area is expanding, and the frequency of updates is increasing.

1.2. Attitudes toward Fish Purchase and Consumption Intentions

In marketing, understanding attitudes can help marketers predict consumer behavior [14]. The differences in attitudes are reflected in consumers' purchase decisions and consumption practices [15]. For this reason, marketing managers spend millions of dollars researching consumer attitudes toward products and brands and spend many more trying to influence these attitudes. By influencing consumers' attitudes, marketers hope to influence consumers' purchase behavior [16]. Attitudes are also suggested to be one of the main determinants explaining food consumption behavior, including seafood consumption behavior [17].

Since fish consumption has been associated with the healthy life concept and a balanced diet during the last decades, authorities have encouraged fish consumption [2,3,8,18–27]. The World Health Organization recommends eating at least two servings of fish per week [17,21,28].

From 1961 to 2017, global fish consumption increased by 3.1% at a rate nearly twice that of the annual world population growth (1.6%) and faster than all other animal protein foods (meat, dairy, milk, etc.), which increased at a rate of 2.1% per year. Per capita, fish consumption increased by about 1.5% per year from 9.0 kg (live weight equivalent) in 1961 to 20.5 kg in 2018. In 2017, fish consumption accounted for 17% of the global population's animal protein consumption and 7% of the total protein consumption [19,29,30]. The increase in the world population with higher living standards, efforts of consumers in healthy nutritional consumption, improvements in seafood processing technologies, and promotion campaigns, which changed consumption habits, have led to an increase in fish consumption in recent years [31–35]. Thus, the increased fish consumption has directed marketers and researchers to focus more on consumers' fish consumption behavior.

Consumer attitudes significantly impact fish consumption and buying decisions, so there is significant interest in learning more about the dynamics of choice behavior for seafood [20,32]. Seafood consumption is influenced by many interrelating factors, including attitudes, personal factors, and the cultural and social environment [36]. The attitudes that may affect individuals' fish purchase and consumption behavior have been studied previously. In this context, the attributes and patterns for fish consumption vary across cultures, countries, and individuals. The most commonly studied attributes are the taste, texture, nutritional value, price, familiarity, health, convenience, availability, season, geographical origin, freshness, and obtainment method [19,25,30,34,36–40]. In addition to those, consumers' sociodemographic and economic characteristics are also accepted as impacting fish consumption behavior [25,34,39,41].

Attitudes act as a mediator in consumers' food purchasing behavior and affect the intention and consumption of a food product [42,43]. Thus, it is important to understand the relationship between attitudes and choice behavior. The demand for fish has increased in the last few decades due to the high nutritional value, increase in health problems based on nutrition in society, easier access to fish (sales to supermarkets), increase in the consumption behavior outside the home, promotional campaigns, and innovations in the form of fish consumption (such as fish fingers) [23,25,31]. In this context, understanding consumer variable demand structures in the face of increasing demand is very important for all parties involved (fish sector, policy makers, relevant non-governmental organizations, and consumers). The main focus of this research was to identify and explain the factors that affect consumers' choice behavior towards fish within the framework of attitudes and sociodemographic characteristics. For this purpose, the fish consumption pattern and attitudes and socio-demographic characteristics that determine and affect the fish consumption and purchasing preference behaviors in Turkey were analyzed using the regression model and descriptive statistics. In the model, attitudes and sociodemographic features were utilized as a proxy for an overall assessment of fish and they referred to the degree to which consumers positively or negatively evaluate. The results provide information to researchers, policy makers, NGOs, and fish industry stakeholders by understanding and explaining the fish choice behavior [19,44,45]. Thus, they can create new effective strategies and marketing plans to create a sustainable value chain for fish consumption.

2. Material and Methods

2.1. Data Description

The quantitative data required to present the consumers' fish purchase and consumption patterns and explain the relationship between attitudes and fish choice behavior were gathered by a cross-sectional survey. This survey was conducted in the major cities (Adana, Ankara, Gaziantep, İstanbul, İzmir, Trabzon, and Van) of the seven geographical regions of Turkey during August–October 2021 by a commercial marketing agency (Ayna Market Research Co.). These cities were chosen due to their geographical dispersion and population sizes, making them a good representation of the entire country. The sample size allocated to each city was determined according to their respective populations. The questionnaire was applied to 450 consumers who were responsible for their households' food purchases. After the preliminary examination, 421 were chosen for evaluation and used in the analysis. The final sample size was extracted from an infinite population with a confidence level of 95.5% (p = 0.05). The sociodemographic characteristics of the sample are given in Table 1.

Table 1. Sociodemographic characteristics of the sample.

Gender	% *	*f*	Age	% *	*f*	Education	% *	*f*
Female	52.5	221	≤34	42.1	177	Uneducated	5.0	21
Male	47.5	200	35–49	35.4	148	Primary	32.8	138
			50–69	21.4	90	High School	33.7	142
			≥70	1.4	6	University	28.5	120
Income	**% ***	***f***	**H. Size**	**%***	***f***			
No Income	9.7	41	1	8.3	35			
<3000 TL	21.4	90	2	12.4	52			
3001–4500 TL	36.8	155	3	25.9	110			
4501–6000 TL	20.0	84	4	28.7	120			
6001–7500 TL	7.4	31	≥5	24.7	104			
>7501 TL	4.7	20						

* Respondent percentages for each category. H. size: Household size. TL: Turkish Lira. *f*: frequency of the values

2.2. Survey Design

The questionnaire was divided into two parts. The first part was designed in four sub-sections to gather information related to the consumption patterns, attitudes toward fish choice behavior, and sociodemographic characteristics of the participants, which later formed the independent variables of the analysis. The participants were asked to score the importance of each item on a four- or five-point Likert scale to express how strongly they agreed or disagreed with each statement in this part.

The second part consisted of the question 'How often do you consume fish?', which constituted the model's dependent variable. The dependent variable, which is expressed by the consumption frequency, was formed to take the value of '1' if fish consumption was every day or 2–3 times a week, '2' if it was once a week or 2–3 times a month, and '3' if it was once a month or less than once a month (Table 2).

Table 2. Categorization of the fish consumption frequencies.

Dependent Variable	f	% *
Y_1 = often	every day or 2–3 times a week	11.6
Y_2 = sometimes	once a week or 2–3 times a month	56.8
Y_3 = rarely	once a month or less	31.6

Description of each dependent variable. Answers for dependent variables were categorized into three levels: Y_1, Y_2, and Y_3. f: frequency of fish consumption. * Respondent percentages for each category.

2.3. Analytical Framework and the Model

The data gathered through the survey application were analyzed using descriptive statistics and a regression model. The impacts of attitudes on consumer choice behavior have been frequently analyzed by regression models in previous studies [42,43,46,47].

In the model, to determine the factors affecting the frequency of individuals' fish consumption, the answers given to the question 'How often do you consume fish?' in the survey were determined as the dependent variable. In line with the answers given by the individuals, the dependent variable was classified into three categories: 'often' if it was every day or 2–3 times a week, 'sometimes' if it was once a week or 2–3 times a month, and 'rarely' if it was once a month or less than once a month (see Table 2). In this context, since the dependent variable consists of more than two values and has a natural ordering, it is appropriate to apply the ordered probit model to the variables [48].

Generally, the ordered probit model can be specified from the unobserved latent variable (Y_i^*):

$$Y_i^* = x_i'\beta + \varepsilon_i \tag{1}$$

where i ($i = 1, 2, \ldots, N$) denotes consumers, x_i is the vector of independent explanatory variables, β is the vector of unknown slope parameters, and ε_i is an unobserved individual error term with a standard logistic distribution. The answers given to the question 'How often do you consume fish?', which is the dependent variable expressed by the consumption frequency in the model, were coded as shown in Table 2 and used in this form in the analysis. So, lower levels of Y_i^* indicated that the individual has a higher consumption frequency. In comparison, higher levels of Y_i^* indicated that an individual has a lower consumption frequency. In short, in the ordered probit model, the dependent variable (i.e., the consumption frequency of individuals) was classified according to the values that the unobserved latent variable (Y_i^*) takes within certain threshold value ranges τ_k:

$$Y_i = \begin{cases} 1, & 'often' \quad if \;\; \tau_0 < Y_i^* \leq \tau_1 \\ 2, & 'sometimes' \quad if \;\; \tau_1 < Y_i^* \leq \tau_2 \\ 3, & 'rarely' \quad if \;\; \tau_2 < Y_i^* \leq \tau_3 \end{cases} \tag{2}$$

where τ_k $(k = 1, 2, 3)$ indicates the threshold level according to the consumption frequency of each individual. In addition, the probability of choosing a particular consumption frequency of each individual is as follows:

$$\begin{aligned} P(Y_i = j | x_i, \beta) &= P(\tau_{j-1} < Y_i^* \leq \tau_j) \\ &= P(\tau_{j-1} < x_i'\beta + \varepsilon_i \leq \tau_j) \\ &= P(\tau_{j-1} - x_i'\beta < \varepsilon_i \leq \tau_j - x_i'\beta) \\ &= F(\tau_j - x_i'\beta) - F(\tau_{j-1} - x_i'\beta) \end{aligned} \tag{3}$$

where F is the cumulative distribution function of ε_i [49]. To calculate the explanatory variables' effects on each individual's preference probability, the derivative of the cumulative distribution specified in Equation (3) was taken according to the explanatory variables. The model estimation was carried out using the maximum likelihood method, and the logarithmic likelihood function is expressed by Equation (4) [50,51]:

$$lnL(\beta,\tau|y,x) = \sum_{j=1}^{J}\sum_{Y_i=j} ln[F(\tau_j - x_i'\beta) - F(\tau_{j-1} - x_i'\beta)] \quad (4)$$

A principal component analysis (PCA) with a varimax rotation was applied to dimensionality reduction in order to reduce the complexity of the model. Thus, it was possible to determine the basic dimensions of consumer participation in fish selection behavior [31,52–54]. Variables with factor loadings of ≥0.4 were removed from the factor set [55]. With the help of this dimensional reduction with PCA, 23 attitudes in the survey were reduced to 15.

The level of significance was set at $p < 0.05$. To perform the data analysis, ordered probit regression was conducted using StataCorp. 2013. Stata Statistical Software (STATA) v13 (College Station, TX, USA: StataCorp LP.) and PCA, and descriptive statistics (mean values) were reached using Social Sciences (SPSS) v21.0 software (Armonk, NY, USA: IBM Corp.) [56].

3. Results

3.1. Descriptive Statistics Results

According to Table 2, 11.6% of the individuals in our sample consumed fish daily or 2–3 times a week, 56.8% once a week or 2–3 times a month, and 31.6% once a month or less than once a month.

In this study, the general meat consumption distribution of the individuals was determined, and it was concluded that the average fish consumption was less than the consumption of red meat and more than the consumption of poultry (3.5 kg per month) (Figure 1).

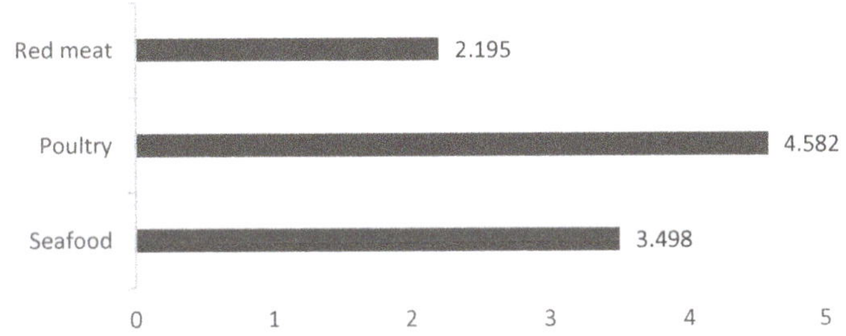

Figure 1. Monthly household meat consumption distribution (kg).

In this research, consumers' purchasing place preferences for where to buy fish were also determined (Figure 2). Accordingly, it is understood that consumers in Turkey mostly buy fish from fish markets. Street markets, supermarkets, and street vendors came after fish markets in the purchasing place preferences.

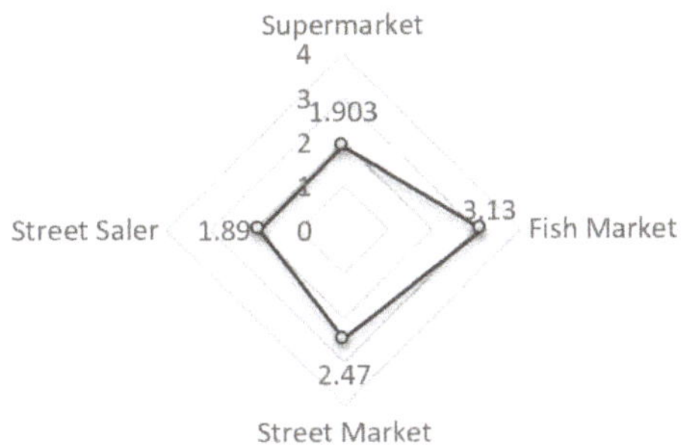

Figure 2. Fish purchase location preferences.

Figure 3 shows consumers' fish consumption preferences. Accordingly, consumers mostly preferred to consume fresh fish instead of processed options (canned, frozen, dried, salted, or smoked).

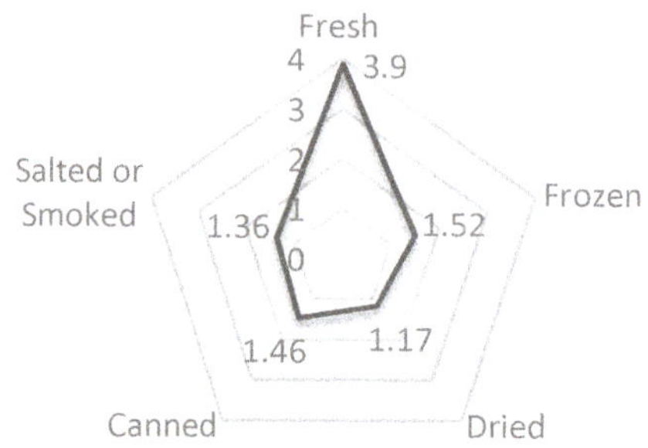

Figure 3. Fish consumption preferences.

The results obtained within the scope of this research show that more than half of the consumers consume fish once a week or 2–3 times a month. Consumers prefer fish markets for their fish purchases, and their consumption preferences are mostly for fresh fish.

The description and mean values of the socio-economic characteristics of the sampled individuals are given in Table 3. According to Table 3, the mean value of the gender variable is 0.524, indicating that 52% of the individuals in this study were female. The mean value of the education variable is 2.857. Accordingly, most individuals in the sample were primary and high school graduates. The average value of the income variable is 3.080, which indicates that the monthly income of the individuals in the sample generally varies between 3000 and 4500 TL. The average household size was found to be three to four individuals.

Table 3. Categorization and mean values of the socio-demographic variables.

		Mean	SE
Gender	female: 1; male: 0	0.524	0.024
Age	<34: 1; 35–49: 2; 50–69: 3; >70: 4	1.817	0.039
Education	Uneducated: 1; primary school: 2; high school: 3; university: 4	2.857	0.043
Income	no income: 1; <3000 TL: 2; 3001–4500 TL: 3; 4501–6000 TL: 4; 6001–7500 TL: 5; >7501 TL: 6	3.080	0.060
H. size	single household: 1; $n = 2$: 2; $n = 3$: 3; $n = 4$: 4; $n \geq 5$: 5	3.491	0.059

Categorization of sociodemographic variables (gender, age, education, income, and household size). SE = standard error. Values are the mean.

Table 4 presents the definitions, mean, and standard errors of this study's explanatory variables (attitudes). Accordingly, the average values of the variables of taste, price, accessibility, traditional consumption, variety, seller trust, production method, and physical appearance were found to be above a score of four. According to this, the individuals in the sample generally showed a traditional approach to fish consumption by finding the fish meat delicious, the price of the fish high, and the fish easy to access or buy. Trust in the seller, production method, and the physical appearance of the fish they buy or consume were also important for the individuals participating in the sampling.

Table 4. Descriptive statistics on fish choices.

V.	Definitions	Mean	SE
Taste	I find fish delicious	4.323	0.034
Price	I find fish prices high	4.095	0.045
Accessibility	Accessibility is important in the fish supply	4.118	0.039
Health	Fish consumption is healthy	4.35	0.031
Traditionality	I perceive fish as a traditional food	4.066	0.042
Convenience	Convenience is important for fish choice	3.681	0.052
Fishing method	Fishing method is important for fish choice	3.888	0.048
Species	Species of fish is important for fish choice	4.040	0.036
Unnatural sub.	Concerned about use of unnatural sub.	3.672	0.049
Location	Location where fish is caught is important	3.94	0.041
Seller trust	trust the seller is important	4.090	0.037
Wild fish	I prefer wild fish	4.010	0.041
Freshness concern	Concerned about the freshness of the fish	3.980	0.041
Phy. appearance	I pay attention to the phy. app. of the fish	3.757	0.046

V.: Variables. SE = standard error. 5-point Likert scales: Strongly disagree (1), Disagree (2), Undecided (3), Agree (4), Strongly agree (5). Values are the mean.

3.2. Regression Model Results

Table 5 shows the results of the ordered probit analysis regarding the relationships between the attitudes toward fish choice behavior, sociodemographic characteristics of the participants, and frequency of fish consumption. Table 5 also demonstrates the coefficient estimates of the explanatory variables in the ordered probit model. In this study, the dependent variable, which expresses the fish consumption frequency of individuals, takes the values of 1, 2, and 3, where 1 represents the situations where the frequency of fish consumption is the least, 2 represents the medium level, and 3 represents the situations where it is the most.

Table 5. Ordered probit model coefficient estimates.

V.	Coeff.	SE	p
Taste	0.203	0.084	0.016 **
Price	−0.121	0.059	0.043 **
Accessibility	0.106	0.071	0.134
Health	0.069	0.087	0.425
Traditionality	−0.048	0.064	0.454
Convenience	0.109	0.051	0.035 **
Fishing method	0.032	0.061	0.594
Species	−0.117	0.073	0.111
Use of unnatural sub.	0.0431	0.057	0.434
Location	−0.043	0.068	0.526
Seller trust	0.153	0.076	0.044 **
Wild fish	−0.113	0.063	0.074 *
Freshness concern	−0.095	0.068	0.164
Physical appearance	0.182	0.058	0.002 ***
Gender	0.081	0.1050	0.439
Age	0.027	0.073	0.703
Education	0.190	0.069	0.006 ***
Income	0.047	0.0.045	0.306
Household size	−0.058	0.043	0.226
Cut1	1.556	0.667	
Cut2	3.446	0.676	

V.: independent variables of the model. Coeff: coefficients. p: prob value. SE = standard error. Level of statistical significance: * $p < 0.01$, ** $p < 0.05$, *** $p < 0.010\%$.

The answers given to the questions regarding the attitudes toward fish consumption of the individuals are presented in Tables 4 and 5, where strongly disagree = 1, disagree = 2, undecided = 3, agree = 4, and completely agree = 5. When the regression results were examined, statistically significant relationships were found between the attitudes, taste, price, convenience, seller trust, wild fish, physical appearance, and fish consumption frequency. In addition, a significant relationship between the education variable, one of the sociodemographic variables, and the frequency of fish consumption was found. Other attitudes about fish consumption and sociodemographic variables were not found to be statistically significant with the frequency of fish consumption.

4. Discussion

Within the scope of this research, all attitudes that may be effective in fish consumption were evaluated, and those related to consumers' purchase or consumption were determined with the help of the established probit model. Accordingly, taste, convenience, seller confidence, wild fish, and physical appearance attitudes were found to be positively related to fish purchase and consumption while the price was negatively related. In addition, the effect of sociodemographic characteristics on fish consumption or purchase was also included in this research model. In this context, a positive relationship between education level and fish consumption behavior was revealed.

According to the results of the descriptive statistics, monthly fish consumption was determined to be higher than red meat and lower than poultry consumption. Middle-income families in Turkey mostly consume poultry for their meat requirements. The relatively low price of poultry is the main reason for this preference, especially in times of high inflation such as now. In addition, fish are mostly preferred by consumers with higher education levels [25,31,57,58]. Due to its high price, red meat consumption is lower than the consumption of fish and poultry.

The descriptive statistics results also show that consumers mostly prefer fish markets for their fish purchases. Among the factors that cause the preference for purchasing in fish markets is the abundance of fish varieties, high probability of finding fresh fish, and relatively lower prices. The fact that fresh fish is the first preference supports the finding that individuals prefer fish markets, which are one of the first links in the supply chain and provide access to fresh fish at lower prices. This is also supported by the correlation between the purchase frequency and the price factor in the regression results.

The majority of consumers preferred fresh fish instead of processed options. Considering the consumption habits of Turkish consumers, dried, salted, and smoked fish are much less preferred than fresh, canned, and frozen fish. Fresh fish is cheaper than processed fish, and habits and cultural factors are effective in the prominence of fresh fish as a fish consumption preference. In their studies, [8] also stated that freshness is one of the most important determinants affecting the level of fish consumption.

According to the ordered probit regression results revealing the relationships between attitudes toward fish and the probability of fish consumption (see Table 5), the attitude with the highest significance with the frequency of fish consumption was physical appearance. The greater the importance given to the physical appearance of the fish, the more likely an individual is to consume or purchase fish more often. In their study, [39] established a relationship between the physical appearance of fish and consumer confidence in fresh seafood. Moreover, [36] reported that one of the attitudes that has the highest impact on the measured perceived quality is the physical appearance of the fish, along with taste and texture. The study of [59] also stated that the indicators of freshness, physical appearance, and color were found to be significant in terms of consumer attitudes.

There is strong evidence that taste is a good predictor of consumers' food choices as a sensory attribute. Consumers often mention taste as a major factor in their food preference decisions [60–63]. The research of [64] was one of the first to reveal the importance of taste attitudes by concluding that husbands, rather than wives, priorities the taste factor in their food preferences.

In many studies on fish consumption preferences, the importance of the taste variable for consumers has also been revealed. In their studies, [3,18,20,27,31,36,37,39,65–69] found that taste is related to fish or other types of seafood purchase and consumption behaviors.

The probit model results of the current study also show that the more delicious consumers find the fish, the more they increase their consumption.

In many studies, price was determined as the leading factor among the barriers to fish consumption, together with the trust of the seller, presence of fishbone, availability, and smell [18,20,27,31,36,68–71]. Fish is perceived to be expensive in most countries, and if consumers indicate that the price level is important, their intention to purchase fish may decrease [37,57]. The low price of fresh fish in the market is probably an advantage for the product to be selected or substituted for other foods, especially in developing countries where high inflation rates are effective such as Turkey [35,36]. Additionally, in our study, price was a significant factor in fish choice behavior. In their study, [72] reached similar results: the probability of consuming fish decreased by 3.4% if the consumers perceived the price as being high.

As previously mentioned, trust is another barrier to consuming fish or seafood products. Lack of trust in choosing and preparing seafood is one of the biggest barriers to seafood consumption [20,34]. Consumer confidence in products moderates the relationship between attitudes and behavioral intentions [73]. In Turkey, consumers prefer fish markets instead of organized retail stores such as supermarkets as the place to buy fish, which reveals the trust factor. In fish markets, on the other hand, there is no intermediary assurance mechanism in determining the qualities of fish, such as freshness and being natural. The seller's declaration is the only element of trust. Therefore, seller reliability is very important. The research results also show a negative relationship between consumers giving importance to the seller's reliability when purchasing fish and fish consumption. As the trust in the seller decreases, the probability of purchasing decreases.

In fish choice behavior, convenience is accepted as another important factor and can determine whether an individual consumes fish or not [39]. The convenience attitude was detected as being significant in previous studies and accepted as a controlling factor determining the seafood choice [2,34,36,37,39,65,74]. The authors of [67] found that convenience is the second most important attitude for fish consumption after shelf life. The authors of [20] indicated that convenience is one of the leading drivers of seafood consumption in Australia. Due to the role of the convenience attitude, there is a demand for fish consumption in a frozen or canned form, which is ready to be cooked or eaten [34]. In our study, the convenience attitude is significantly related to fish consumption. As the consumers perceive that the preparation or consumption fish to be convenient, the probability of consuming or purchasing fish will increase.

In the seafood sector, fish are sourced from aquaculture farms or caught in the wild [2]. In many fish markets, wild fish have a more favorable image regarding taste, health, and nutritional value than aquaculture fish [2,26,40,75]. The quality perception of wild fish is also higher than aquaculture fish. Consumers have more positive attitudes toward wild fish and are willing to pay a premium for wild fish [2,26]. The probit model of this study also established a relation between wild fish preferences and an increase in consumption. According to the model results, the fact that the fish is wild rather than from aquaculture increases the probability of consumption.

The effects of sociodemographic variables on the frequency of consumption were also evaluated within the scope of this study. The results show that among the variables of gender, age, education, income, and household size, education is the only variable associated with the consumption frequency. The positive relationship between education and consumption frequency shows that the higher the education level of the consumers, the higher the probability of an increase in the fish consumption frequency. This is in line with previous studies of [25,28,57,69,76,77], which all found a positive association between education and fish consumption behavior.

5. Conclusions

It can be concluded from the results of this study that greater importance should be given to the modernization of fish markets and more efficient supply chains should be established to deliver fresh fish products to consumers in Turkey. Promotional activities should be carried out for canned and frozen products, which have important industrial infrastructure opportunities in Turkey. Thus, convenience, an important attitude regarding fish, can also be achieved. Fish should be made to be a more accessible and consumable product by applying policies to reduce consumer prices. In Turkey, fish is usually cooked and served without using diverse receipts and culinary arts. We can state that applications that enhance the flavor of fish (such as new cooking methods, sauces, or recipes) will positively affect fish consumption. Diversity and value should be added to fish presentation by researching fish dishes and serving methods in countries with high fish consumption habits. Since consumers give importance to the physical appearance of fish, and this attitude is related to consumption, the appearance characteristics of fresh fish should be highlighted in the market.

One of the main determinants of the pricing of fresh fish in Turkey is whether the fish is wild or from aquaculture. Consumers generally prefer wild fish to aquaculture fish, but wild fish costs about 20% more. There is no standard-setting application that enables them to distinguish whether the fish is wild or farmed, and they generally have to rely on the seller's statement. Seafood fraud has recently increased by 60%, and the seafood industry has been hit particularly hard. Salmon, tuna, and halibut are just a few varieties that are commonly mislabeled in restaurants and grocery stores. A new blockchain network aims to eliminate fraud by tracking seafood from the point of capture to the kitchen. With the knowledge of where and how the fish are produced using QR code applications, the consumer can better know what they are buying. With the increase in these applications, the complexity of this matter may be solved to a great extent. However, the absence of QR codes and blockchain applications creates doubt in consumers, and their trust in the seller becomes very important. For this reason, consumers generally buy fish from people or institutions they know. Features such as the freshness or wildness of fish should be positioned according to market standards and consumer perceptions.

With the help of these results, efficient marketing strategies and promotional practices can be created, and inefficient ones can be avoided. Thus, the research findings have important managerial and policy implications that concern various stakeholders.

Limitations

Some limitations of this study should be acknowledged. First, this study only focused on individuals' fish purchasing behavior for their home consumption. However, it is known that there are options for away-from-home fish consumption such as fish restaurants, fast food, and cafes. Research can also be applied to the away-from-home consumption behaviors of consumers regarding fish. Second, continuing pandemic conditions during the research process forced this research to be conducted with a limited number of samples to prevent an impact on public health. Last, to avoid confusing results, only fish were focused on, neglecting other aquatic products. Similar studies can also be carried out for other types of seafood.

Author Contributions: Conceptualization, O.I.G. and L.S.; methodology, O.I.G. and L.S.; software, O.I.G. and L.S.; validation, O.I.G., L.S., Ç.D. and I.Ö.; formal analysis, O.I.G. and L.S.; investigation, O.I.G., Z.E., I.Ö., H.Y. and Ç.D.; resources, Z.E., H.Y., I.Ö. and Ç.D.; data curation, O.I.G. and L.S.; writing—original draft preparation, O.I.G., Z.E., I.Ö., Ç.D. and H.Y.; writing—review and editing Z.E., I.Ö., Ç.D. and H.Y.; visualization, Z.E., I.Ö., Ç.D. and H.Y.; supervision, O.I.G. and L.S. All authors have read and agreed to the published version of the manuscript.

Funding: This research received no external funding.

Data Availability Statement: Data is contained within the article.

Conflicts of Interest: The authors declare no conflict of interest.

References

1. Sobal, J.; Bisogni, C.A. Constructing Food Choice Decisions. *Ann. Behav. Med.* **2009**, *38* (Suppl. 1), 37–46. [CrossRef] [PubMed]
2. Thong, N.T.; Solgaard, H.S. Consumer's Food Motives and Seafood Consumption. *Food Qual. Prefer.* **2017**, *56*, 181–188. [CrossRef]
3. Tomi, M.; Matuli, D.; Jeli, M. What Determines Fresh Fish Consumption in Croatia. *Appetite* **2016**, *106*, 13–22. [CrossRef] [PubMed]
4. Saba, A.; Di Natale, R. A Study on the Mediating Role of Intention in the Impact of Habit and Attitude on Meat Consumption. *Food Qual. Prefer.* **1999**, *10*, 69–77. [CrossRef]
5. Cardello, A.V. The Role of the Human Senses in Food Acceptance. In *Food Choice, Acceptance and Consumption*, 1st ed.; Meiselman, H.L., Macfie, H.J.H., Eds.; B1ackie Academic & Professional, an Imprint of Chapman & Hall, 2–6 Boundary Row: London, UK, 1996; pp. 1–82.
6. Wansink, B.; Sobal, J. Mindless Eating The 200 Daily Food Decisions We Overlook. *Environ. Behav.* **2007**, *39*, 106–123. [CrossRef]
7. Shepherd, R.; Stockley, L. Fat Consumption and Attitudes Towards Food with a High-Fat Content. *Hum. Nutr. Appl. Nutr.* **1985**, *39A*, 431–442.
8. Honkanen, P.; Olsen, S.O.; Verplanken, B. Intention to Consume Seafood—The Importance of Habit. *Appetite* **2005**, *45*, 161–168. [CrossRef] [PubMed]
9. Thomas, P.R. Improving America's Diet and Health: From Recommendations to Action. In *Committee on Dietary Guidelines Implementation*; Institute of Medicine: Washington, DC, USA, 1991; pp. 1–256.
10. Kearney, J. Food Consumption Trends and Drivers. *Phil. Trans. R. Soc. B* **2010**, *365*, 2793–2807. [CrossRef] [PubMed]
11. Szakaly, Z.; Kontor, E.; Kovacs, S.; Popp, J.; Peto, K.; Polereczki, Z. Adaptation of the Food Choice Questionnaire: The case of Hungary. *Br. Food J.* **2018**, *120*, 1474–1488. [CrossRef]
12. Matsumoto, S.; Yamamoto, M. Empirical Analysis of the Vegetable Wholesale Market in Kanto Region. In *Consumer Concerns about Radioactive Contamination*; Matsumoto, S., Otsuki, T., Eds.; CRC Press: Boca Raton, FL, USA; Taylor & Francis Group: London, UK, 2018; pp. 1–298.
13. Brunso, K.; Verbeke, W.; Olsen, S.O.; Jeppesen, L.F. Motives, Barriers and Quality Evaluation in Fish Consumption Situations Exploring and Comparing Heavy and Light Users in Spain and Belgium. *Br. Food J.* **2009**, *111*, 699–716. [CrossRef]
14. Solomon, M.; Bamossy, G.; Askegaard, S.; Hogg, M.K. *Consumer Behaviour: A European Perspective*, 3rd ed.; Prentica Hall, Pearson, Education Limited: London, UK, 2006.
15. Mothersbaugh, D.L.; Hawkins, D.I. *Consumer Behavior Building Marketing Strategy*, 13th ed.; Mothersbaugh, L.L., Tom, G., Eds.; McGraw-Hill Education: New York, NY, USA, 2016.
16. Peter, J.P.; Olson, J.C. *Consumer Behavior & Marketing Strategy*, 9th ed.; McGraw-Hill/Irwin: New York, NY, USA, 2010.
17. Tomic, M.; Kovacicek, T.; Matulıc, D. Attitudes as Basis for Segmenting Croatian Fresh Fish Consumers. *New Medit.* **2016**, *4*, 63–71.
18. Fiandari, Y.R.; Surachman, S.; Rohman, F.; Hussein, A.S. Perceived Value Dimension in Repetitive Fish Consumption in Indonesia by Using an Extended Theory of Planned Behavior. *Br. Food J.* **2019**, *121*, 1220–1235. [CrossRef]
19. Birch, D.; Dean, D.; Fazal-e-Hasan, S.M.; Lawley, M. Train the Child and Teach the Adult: Developing Intervention Strategies for Increasing Seafood Consumption. *J. Consum. Behav.* **2018**, *17*, 426–438. [CrossRef]
20. Christenson, J.K.; O'kane, G.M.; Farmery, A.K. The Barriers and Drivers of Seafood Consumption in Australia: A narrative literature review. *Int. J. Consum. Stud.* **2017**, *41*, 299–311. [CrossRef]
21. Jacobs, S.; Sioen, I.; Marques, A.; Verbeke, W. Consumer Response to Health and Environmental Sustainability Information Regarding Seafood Consumption. *Environ. Res.* **2018**, *161*, 492–504. [CrossRef]
22. Baptista, R.C.; Rodrigues, H.; Sant'Ana, A.S. Consumption, Knowledge, and Food Safety Practices of Brazilian Seafood Consumers. *Food Res. Int.* **2020**, *132*, 109084. [CrossRef]
23. Hoque, M.Z.; Alam, M.N. Consumers' knowledge Discrepancy and Confusion in Intent to Purchase Farmed fish. *Br. Food J.* **2020**, *122*, 3567–3583. [CrossRef]
24. Rebecca, W.; Ross, R.P.; Fitzgerald, G.F.; Stanton, C. Fatty Acids from Fish: The anti-Inflammatory Potential of Long-chain Omega-3 Fatty Acids. *Nutr. Rev.* **2022**, *68*, 280–289.
25. Myrland, O.; Trondsen, T.; Johnston, R.S.; Lund, E. Determinants of Seafood Consumption in Norway: Lifestyle, Revealed Preferences, and Barriers to Consumption. *Food Qual. Prefer.* **2000**, *11*, 169–188. [CrossRef]
26. Altintzoglou, T.; Verbeke, W.; Filiep, V.; Luten, J. The Image of Fish from Aquaculture Among Europeans: Impact of Exposure to Balanced Information. *J. Aquat. Food Prod. Technol.* **2010**, *19*, 103–119. [CrossRef]
27. Badr, L.M.; Salwa, O.; Ahmed, Y. Perceived Barriers to Consumption of Freshwater Fish in Morocco. *Br. Food J.* **2015**, *117*, 274–285. [CrossRef]
28. Grieger, J.A.; Miller, M.; Cobiac, L. Knowledge and Barriers Relating to Fish Consumption in Older Australians. *Appetite* **2012**, *59*, 456–463. [CrossRef]
29. FAO. The State of World Fisheries and Aquaculture. In *Sustainability in Action*; FAO: Rome, Italy, 2020.
30. Richter, I.G.M.; Klöckner, C.A. The Psychology of Sustainable Seafood Consumption: A Comprehensive Approach. *Foods* **2017**, *6*, 86. [CrossRef]
31. Verbeke, W.; Vackier, I. Individual Determinants of Fish Consumption: Application of the Theory of Planned Behaviour. *Appetite* **2005**, *44*, 67–82. [CrossRef]
32. Santeramo, F.G.; Carlucci, D.; de Devitiis, B.; Nardone, G.; Viscecchia, R. On consumption patterns in oyster markets: The role of attitudes. *Mar. Policy* **2017**, *79*, 54–61. [CrossRef]

33. Cahu, C.; Salen, P.; de Lorgeril, M. Farmed and Wild Fish in the Prevention of Cardiovascular Diseases: Assessing Possible Differences in Lipid Nutritional Values. *Nutr. Metab. Cardiovasc. Dis.* **2004**, *14*, 34–41. [CrossRef]
34. Claret, A.; Guerrero, L.; Aguirre, E.; Rincón, L.; Hernández, M.D.; Martínez, I.; Peleteiro, J.B.; Grau, A.; Rodríguez-Rodríguez, C. Consumer Preferences for Sea Fish Using Conjoint Analysis: Exploratory Study of the Importance of Country of Origin, Obtaining Method, Storage Conditions and Purchasing Price. *Food Qual. Prefer.* **2012**, *26*, 259–266. [CrossRef]
35. Eygi Erdoğan, B.; Mol, S.; Coşansu, S. Factors Influencing the Consumption of Seafood in Istanbul, Turkey. *Turk. J. Fish. Aquat. Sci.* **2011**, *11*, 631–639.
36. Thong, N.T.; Olsen, S.O. Attitude toward and Consumption of Fish in Vietnam. *J. Food Prod. Mark.* **2012**, *18*, 79–95. [CrossRef]
37. Olsen, S.O. Antecedents of Seafood Consumption Behavior. *J. Aquat. Food Prod. Technol.* **2004**, *13*, 79–91. [CrossRef]
38. Steptoe, A.; Pollard, T.M. Development of a Measure of the Motives Underlying the Selection of Food: The Food Choice Questionnaire. *Appetite* **1995**, *25*, 267–284. [CrossRef]
39. Olsen, S.O. Understanding the Relationship Between Age and Seafood Consumption: The Mediating Role of Attitude, Health Involvement and Convenience. *Food Qual. Prefer.* **2003**, *14*, 199–209. [CrossRef]
40. Carlucci, D.; Nocella, G.; De Devitiis, B.; Viscecchia, R.; Bimbo, F.; Nardone, G. Consumer Purchasing Behavior Towards Fish and Seafood Products. Patterns and Insights from a Sample of International Studies. *Appetite* **2015**, *84*, 212–227. [CrossRef] [PubMed]
41. Smith, S.; Varble, S.; Secchi, S. Fish Consumers: Environmental Attitudes and Purchasing Behavior. *J. Food Prod. Mark.* **2017**, *23*, 267–282. [CrossRef]
42. Wang, O.; Somogyi, S.; Charlebois, S. A Comparison Between Business-to-consumer (B2C), Online-to-offline (O2O) and New Retail. *Br. Food J.* **2020**, *122*, 1215–1237. [CrossRef]
43. Schäufele, I.; Janssen, M. How and Why Does the Attitude-Behavior Gap Differ Between Product Categories of Sustainable Food? Analysis of Organic Food Purchases Based on Household Panel Data. *Front. Psychol.* **2021**, *12*, 1901–1914. [CrossRef] [PubMed]
44. Skallerud, K.; Armbrecht, J.; Tuu, H.H. Intentions to Consume Sustainably Produced Fish: The Moderator Effects of Involvement and Environmental Awareness. *Sustainability* **2021**, *13*, 946. [CrossRef]
45. Zander, K.; Feucht, Y. How to Increase Demand for Carp? Consumer Attitudes and Preferences in Germany and Poland. *Br. Food J.* **2020**, *122*, 3267–3282. [CrossRef]
46. Aikman, S.N.; Min, K.E.; Graham, D. Food Attitudes, Eating Behavior, and the Information Underlying Food Attitude. *Appetite* **2006**, *4*, 111–114. [CrossRef]
47. Ham, M.; Jeger, M.; Ivković, A.F. The Role of Subjective Norms in Forming the Intention to Purchase Green Food. *Econ. Res. Ekon. Sstraživanja* **2015**, *28*, 738–748. [CrossRef]
48. McKelvey, R.D.; Zavoina, W.A. Statistical Model for the Analysis of Ordinal Level Dependent Variables. *J. Math. Sociol.* **1975**, *4*, 103–120. [CrossRef]
49. Cameron, A.C.; Trivedi, P.K. *Microeconometrics: Methods and Applications*; Cambridge University Press: Cambridge, UK, 2005.
50. Long Scott, J.; Freese, J. Regression Models for Categorical and Limited Dependent Variables. In *Advanced Quantitative Techniques in the Social Sciences. A Stata Press Publication*, 2nd ed.; Sage Publications: Austin, TX, USA, 2006.
51. Greene, W.H. *Econometric Analysis*, 5th ed.; Prentice Hall: Hoboken, NJ, USA, 2003; pp. 1–1168.
52. Heise, H.; Theuvsen, L. What Do Consumers Think About Farm Animal Welfare in Modern Agriculture? Attitudes and Shopping Behaviour. *Int. Food Agribus. Manag. Rev.* **2017**, *20*, 379–399. [CrossRef]
53. Franz, A.; Deimel, I.; Spiller, A. Concerns About Animal Welfare: A Cluster Analysis of German Pig Farmers. *Br. Food J.* **2012**, *114*, 1445–1462. [CrossRef]
54. Kraus, D. Components and Completion of Partially Observed Functional Data. *J. R. Stat. Soc.* **2015**, *77*, 777–801. [CrossRef]
55. Backhaus, K.; Erichson, B.; Weiber, R. *Fortgeschrittene Multivariate Analysemethoden: Eine Anwendungsorientierte Einführung, Lehrbuch*, 2nd ed.; Springer Gabler: Berlin/Heidelberg, Germany, 2013.
56. Benam, N.S.; Goli, M.; Ardebili, S.M.S.; Vaezshoushtari, N. The quality characteristics of dough and toast bread prepared with wheat flour containing different levels of Portulaca oleracea leaf powder. *Food Sci. Technol.* **2022**, *42*, 1–7.
57. Rahman, M.N.; Islam, A.R.M.T. Consumer Fish Consumption Preferences and Contributing factors: Empirical Evidence from Rangpur City Corporation, Bangladesh. *Heliyon* **2020**, *6*, e05864. [CrossRef]
58. Pieniak, Z.; Kołodziejczyk, M.; Kowrygo, B.; Verbeke, W. Consumption patterns and labelling of fish and fishery products in Poland after the EU accession. *Food Control* **2011**, *22*, 843–850. [CrossRef]
59. Hall, E.T.; Amberg, S.M. Factors Influencing Consumption of Farmed Seafood Products in the Pacific Northwest. *Appetite* **2013**, *66*, 1–9. [CrossRef] [PubMed]
60. Luomala, H.H.; Jokitalo, M.; Karhu, H. Perceived Health and Taste Ambivalence in Food Consumption. *J. Consum. Mark.* **2015**, *32*, 290–301. [CrossRef]
61. Roininen, K.; Tuorila, H.; Zandstrab, E.H.; de Graaf, C.; Vehkalahtic, K.; Stubenitsky, K.; Melad, D.J. Differences in Health and Taste Attitudes and Reported Behaviour Among Finnish, Dutch and British Consumers: A Cross-national Validation of the Health and Taste Attitude Scales (HTAS). *Appetite* **2001**, *37*, 33–45. [CrossRef] [PubMed]
62. Zandstra, E.H.; de Graaf, C.; Staveren, V.W.A. Influence of Health and Taste Attitudes on Consumption of Low- and High-fat Foods. *Food Qual. Prefer.* **2001**, *12*, 75–82. [CrossRef]
63. Chen, M.F. Consumers' Health and Taste Attitude in Taiwan The Impacts of Modern Tainted Food Worries and Gender Difference. *Br. Food J.* **2013**, *115*, 526–540. [CrossRef]

64. Schafer, R.B. Factors Affecting Food Behavior and the Quality of Husbands' and Wives' Diets. *J. Am. Diet. Assoc.* **1978**, *72*, 138–143. [CrossRef]
65. Higuchi, A.; Dávalos, J.; Hernani-merino, M. Theory of Planned Behavior Applied to Fish Consumption in Modern Metropolitan Lima. *Food Sci. Technol.* **2017**, *37*, 202–208. [CrossRef]
66. Zheng, Q.; Wang, H.H.; Lu, Y. Consumer Purchase Intentions for Sustainable Wild Salmon in the Chinese Market and Implications for Agribusiness Decisions. *Sustainability* **2018**, *10*, 1377. [CrossRef]
67. Heide, M.; Olsen, S.O. Influence of Packaging Attributes on Consumer Evaluation of Fresh Cod. *Food Qual. Prefer.* **2017**, *60*, 9–18. [CrossRef]
68. Hermida, M.; Costa, S. Between Tradition and Taste: Fish Consumption Habits in a Small Portuguese Archipelago. *J. Aquat. Food Prod. Technol.* **2020**, *29*, 335–349. [CrossRef]
69. Akuffo, A.S.; Quagrainie, K.K.; Obirikorang, K.A. Analysis of the Determinants of Fish Consumption by Households in Ghana. *Aquac. Econ. Manag.* **2020**, *24*, 294–309. [CrossRef]
70. Kitano, S.; Yamamoto, N. The Role of Consumer Knowledge, Experience, and Heterogeneity in Fish Consumption: Policy Lessons From Japan. *J. Retail. Consum. Serv.* **2020**, *56*, 102151. [CrossRef]
71. Brayden, W.C.; Noblet, L.C.; Evans, S.K.; Rickard, L. Consumer Preferences for Seafood Attributes of Wild-harvested and Farm-raised Products. *Aquac. Econ. Manag.* **2018**, *22*, 362–382. [CrossRef]
72. Lee, M.K.; Nam, J. The Determinants of Live Fish Consumption Frequency in South Korea. *Food Res. Int.* **2019**, *120*, 382–388. [CrossRef] [PubMed]
73. Bennett, P.D.; Harrell, G.D. The Role of Confidence in Understanding and Predicting Buyers' Attitudes and Purchase Intentions. *J. Consum. Res.* **1975**, *2*, 110–117. [CrossRef]
74. Leek, S.; Maddock, S.; Foxall, G. Situational Determinants of Fish Consumption. *Br. Food J.* **2000**, *102*, 18–39. [CrossRef]
75. Vanhonacker, F.; Pieniak, Z.; Verbeke, W. Fish Market Segmentation Based on Consumers' Motives, Barriers and Risk Perception in Belgium. *J. Food Prod. Mark.* **2010**, *16*, 166–183. [CrossRef]
76. Onumah, E.E.; Quaye, E.A.; Ahwireng, A.K.; Campion, B.B. Fish Consumption Behaviour and Perception of Food Security of Low-Income Households in Urban Areas of Ghana. *Sustainability* **2020**, *12*, 7932. [CrossRef]
77. Abdikoğlu, D.İ.; Azabağaoğlu, M.Ö.; Unakıta, G. An Econometric Analysis of Factors Affecting Fish Consumption: The Case of Tekirdag, Turkey. *KSU J. Agric. Nat.* **2020**, *23*, 446–452.

foods

MDPI

Article

Comparative Study of the Nutritional Composition and Antioxidant Ability of Soups Made from Wild and Farmed Snakehead Fish (*Channa Argus*)

Mengting Ren [1,2], Tao Yin [1,2,*], Juan You [1,2], Ru Liu [1,2], Qilin Huang [1,2] and Shanbai Xiong [1,2]

1 Key Laboratory of Environment Correlative Dietology (Ministry of Education), College of Food Science and Technology, Huazhong Agricultural University, Wuhan 430070, China
2 National R & D Branch Center for Conventional Freshwater Fish Processing, Wuhan 430070, China
* Correspondence: yintao@mail.hzau.edu.cn; Tel.: +86-27-8728-8375

Abstract: In recent years, fish soup has become an important product for commercial processing of fish due to its health effects. In this study, nutritional composition and antioxidant ability of soups prepared from farmed and wild snakehead fish were analyzed (hereafter FS and WS soup, respectively). For the FS soup, the proximate composition of protein, fat, ash, free amino acids, and soluble peptides were 2.55%, 0.89%, 0.92%, 0.47%, and 0.62%, respectively. The total amino acid was 390.11 mg/ g, and the proportion of essential amino acid was 27.59%. The total fatty acid was 13.64 g/100 g, of which monounsaturated fatty acid was 5.78 g/100 g, n-6 polyunsaturated fatty acid 3.50 g/100 g, and n-3 polyunsaturated fatty acid 0.41 g/100 g, respectively. The contents of Zn and Ca were 9.04 mg/ kg and 1.13 mg/ g, respectively. The DPPH radical-scavenging ability, Fe^{2+} chelating ability, and hydroxyl radical-scavenging ability was 57.89%, 21.21%, and 25.61%, respectively. Overall, there was no obvious difference in the nutritional composition and antioxidant activity between the FS and WS soups. The protein content (1.90%) of the WS soup was relatively lower, but the total fatty acid (16.22 g/100 g), MUFA (7.17 g/100 g), and Zn (12.57 mg/ kg) contents were significantly higher.

Keywords: growth environment; snakehead fish; fish soup; amino acid composition; fatty acid composition

Citation: Ren, M.; Yin, T.; You, J.; Liu, R.; Huang, Q.; Xiong, S. Comparative Study of the Nutritional Composition and Antioxidant Ability of Soups Made from Wild and Farmed Snakehead Fish (*Channa Argus*). *Foods* **2022**, *11*, 3294. https://doi.org/10.3390/foods11203294

Academic Editor: Isabel Castanheira

Received: 13 September 2022
Accepted: 18 October 2022
Published: 21 October 2022

1. Introduction

Snakehead fish (*Channa argus*), commonly known as black fish, fortune fish, etc., belongs to the family Channidae, and is a ferocious carnivorous freshwater fish [1]. It is widely distributed in rivers, lakes, ponds, and swamps in China, Korea, Japan, Southeast Asia, India, and the Russian Far East [1,2]. The volume of snakehead fish reached 540,000 tons in 2021 in China [3]. Compared to other freshwater fish, snakehead fish has the advantages of high meat yield, high protein and other nutrient content, and various therapeutic effects [4].

Snakehead fish is mainly farmed in high density or grows wildly in fish ponds. The findings of Fuentes, et al. [5] showed differences in proximate composition, color, texture (especially), fatty acids, and free amino acids between the farmed and wild sea bass. O'Neill, et al. [6] did not observe differences in proximate composition between the farmed and wild yellowtail fish. However, from a nutritional point of view, they suggested that the wild yellowtail fish contained better fatty acid complexes and thus might have better nutritional value. Cahu, et al. [7] claimed that the fish farmed under suitable conditions was at least as beneficial in nutrients as wild fish. It can be seen that the influence of growth environment (farm versus wild) on the nutritional composition of fish is controversial, and further research is needed. The major nutritional composition of snakehead fish muscle is water (74.65–79.95%), protein (17.63–20.03%), fat (0.10–1.24%), and ash (1.09–2.16%) [8]. The nutritional composition of snakehead fish muscle may be affected by factors such as origin, food, growth environment, as well as fish size, species, farming time, and processing [6,9–11].

In recent years, fish soup has become an important product for commercial processing of fish due to its health effects. The soup prepared from the snakehead fish can promote the healing of wounds and burns [12]. Yuan, et al. [13] reported that snakehead soup had an anti-fatigue effect. In recent years, according to the research results of Zhang, et al. [4], snakehead soup also possessed antioxidant effects. The growth environment may affect the muscle composition of snakehead fish [14], which in turn affects the nutritional composition and functionality of the soup made from the snakehead fish. However, there is no report on the effects of growth environment on the nutritional composition and functionality of snakehead fish soup.

Therefore, the nutritional composition (including proximate composition, amino acid composition, fatty acid composition, and mineral composition) and antioxidant ability of the two soups prepared from the farmed and wild snakehead fish were comparatively analyzed in this study, in order to provide a basis for the nutritional evaluation of the snakehead fish.

2. Materials and Methods

2.1. Materials and Reagents

In this study, the farmed snakehead fish grew in the fish ponds in Guangzhou. It was farmed alone in high density and fed regularly. The wild snakehead fish of single or very little popularity (<5 tails per pond) grew in the fish ponds in Guangzhou with other farmed fish, and hunted small trash fish, crabs, frogs, etc., for food. The morphology of the two kinds of fish is shown in Figure S1. The comparison of the growth environment and food nutrient composition is shown in Table S1. The length-to-height ratio and BMI (Body Mass Index) were 4.74 and 5.20, respectively. For the wild snakehead fish, they were 5.81 and 4.24, respectively. Approximate 15 kg raw fish each was taken for the study.

Refined cooking salt (China Salt Industry Group Co., Beijing, China) and peanut oil (Shandong Luhua Group Co., Shandong, China) are commercially available. Copper sulfate, potassium sulfate, concentrated sulfuric acid, sodium hydroxide, hydrochloric acid, ethanol, petroleum ether, zinc acetate, concentrated nitric acid, 1,10-phenanthroline monohydrate, ferrous sulfate, Trichloroacetic acid (TCA), L-leucine (L-Leu), bovine serum albumin (BSA) and other reagents were purchased from China National Pharmaceutical Co., Ltd. (Beijing, China). 1,1-Diphenyl-2-picrylhydrazine (DPPH), mixed standard of 35 fatty acids was obtained from Sigma-Aldrich (St. Louis, MO, USA).

2.2. Preparation of Snakehead Fish Soup

We adopted a traditional method to prepare snakehead soup (according to the method of Zhang, et al. [15]), with some improvements. Fresh fish meat (about 700 g) with head and bones were washed with tap water, and then the water was drained off. Luhua peanut oil (3% of the weight of the fish) was added into an iron pot and heated by an induction cooker (RT2134, Midea group, Foshan, China). After boiling, the fish meat was added into to the pan and stir-fried until golden brown. Water (3 times the weight of fish) was poured into the pan. The mixture was heated up to boiling temperature under 2100 W power, and kept boiling for 15 min. Foam in the surface was removed during the process. After the pan was covered by a lid, the mixture was heated under mild power (300 W) for 30 min. Salt (1.2% of the weight of the fish) was added right before turning the induction cooker off. The fish soup was separated by filtering through single layer cheese cloth.

With reference to the method of Guo, et al. [16], snakehead fish soup was dried in a pilot-scale vacuum freeze dryer (102241, Martin Christ, Osterode, Germany). The dryer was equipped with 10 stainless steel pans (50 cm diameter). The wild and farmed snakehead soup were transferred to plates, cooled to room temperature, and then frozen at −18 °C. The frozen fish soup was dried under the conditions of shelf temperature at 20 °C and a vacuum of 338 Pa for 4 h, followed by drying at −5 °C for 6 h, to remove most of the water. After adjusting the temperature and vacuum to −1 °C and 338 Pa, respectively, the fish soup was further dried for 10 h. The dried snakehead fish soup was pulverized into powder, placed in a plastic bag, vacuum-packed, and stored in a desiccator at room temperature. The preparation of snakehead soup powder was repeated twice. The yield of soup powder made from farmed snakehead was about 14.97%, and that of wild snakehead 12.12%.

2.3. Proximate Composition

The moisture content of the samples was determined by drying the samples in an oven at 105 °C for 4–6 h [17]. Crude protein was determined by Kjeldahl method [17,18]. During the analysis, an automatic Kjeldahl analyzer (K9840, Haineng, Shandong, China) was used, and a factor of 6.25 was used to convert nitrogen content to protein content. Ash was determined by incinerating the sample in a muffle furnace (BF51794C-1, Thermo Scientific, Waltham, MA, USA) at 550 °C for 18 h [19]. Crude fat was extracted according to the method of Bukhanko, et al. [20].

The free amino acid content was determined with reference to the method of Bordons [21] with minor modifications. The lyophilized powders of wild and farmed snakehead soup were weighed and dissolved in distilled water to constant volume. The soup was put in a centrifuge tube. Two milliliters of 10% TCA solution was added to precipitate the protein for 10 min. Then, the sample was centrifuged at 4000 r/min for 15 min. Finally, the supernatant was diluted by 15 times. The absorbance was measured at 570 nm using a spectrophotometer (UV-1750, Shimadzu, Kyoto, Japan) using a ninhydrin colorimetric method.

The determination of TCA-soluble peptide content was performed with reference to the method of Rawdkuen, et al. [22]. About 0.5 g of fish broth freeze-dried powder was mixed in 18 mL of 5% TCA solution, homogenized at 11,000 r/min for 2 min, then kept at 4 °C for 1 h after homogenization, and then centrifuged at 8500 r/min for 5 min. The supernatant was assayed for TCA-soluble peptide concentration by Lowry's method [23].

2.4. Total Amino Acid Composition Analysis

Referring to the method of Fuentes, et al. [5], the amino acid composition was determined by HPLC method. The sample was weighed into a hydrolysis tube for hydrolyzing. After this step, it was taken out and cooled. Then it was transferred to a colorimetric tube to constant volume and put it in a vacuum drying oven for drying. The centrifuge tube was filled with nitrogen, then reagents were added to derivatize at room temperature. Mobile phase was added and mixed well. The processed sample was passed through the filter membrane and entered the instrument (1525 + 717 + 2998PDA + RI, Waters, MA, USA) for analysis. The amino acid content was calculated according to the following formula

$$X = \frac{c \times V}{m}$$

X: the content of amino acids in the sample, in mg/ kg; m: the sample weight, in g; V: the constant volume of the sample after hydrolysis, in mL; c: the calculated concentration of each amino acid on the standard curve, in μg/ mL.

Amino acid score (AAS), chemical score (CS) and the Essential Amino Acid Index (EAAI) were calculated by the following equations.

$$AAS = \frac{\text{Sample amino acid content}/\%}{\text{Content of the same amino acid in the FAO/WHO scoring standard model}/\%}$$

$$CS = \frac{\text{Sample amino acid content}/\%}{\text{Identical Amino Acid Content in Whole Egg Protein}/\%}$$

$$EAAI = \sqrt[n]{\frac{100a}{ae} \times \frac{100b}{be} \times \cdots \times \frac{100j}{je}}$$

Among them, n is the number of essential amino acids compared; a, b ... j are the essential amino acid content of the protein /%; ae, be ... je is the essential amino acid content of whole egg protein /%.

2.5. Fatty Acid Composition Analysis

Referring to the method of Zhang, et al. [24], the fatty acid composition was determined by GC-MS method. An appropriate amount of the sample was weighed. Pyrogallic acid (100 mg), 95% ethanol (2 mL) hydrochloric acid (10 mL), and zeolite were added to the flask with the sample. Then the flask was soaked in water at 70~80 °C for 40 min.

After hydrolysis, 95% ethanol (10 mL) was added to the sample, and 50 mL diethyl ether petroleum ether mixture and hydrolysate were merged into a separating funnel. After shaking and standing still, the ether layer extract was collected. The above steps were repeated 3 times. Finally, it was collected into a flask of constant weight. The flask was dried in a water bath and then dried at 100 °C ± 5 °C for 2 h.

The extracted fat was subjected to fat saponification and fatty acid methylation. The processed sample was passed through the filter membrane (0.45 μm) and entered the instrument (Trace1310 ISQ, Thermo Scientific, Waltham, MA, USA) for analysis.

Chromatographic column: TG-5MS (30 m × 0.25 mm × 0.25 μm); heating procedure: keep at 80 °C for 1min, then rise to 200 °C at the rate of 10 °C /min, continue to rise to 250 °C at the rate of 5 °C /min, and finally rise to 270 °C at the rate of 2 °C /min, and keep for 3 min; inlet temperature: 290 °C; carrier gas flow rate: 1.2 mL/min; mass spectrometry conditions: ion source temperature: 280 °C, transmission line temperature: 280 °C, solvent delay time: 5 min, scanning range: 30~400 amu, ion source: EI source 70 eV.

2.6. Mineral Element Composition Analysis

With reference to the method of Leme, et al. [25], the mineral composition was determined by the ICP-MS method. An appropriate amount of sample was weighed into the digestion tank and then mixed with nitric acid. After the reaction, it was put into a microwave digestion apparatus for digestion. After the temperature was cooled, the digestion tank was taken out and put in a fume hood. Then the digestion tank was opened, and it was diluted with ultrapure water to a certain concentration. The blank control was treated in the same way. Samples were tested on an ICP-MS instrument (iCAPQ, Thermo Scientific, Waltham, MA, USA).

2.7. Measurement of Antioxidant Ability

2.7.1. DPPH Radical-Scavenging Ability

Measurement of DPPH radical-scavenging ability was carried out referring to the method of Zhang, et al. [24]. Fish soup powder was weighed and dissolved in distilled water and diluted. Then it was transferred into a centrifuge tube and centrifuged. A 4.0 mL supernatant was mixed with 1.0 mL DPPH solution (0.1 mmol/L in 95% ethanol). The mixture was shaken and left for 30 min at room temperature, and the absorbance of the resulting solution was measured at 517 nm using a spectrophotometer (UV-1750, Shimadzu, Kyoto, Japan). Distilled water was used as the control group instead of the sample, and ethanol was used as the blank group. The DPPH radical-scavenging ability was calculated according to the following formula.

$$\text{DPPH Radical-Scavenging (\%)} = [(OD_1 - (OD_2 - OD_3))/OD_1] * 100$$

Among them, OD_1, OD_2, and OD_3 represent the absorbance values of the control group, sample group and blank group, respectively.

2.7.2. Fe^{2+} Chelating Ability

We referred to the method of Zhang, et al. [24] for the determination of Fe^{2+} chelating ability with some modifications. Fish soup powder sample was weighed and dissolved in distilled water and diluted. Then it was transferred into a centrifuge tube and centrifuged. The supernatant (1 mL) was mixed with 3.85 mL of distilled water, ferrous chloride (0.05 mL, 2 mmol/L) and ferrozine (0.1 mL, 5 mmol/L). After vigorous mixing, the solution was left to stand at room temperature for 20 min, and the absorbance of the resulting solution was measured at 562 nm with a spectrophotometer (UV-1750, Shimadzu, Kyoto, Japan). The samples and chemical reagents were replaced by distilled water as the control and blank groups, respectively. The Fe^{2+} chelating ability was calculated according to the following formula.

$$Fe^{2+}\text{ Chelating (\%)} = [(OD_1 - (OD_2 - OD_3))/OD_1] * 100$$

Among them, OD_1, OD_2 and OD_3 represent the absorbance values of the control group, the sample group, and the blank group, respectively.

2.7.3. Hydroxyl Radical-Scavenging Ability

It was referred to the method of Zhang, et al. [24] with some modifications. Fish soup powder sample was weighed and dissolved in distilled water and diluted. Then it was transferred into a centrifuge tube and centrifuged. The supernatant (2 mL) mixed with 0.75 mM 1,10-phenanthroline (2 mL), 0.75 mM $FeSO_4$ (2 mL), and 0.2 M phosphate buffer (pH = 7.4, 2 mL). Then 0.12% H_2O_2 (1 mL) solution was added to the mixture and incubated at 37 °C for 1 h, and the absorbance of the resulting solution was measured at 536 nm with a spectrophotometer (UV-1750, Shimadzu, Kyoto, Japan). Control group 1 consisted of the same solution as the sample group, except that equivalent deionized water was used instead of the sample solution; control group 2 was based on control group 1 by replacing H_2O_2 with distilled water. The composition of the blank group was the same as that of the sample group, except that equivalent deionized water was used instead of 1,10-phenanthroline monohydrate and $FeSO_4$ solution. The hydroxyl radical-scavenging ability was calculated according to the following formula.

$$\text{Hydroxyl Radical-Scavenging (\%)} = (OD_s - OD_b - OD_1)/(OD_2 - OD_1) * 100\%$$

Among them, OD_s, OD_b, OD_1 and OD_2 represent the absorbance of the sample group, blank group, control group 1 and control group 2, respectively.

2.8. Statistical Analysis

The experimental data were expressed as mean ± standard deviation (mean ± SD). Multiple comparison analysis was performed using SPSS 26.0. Each index was repeated three times, and the minerals were repeated two times.

3. Results

3.1. Proximate Composition

It can be seen from Table 1 that the proximate composition of moisture, protein, fat, ash, free amino acids, and soluble peptides in the farmed snakehead fish soup (wet base) were 95.01%, 2.55%, 0.89%, 0.92%, 0.47%, and 0.62%, respectively. Zhu, et al. [26] reported that the proximate composition of protein, fat, ash in snakehead fish soup were 0.5–0.9%, 0.6%, and 1.4%, respectively. With respect to the soup prepared by Zhu, et al. [26], the ash content of the snakehead fish soup in this study (Table 1) was relatively lower, but the crude protein and crude fat content were relatively higher. Compared with the soup making method by Zhu, et al. [26], the heating power used in this study is stronger, and therefore the protein and fat were more easily dissolved from the fish meat into the soup. Zhu, et al. [26] chosen a fish to water mass ratio of 1:6 to make soup. The fish to water mass ratio (1:3) was smaller in our study (Method 2.2). The study of Xu [27] showed that when the fish-water mass ratio was 1:4, the protein concentration was the highest in the soup; when the ratio was 1:10, the concentration was the lowest. This was because a large amount of water diluted the concentration of dissolved protein. Therefore, the fish soup prepared in our study (Table 1) contained more protein and crude fat. Zhu, et al. [26] used the fish head as the raw material to make fish soup. We used the descaled and gutted snakehead fish as the raw material to make fish soup. The fish head contains higher ash content. As a result, the ash content of our soup was much lower.

Table 1. Proximate composition of soup made from farmed and wild snakehead fish.

Composition	Farmed Snakehead Fish Soup		Wild Snakehead Fish Soup	
	Wet Base	Dry Base	Wet Base	Dry Base
Moisture	95.01 ± 0.32 [a]	-	95.96 ± 0.90 [a]	-
Crude fat	0.89 ± 0.09 [a]	17.86 ± 1.82 [a]	0.79 ± 0.03 [a]	19.65 ± 2.42 [a]
Crude protein	2.55 ± 0.08 [a]	51.20 ± 1.62 [a]	1.90 ± 0.01 [b]	47.03 ± 0.36 [a]
Crude ash	0.92 ± 0.12 [a]	18.41 ± 2.47 [a]	0.71 ± 0.11 [a]	17.66 ± 2.80 [a]
Free amino acids	0.47 ± 0.05 [a]	9.49 ± 0.99 [a]	0.30 ± 0.03 [b]	7.30 ± 0.74 [a]
Soluble peptide	0.62 ± 0.04 [a]	12.47 ± 0.74 [a]	0.47 ± 0.08 [a]	11.69 ± 1.93 [a]

Note: Proximate composition is expressed as %. Mean ± SD (standard deviation) from three replicates. Different lowercase letters in the same line indicate significant differences in composition between soup made from farmed and wild snakehead fish ($p < 0.05$).

In China, in addition to snakehead fish, crucian carp, bighead carp, etc. are also freshwater fish commonly used in making soups. According to the reports of Xia and Xu [28] and Xu [27], the protein, fat, and ash in the crucian carp (*Carassius auratus*) soup and bighead carp (*Aristichthys nobilis*) soup (wet base) were in the range of 0.70%~1.11%, 0.70%~1.11% and 1.2%~1.6%, respectively. By comparing with the basic nutrients of the fish soups mentioned above, it might be concluded that the snakehead fish soup possesses the characteristic of high protein and fat contents, but low content of ash.

The proximate composition of moisture, protein, fat, ash, free amino acids, and soluble peptides in the wild snakehead fish soup (wet base) were 95.96%, 1.90%, 0.79%, 0.71%, 0.30%, and 0.47%, respectively. Compared with the farmed snakehead soup, its crude fat, crude ash, free amino acid, and soluble peptide content were slightly lower ($p > 0.05$); but crude protein content was significantly lower ($p < 0.05$). The total amount of protein and its degradation products in the farmed snakehead soup was higher than that of wild snakehead soup. It might be related to the relatively higher protein content in the muscle of farmed snakehead [14]. Yin, et al. [29] and O'Neill, et al. [6] and Cahu, et al. [7] reported that the mass fractions of protein and fat in muscles of fish in different growth environments varied greatly. The nutritional content of fish is closely related to its living environment (farmed or wild growth), feed composition, and growth period (larvae or adults, etc.). The fat content of feed for farmed snakehead fish is higher than that in the food of wild snakehead fish (Table S1). Increasing the fat level in the feed could increase the energy of farmed snakehead fish, which might reduce the energy supply proportion of protein and thus increase the protein content in fish meat [30]. As a result, the protein content of the soup made from the farmed snakehead fish was higher.

3.2. Total Amino Acid Composition

It can be seen from Table 2 that the amino acid of the highest content in the farmed snakehead soup (dry base) is glycine (82.56 mg/g, 21.16%), followed by arginine (42.79 mg/g, 10.97%), alanine (39.13 mg/g, 10.03%), proline (39.05 mg/g, 10.01%) and glutamic acid (36.20 mg/g, 9.28%), the lowest content was tyrosine (3.81 mg/g, 0.98%). Correspondingly, the content of glycine in farmed snakehead soup (wet base) was 4.12 mg/g, followed by arginine 2.13 mg/g, alanine 1.95 mg/g, proline 1.95 mg/g and glutamic acid 1.81 mg/ g, the lowest content was tyrosine 0.19 mg/g. Zhu, et al. [26] reported that the highest content of snakehead soup (wet base) was glycine (2.011 mg/g), which was consistent with our result.

Table 2. Amino acid composition of soup made from farmed and wild snakehead fish.

Amino Acids	Farmed Snakehead Fish Soup		Wild Snakehead Fish Soup	
	mg/g	%	mg/g	%
Aspartic acid	17.20 ± 6.34 [a]	4.41	19.50 ± 5.49 [a]	5.43
Glutamic acid	36.20 ± 10.03 [a]	9.28	36.80 ± 8.06 [a]	10.25
Serine	14.25 ± 1.68 [a]	3.65	13.61 ± 1.18 [a]	3.79
Glycine	82.56 ± 1.53 [a]	21.16	76.41 ± 1.44 [b]	21.29
Histidine	7.48 ± 0.97 [a]	1.92	6.78 ± 0.86 [a]	1.89
Arginine	42.79 ± 7.10 [a]	10.97	33.77 ± 2.66 [a]	9.41
Threonine *	14.59 ± 0.96 [a]	3.74	13.93 ± 0.56 [a]	3.88
Alanine	39.13 ± 0.45 [a]	10.03	34.74 ± 0.21 [b]	9.68
Proline	39.05 ± 3.04 [a]	10.01	34.30 ± 3.01 [a]	9.55
Tyrosine *	3.81 ± 0.02 [a]	0.98	3.64 ± 0.05 [b]	1.01
Valine *	10.98 ± 0.15 [a]	2.81	10.07 ± 0.14 [b]	2.80
Methionine *	7.92 ± 0.32 [a]	2.03	7.58 ± 0.30 [a]	2.11
Cysteine	-	-	-	-
Isoleucine *	8.68 ± 0.40 [a]	2.22	7.98 ± 0.11 [a]	2.22
Leucine *	19.36 ± 0.74 [a]	4.96	18.28 ± 0.09 [a]	5.09
Phenylalanine *	15.95 ± 0.82 [a]	4.09	14.35 ± 0.40 [a]	4.00
Lysine *	30.15 ± 0.97 [a]	7.73	27.23 ± 0.60 [b]	7.59
TAA	390.11 ± 10.38 [a]	100%	358.97 ± 15.92 [a]	100%
EAA	107.63 ± 3.71 [a]	27.59%	99.42 ± 0.32 [a]	27.70%
NEAA	282.48 ± 14.08 [a]	72.41%	259.55 ± 15.92 [a]	72.30%
E/N		38.10%		38.31%

Mean ± SD (standard deviation) from two replicates. * indicates essential amino acids. "-" indicates that the amino acid is not detected. TAA indicates total amino acids, EAA essential amino acids, NEAA non-essential amino acids, E/N the ratio of essential amino acids to non-essential amino acids. Different lowercase letters in the same line indicate significant differences in the amino acid content between soup made from farmed and wild snakehead fish ($p < 0.05$).

The total amino acid content of the farmed snakehead soup (dry base) was 390.11 mg/g, and the corresponding total amino acid content in wet basis was 19.46 mg/g. The lysine content of the farmed snakehead soup (dry base) was 30.15 mg/g, the corresponding wet basis content was 1.50 mg/g, and the percentage was 7.73%. Zhu, et al. (2017) reported that the total amino acid content of snakehead fish soup was 10.177 mg/g, the lysine content was 0.635 mg/g, and the percentage was 6.24%. Compared with the snakehead fish soup reported by Zhu, et al. [26], the total amino acid content and lysine content of the farmed snakehead soup in our study (Table 2) were significantly higher, which were the same as the differences in free amino acids in the proximate composition (Table 1).

Zhang, et al. [31] reported that the total amino acid content of crucian carp soup was 0.66–0.70 mg/g. The essential amino acid content of the soup in this study (Table 2) was 107.63 mg/g, and the corresponding wet basis content was 5.37 mg/g, accounting for 27.59%. The essential amino acid content of crucian carp soup reported by Zhang, et al. [31] was 24.97% to 25.85%. The total amino acid and essential amino acid content of snakehead fish soup in this study were higher than those reported by Zhang, et al. [31] in crucian carp soup. Therefore, compared with crucian carp soup, snakehead fish soup's nutritional value of amino acids was higher.

Compared with farmed snakehead fish soup, the content of serine, histidine, arginine, threonine, proline, methionine, isoleucine, leucine, and phenylalanine in wild snakehead soup was relatively lower ($p > 0.05$). The contents of aspartic acid and glutamic acid ($p > 0.05$) were relatively higher. The contents of glycine, alanine, tyrosine, valine and lysine ($p < 0.05$) were significantly lower. In terms of percentages, except for aspartic acid, glutamic acid, and arginine, there was no significant difference. The total amino acid content (dry basis) of wild snakehead fish soup was 358.97 mg/g, which was slightly lower than that of the farmed snakehead fish soup ($p > 0.05$). The result was consistent with the protein content (Table 1). The essential amino acid content of wild snakehead soup was

99.42 mg/g, the percentage was 27.70%. There was no significant difference between the wild and farmed snakehead fish soup ($p > 0.05$). Zhao, et al. [32] reported that there was no significant difference in the total amino acid content and composition of farmed and wild snakehead fish meat, which was consistent with the results of this study. The results of Table 2 showed that the growth environment had no significant effect on the total amino acid content and composition of snakehead fish soup.

It can be seen from Table 3 that the AAS and CS of the lysine of the two kinds of snakehead soup were greater than 1. Lysine is the first limiting amino acid of cereals, which is the staple food of Asian people [33]. Eating fish soup and grains at the same time can achieve nutritional balance. The AAS close to 1 were threonine, phenylalanine + tyrosine, and leucine. Those below 1 were valine, isoleucine, methionine + cysteine, and the lowest among them were valine and isoleucine, which were the first type of restrictive amino acids. Compared with the farmed snakehead soup, except the AAS and CS of methionine + cysteine were significantly higher ($p < 0.05$), the AAS and CS of essential amino acids in the wild snakehead soup had no significant difference ($p > 0.05$).

Table 3. Scoring AAS, CS and EAAI of protein in soup made from farmed and wild snakehead fish.

Amino Acid	FAO/WHO Amino Acid Scoring Standard Model/%	Whole Egg Protein Scoring Mode/%	Farmed Snakehead Fish Soup		Wild Snakehead Fish Soup	
			AAS	CS	AAS	CS
Threonine	4.00	4.98	0.94 ± 0.09 [a]	0.75 ± 0.07 [a]	0.97 ± 0.08 [a]	0.78 ± 0.06 [a]
Valine	5.00	7.42	0.56 ± 0.02 [a]	0.38 ± 0.01 [a]	0.56 ± 0.02 [a]	0.38 ± 0.01 [a]
Isoleucine	4.00	6.60	0.56 ± 0.04 [a]	0.34 ± 0.02 [a]	0.56 ± 0.03 [a]	0.34 ± 0.02 [a]
Leucine	7.00	8.80	0.71 ± 0.04 [a]	0.56 ± 0.04 [a]	0.73 ± 0.03 [a]	0.58 ± 0.02 [a]
Lysine	5.50	6.40	1.41 ± 0.08 [a]	1.21 ± 0.07 [a]	1.38 ± 0.04 [a]	1.19 ± 0.03 [a]
Methionine + Cysteine	3.50	5.48	0.58 ± 0.01 [b]	0.37 ± 0.01 [b]	0.60 ± 0.00 [a]	0.39 ± 0.00 [a]
Phenylalanine + Tyrosine	6.00	10.08	0.85 ± 0.06 [a]	0.50 ± 0.03 [a]	0.84 ± 0.05 [a]	0.50 ± 0.03 [a]
Tryptophan	1.00	1.70	-	-	-	-
Total	36.00	51.46				
EAAI			53.25			53.76

Note: AAS indicates amino acid scores, CS chemical scores, EAAI essential amino acid indices. "-" indicates the amino acid was not detected. Different lowercase letters in the same line indicate significant differences in amino acid scores and chemical scores between soup made from farmed and wild snakehead fish ($p < 0.05$).

3.3. Fatty Acid Composition

The GC-MS chromatogram of fatty acid analysis is shown in Figure S2, and the calculated individual fatty acid content is in Table 4. It can be seen from Table 4 that a total of 25 kinds of fatty acids were detected in the farmed snakehead soup, including 20 kinds of saturated fatty acids (SFA), 5 kinds of monounsaturated acids (MUFA) and 8 kinds of polyunsaturated acids (PUFA). The highest content was C18:1n9c, which was 5.29 g/100 g. The relatively low content was C15:0, C17:0, C18:3n6, C21:0, C20:2, C20:3n6, C20:3n3, C22:1n9, C20:4n6, C23:0, C22:2, C20:5n3 and C24:1. Their content was in the range of 0.01~0.06 g/100 g. The content of EPA+DHA was 0.24 g/ 100 g, accounting for 1.77%. The total amount of MUFA was 5.78 g/100 g, accounting for 42.33%. The total amount of n-6 PUFA and n-3 PUFA were 3.50 g/100 g and 0.41 g/100 g, respectively, and the total proportion were 25.67% and 2.98%, respectively.

Table 4. Fatty acid composition in soup made from farmed and wild snakehead fish.

Fatty Acids	Farmed Snakehead Fish Soup		Wild Snakehead Fish Soup	
	g/100 g	%	g/100 g	%
C8:0	-	-	-	-
C10:0	-	-	-	-
C11:0	-	-	-	-
C12:0	-	-	-	-
C13:0	-	-	-	-
C14:0	0.12 ± 0.02 [a]	0.85	0.11 ± 0.09 [a]	0.65
C14:1	-	-	-	-
C15:0	0.02 ± 0.00 [a]	0.13	0.02 ± 0.01 [a]	0.11
C15:1	-	-	-	-
C16:0	2.46 ± 0.13 [a]	18.00	2.88 ± 0.38 [a]	17.75
C16:1	0.23 ± 0.03 [a]	1.66	0.22 ± 0.08 [a]	1.38
C17:0	0.03 ± 0.01 [a]	0.23	0.03 ± 0.01 [a]	0.18
C17:1	-	-	-	-
C18:0	0.76 ± 0.01 [b]	5.58	1.00 ± 0.05 [a]	6.16
C18:1n9t	-	-	-	-
C18:1n9c	5.29 ± 0.15 [b]	38.77	6.44 ± 0.10 [a]	39.69
C18:2n6t	-	-	-	-
C18:2n6c	3.41 ± 0.18 [a]	24.98	3.67 ± 0.28 [a]	22.64
C20:0	0.16 ± 0.02 [a]	1.16	0.19 ± 0.00 [a]	1.17
C18:3n6	0.03 ± 0.00 [a]	0.20	0.03 ± 0.01 [a]	0.17
C18:3n3	0.15 ± 0.00 [a]	1.07	0.07 ± 0.03 [a]	0.42
C20:1	0.20 ± 0.02 [a]	1.49	0.19 ± 0.04 [a]	1.15
C21:0	0.01 ± 0.00 [a]	0.04	0.01 ± 0.00 [a]	0.03
C20:2	0.06 ± 0.00 [a]	0.44	0.05 ± 0.01 [a]	0.29
C22:0	0.23 ± 0.06 [a]	1.69	0.30 ± 0.03 [a]	1.83
C20:3n6	0.02 ± 0.00 [b]	0.17	0.04 ± 0.00 [a]	0.24
C20:3n3	0.019 ± 0.000 [a]	0.14	0.003 ± 0.000 [b]	0.02
C22:1n9	0.04 ± 0.01 [a]	0.32	0.31 ± 0.18 [a]	1.93
C20:4n6	0.04 ± 0.02 [a]	0.32	0.13 ± 0.15 [a]	0.80
C23:0	0.01 ± 0.00 [a]	0.04	0.01 ± 0.00 [a]	0.05
C22:2	0.01 ± 0.00 [a]	0.07	-	-
C20:5n3	0.04 ± 0.01 [a]	0.27	0.08 ± 0.10 [a]	0.47
C24:0	0.11 ± 0.03 [a]	0.80	0.14 ± 0.01 [a]	0.87
C24:1	0.01 ± 0.00 [a]	0.09	0.01 ± 0.00 [b]	0.05
C22:6n3	0.21 ± 0.05 [a]	1.50	0.31 ± 0.39 [a]	1.93
Total fatty acid	13.64 ± 0.04 [b]		16.22 ± 0.73 [a]	
EPA+DHA	0.24 ± 0.06 [a]	1.77	0.39 ± 0.49 [a]	2.39
∑SATD	3.89 ± 0.01 [a]	28.52	4.67 ± 0.49 [a]	28.82
∑MUFA	5.78 ± 0.14 [b]	42.33	7.17 ± 0.16 [a]	44.20
∑PUFAω6	3.50 ± 0.16 [a]	25.67	3.87 ± 0.13 [a]	23.85
∑PUFAω3	0.41 ± 0.05 [a]	2.98	0.46 ± 0.52 [a]	2.84

Mean ± SD (standard deviation) from three replicates. EPA indicates eicosapntemacnioc acid, DHA docosahexaenoic acid, SATD saturated fatty acid, MUFA monounsaturated fatty acid, PUFAω6 omega 6 polyunsaturated fatty acids, PUFAω3 omega 3 polyunsaturated fatty acids. "-" indicates the fatty acid was not detected. Different lowercase letters in the same line indicate significant differences in the fatty acid content soup made from farmed and wild snakehead fish ($p < 0.05$).

Twenty-four kinds of fatty acids were detected in wild snakehead soup. Compared with the fatty acids detected in farmed snakehead soup, the undetected fatty acid was C22:2. The contents of C18:0, C18:1n9c and C20:3n6 in wild snakehead soup were significantly higher than those in farmed snakehead soup ($p < 0.05$); while the contents of C20:3n3 and C24:1 were significantly lower than those in farmed snakehead soup ($p < 0.05$). There was no significant difference in other fatty acids content ($p > 0.05$). The total amount of fatty acids in wild snakehead soup was 16.22 g/100 g, which was 1.19 times that of farmed snakehead fish soup. The difference was the same as the fat content in dry basis (Table 1).

The absolute contents of EPA+DHA and ∑MUFA In wild snakehead soup were 0.39 g/100 g and 7.17 g/100 g, respectively. The absolute content of EPA+DHA was higher than that of farmed snakehead fish soup ($p > 0.05$), and the absolute content of MUFA was significantly higher than that of farmed snakehead fish soup ($p < 0.05$). Compared with farmed fish, wild fish usually contains higher levels of eicosapentaenoic acid (EPA) and docosahexaenoic acid (DHA). Several studies have pointed out that the content of MUFA in wild fish is higher [6]. The results of this study (Table 3) were consistent with the above results. According to the reported research differences in diet and growth environment lead to differences in the fatty acid composition of fish muscle between the wild and farmed fish [6,34,35]. Farmed snakehead fish swims on the surface layer of fish ponds with high temperature. However, the wild snakehead fish stay at the bottom layer of water with low temperature and high activity. Studies have shown that the greater the depth, the more total polyunsaturated fatty acids [36]. It was speculated that the different growth environment might change the fatty acid composition of snakehead fish muscles, thereby resulting in different fatty acid composition of the soups made from farmed and wild snakehead fish.

n-3 polyunsaturated fatty acids (n-3 PUFAs) have good anti-inflammatory properties [37]; some n-6 polyunsaturated fatty acids are required for normal human metabolism and human health, while n-6 polyunsaturated fatty acids have pro-inflammatory activity and play an important role in immune function [38]. Studies have shown that a lower n-6 PUFAs to n-3 PUFAs ratio is beneficial to human health [39,40]. The proportion of n-6 PUFAs to n-3 PUFAs in the diet of the population in developed countries is high (15:1-20:1) [40]. The ratio of n-6 polyunsaturated fatty acids and n-3 polyunsaturated fatty acids in the farmed fish snakehead soup in this study is 8.61, and 8.40 in wild snakehead fish soup. The results indicated that snakehead fish soup is a good source of polyunsaturated fatty acids.

3.4. Mineral Composition

Calcium is essential for the various basic physiological functions such as bone mineralization, blood coagulation, neuronal transmission, muscle contraction, and intracellular signaling [41]; Iron is an important part of human hemoglobin, myoglobin, which are involved in the transportation and storage of human oxygen and the synthesis of various metalloenzymes [42]. Zinc deficiency can cause metabolic disorders and decreased immune function, leading to bacterial, viral, and fungal infections, growth retardation, premature and poor wound healing [42]. It can be seen from Table 4 that the contents of the four trace elements Cu, Fe, Zn and Mn in the farmed snakehead soup were 0.55 mg/kg, 13.00 mg/kg, 9.04 mg/kg and 0.66 mg/kg, respectively, and the content of the major element Ca was 1.13 mg/g. Zhang, et al. [24] reported that the content of Ca, Zn, and Fe in snakehead fish soup were 0.81 mg/kg, 0.25 mg/kg and 0.15 mg/kg, respectively, which were lower than the mineral contents detected in this study. This might be due to the higher heating power during the processing of snakehead soup in this study, which was beneficial to the dissolution of minerals. At the same time, the choice of pot and boiling water may have a great impact on the mineral content. Tang, et al. [43] reported that the Ca, Fe and Zn contents in crucian carp soup were 0.08 mg/g, 1.41 mg/kg and 4.81 mg/kg, respectively. Through the comparative analysis with the above experimental results, it can be seen that the mineral content of the snakehead fish soup was high. Experience and experiments have proved that the snakehead fish soup has the effects of promoting wound healing [12,13]. Some scholars infer that the high Zn content in snakehead soup is one of the main factors for its outstanding wound-healing effect [13].

Compared with farmed snakehead soup, the contents of Cu, Fe, Mn, and Ca in wild snakehead soup were 0.43 mg/kg, 11.00 mg/kg, 0.32 mg/kg and 0.83 mg/g respectively, which were all significantly lower ($p < 0.05$). However, the Zn content was significantly higher ($p < 0.05$). O'Neill, et al. [6] showed that compared with farmed yellowtail, the Ca content of wild yellowtail was significantly lower ($p < 0.05$) but the Zn content was significantly higher ($p < 0.05$). Their findings were consistent with this study (Table 5).

Table 5. Mineral composition of soup made from farmed and wild snakehead fish.

Minerals	Farmed Snakehead Fish Soup	Wild Snakehead Fish Soup
Cu * (mg/kg)	0.55 ± 0.00 ^{a}	0.43 ± 0.02 ^{b}
Fe * (mg/kg)	13.00 ± 0.08 ^{a}	11.00 ± 0.57 ^{b}
Zn *(mg/kg)	9.04 ± 0.14 ^{b}	12.57 ± 0.53 ^{a}
Mn *(mg/kg)	0.66 ± 0.00 ^{a}	0.32 ± 0.02 ^{b}
Ca (mg/g)	1.13 ± 0.03 ^{a}	0.83 ± 0.02 ^{b}

* indicates trace elements. Mean ± SD (standard deviation) from two replicates. Different lowercase letters in the same line indicate significant differences in the mineral content between soup made from farmed and wild snakehead fish ($p < 0.05$).

3.5. Antioxidant Ability

The wound repair process can induce cellular oxidative stress and generate various free radicals. In turn, the oxidative stress may seriously interfere with wound healing through skin damage, neuropathy, and local infection [44,45]. The anti-oxidative ability of snakehead soup mainly comes from antioxidant peptides [46], which can reduce oxidative stress by scavenging reactive oxygen species and chelating transition metals, thereby accelerating wound healing [47].

The antioxidant ability of soup made from farmed and wild snakehead fish was shown in Figure 1. The DPPH radical-scavenging ability of the farmed snakehead soup was 57.89%, the hydroxyl radical-scavenging ability was 25.61%, and the Fe^{2+} chelating ability was 21.21%. Zhang, et al. [24] reported that the DPPH radical-scavenging ability of snakehead soup was 75%, the hydroxyl radical-scavenging ability was 26%, and the Fe^{2+} chelating ability was 26%. The research results of Zhang, et al. [24] showed that the DPPH radical-scavenging ability was 1.3 times that of farmed snakehead soup in this study, and the hydroxyl radical scavenging ability and Fe^{2+} chelating ability were slightly lower. The lower antioxidant capacity of the fish soup in this study might be due to the fact that the heating power of the fish soup in this subject was higher, so the biologically active substances in the fish soup might be decomposed, resulting in lower antioxidant capacity [48].

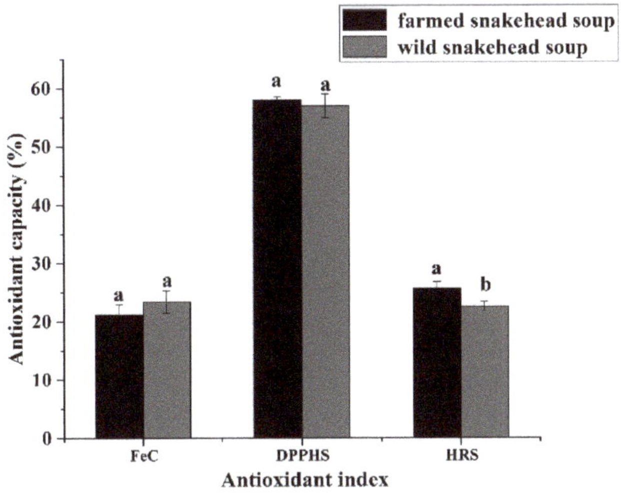

Figure 1. Antioxidant ability of soup made from farmed and wild snakehead fish. FeC: Fe2+ chelating ability, DPPHS: DPPH scavenging ability, HRS: hydroxyl radical scavenging ability. Different lowercase letters indicate significant difference in antioxidant capacity between the soup made from farmed and wild snakehead fish ($p < 0.05$).

The DPPH radical-scavenging ability of wild snakehead soup was 56.90%, the hydroxyl radical-scavenging ability was 22.54%, and the Fe^{2+} chelating ability was 23.36%. There was no significant difference in Fe^{2+} chelation ability and DPPH radical-scavenging ability of wild snakehead soup ($p > 0.05$), while the hydroxyl radical-scavenging ability was slightly lower ($p < 0.05$). The results showed that there was no obvious difference in the antioxidant capacity of the two soups.

4. Conclusions

Snakehead fish soup was rich in protein, lysine, and trace element Zn, and has high antioxidant activity. There was no obvious difference in the nutritional composition and antioxidant activity of soup made from snakehead fish in different growth environments. However, the crude protein content of farmed snakehead soup was higher, while the total fatty acid, MUFA and Zn content of wild snakehead soup were significantly higher, which may be related to the differences in food and environment. Further research is needed on the effect of growth environment on the health effects of snakehead fish soup, especially the functionality of wound healing.

Supplementary Materials: The following supporting information can be downloaded at: https://www.mdpi.com/article/10.3390/foods11203294/s1, Figure S1: Appearance and morphological parameters of farmed and wild snakehead fish. Different lowercase letters indicate significant difference in the body mass index between the farmed and wild snakehead fish ($p < 0.05$); Figure S2: The GC-MS chromatogram of fatty acid analysis. a: farmed snakehead fish soup; b: wild snakehead fish soup; Table S1: Food nutritional composition and growth environment of farmed and wild snakehead fish [49–59].

Author Contributions: Conceptualization, M.R. and T.Y.; methodology, M.R. and T.Y.; validation, M.R. and T.Y.; data curation, M.R. and T.Y.; writing—original draft preparation, M.R., T.Y. and J.Y.; writing—review and editing, M.R., T.Y. and J.Y.; project administration, T.Y., J.Y., R.L., Q.H. and S.X. All authors have read and agreed to the published version of the manuscript.

Funding: This research was funded by the Fundamental Research Funds for the Central University (2662020SPPY007) and China Agricultural Research System (CARS-45).

Institutional Review Board Statement: Not applicable.

Informed Consent Statement: Not applicable.

Data Availability Statement: Not applicable.

Acknowledgments: Authors gratefully acknowledge financial support from the Fundamental Research Funds for the Central University (2662020SPPY007) and earmarked fund for CARS (CARS-45).

Conflicts of Interest: The authors declare no conflict of interest.

References

1. Lam, T.J. Fish culture in southeast Asia. *Can. J. Fish. Aquat. Sci.* **1982**, *39*, 138–142. [CrossRef]
2. Li, M.-D. *Chinese Economic Fish Ecology*, 2nd ed.; Tianjin Science and Technology Press: Tianjin, China, 2012.
3. Ministry of Agriculture and Rural Affairs; Fisheries and Fisheries Administration; National Aquatic Technology Promotion Station; China Fisheries Society. *2022 China Fisheries Statistical Yearbook*; China Agricultural Press: Beijing, China, 2022.
4. Zhang, J.; Li, M.; Zhang, G.; Tian, Y.; Kong, F.; Xiong, S.; Zhao, S.; Jia, D.; Manyande, A.; Du, H. Identification of novel antioxidant peptides from snakehead (*Channa argus*) soup generated during gastrointestinal digestion and insights into the anti-oxidation mechanisms. *Food Chem.* **2021**, *337*, 127921. [CrossRef] [PubMed]
5. Fuentes, A.; Fernández-Segovia, I.; Serra, J.A.; Barat, J.M. Comparison of wild and cultured sea bass (*Dicentrarchus labrax*) quality. *Food Chem.* **2010**, *119*, 1514–1518. [CrossRef]
6. O'Neill, B.; Le Roux, A.; Hoffman, L.C. Comparative study of the nutritional composition of wild versus farmed yellowtail (Seriola lalandi). *Aquaculture* **2015**, *448*, 169–175. [CrossRef]
7. Cahu, C.; Salen, P.; de Lorgeril, M. Farmed and wild fish in the prevention of cardiovascular diseases: Assessing possible differences in lipid nutritional values. *Nutr. Metab. Cardiovasc. Dis.* **2004**, *14*, 34–41. [CrossRef]
8. Zhang, J.; Yang, X.; Zhang, C. Nutritional composition and nutritional value evaluation of snakehead muscle. *Food Res. Dev.* **2020**, *41*, 192–197.

9. Zou, Y.; Wen, Z.; Qin, C.; Li, H.; Wu, J.; Xie, B. Analysis of Muscle Nutritional Components of Snakehead Snails at Different Ages. *J. Nutr.* **2017**, *39*, 616–618. [CrossRef]
10. Zhou, C.; Lei, L.; Deng, X.; Zheng, Z.; Zheng, Y.; Wu, J.; Su, J.; Fan, W.; Cao, G.; Li, Y. Analysis and Evaluation of Muscle Nutritional Components of Snakehead and White Snakehead. *Freshw. Fish.* **2018**, *48*, 83–89. [CrossRef]
11. Deng, X.; Lei, L.; Yang, H.; He, Y.; Wu, J.; Su, J.; Cao, G.; Li, Y.; Lu, G.; Zhou, C. Effects of different storage temperature and time on muscle quality and nutritional components of white snakehead snakehead. *Food Ferment. Ind.* **2019**, *45*, 170–176. [CrossRef]
12. Sahid, N.A.; Hayati, F.; Rao, C.V.; Ramely, R.; Sani, I.; Dzulkarnaen, A.; Zakaria, Z.; Hassan, S.; Zahari, A.; Ali, A.A. Snakehead Consumption Enhances Wound Healing? From Tradition to Modern Clinical Practice: A Prospective Randomized Controlled Trial. *Evid. Based Complement. Altern. Med.* **2018**, *2018*, 3032790. [CrossRef] [PubMed]
13. Yuan, S.; Huo, J.; Cai, Z. Experimental study on the anti-fatigue and muscle-building effects of snakehead and moon snakehead in mice. *Youjiang Med.* **2005**, *33*, 109–111.
14. Periago, M.J.; Ayala, M.D.; López-Albors, O.; Abdel, I.; Martínez, C.; García-Alcázar, A.; Ros, G.; Gil, F. Muscle cellularity and flesh quality of wild and farmed sea bass, *Dicentrarchus labrax* L. *Aquaculture* **2005**, *249*, 175–188. [CrossRef]
15. Zhang, J.; Du, H.; Zhang, G.; Kong, F.; Hu, Y.; Xiong, S.; Zhao, S. Identification and characterization of novel antioxidant peptides from crucian carp (Carassius auratus) cooking juice released in simulated gastrointestinal digestion by UPLC-MS/MS and in silico analysis. *J. Chromatogr. B Anal. Technol. Biomed. Life Sci.* **2020**, *1136*, 121893. [CrossRef] [PubMed]
16. Guo, X.; Shi, L.; Xiong, S.; Hu, Y.; You, J.; Huang, Q.; Yin, T. Gelling properties of vacuum-freeze dried surimi powder as influenced by heating method and microbial transglutaminase. *LWT* **2019**, *99*, 105–111. [CrossRef]
17. *ISO-1442:1997*; Meat and Meat Products—Determination of moisture Content. International Organization for Standardization: Geneva, Switzerland, 1997.
18. *ISO-5983:1997*; Animal Feeding Stuffs—Determination of Nitrogen Content and Calculation of Crude Protein Content-Kjeldahl Method. International Organization for Standardization: Geneva, Switzerland, 1997.
19. *SO-5984:2002*; Animal Feeding Stuffs—Determination of Crude Ash. International Organization for Standardization: Geneva, Switzerland, 2002.
20. Bukhanko, N.; Attard, T.; Arshadi, M.; Eriksson, D.; Budarin, V.; Hunt, A.J.; Geladi, P.; Bergsten, U.; Clark, J. Extraction of cones, branches, needles and bark from Norway spruce (Picea abies) by supercritical carbon dioxide and soxhlet extractions techniques. *Ind. Crops Prod.* **2020**, *145*, 112096. [CrossRef]
21. Bordons, A. Automated determination of lysine by colorimetric method with ninhydrin. *Biotechnol. Lett.* **1986**, *8*, 411–414. [CrossRef]
22. Rawdkuen, S.; Jaimakreu, M.; Benjakul, S. Physicochemical properties and tenderness of meat samples using proteolytic extract from Calotropis procera latex. *Food Chem.* **2013**, *136*, 909–916. [CrossRef]
23. Lowry, O.; Rosebrough, N.; Farr, A.L.; Randall, R. Protein Measurement with the Folin Phenol Reagent. *J. Biol. Chem.* **1951**, *193*, 265–275. [CrossRef]
24. Zhang, G.; Zheng, S.; Feng, Y.; Shen, G.; Xiong, S.; Du, H. Changes in Nutrient Profile and Antioxidant Activities of Different Fish Soups, Before and After Simulated Gastrointestinal Digestion. *Molecules* **2018**, *23*, 1965. [CrossRef]
25. Leme, A.B.P.; Bianchi, S.R.; Carneiro, R.L.; Nogueira, A.R.A. Optimization of Sample Preparation in the Determination of Minerals and Trace Elements in Honey by ICP-MS. *Food Anal. Methods* **2013**, *7*, 1009–1015. [CrossRef]
26. Zhu, L.; Chen, L.; Yuan, M.; Bai, C.; Jiang, Y.; Zhao, L. Effects of processing methods on nutritional components of mullet soup. *Chin. Condiments* **2017**, *42*, 61–64+72.
27. Xu, H. Study on the Effect of Thermal Processing on the Quality of Bighead Carp Soup. Master's Thesis, Jiangnan University, Wuxi, China, 2008.
28. Xia, Q.; Xu, H. Production technology and nutritional analysis of crucian carp soup. *Yangzhou Univ. Culin. J.* **2005**, *22*, 25–27. [CrossRef]
29. Yin, H.; Sun, Z.; Sun, D.; Qiu, L. Comparative analysis of nutritional components in muscle of 6 species of cultured sturgeon. *J. Dalian Fish. Inst.* **2004**, *19*, 92–96. [CrossRef]
30. Bendiksen, E.Å.; Berg, O.K.; Jobling, M.; Arnesen, A.M.; Måsøval, K. Digestibility, growth and nutrient utilisation of Atlantic salmon parr (*Salmo salar* L.) in relation to temperature, feed fat content and oil source. *Aquaculture* **2003**, *224*, 283–299. [CrossRef]
31. Zhang, M.; Chen, M.; Xing, S. Characterization of the key odorants of crucian carp soup and flavour improvement by modulated temperature mode in electrical stewpot. *Flavour Fragr. J.* **2021**, *36*, 637–651. [CrossRef]
32. Zhao, L.; Chen, J.; Zhao, C.; Bai, Q.; Bi, Y.; Wang, H. Component analysis and nutritional evaluation of wild and cultured snakehead muscle. *Mod. Food Technol.* **2015**, *31*, 244–249. [CrossRef]
33. Matthews, D.E. Review of Lysine Metabolism with a Focus on Humans. *J. Nutr.* **2020**, *150*, 2548S–2555S. [CrossRef]
34. Ljubojević, D.; Radosavljević, V.; Puvača, N.; Živkov Baloš, M.; Đorđević, V.; Jovanović, R.; Ćirković, M. Interactive effects of dietary protein level and oil source on proximate composition and fatty acid composition in common carp (*Cyprinus carpio* L.). *J. Food Compos. Anal.* **2015**, *37*, 44–50. [CrossRef]
35. Ljubojević, D.; Ćirković, M.; Novakov, N.; Jovanović, R.; Janković, S.; Đorđević, V.; Mašić, Z. Productivity and Meat Nutrient in Fish: The Diet Effect. *Kafkas Univ. Vet. Fak. Derg.* **2013**, *19*, 43–49. [CrossRef]
36. Garcia-Esquinas, E.; Ortola, R.; Banegas, J.R.; Lopez-Garcia, E.; Rodriguez-Artalejo, F. Dietary n-3 polyunsaturated fatty acids, fish intake and healthy ageing. *Int. J. Epidemiol.* **2019**, *48*, 1914–1924. [CrossRef] [PubMed]

37. Yates, C.M.; Calder, P.C.; Ed Rainger, G. Pharmacology and therapeutics of omega-3 polyunsaturated fatty acids in chronic inflammatory disease. *Pharmacol. Ther.* **2014**, *141*, 272–282. [CrossRef]
38. Patterson, E.; Wall, R.; Fitzgerald, G.F.; Ross, R.P.; Stanton, C. Health implications of high dietary omega-6 polyunsaturated Fatty acids. *J. Nutr. Metab.* **2012**, *2012*, 539426. [CrossRef]
39. Simopoulos, A.P. The importance of the ratio of omega-6/omega-3 essential fatty acids. *Biomed. Pharmacother.* **2002**, *56*, 365–379. [CrossRef]
40. Simopoulos, A.P. Omega-6/Omega-3 Essential Fatty Acid Ratio and Chronic Diseases. *Food Rev. Int.* **2004**, *20*, 77–90. [CrossRef]
41. Song, L. Calcium and Bone Metabolism Indices. *Adv. Clin. Chem.* **2017**, *82*, 1–46. [CrossRef] [PubMed]
42. Gregory, P.J.; Wahbi, A.; Adu-Gyamfi, J.; Heiling, M.; Gruber, R.; Joy, E.J.M.; Broadley, M.R. Approaches to reduce zinc and iron deficits in food systems. *Glob. Food Secur.* **2017**, *15*, 1–10. [CrossRef]
43. Tang, X.; Chen, J.; Li, G.; Zhu, L.; Dai, Y. Effects of Processing Methods on Nutritional Components of Fish Soup. *Food Ind. Technol.* **2008**, *29*, 248–251. [CrossRef]
44. Deng, L.; Du, C.; Song, P.; Chen, T.; Rui, S.; Armstrong, D.G.; Deng, W. The Role of Oxidative Stress and Antioxidants in Diabetic Wound Healing. *Oxid. Med. Cell. Longev.* **2021**, *2021*, 8852759. [CrossRef] [PubMed]
45. Schafer, M.; Werner, S. Oxidative stress in normal and impaired wound repair. *Pharm. Res.* **2008**, *58*, 165–171. [CrossRef] [PubMed]
46. Wang, C.-H.; Doan, C.T.; Nguyen, V.B.; Nguyen, A.D.; Wang, S.-L. Reclamation of Fishery Processing Waste: A Mini-Review. *Molecules* **2019**, *24*, 2234. [CrossRef]
47. Orsini Delgado, M.C.; Nardo, A.; Pavlovic, M.; Rogniaux, H.; Anon, M.C.; Tironi, V.A. Identification and characterization of antioxidant peptides obtained by gastrointestinal digestion of amaranth proteins. *Food Chem.* **2016**, *197 (Pt B)*, 1160–1167. [CrossRef]
48. Hwang, E.-S. Bioactive Compounds and Antioxidant Activity of Cauliflower According to Heat Treatment Method (P06-074-19). *Curr. Dev. Nutr.* **2019**, *3* (Suppl. 1), 586. [CrossRef]
49. Feed A Consists of Imported Fish Meal, Starch, Soybean Meal, Fish Oil, Calcium Dihydrogen Phosphate, Vitamins, Multivitamins and Organic Trace Elements. The ingredients refer to the nutrient composition table of snakehead feed, which comes from Zhejiang Dongyu Biotechnology Co., Ltd. Available online: http://www.zjdjsw.com/ (accessed on 1 September 2022).
50. Feed B Consists of Imported High-Quality Fish Meal, Flour, Soybean Meal, Yeast Powder, Organic Chelated Minerals, Multiple Vitamins, Trace Elements, etc. The Ingredients Refer to the Nutrient Composition Table of Snakehead Feed, which Comes from Jiangmen Hengsheng Industrial Co., Ltd. Available online: http://www.hengshengfeed.com/ (accessed on 1 September 2022).
51. Zhejiang Lianxing Feed Technology Co., Ltd. An Extruded Compound Feed for Blackfish: CN201510830228.8.2016-03-30. Available online: http://lianxingsiliao.cn.tonbao.com/ (accessed on 1 September 2022).
52. Wang, Y.H.; Ding, W.; Chen, J.; Wang, H.C.; Luo, G.L.; Xing, J.; Xin, J.Q. Effects of feeding compound feed and chilled trash fish on the growth rate and feed cost of turbot. *Jiangsu Agric. Sci.* **2016**, *44*, 282–285.
53. Li, J.Q.; Lin, J.B.; Zhu, Q.G.; Li, C.L. A comparative experiment on the effect of artificial compound feed and small trash fish on raising red-spotted grouper. *Taiwan Strait* **2004**, *2*, 167–173.
54. Huang, Z.C.; Chen, D.H.; Lin, J.B.; Lin, K.B.; Zhu, Q.G.; Liang, P.; Zheng, L.Y.; Qiu, F.Y. Effects of compound feed and small trash fish on the growth performance of grouper oblique. *Feed. Res.* **2012**, *9*, 1–4.
55. Chen, D.H.; Zheng, L.Y.; Lin, J.B.; Zhu, Q.G.; Liang, P.; Lin, K.B.; Huang, Z.C.; Qiu, M.L. Study on the effects of different feeds and small trash fish on the growth and immunity of grouper serrata. *J. Fujian Agric.* **2013**, *28*, 309–314.
56. Cui, J.; Meng, C.; Liu, B.L.; Li, W.D. Nutrient content and evaluation of five main species of freshwater shrimp in Anhui Province. *Health Res.* **2020**, *49*, 962–968.
57. Zhuang, P.; Song, C.; Zhang, L.Z. Nutrient composition comparison between white shrimp and Macrobrachium japonicus in the Yangtze Estuary. *J. Zool.* **2008**, *4*, 822–829.
58. He, X.R.; Wang, X.Q.; Fan, W.J.; Zhou, X.W. Biological characteristics and nutritional composition analysis of Sangzhi toe ditch frog. *J. Hunan Agric. Univ.* **2008**, *4*, 482–484.
59. Ouyang, F.; Chen, Z.; Shi, L.; Chen, X.H. Analysis of the nutritional components of the anal frog of Taihang longa. *J. Tianjin Norm. Univ.* **2015**, *35*, 128–129+136.

foods

MDPI

Article

Proximate Composition, Retained Water, and Bacterial Load for Two Sizes of Hybrid Catfish (*Ictalurus furcatus* × *Ictalurus punctatus*) Fillets at Different Process Steps

Manirul Haque [1,2] and Juan L. Silva [1,*]

[1] Department of Food Science, Nutrition and Health Promotion, Mississippi State University, Mississippi State, MS 39762, USA
[2] Department of Food Science and Technology, University of Nebraska-Lincoln, Lincoln, NE 68588, USA
* Correspondence: jsilva@foodscience.msstate.edu

Abstract: The catfish processors in the US are required to state the maximum percentage of retained water content (RWC) on the product label. The objectives of our study were to quantify the RWC of processed hybrid catfish fillets from proximate composition and the bacterial load at different processing points. Water content was determined using oven-dry (AOAC950.46,1990) and Near-infrared (NIR) spectroscopy. Protein and fat content were determined by NIR spectrometer. Psychrotrophic (PPC) and Total Coliform (TCC) counts were enumerated using 3MPetrifilm^{TM}. The fillets' overall baseline water, protein, and fat content were 77.8, 16.7 and 5.7%, respectively. The RWC of final fresh and frozen fillets were ~1.1=/- 2.0% (not significant) and ~4.5%, respectively, and was not fillet size or harvest season dependent. Baseline water content (78.0 vs. 76.0%) was higher ($p \leq 0.05$), and fat content (6.0% vs. 8.0%) was lower ($p \leq 0.05$) for small (50–150 g) compared to large fillets (150–450 g). Higher ($p \leq 0.05$) baseline PPC (~4.2 vs. ~3.0) and TCC (~3.4 vs. ~1.7) were observed for the warm season (April–July) fillets compared to the cold season (Feb–April). This study provides information to processors and others on estimating retained water and microbiological quality of the hybrid catfish fillets over the process line.

Keywords: retained water; catfish; proximate composition; water content; processing; bacterial load

Citation: Haque, M.; Silva, J.L. Proximate Composition, Retained Water, and Bacterial Load for Two Sizes of Hybrid Catfish (*Ictalurus furcatus* × *Ictalurus punctatus*) Fillets at Different Process Steps. *Foods* **2023**, *12*, 1112. https://doi.org/10.3390/foods12051112

Academic Editors: Tao Yin and Liu Shi

Received: 20 October 2022
Revised: 14 November 2022
Accepted: 28 November 2022
Published: 6 March 2023

1. Introduction

Farm-raised catfish is a significant commercial food commodity in the United States, contributing around $4 billion to the US economy each year [1,2]. The commercial catfish processors produce the hybrid catfish for greater fillet yield [3,4]. Catfish growers directly sell food-size (0.3 to 1.5 kg) fish to processors who process them into whole fish, dressed fish, fillets, fillet strips, nuggets, and steaks [2]. These products are usually sold as either iced, frozen, battered, breaded, or fresh [5,6]. The United States Department of Agriculture-Food Safety Inspection Service (USDA-FSIS) started inspecting Siluriformes including catfish in September 2017 [7]. Fresh or fresh-frozen packages of catfish or parts must be labeled to reflect 100% net weight after thawing. Processors must state the maximum percentage of retained water (if any) on the product label [8]. The USDA-FSIS adopted existing meat and poultry net weight and retained water regulations (9 CFR Parts 381 and 441) for Siluriformes' products (USDA, 2001) without modification. A significant number of studies reported that poultry carcasses retained 4 to 11% water after immersion chilling which may be different from catfish retained water [9–12]. Arbitrary adoption of poultry retained water regulations in the Siluriformes industry could be misleading. A data gap exists in estimating catfish retained water during and after processing. One of our goals of this study is to estimate the fresh and processed catfish (final product) retained water for filling out the existing data gap. The catfish industry also needs to identify the key variables that affect the water uptake or loss in catfish products during or after processing, which may contribute to better process control.

During commercial processing, a chilling procedure may contribute to 'added water' in the final product. The added water could be determined using catfish muscle water-protein ratio as protein content typically remains constant with the process steps [13–15]. Water content has an inverse relationship with fat content of fish [15–17]. This relationship could be used to predict the water content from fat or protein content of the processed catfish [11]. However, proximate composition of fish differs from species to species, individual to individual, considering size, sex, season, feeding, and processing stress [18–21]. Many studies have reported channel catfish baseline proximate composition [22–24], while data for hybrid catfish proximate composition in the process lines are scarce.

In a fish processing facility, indicator bacterial counts could reveal temperature abuse, cross-contamination, and mishandling during processing [21,25]. However, catfish's bacterial counts (APC, PPC, TCC, and *E. coli*) differ with harvesting season, size of the processing plant, and processing methods [26,27]. Although some studies [26–29] reported bacterial load in the different process steps for channel catfish (*Ictalurus punctatus*) fillets; data for hybrid catfish fillets bacterial load are insufficient to control the existing catfish processing lines. The experimental data from our study will help identify the critical process steps for controlling the bacterial load in catfish fillets.

This study aimed to quantify the retained water content from the estimated proximate composition of hybrid catfish fillets and the bacterial load at different process steps. The purpose of determining the proximate composition and bacterial load at each process step was to identify the critical process step where fillets uptake the highest amount of the water content and had the highest bacterial load. Because the baseline water (natural or as received in the plant) content dictates the amount of retained water with each process step, fillet sizes and harvesting seasons were included to observe any impact of these parameters on the fillet's water content and bacterial load as received. The natural composition, especially water content of catfish products before and during processing, can provide information to processors and inspection authorities regarding regulatory compliance and labeling requirements.

2. Materials and Methods

2.1. Sample Collection and Treatment

A total of 228 hybrid catfish [Blue (*Ictalurus furcatus*) × Channel (*Ictalurus punctatus*)] fillets were collected from two local catfish processing plants in Mississippi between February to April and May to July 2018. Three fillet samples (one for microbiological analysis and two for proximate analysis) of two sizes (small: 50 g to 150 g and large: 150 g to 450 g) from seven process steps; Fillets before trimming (BT): assumed to have similar proximate composition as received fish in the processing plant, after trimming/before water chilling (BC), after water chilling (AC), after ice slush chilling (AS), before ice packing (BIP): fresh scalded fillets; fillets after injection (AI), and after freezing (AF) (Figure 1), were randomly picked and placed into quart size Ziplock® bags (GreatValue™ Slide Zipper 7in'8 in). The temperature of the BT, BC, AC, AS, BIP, AI, and AF fillets during sampling averaged 21 °C, 20.6 °C, 6.2 °C, 0 °C, 3.7 °C, 4.6 °C, and −2.6 °C, respectively. The sampled catfish fillets were kept in an ice chest with ice and transported within 40 min to the Food Safety and Processing Laboratory of the Department of Food Science, Nutrition and Health Promotion at Mississippi State University. Microbiological analysis was performed within six hours of sampling. Collected catfish fillets (placed in bags) kept in the ice chest covered with ice were placed at 4 °C in a refrigerator (Isotemp Plus Laboratory Refrigerator, Fisher Scientific, Pittsburg, PA, USA) for 22 to 24 h before proximate analysis.

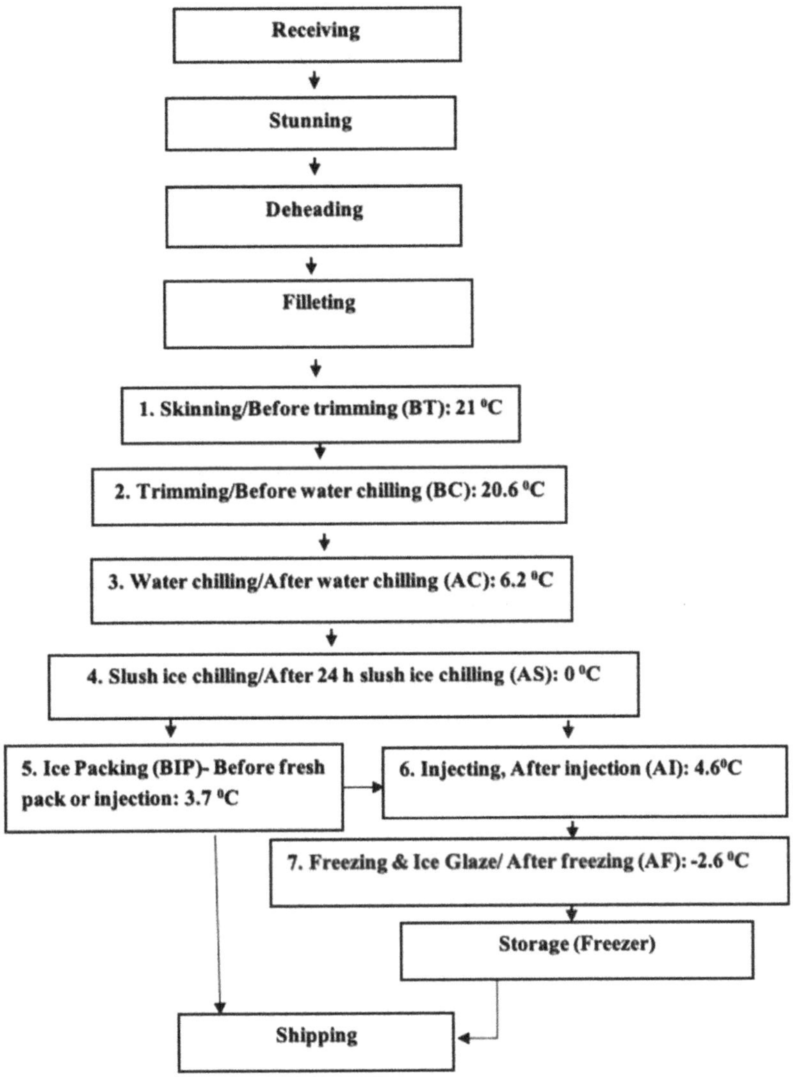

Figure 1. Typical process flow for catfish fillet showing sampling points.

2.2. Proximate Analysis

The weight and length of fillets were measured before proximate analysis. Ice glaze of the frozen fillets was removed by spraying cold water and draining the water for two minutes and immediately transferred to the refrigerator (4 °C) for further proximate analysis (AOAC 963.18). The water content was measured following AOAC 950.46,1990 procedure. In brief, the whole fillet was homogenized with a food chopper (Black & Decker® Handy Chopper Plus™, Towson, MD, USA) and transferred to a large (150′15 mm) Petri dish (Falcon 35 1058 PetriDish Style Sterile, Oxnard, CA, USA). An aliquot of 5 g of homogenized sample was evenly distributed into the weighing dish (Fisher Scientific, 08732101, Houston, TX, USA). Dishes (with sample) were weighed and dried at 105 ± 2 °C in an ISOTEM OVEN 300 (300 series Model 318, Fisher Scientific, Houston, TX, USA) for 5 ± 2 h or until a constant weight was achieved and were placed in a desiccator (Sanplatec

Corporation, Japan) for 15 ± 5 min to cool after drying. Water content was calculated on a wet basis as follows:

$$\text{Water content} = \frac{(\text{W2} - \text{W3}) * 100}{\text{W2} - \text{W1}} \tag{1}$$

where,

W1 = weight of dish (without sample)
W2 = weight of dish (with sample) before drying
W3 = weight of dish (with sample) after drying

Proximate composition (protein, fat, collagen, and water content) of the fish fillets was analyzed on a wet basis using a Near-infrared (NIR) spectrometer (Food Scan Lab Analyzer Model 78,800, Foss Analytical, Eden Prairie, MN, USA).

2.3. Retained Water Calculation

A significant correlation (r = 0.90, $p \leq 0.05$) was obtained between water determined by NIR and the oven method. To establish a prediction model, water content determined by NIR was fitted using simple linear regression. The water-protein ratio (wet basis) was calculated as follows

$$\text{Water} - \text{protein ratio (M : P)} = \frac{\text{Fitted water content}}{\text{protein content determined by NIR}} \tag{2}$$

Retained water (%) was calculated based on the water retention/loss at each point of the processing as follows:

Retained water (%) = water (%) at any process point (e.g., AC, BIP, AF)—Water (%) at baseline (BT).

2.4. Microbiological Analysis

A 25 g fillet sample was aseptically cut with a sterile stainless-steel knife, weighed, and placed in a stomacher bag (Nasco, Whirl-Pak, 19 × 30 cm: Fort Atkinson, WI, USA). A 225 mL of 0.1% sterilized buffer peptone water (BPW) solution (Difco, Detroit, MI, USA) was added and stomached for two minutes in a laboratory blender stomacher 400 (A. J. Seward and Co., Ltd., London, England). Dilutions were made by transferring 1 mL of the homogenate into dilution tubes with 9 mL of 0.1% sterilized peptone solution. Plating was conducted on aerobic (APC) count 3M^{TM} Petrifilm (3M Co., St. Paul, MN, USA) in duplicate, and these were incubated for 72 h at 20 ± 2 °C [27] for psychrotrophic counts (PPC). *E. coli* plates were incubated for 24 to 48 h at 35 ± 2 °C on 3M^{TM} Petrifilm *E. coli* (3M Co., St. Paul, MN, USA) in duplicate for the enumeration of *E. coli* and total coliform counts (TCC) [26,30]. According to the manufacturer's instructions, colonies were identified and enumerated using 3M™ PetrifilmPlate Reader (3M Company, Technopath, St. Paul, MN, USA). Selected plate counting was verified by the conventional (visual) counting method.

2.5. Experimental Design and Statistical Analysis

Data were arranged in a 2-way factorial [2 sizes of the fillets (small: 50 to 150 g; large: 150 g to 450 g) × 7 process points] randomized complete block (RCB) design with 12 replications (blocks) based on the availability of the fillets from each process point [BT: 15 fillets (small = 7, large = 8); BC:16 fillets (small = 9, large = 6); AC = 10 fillets (small = 5, large = 5); AS: 14 fillets (small = 8, large = 6); BIP: 9 (small = 3, large = 6); AI: 9 fillets (small = 5, large = 4); AF: 7 fillets (small = 3, large = 4)]. Data were unbalanced in the blocks due to the unavailability of the fillets for some replications. The General Linear and mixed Model procedure (PROC GLMMIX) of the Statistical Analysis System (SAS studio edition, 2020) was used to examine the interaction of sizes, process steps, and blocks. There was no interaction found between size and process steps and between size, process steps and blocks. Tukey's honest significant difference (HSD) was used for the mean separation of the measurements of the fillets ($p \leq 0.05$). Simple linear regression (SLR) and multiple linear regression (MLR) [31] models were used to calculate the correlation of the variables. All Statistical analysis were performed using SAS universal edition (2020).

3. Results and Discussion

3.1. Proximate Composition and Microbial Load of the Hybrid Catfish Fillets by Fillet Size and Harvesting Season

The proximate composition especially water content of the receiving fillets (BT fillets) as well as microbial load dictates the final products' water retention and microbial quality during catfish processing. However, initial proximate composition and bacterial load could differ for different sizes of fillets and harvesting season. Our study estimated the initial proximate composition and microbial load of hybrid catfish fillets and presented in Table 1.

Small-sized fillets had higher ($p \leq 0.05$) water (78.0 vs. 76.8) and lower ($p \leq 0.05$) fat content (6.0 vs. 8.0%) (Table 1). Conversion of water into fat as fish grows may contribute to higher fat retention content in a larger fish [16,32,33]. Silva and Ammerman (1993) [34] reported higher water content (70.8% vs. 68.1%) and lower fat content (10.8 vs. 13.2%) for small (0.3 kg) sized channel catfish (*Ictalurus punctatus*) fillets than larger ones (1.0 kg). However, protein content and bacterial load (PPC and TCC) did not differ ($p > 0.05$) between sizes.

Proximate composition and bacterial load of warm weather (May to July) fish were different ($p \leq 0.05$) from cold weather (February to April) fish. BT fillets' water content (77.8 vs. 75.6%) was higher ($p \leq 0.05$), and fat content (5.8 vs. 8.7%) was lower ($p \leq 0.05$) for those collected in the cold weather than that of the warmer weather (Table 1). Some studies [35,36] reported that low mean energy value for muscle tissue during winter are related to the lower level of fat within the muscle. Nettleton et al. (1990) also reported higher water content for fillets collected in the winter than in summer (77.4 vs. 76%). However, Robinson et al. (2001) reported no seasonal (May, October, and February) impact on the water content of large sizes (0.23 to 0.45 kg) of channel catfish (*Ictalurus punctatus*) fillets.

BT fillets' PPC was higher ($p \leq 0.05$) for fillets collected in the cold season compared to those from the warm season (4.2 vs. 3.0 log CFU/g); however, TCC was higher ($p \leq 0.05$) for warmer season fillets (3.4 vs. 1.7 log CFU/g). Huang and Leung (1993) [37] reported similar PPC (2.8 to 3 log CFU/mL) and TCC (1.48 log CFU/mL) in dressed channel catfish harvested from southern Georgia during the spring season. Nunez et al. (2003) also reported higher TCC in channel catfish processed in the spring than those processed in the fall or winter. Fernandes et al., (1997) reported higher counts of *E. coli* and *S. aureus* in catfish fillets collected in summer compared to those collected in winter.

Table 1. Proximate composition and bacterial count of baseline (BT) hybrid catfish fillets.

		Moisture (%) (Oven)		Fat (%) (NIR)		Protein (%) (NIR)		Psychrotrophic Counts (PPC) (log CFU/g)		Total Coliform Counts (TCC) (log CFU/g)		*E. coli*
		(mean ± SD)	Range	(mean ± SD)	Range	(mean ± SD)	Range	(mean ± SD)	Range	(mean ± SD)	Range	
Fillet Size (Weight)	Small (50–150 g)	78.0 ± 1.1 ^{a}	76.4–80.0	6.0 ± 1.1 ^{a}	4.0–7.2	17.0 ± 0.5 ^{a}	16.0–17.4	3.8 ± 1.4 ^{a}	0–5.4	2.3 ± 1.8 ^{a}	0.0–5.4	ND
	Large (150–450 g)	76.0 ± 1.7 ^{b}	73.2–78.0	8.0 ± 2.8 ^{b}	4.8–12.8	16.6 ± 0.4 ^{a}	15.5–17.2	3.5 ± 1.5 ^{a}	0–5.5	2.6 ± 1.8 ^{a}	0.0–5.5	ND
Harvest season	Cold (Feb–April)	77.8 ± 1.3 ^{a}	74.5–80.0	5.8 ± 1.6 ^{a}	4.0–10.3	16.7 ± 0.5 ^{a}	15.5–17.4	4.2 ± 0.6 ^{a}	3.2–5.3	1.7 ± 1.3 ^{a}	0.0–3.1	ND
	Warm (April–July)	75.6 ± 1.7 ^{b}	73.2–77.5	8.7 ± 2.7 ^{b}	5.7–12.8	16.4 ± 0.2 ^{a}	16.1–16.7	3.0 ± 2.0 ^{b}	0–5.5	3.4 ± 2.0 ^{b}	0.0–5.5	ND

^{a,b} Means in same column not followed by same letter differ ($p \leq 0.05$); SD = Standard Deviation; ND = Not detected; NIR = Near infrared spectrometer method.

3.2. Water and Retained Water Content of the Hybrid Catfish Fillet at Several Process Steps

The USDA-FSIS stated that 'retained moisture' should be documented to provide consumers with the information necessary to make reasonable purchase decisions. In 2017, the USDA-FSIS started inspecting 'Siluriformes' and stated that processors must state the maximum percentage of retained water (if any) on the product label. No published study reported the retained water content for the catfish's final commercial product. We have determined the fillets' retained water content from the proximate composition using a systematic approach so that the authority and the catfish processors could use the data for better process control.

There was no interaction ($p > 0.05$) between fillets' sizes, harvesting season, and processing steps, indicating fillets' sizes or seasons did not affect the retained water (water difference from baseline; BT) at any process step. Retained water was proportional to water content at every process step (Figure 2). Retained water content (RWC) differed ($p \leq 0.05$) at some process steps. Final fresh and frozen fillets' RWC were 1.1% with a range of 0 to 5% and 4.5% with a range of 2 to 8%, respectively (Figure 2).

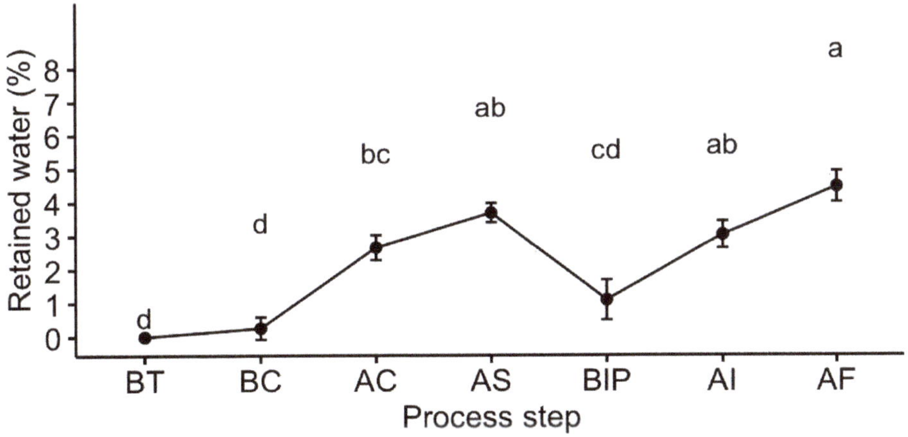

Figure 2. Retained water content (%) of hybrid catfish fillets at different catfish process steps regardless of size; ^{abcd} Means not followed by the same letter differ ($p \leq 0.05$); BT = Before Trimming (Baseline; assumed to have the same proximate composition as received fish at processing plant); BC = After trimming/before chilling; AC = After water chilling; AS = After slush ice chilling; BIP = Before ice packing (Fresh fillets); AI = After injecting (polyphosphate injection), AF = After freezing (Frozen fillets).

Baseline (BT) retained water content (RWC) was assumed to be zero, where water content depends on the size of the fillets and seasonal changes. BC fillets' RWC was similar ($p > 0.05$) to that of BT fillets. After water chilling, RWC for AC fillets was 2.7 ± 1.5% (with a range of 0.7 to 5%) possibly due to the adsorbance of chilled water in the subcutaneous layer of the muscle tissue [10]. After 24 h ice slush chilling, AS fillets retained more ($p \leq 0.05$) water (3.7 ± 1.4%, with a range of 0.3 to 6.2%) than other fillets (except for AF fillets) may be due to the immersion of the fillets for a more extended period in the slush ice, where fillets trap more water in the intercellular space of the muscle tissues [10,12]. Carciofi & Laurindo (2007) [38] reported that water absorption of poultry depends on immersion time, water temperature, and water stirring conditions during chilling. Afterwards, fillets lost around 2.6% of this water content before ice packing (BIP). Retained water of the BIP fillets was 1.1 ± 2.0 with a range of −2.1 to 5.0%. Klose et al. (1960) [39] reported that most of the absorbed water is loosely held (unbound water, adsorbed) in pockets between the tissues of the muscle during immersion chilling, and most could be lost afterwards. Silva et al. (2001) supported these results stating that fillets could gain weight due to water adsorption during chilling but lose most of it before ice packing. The ranges of RWC (−0.7 to 6.3%) of hybrid catfish fillets after water and ice slush chilling were lesser than reported retained water (6 to 12%) of poultry carcass after immersion chilling [10,12]. Retained water was not different ($p > 0.05$) for fillets after injection (3.0 ± 1.6% with a range of 0.3 to 5.1%) and freezing (4.5 ± 2.0% with a range of 1.8 to 8.0%). This result might be due to the injection of polyphosphate or other chemical agents in the fillets before freezing, which increased the water binding capacity of the muscle's myofibrillar protein and protected water loss during freezing [40–42].

3.3. Factors Affecting the Retained Water Content of Catfish Fillets

NIR spectroscopy is fast, noninvasive, and more economical for determining the proximate composition of the muscle food in comparison to other conventional (oven dry, Kjeldahl) methods [43,44]. Multiple linear regression analysis was used to analyze the factors affecting retained water of the hybrid catfish fillets during processing. Several studies [14,45] reported that the water-protein ratio could be used to determine added water during seafood processing. Breck (2014) reported that the relationship between water and protein is size-dependent, and that fat content is inversely correlated to the water content of the fish. Thus, the catfish's water-protein ratio (M:P), weight (g), and fat content were examined by multiple linear regression analysis to predict the retained water of the catfish fillets during processing.

Water content determined by NIR spectrometer was fitted based on water content determined by oven method (AOAC approved method) using a simple linear regression model [46]. A significant correlation ($F (1.74) = 513.97$, $p < 0.0001$), $R^2 = 0.87$) was obtained between water content determined by NIR and water content determined by the oven method. Fitted water was equal to 14.7 + 0.80 (water content determined by oven method) %.

The retained water calculated from water determined by NIR was fitted based on retained water calculated from water determined by the oven method using a simple linear regression model. The fitted retained water was equal to 3.0 + 1.10 (calculated retained water from water determined by oven) %, [($F (1.56) = 255.93$, $p < 0.0001$), $R^2 = 0.82$). This fitted retained water was used as a dependent variable (Y), and water-protein ratio (M:P), fat content (%), and weight (g) of the catfish were used as independent variables (X) in the prediction models. A stepwise regression analysis was conducted with backward elimination of the independent variables to fit the models. At first, all the independent variables (M:P, fat content, and weight) were used for the model establishment. The regression equation of this model ($F (3, 57) = 419.36$, $p < 0.0001$, $R^2 = 0.96$) was as follows:

$$\text{Retained water (\%)} = -5.6 + 2.1\ (\text{M:P}) - 0.13\ (\text{Fat}) + 0.0004\ (\text{weight})\ (\text{Model 1}) \quad (3)$$

Both M:P and fat were significant ($p \leq 0.05$) predictors for retained water; however, weight was not a significant ($p > 0.05$) predictor for retained water in this model.

Thus, weight was excluded from the model, and a reduced model ($F (2, 58) = 635.59$, $p < 0.0001$), $R^2 = 0.96$) was established. Adjusted R-square was not different ($p > 0.05$) for this reduced model after excluding weight, indicating that weight was not a significant predictor along with water-protein ratio and fat content. The regression equation of this reduced model was as follows:

$$\text{Retained water (\%)} = -5.6 + 2.13\ (\text{M:P}) - 0.70\ (\text{Fat})\ (\text{Model 2}) \quad (4)$$

Both M:P and fat were significant predictors of retained water in this model 2. However, when fat content was excluded from model 2, adjusted R^2 (0.58) was different ($p \leq 0.05$) for the reduced model 3. This indicated that fat content was a significant predictor for retained water in model 2. The regression equation of this reduced model ($F (1, 59) = 84.84$, $p < 0.0001$) was as follows:

$$\text{Retained water (\%)} = -12.2 + 2.8\ (\text{M:P})\ (\text{Model 3}) \quad (5)$$

However, when weight (g) was added excluding fat content in this reduced model 3, adjusted R^2 (0.73) increased ($p \leq 0.05$), which indicated that weight was a significant predictor for retained water excluding fat in model 4. The regression equation of this model ($F (2, 58) = 79.78$, $p < 0.0001$, $R^2 = 0.73$) was as follows.

$$\text{Retained water (\%)} = -12.3 + 3.0\ (\text{M:P}) - 0.007\ (\text{weight})\ (\text{Model 4}) \quad (6)$$

Both M:P and weight were significant predictors of model 4.

Model 2 fulfilled the goodness of fit criteria of a multiple linear regression model [46]. This model contained 76 observations and 3 parameters. The coefficient of multiple determination (R^2) was 0.96, indicating that this model accounted for the more significant proportion of variation. The residual of M:P and fat content followed random distribution (Table 2). The value of residual degrees of freedom adjusted R square (Adj. $R^2 = 0.96$) and means square error (MSE = 0.105) also exhibited a good fit of this model for the prediction of retained water based on water-protein ratio and fat content. Thus, water-protein ratio and fat content were significant predictors for retained water during processing of hybrid catfish fillets.

Table 2. Regression analysis of model 2 for the prediction of retained water (%) of hybrid catfish fillets during processing.

Variable	Coefficient	Std. Error	T-Statistic	Pr > |t|
Intercept	−5.734	0.590	−9.780	<0.0001
Moisture-protein ratio	2.140	0.103	20.700	<0.0001
Fat (%)	−0.670	0.030	−22.070	<0.0001
R-Squared	0.960	MSE	0.105	
Adjusted R-Squared	0.954	F-statistics	635.590	
No. of observations	61	Pr (F-statistics)	<0.0001	

Dependent variable = Fitted retained water (%) (Calculated from regression analysis between retained water from water determined by oven and NIR method; Independent variable = water-protein ratio, weight (g) and fat content (%) of the catfish fillets.

3.4. Proximate Compositions of the Hybrid Catfish Fillet at Several Process Steps

There was no interaction ($p \leq 0.05$) between fillet size, harvest season, and processing step on proximate composition at any process step. Fat content differed when water content differed ($p \leq 0.05$) at similar process steps (Figure 3). For instance, fillets after ice slush (AS) had less ($p \leq 0.05$) fat content in comparison to BT and BC fillets (4.7 ± 1.6 vs. 7.0 ± 2.5 and 5.6 ± 2.0%), respectively (Figure 3). An inverse correlation was reported between fat and water content of the fish fillets [15–17]. However, fat content of the fillets was not different ($p > 0.05$) at any of the process steps when measured on a dry basis.

Protein content was not different ($p > 0.05$) for BT (17.0 ± 0.5%), BC (17.0 ± 0.7%) and BIP (16.3 ± 0.6%) fillets (Figure 3). However, when AC and AS fillets' water content increased (2 to 4%) due to water absorption during chilling (both water and slush ice), the percentage of protein content was lower ($p \leq 0.05$) for these fillets (AC: 16.0 ± 0.7%; AS: 15.0 ± 0.7) in comparison to BT, BC and BIP fillets. AI and AF fillets also resulted in lower ($p \leq 0.05$) protein content (AI: 14.5 ± 0.5; AF: 14.5 ± 0.5), where the water content of these (AI and AF) fillets was higher ($p \leq 0.05$) in comparison to BT, BC and BIP fillets (Figure 3). An inverse relationship between protein and water content was also reported by Breck et al. (2014). Fillets' protein content on a dry basis also differed ($p \leq 0.05$) with process steps.

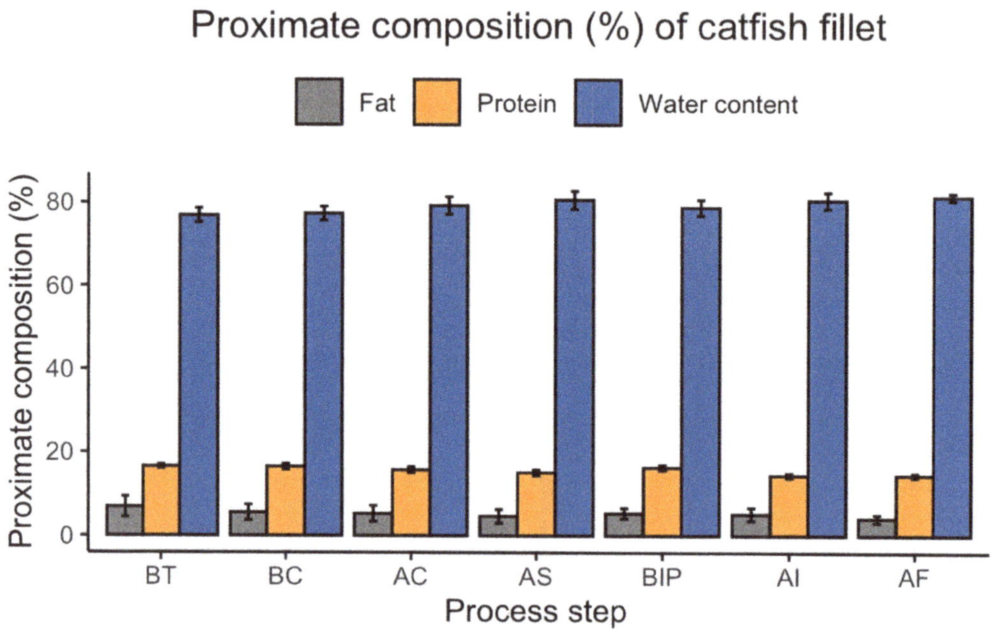

Figure 3. Proximate composition (%) of hybrid catfish fillets at different catfish process steps regardless of size; BT = Before Trimming (Baseline; assumed to have the same proximate composition as received fish at processing plant); BC = After trimming/before chilling; AC = After water chilling; AS = After slush ice chilling; BIP = Before ice packing (Fresh fillets); AI = After injecting (polyphosphate injection), AF = After freezing (Frozen fillets).

3.5. Bacterial Load of the Hybrid Catfish Fillets at Several Process Steps

Indicator bacteria may reveal temperature abuse [47] and cross-contamination [26] during fish handling and storage. We have determined psychrotrophic plate counts (PPC) throughout the process line because of the low temperature (4.6 to 21 °C) of the processing environment where PPC could be the indicator microorganisms [29]. Total coliform plate counts (TCC) was determined to investigate cross-contamination in the processing environment. Fillets' sizes and harvesting seasons were considered to investigate any influence of these two parameters on the final bacterial count [27]. The bacterial load data are presented in Table 1 and Figure 4.

There was no interaction ($p > 0.05$) between fillet size, harvest season, and process steps for psychrotrophic plate counts (PPC) and total coliform plate counts (TCC). The average PPC of BT fillets was ~4 log CFU/g, with a range of ND (Not detected) to ~5 log CFU/g for both sizes and seasons (Table 1). Fernandes et al. (1997) and Watchalotone et al. (1996) reported similar ranges of PPC (3.5 to 5.5 log CFU/g); however, Huang & Leung (1993) and Nunez et al. (2003) reported lower PPC (2 to 3 log CFU/g) for channel catfish (*Ictalurus punctatus*) fillets collected from processing plants. The average TCC of BT fillets was 2.4 log CFU/g, with a range of ND to ~5 log CFU/g for both sizes and seasons (Table 1).

Previous studies also reported a similar range of TCC (1 to ~2.7 log CFU/g) for channel catfish (*Ictalurus punctatus*) fillets collected from catfish processing plants [26,28,48]. The reported higher TCC results might be due to temperature abuse and mishandling, as these fillets were collected from a manual catfish processing scheme [48]. *E. coli* was not detected in any sample at any process step which could be found in other meat products [49]. The maximum acceptable limits of aerobic plate counts (APC) at 20 to 25 °C and *E. coli* in fresh and frozen fish are 5.7 log CFU/g and 1.0 log CFU/g, respectively, as specified by ICMSF (International Commission on Microbiological Specifications for Foods) [25]. However,

Watchalotone et al. (2001) suggested that PPC and TCC of catfish fillets during processing should not be more than 3–4 log CFU/g and 2 log CFU/g, respectively.

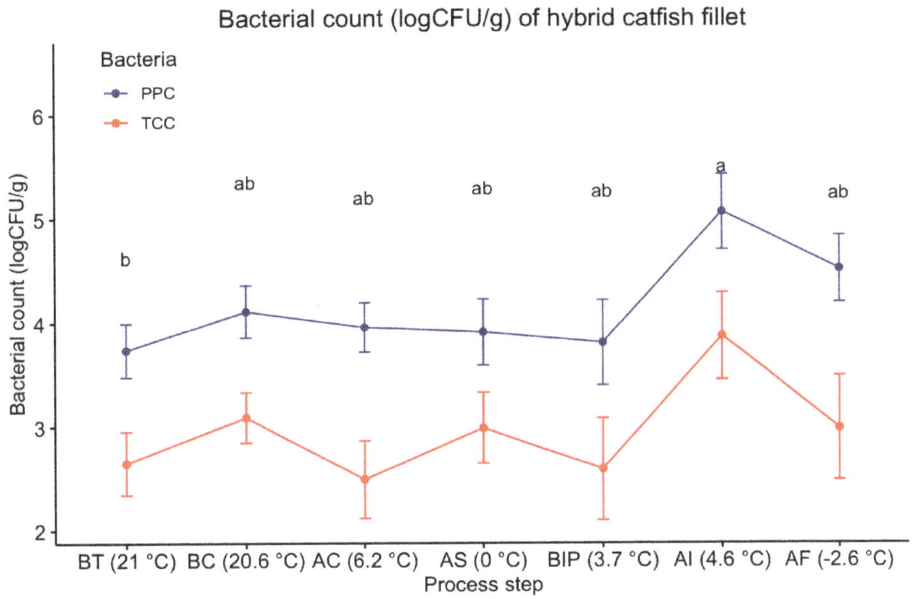

Figure 4. Bacterial count (log CFU/g) of hybrid catfish fillets at different process steps regardless of sizes; PPC = Psychrotrophic plate counts; TCC = Total Coliform plate counts. [ab] Means not followed by the same letter differ ($p \leq 0.05$); BT = Before Trimming (Baseline; assumed to have the same proximate composition as received fish at processing plant); BC = After trimming/before chilling; AC = After water chilling; AS = After slush ice chilling; BIP = Before ice packing (Fresh fillets); AI = After injecting (polyphosphate injection), AF = After freezing (Frozen fillets). Temperature in the parenthesis denotes the fillet temperature at the designated process step.

PPC and TCC of fillets were not different ($p > 0.05$) at any process steps (Figure 4) except at AI. The highest PPC and TCC was observed at this point (AI). The temperature at this step was slightly increased (~1 °C) from the previous step (BIP), which may explain the reason of higher count after injection. No differences in PPC and TCC ($p > 0.05$) for AC and AS fillets indicated that 24 h slush ice chilling could not reduce the bacterial load (PPC and TCC) in comparison to water chilling.

4. Conclusions

This study provided extensive data on baseline proximate composition and bacterial load of hybrid catfish fillets. As the baseline water content dictated the amount of retained water of the catfish fillets at the process steps, data (water-protein ratio and fat content) from this study could be used to estimate the fillets' retained water at any process step which would further help the processor label their final products' retained water. Data showed that there was no water gain in fresh fillets and minimal water gain in frozen fillets. Nevertheless, processors should be careful to estimate the retained water using baseline water content from one season to another or big fish to small ones since baseline water content is fillet size and harvest season dependent. It is worth noting that the processor should be more cautious at the injection step keeping temperature low, where the bacterial load was considerably higher than in other steps. Overall, the catfish industry and others will be benefitted using our data on estimating retained water and microbiological quality of the hybrid catfish fillets for stringent process control.

Author Contributions: Conceptualization, J.L.S. and M.H.; methodology, J.L.S. and M.H.; software, M.H.; validation, J.L.S. and M.H.; formal analysis, M.H.; investigation, M.H.; resources, J.L.S. and M.H.; data curation, M.H.; writing—original draft preparation, M.H.; writing—review and editing, J.L.S. and M.H.; visualization, M.H.; supervision, J.L.S.; project administration, J.L.S.; funding acquisition, J.L.S. All authors have read and agreed to the published version of the manuscript.

Funding: This publication is a contribution of the Mississippi Agricultural and Forestry Experiment Station (MAFES), Mississippi State University, Mississippi State, MS, USA.

Institutional Review Board Statement: Not applicable.

Informed Consent Statement: Not applicable.

Data Availability Statement: Data are contained within the article.

Conflicts of Interest: The authors declare no conflict of interest.

References

1. TCI, The Catfish Institute, FAQs. 2022. Available online: http://uscatfishhttps://www.uscatfish.com/faqs.com/faqs/ (accessed on 31 July 2022).
2. USDA, National Agricultural Statistics Service Catfish Production (July 2022). 2022. Available online: https://downloads.usda.library.cornell.edu/usda-esmis/files/bg257f046/nc581v125/gt54mt95t/cfpd0722.pdf (accessed on 31 October 2022).
3. Gosh, K.; Hanson, T.R.; Drescher, D.; Robinson, D.; Bugg, W.; Chatakondi, N.; Kumar, G.; Jeffers, C.; Dunham, R.A. Economic Effect of Hybrid Catfish (Channel Catfish ♀× Blue Catfish ♂) Growth Variability on Traditional and Intensive Production Systems. *N. Am. J. Aquac.* **2022**, *84*, 25–41. [CrossRef]
4. Green, B.W.; Rawles, S.D. Comparative Growth and Yield of Channel Catfish and Channel × Blue Hybrid Catfish Fed a Full or Restricted Ration. *Aquac. Res.* **2010**, *41*, e109–e119. [CrossRef]
5. Ammerman, G.R. Processing. In *Channel Catfish Culture*; Tucker, C.S., Ed.; Elsevier Science Publishing Co., Inc.: New York, NY, USA, 1985; pp. 594–598.
6. Silva, J.L.; Ammerman, G.R.; Dean, S. *Processing Channel Catfish*; SRAC Publication Number 183; SRAC Publication: Beltsville, MD, USA, 2001.
7. United States Department of Agriculture, Food Safety and Inspection Service. Mandatory Inspection of Fish of the Order Siluriformes and Products Derived from Such Fish; Final Rule. 9 CFR part 441. Docket No. FSIS-2008-0031. Federal Register. 80:231. 2015. Available online: https://www.fsis.usda.gov/wps/wcm/connect/45f61995-b867-4a5b-a4e0-064ad052725c/2008-0031F.htm?MOD=AJPERES (accessed on 25 November 2021).
8. United States Department of Agriculture, Food Safety and Inspection Service. FSIS Compliance Guideline for Establishments That Slaughter or Further Process Siluriformes Fish and Fish Products. 2017. Available online: https://www.fsis.usda.gov/wps/wcm/connect/8ec92a7f-8f9b-45ae-b80f-7c336f7d6ff5/Compliance-Guideline-Siluriformes-Fish.pdf?MOD=AJPERES (accessed on 25 November 2021).
9. Bigbee, D.G.; Dawson, L.E. Some Factors That Affect Change in Weight of Fresh Chilled Poultry: 1. Length of Chill Period, Chilling Medium and Holding Temperature. *Poult. Sci.* **1963**, *42*, 457–462. [CrossRef]
10. James, C.; Vincent, C.; Andrade Lima, T.I.; James, S.J. The Primary Chilling of Poultry Carcasses—A Review. *Int. J. Refrig.* **2006**, *29*, 847–862. [CrossRef]
11. Jeong, J.Y.; Janardhanan, K.K.; Booren, A.M.; Karcher, D.M.; Kang, I. Moisture Content, Processing Yield, and Surface Color of Broiler Carcasses Chilled by Water, Air, or Evaporative Air. *Poult. Sci.* **2011**, *90*, 687–693. [CrossRef] [PubMed]
12. Young, L.L.; Smith, D.P. Moisture Retention by Water-and Air-Chilled Chicken Broilers during Processing and Cutup Operations. *Poult. Sci.* **2004**, *83*, 119–122. [CrossRef]
13. Botta, J.R.; Cahill, F.M. Moisture Content of Scallop Meat: Effect of Species, Time and Season and Method of Determining "Added Water". In Proceedings of the Annual Conference Tropical and Subtropical Fisheries Technological Conference of the Americas, Williamsburg, VA, USA, 4–6 November 1992; pp. 43–50.
14. Breck, J.E. Body Composition in Fishes: Body Size Matters. *Aquaculture* **2014**, *433*, 40–49. [CrossRef]
15. Yeannes, M.I.; Almandos, M.E. Estimation of Fish Proximate Composition Starting from Water Content. *J. Food Compos. Anal.* **2003**, *16*, 81–92. [CrossRef]
16. Karl, H.; Numata, J.; Lahrssen-Wiederholt, M. Variability of Fat, Water and Protein Content in the Flesh of Beaked Redfish (*Sebastes mentella*) and Greenland Halibut (*Reinhardtius hippoglossoides*) from Artic Fishing Grounds. *J. Consum. Prot. Food Saf.* **2018**, *13*, 383–389. [CrossRef]
17. Linhartová, Z.; Krejsa, J.; Zajíc, T.; Másílko, J.; Samples, S.; Mráz, J. Proximate and Fatty Acid Composition of 13 Important Freshwater Fish Species in Central Europe. *Aquac. Int.* **2018**, *26*, 695–711. [CrossRef]
18. Emre, Y.; Uysal, K.; Emre, N.; Pak, F.; Oruç, H.; Yetek, İ. Seasonal Variations of Fatty Acid Profiles in the Muscle of *Capoeta angorae*. *Turk. J. Fish. Aquat. Sci.* **2015**, *15*, 103–109. [CrossRef]

19. FAO. *FAO Fisheries and Aquaculture Fact Sheets 2015*; FAO: Rome, Italy, 2015.
20. Huss, H.H. *Fresh Fish–Quality and Quality Changes: A Training Manual Prepared for the FAO/DANIDA Training Programme on Fish Technology and Quality Control*; Food & Agriculture Organization: Rome, Italy, 1988.
21. Huss, H.H. *Quality and Quality Changes in Fresh Fish*; FAO: Rome, Italy, 1995.
22. Banrie Nutrition and Feeding Studies for Pond-Raised Hybrid Catfish. Available online: https://thefishsite.com/articles/nutrition-and-feeding-studies-for-pondraised-hybrid-catfish (accessed on 24 August 2022).
23. Effiong, M.U.; Peter, N.E. Proximate Compositions and Length-Weight Relationships of Hybrid Catfish, Clarias Gariepinus (♂) X Heterobranchus Bidorsalis (♀) under Various Densities. *J. Aquat. Sci.* **2019**, *34*, 153–160. [CrossRef]
24. Li, M.H.; Wise, D.J.; Kumar, G.; Bosworth, B.G.; Mischke, C.C.; Aarattuthodiyil, S.; Rutland, W.; Lucas, P.M. Effects of Long-Term Restricted Feeding Followed by Full Feeding on Growth, Processing Yield, Fillet Proximate Composition, and Economics of Market-Size Hybrid Catfish, ♀Ictalurus Punctatus × ♂Ictalurus Furcatus. *J. World Aquac. Soc.* **2020**, *51*, 931–943. [CrossRef]
25. Gould, G.W. *Micro-Organisms in Foods 2. Sampling for Microbiological Analysis: Principles and Specific Applications*, 2nd ed.; ICMSF, Blackwell Scientific Publications: Oxford, UK, 1990; ISBN 0-632-01567-5.
26. Fernandes, C.F.; Flick, G.J.; Silva, J.L.; McCaskey, T.A. Influence of Processing Schemes on Indicative Bacteria and Quality of Fresh Aquacultured Catfish Fillets. *J. Food Prot.* **1997**, *60*, 54–58. [CrossRef]
27. Marroquin, E.; Silva, J.L.; Wannapee, B.; Kim, T. Processing Method Effect on Texture, Color, and Microbial Load of Channel Catfish Fillets. *J. Aquat. Food Prod. Technol.* **2004**, *13*, 101–110. [CrossRef]
28. Nunez, A.L.; Silva, J.L.; Hood, A.F.; Chamul, R.S. Variations in Microbial Contamination Through the Process in Three Typical Catfish Operations. *Bull. Miss. Agric. For. Exp. Stn.* **2003**, *1121*, 28.
29. Watchalotone, S.; Silva, J.L.; Chen, T.C.; Handumrongkul, C. Influence of Process Flow on Microbial Profile of Channel Catfish Fillets. In *Proceedings of the 1998 Catfish Processors' Workshop, Starkville, MS, USA, 1998*; Silva, J.L., Hood, A.F., Dean, S., Eds.; Office of Agricultural Communications, Division of Agriculture, Forestry, and Veterinary Medicine: Starkville, MS, USA, 2001; pp. 14–16.
30. Swanson, K.M.J.; Busta, F.F.; Peterson, E.H.; Johnson, M.G. Colony Count Methods. *Compend. Methods Microbiol. Exam. Foods* **1992**, *3*, 75–95.
31. Lai, T.L.; Robbins, H.; Wei, C.Z. Strong Consistency of Least Squares Estimates in Multiple Regression II. *J. Multivar. Anal.* **1979**, *9*, 343–361. [CrossRef]
32. Boggess , T.S., Jr.; Heaton, E.K.; Shewfelt, A.L. Storage Stability of Commercially Prepared And Frozen Pond-Raised Channel Catfish (Ictalurus Punctauts, Rafinisque). *J. Food Sci.* **1971**, *36*, 969–973. [CrossRef]
33. Shearer, K.D. Factors Affecting the Proximate Composition of Cultured Fishes with Emphasis on Salmonids. *Aquaculture* **1994**, *119*, 63–88. [CrossRef]
34. Silva, J.L.; Ammerman, G.R. Composition, Lipid Changes, and Sensory Evaluation of Two Sizes of Channel Catfish during Frozen Storage. *J. Appl. Aquac.* **1993**, *2*, 39–50. [CrossRef]
35. Robinson, E.H.; Li, M.H.; Oberle, D.F. *Nutrient Characteristics of Pond-Raised Channel Catfish*; Mississippi Agricultural & Forestry Experiment Station: Beaumont, MS, USA, 2001.
36. Nettleton, J.A.; Allen, W.H.; Klatt, L.V.; Ratnayake, W.M.N.; Ackman, R.G. Nutrients and chemical residues in one- to two-pound Mississippi farm-raised channel catfish (Ictalurus punctatus). *J. Food Sci.* **1990**, *55*, 954–958. [CrossRef]
37. Huang, Y.-W.; Leung, C.-K. Microbiological Assessment of Channel Catfish Grown in Cage and Pond Culture. *Food Microbiol.* **1993**, *10*, 187–195. [CrossRef]
38. Carciofi, B.A.; Laurindo, J.B. Water Uptake by Poultry Carcasses during Cooling by Water Immersion. *Chem. Eng. Process. Process Intensif.* **2007**, *46*, 444–450. [CrossRef]
39. Klose, A.A.; Pool, M.F.; De Fremery, D.; Campbell, A.A.; Hanson, H.L. Effect of Laboratory Scale Agitated Chilling of Poultry on Quality. *Poult. Sci.* **1960**, *39*, 1193–1198. [CrossRef]
40. Kin, S.; Wes Schilling, M.; Silva, J.L.; Smith, B.S.; Jackson, V.; Kim, T. Effects of Phosphate Type on the Quality of Vacuum-Tumbled Catfish Fillets. *J. Aquat. Food Prod. Technol.* **2009**, *18*, 400–415. [CrossRef]
41. Walayat, N.; Wang, X.; Liu, J.; Nawaz, A.; Zhang, Z.; Khalifa, I.; Rincón Cervera, M.Á.; Pateiro, M.; Lorenzo, J.M.; Nikoo, M.; et al. Kappa-Carrageenan as an Effective Cryoprotectant on Water Mobility and Functional Properties of Grass Carp Myofibrillar Protein Gel during Frozen Storage. *LWT* **2022**, *154*, 112675. [CrossRef]
42. Walayat, N.; Tang, W.; Wang, X.; Yi, M.; Guo, L.; Ding, Y.; Liu, J. Effective Role of Konjac Oligosaccharide against Oxidative Changes in Silver Carp Proteins during Fluctuated Frozen Storage. *Food Hydrocoll.* **2022**, *131*, 107761. [CrossRef]
43. Hirose, A.; Yoshitake, M.; Onodera, J.; Ooba, K.; Sakakibara, T.; Ito, A.; Shiina, Y. Measurement of fat content of cultured Pacific bluefin tuna Thunnus orientalis by near-infrared spectroscopy. *Nippon Suisan Gakkaishi* **2016**, *82*, 753–762. [CrossRef]
44. Xiccato, G.; Trocino, A.; Tulli, F.; Tibaldi, E. Prediction of Chemical Composition and Origin Identification of European Sea Bass (*Dicentrarchus labrax* L.) by near Infrared Reflectance Spectroscopy (NIRS). *Food Chem.* **2004**, *86*, 275–281. [CrossRef]
45. Ruth, S.M.; Brouwer, E.; Koot, A.; Wijtten, M. Seafood and Water Management. *Foods* **2014**, *3*, 622–631. [CrossRef] [PubMed]
46. Kutner, M.H.; Nachtsheim, C.; Neter, J. *Applied Linear Regression Models*; McGraw-Hill/Irwin: New York, NY, USA, 2004.
47. Mayer, B.K.; Ward, D.R. Microbiology of Finfish and Finfish Processing. In *Microbiology of Marine Food Products*; Springer: Boston, MA, USA, 1991; pp. 3–17.

48. Watchalotone, S. Influence of Process Flow on Microbial Profile of Channel Catfish and of Selected Antimicrobials on Its Shelf Life. Mater's Thesis, Mississippi State University, Starkville, MS, USA, 1996.
49. Haque, M.; Bosilevac, J.M.; Chaves, B.D. A Review of Shiga-Toxin Producing Escherichia Coli (STEC) Contamination in the Raw Pork Production Chain. *Int. J. Food Microbiol.* **2022**, *377*, 109832. [CrossRef] [PubMed]

foods

Article

Fattening in Saline and Alkaline Water Improves the Color, Nutritional and Taste Quality of Adult Chinese Mitten Crab *Eriocheir sinensis*

Shihui Wang [1,2], Kun Guo [1,2], Liang Luo [1,2], Rui Zhang [1,2], Wei Xu [1,2], Yingying Song [1,3] and Zhigang Zhao [1,2,*]

1 Key Open Laboratory of Cold Water Fish Germplasm Resources and Breeding of Heilongjiang Province, Heilongjiang River Fisheries Research Institute, Chinese Academy of Fishery Sciences, Harbin 150070, China
2 Engineering Technology Research Center of Saline-Alkaline Water Fisheries (Harbin), Chinese Academy of Fishery Sciences, Harbin 150070, China
3 College of Animal Science and Technology, Northeast Agricultural University, Harbin 150030, China
* Correspondence: zhaozhigang@hrfri.ac.cn; Tel.: +86-451-8793-0948

Abstract: The majority of pond-reared Chinese mitten crab (*Eriocheir sinensis*) grow and fatten in freshwater. Previous studies illustrated that *E. sinensis* cultured in saline-alkaline water in outdoor environments showed a higher quality than that cultured in freshwater. However, it is still unclear whether salinity or alkalinity has an important positive effect on the quality of *E. sinensis*. This study aimed to investigate the gonadal development, edible yield, coloration, and nutritional and flavor quality of *E. sinensis* fattening in saline and alkaline water indoors. Results showed that there were no significant changes observed in gonadosomatic index (GSI) and other edible parameters among freshwater (FW), saline water (SW), and alkaline water (AW) during the 55-day fattening period ($p > 0.05$). Significantly higher a^* and b^* values of freeze-dried female carapace were observed fattening in SW and AW compared with that of FW ($p < 0.05$). The crude protein in gonad and male muscle, moisture in female muscle, and crude lipid in male muscle increased significantly from FW to SW and AW ($p < 0.05$). Better nutritional and flavor values were also detected in male hepatopancreas and muscles. In conclusion, numerous advantages of fattening in SW and AW were observed, including the improvement of carotenoid accumulation in freeze-dried carapace, DHA, EPA, total essential free amino acids (∑EFAA), total free amino acids (∑FAA), and total umami values (∑TUV) contents in male hepatopancreas and muscle.

Keywords: *Eriocheir sinensis*; aquaculture; fatty acid; free amino acid

Citation: Wang, S.; Guo, K.; Luo, L.; Zhang, R.; Xu, W.; Song, Y.; Zhao, Z. Fattening in Saline and Alkaline Water Improves the Color, Nutritional and Taste Quality of Adult Chinese Mitten Crab *Eriocheir sinensis*. *Foods* **2022**, *11*, 2573. https://doi.org/10.3390/foods11172573

Academic Editors: Tao Yin and Liu Shi

Received: 6 August 2022
Accepted: 23 August 2022
Published: 25 August 2022

1. Introduction

The Chinese mitten crab *Eriocheir sinensis*, an important freshwater product, has high economic value and is favored by consumers because of its delicious taste and high polyunsaturated fatty acid (PUFA) and amino acid (AA) contents in China [1]. The aquaculture yield reached 808,274 t in 2021, and the output value has increased to beyond CNY 50 billion [2]. *E. sinensis* has a unique life cycle. Juvenile and adult individuals live in freshwater, while mature or nearly mature individuals migrate to the estuary for reproduction [3]. The culture modes of *E. sinensis* mainly include pond culture, lake, or reed pond proliferation, and rice crab co-culture, which are distributed in different regions of China. With increasing pressure for environmental protection, expanding the aquaculture space of aquatic products has become one of the most important elements affecting the development of the *E. sinensis* industry. The area of *E. sinensis* cultured in saline-alkaline water has gradually increased in recent years [4,5].

Globally, saline-alkaline land covers 0.95 billion hectares, accounting for 1/3 of the total land area [6]. The area of saline-alkaline land in China is approximately 99.13 million hectares, and the low-lying saline-alkaline water is approximately 45.87 million hectares, accounting for approximately 55% of the total lake area [7]. Previous studies have demonstrated that *E. sinensis* cultured in saline-alkaline water in outdoor ponds does not affect growth and gonadal development [8]. Meanwhile, *E. sinensis* cultured in saline-alkaline water in outdoor environments show higher levels of long-chain unsaturated fatty acids (LC-PUFA), especially DHA and EPA, than that cultured in freshwater [4,5]. However, the nutritional quality of *E. sinensis* cultured in outdoor earthen ponds can be affected by numerous factors, such as germplasm [9], culture environment [10], and diet [11]. Salinity and alkalinity are important ecological factors in the culture environment [12,13]. Hence, it is still unclear whether salinity or alkalinity has an effect on the nutritional quality of *E. sinensis*, and the main reason why *E. sinensis* accumulates high levels of DHA and EPA in outdoor saline-alkaline earthen ponds.

Since *E. sinensis* is a migratory aquatic animal, salinity plays an important role in its reproduction. Therefore, there are many studies on salinity [14–17]. The majority of studies have mainly focused on the physiological metabolism and osmoregulation of salinity in *E. sinensis* [14,15]. Only a few studies have paid close attention to the nutritional quality and flavor quality of *E. sinensis* [16,17]. Even though there is little literature, they still focus on brackish water (12 ppt) during the process of reproductive migration or low salinity seawater (7 ppt) for a short time. Previous studies have investigated edible yield and nutritional quality for fattening over 60 days in outdoor ponds [1,11]. Nevertheless, no studies have reported on the gonadal development, color, nutrition, and flavor quality of *E. sinensis* fattening in low salinity water (1.5 ppt) indoors for a long time.

Compared with numerous studies on the salinity of *E. sinensis*, few reports on the alkalinity of *E. sinensis* exist in the literature. More reports have focused on the toxicity of alkalinity on *Tribolodon brandti* [18] and *Macrobrachium nipponense* [19]. A recent study simply illustrated the toxicity of carbonate alkalinity ($NaHCO_3$) on *E. sinensis* [20]. However, no future reports have focused on the quality of *E. sinensis* edible tissues.

During the period of fattening, edible parameters, commonly including the hepatosomatic index (HSI), gonadosomatic index (GSI), meat yield (MY), and total edible yield (TEY), are quite important indicators utilized in the evaluation of fattening performance for *E. sinensis* [11]. Meanwhile, color parameter, proximate composition, fatty acid, and free amino acid are also key indicators for the coloration and nutritional and flavor quality of *E. sinensis* [4,21]. Therefore, based on the above reasoning, this experiment was designed to investigate the effects on the gonadal development, edible yield, coloration, and nutritional and flavor quality of *E. sinensis* fattening in saline or alkaline water.

2. Materials and Methods

2.1. Experimental Design

This experiment was conducted at the Heilongjiang River Fisheries Research Institute, CAFS (Harbin, China). Approximately 140 adult *E. sinensis* post puberty molt (female body weight 80~100 g, male body weight 100~120 g) were obtained from a local crab farm in early August 2021. The crabs were cultured in earth pond outside with *Elodea canadensis* transplanted from 1 May to the sampling time, and fed once a day at 17:00 with a commercial formulated diet (crude protein $\geq$ 36.0%, crude fat $\geq$ 5.0%, moisture $\leq$ 12.0%, ash $\leq$ 18.0%; Nanjing Aohua Biotechnology Co., Ltd., Nanjing, China). Among them, a total of 120 healthy, active, and intact individuals were selected and conducted for the experiment on 5th August according to our own protocol. The fattening trial was conducted in 12 indoor glass tanks (64 $\times$ 38 $\times$ 43 cm) with 60 L water in each glass tank. Four glass tanks as one treatment and three treatments were set in this study: freshwater (FW: salinity 0 ppt, alkalinity 0 mmol/L), saline water (SW: salinity 1.5 ppt, alkalinity 0 mmol/L), and alkaline water (AW: alkalinity 10 mmol/L, salinity 0 ppt). Forty individuals together (half females and half males) were randomly distributed, averaging 10 individuals in each glass tank. FW was sourced from tap water. Analytical pure sodium chloride NaCl (Sinopharm Chemical Reagent Co., Shenyang, China) was used to adjust the salinity to 1.5 ppt, and analytical pure sodium bicarbonate $NaHCO_3$ (Sinopharm Chemical Reagent Co., Shenyang, China) was used to adjust 10 mmol/L alkalinity. The saline-alkaline concentration in this experiment originated from the outdoor earthen pond from Dongying in Shandong Province and Daqing from Heilongjiang Province of China.

2.2. Culture Management

During the fattening period, the water temperature was changed 2 °C every 10 days and presented a downward trend from 25 to 15 °C depending on the season from 5th August to 28th September. In an indoor circulating aquaculture system, around 30% of glass tank water was replaced in each tank every day with dechlorinated tap water at the appropriate temperature and maintained constant water salinity and alkalinity. The ammonia-N, nitrite, salinity, alkalinity, dissolved oxygen (DO), and pH of the water were checked every day. Over the course of the trial, *E. sinensis* was fed daily at 5 pm with a commercial crab diet (crude protein $\geq$ 36.0%, crude fat $\geq$ 5.0%, moisture $\leq$ 12.0%, ash $\leq$ 18.0%; Nanjing Aohua Feed Co., Ltd., Nanjing, China), and food residue was removed next morning. The feeding amount was adjusted according to Zhang's study [16].

2.3. Sample Collection and Dissection

The studies in *E. sinensis* were reviewed and approved by the Committee for the Welfare and Ethics of Laboratory Animals of the Heilongjiang River Fisheries Research Institute (Approval numbers: 20210915-001, approved on 15 September 2021). On 28 September, a total of 40 individuals from each treatment were sampled. The *E. sinensis* surface was wiped with a towel, and digital balance (JA2002, precision = 0.01 g, Shanghai Puchun Measuring Instrument Co., Ltd., Shanghai, China) was used to measure the body weight of each crab. A Vernier caliper (111-101-10G, precision = 0.01 mm, Guilin Guanglu Measuring Instrument Co., Ltd., Guilin, China) was used to measure the carapace length and width parameters. The *E. sinensis* was then dissected to obtain edible tissues (hepatopancreas, gonad, and muscle). The method and steps of dissection were as follows: First, the carapace and breastplate of *E. sinensis* were separated, and then the hepatopancreas and gonads hidden in the carapace were carefully taken out. Second, the hexagonal heart was discarded, and the remaining hepatopancreas and gonads in the chest were carefully taken out. The edible tissues were weighed for calculating hepatosomatic index (HSI, %), gonadosomatic index (GSI, %), meat yield (MY, %), and total edible yield (TEY, %). Subsequently, all edible tissues were stored separately at −40 °C for further biochemical analysis. The HSI, GSI, MY, TEY, and condition factor (CF, g/cm^3) were calculated with the following Formulas (1)–(5):

$$\text{HSI }(\%) = 100 \times \frac{\text{hepatopancreas weight}}{\text{body weight}} \tag{1}$$

$$\text{GSI }(\%) = 100 \times \frac{\text{gonad weight}}{\text{body weight}} \tag{2}$$

$$\text{MY }(\%) = 100 \times \frac{\text{meat weight}}{\text{body weight}} \tag{3}$$

$$\text{TEY }(\%) = 100 \times \frac{\text{hepatopancreas weight} + \text{gonad weight} + \text{meat weight}}{\text{body weight}} \tag{4}$$

$$\text{CF }(\text{g/cm}^3) = \frac{\text{body weight}}{\text{carapace length}^3} \tag{5}$$

2.4. Measurements of Color Parameters

A colorimeter (CR-400, Konica Minolta, Marunouchi, Tokyo, Japan) was used to measure the color values (lightness L^*, redness a^*, and yellowness b^*) of freeze-dried carapace, and female gonad (ovary) of *E. sinensis* from three treatments, respectively. Six relatively smooth points on the carapace surface [21], and three random points on the ovary were selected for L^*, a^*, and b^* measures.

2.5. Proximate Composition

The moisture analysis of experimental *E. sinensis* edible tissues from the three treatments was determined with a vacuum freeze-dryer (FD-1A-50, Biocoll, Beijing, China) at −50 °C to a constant weight [4]. Prior to the future biochemical analysis, five freeze-dried *E. sinensis* tissues were randomly selected to form a replicate. Three replicates were designed in the experiment. The crude protein, crude lipid, and ash contents were separately determined using the Kjeldahl method [22], GB 5009.6-2016 [23], and AOAC procedures [22], respectively.

2.6. Fatty Acid Profile

The peak area percentage method by GB 5009.168-2016 was carried out to measure fatty acids [24]. All samples were repeated three times. Crude lipids extracted from *E. sinensis* edible tissues were further processed for fatty acid analysis. The results are presented as the percentage of each fatty acid with respect to the total fatty acids (%).

2.7. Free Amino Acid Analysis and Taste Activity Value

Freeze-dried *E. sinensis* tissue samples were processed according to the previously described method [16]. A total of 17 amino acids were calculated using this method. The taste activity value (TAV) was calculated as the ratio of the concentration of taste compounds measured above in the *E. sinensis* edible tissue to its threshold value [4].

2.8. Statistical Analysis

The results are presented as mean values ± standard error (SE). SPSS 22.0 software (SPSS Inc., Chicago, IL, USA) was used for statistical analysis. A one-way ANOVA was used to determine the differences among these three treatments, and Duncan's multiple range test was carried out for comparisons. The comparison test $p < 0.05$ was regarded as the statistical significance.

3. Results

3.1. Gonadal Development and Total Edible Yield

The edible yield and condition factors of adult *E. sinensis* from different culture environments are shown in Figure 1. There was no significantly increasing or decreasing trend observed among the FW, SW, and AW during the 55-day fattening period ($p > 0.05$). However, slight numerical differences were found between edible parameters and the condition factor (CF) among the three treatments at the end of the experiment. For females (Figure 1A,C), the HSI, GSI, TEY, and CF parameters of *E. sinensis* from FW were slightly higher than those of SW and AW. For males (Figure 1B,C), the HSI, GSI, and CF parameters of *E. sinensis* fattening in SW were much better than those of FW and AW, while a higher percentage of *E. sinensis* MY and TEY parameters existed in AW.

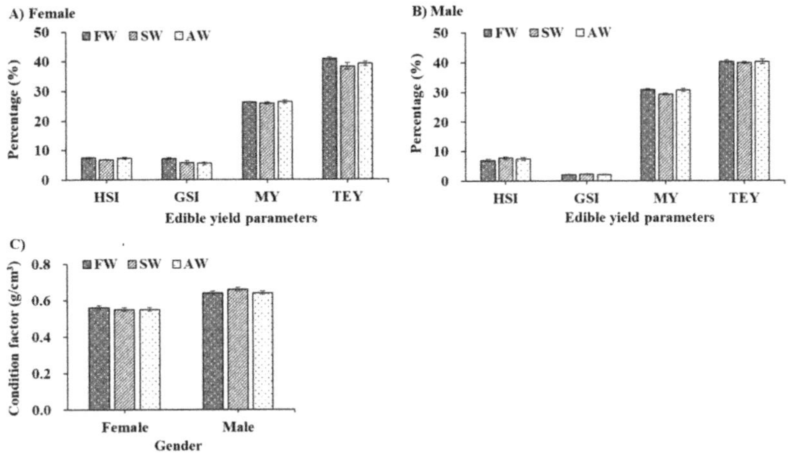

Figure 1. The edible yield (% body weight, (**A**,**B**)) and condition factor (%, (**C**)) of adult *Eriocheir sinensis* from different culture environments. The data are presented as mean ± standard error (SE) (n = 20). FW, freshwater; SW, saline water; AW, Alkaline water; HSI, hepatosomatic index; GSI, gonadosomatic index; MY, meat yield; TEY, total edible yield.

3.2. Color Parameters

Significantly higher a^* (redness) and b^* (yellowness) values of freeze-dried female carapace were observed fattening in SW and AW compared with that of FW ($p < 0.05$) (Figure 2A). No significant differences existed between the freeze-dried female gonad and male carapace among the three treatments (Figure 2B,C) ($p > 0.05$), but the a^* and b^* values of freeze-dried male carapace still showed an increasing trend of fattening in SW and AW. Overall, the effects of fattening in SW and AW on the color quality of the carapace are more obvious than that of the female gonad.

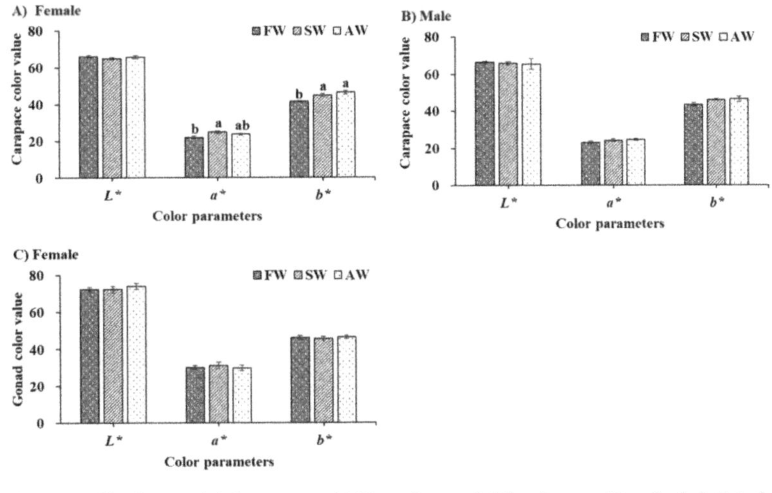

Figure 2. The freeze-dried carapace (**A**,**B**) and gonad (**C**) color quality of adult *Eriocheir sinensis* from different culture environments. The data are presented as mean ± standard error (SE) (n = 20). Different letters indicate a significant difference among different culture environments ($p < 0.05$). FW, freshwater; SW, saline water; AW, Alkaline water; L^*, a^*, and b^* represent the color parameters of lightness, redness, and yellowness, respectively.

3.3. Proximate Composition

The proximate composition of adult *E. sinensis* fattening in different culture environments is presented in Table 1. The crude protein in gonad and male muscle, moisture in female muscle, and crude lipid in male muscle increased significantly from FW to SW and AW ($p < 0.05$), whereas a decreasing trend was detected by crude protein in female muscle from FW to SW and AW ($p < 0.05$). The crude lipid of hepatopancreas exhibited an obviously increasing trend of fattening in SW and AW compared with that of fattening in FW, but no significant difference ($p > 0.05$).

Table 1. The proximate composition (% wet weight) of adult *Eriocheir sinensis* fattening in different culture environments.

Item	Female			Male		
	FW	SW	AW	FW	SW	AW
Hepatopancreas						
moisture	59.50 ± 1.60	59.62 ± 3.39	61.04 ± 2.02	55.29 ± 0.66	50.34 ± 4.16	51.48 ± 3.26
crude protein	10.86 ± 0.51	9.06 ± 0.07	9.94 ± 0.86	8.96 ± 0.11	9.28 ± 2.75	8.01 ± 0.85
crude lipid	24.10 ± 1.01	26.99 ± 0.10	24.17 ± 0.68	30.92 ± 0.07	35.11 ± 3.67	35.76 ± 1.21
ash	1.46 ± 0.16	1.03 ± 0.06	1.19 ± 0.10	1.36 ± 0.07	1.14 ± 0.30	1.07 ± 0
Gonad						
moisture	55.63 ± 1.52	53.68 ± 1.71	55.02 ± 0.53	73.50 ± 0.31	72.76 ± 0.35	73.34 ± 0.36
crude protein	28.84 ± 0.08 ^{b}	29.86 ± 0.25 ^{a}	29.31 ± 0.11 ^{ab}	17.61 ± 0.01 ^{b}	18.01 ± 0.02 ^{a}	17.63 ± 0.05 ^{b}
crude lipid	7.45 ± 0.09	6.72 ± 0.32	7.22 ± 0.11	0.34 ± 0	0.42 ± 0.04	0.35 ± 0
ash	2.17 ± 0.40	1.90 ± 0.05	2.23 ± 0.34	2.46 ± 0.32	2.33 ± 0.04	2.25 ± 0.01
Muscle						
moisture	79.61 ± 0.25 ^{b}	80.88 ± 0.59 ^{ab}	81.01 ± 0.44 ^{a}	79.33 ± 0.64	78.04 ± 0.50	78.96 ± 0.50
crude protein	17.42 ± 0.05 ^{a}	15.71 ± 0.15 ^{b}	15.63 ± 0.30 ^{b}	16.50 ± 0.06 ^{c}	17.98 ± 0.00 ^{a}	17.05 ± 0.11 ^{b}
crude lipid	0.38 ± 0.03	0.46 ± 0	0.37 ± 0.03	0.35 ± 0.02 ^{b}	0.48 ± 0 ^{a}	0.38 ± 0.02 ^{b}
ash	1.51 ± 0.02	1.50 ± 0.03	1.49 ± 0.07	1.35 ± 0.05	1.50 ± 0.01	1.47 ± 0.02

Notes: The data presented as means ± standard error (SE) (n = 20). Values in the same row with different superscripts are significantly different ($p < 0.05$). Forty mature individuals were collected from each culture environment. Abbreviations: FW, freshwater; SW, saline water; AW, alkaline water.

3.4. Fatty Acid Profiles

The fatty acid composition of adult *E. sinensis* fattening in different culture environments is summarized in Table 2. The fatty acids of *E. sinensis* edible tissues contain saturated fatty acids (SFAs), monounsaturated fatty acids (MUFAs), and polyunsaturated fatty acids (PUFAs).

3.5. Free Amino Acid Composition and Taste Activity Value

Concentrations of ∑SFA in male hepatopancreas significantly decreased from FW to SW and AW ($p < 0.05$) (Table 2). This trend was mainly driven by a reduction in 16:0, which decreased from 27.90 ± 0.90% in FW to 23.96 ± 0% in SW. Concentrations of ∑MUFA in male hepatopancreas also decreased from FW to SW and AW, but no significant difference was observed ($p > 0.05$). This trend was generally driven by a reduction in dominant 18:1n9c, which decreased from 46.65 ± 1.61% in FW to 42.76 ± 0.24% in SW. 18:2n6c was present in the highest abundance, followed by 18:3n3, DHA, EPA, and ARA within PUFA. A significantly increasing trend of ∑PUFA in male hepatopancreas was observed from FW to AW and SW ($p < 0.05$), which is generally driven by an increase in the above-mentioned fatty acids. The highest content of DHA in male hepatopancreas was detected in AW compared with that of FW ($p < 0.05$). A significant improvement of ∑HUFA, ∑n-3 PUFA, and ∑DHA + EPA contents was also observed in SW and AW male hepatopancreas ($p < 0.05$).

Table 2. The fatty acid composition in hepatopancreas, gonad, and muscle of adult *Eriocheir sinensis* fattening in different culture environments (% of total fatty acids).

Fatty Acid	Hepatopancreas			Gonad			Muscle		
	FW	SW	AW	FW	SW	AW	FW	SW	AW
Female									
C14:0	1.36 ± 0.12	1.40 ± 0.04	1.62 ± 0.04	0.85 ± 0.02	0.83 ± 0.01	1.01 ± 0.18	—	—	—
C15:0	0.67 ± 0.06	0.68 ± 0.02	0.76 ± 0.01	0.50 ± 0.02	0.53 ± 0.01	0.55 ± 0.08	—	—	—
C16:0	19.68 ± 0.75	19.46 ± 0.22	20.85 ± 1.74	16.03 ± 0.37	15.50 ± 0.07	16.20 ± 0.40	12.53 ± 0.25	12.37 ± 0.32	13.22 ± 0.25
C17:0	0.74 ± 0.03 ^{a}	0.62 ± 0 ^{b}	0.71 ± 0.03 ^{ab}	0.76 ± 0.05	0.63 ± 0.02	0.74 ± 0.11	0.95 ± 0.02 ^{ab}	0.86 ± 0 ^{b}	1.01 ± 0.04 ^{a}
C18:0	3.07 ± 0.02	3.19 ± 0.07	3.42 ± 0.19	3.15 ± 0.17	2.82 ± 0.02	2.87 ± 0.12	9.79 ± 0.02 ^{ab}	9.21 ± 0.21 ^{b}	10.08 ± 0.19 ^{a}
∑SFA	26.54 ± 0.88	26.45 ± 0.04	28.52 ± 1.89	21.38 ± 0.63	20.47 ± 0.08	21.45 ± 0.88	24.01 ± 0.24 ^{ab}	23.12 ± 0.50 ^{b}	25.11 ± 0.15 ^{a}
C15:1n5	0.30 ± 0.03	0.28 ± 0.02	0.32 ± 0.02	—	—	—	0.61 ± 0.03	0.66 ± 0.01	0.79 ± 0.07
C16:1n7	8.70 ± 0.82	9.57 ± 0.92	9.76 ± 1.32	7.80 ± 0.95	10.33 ± 0.13	10.27 ± 0.48	2.97 ± 0.19	2.81 ± 0.05	2.86 ± 0.46
C18:1n9c	38.59 ± 0.27	36.55 ± 1.04	37.94 ± 1.21	34.18 ± 0.31 ^{a}	31.40 ± 0.75 ^{b}	32.55 ± 0.46 ^{ab}	25.27 ± 0.32	24.92 ± 0.51	25.45 ± 0.14
C20:1n9	1.33 ± 0.08	1.04 ± 0.04	1.18 ± 0.14	0.48 ± 0.05 ^{a}	0.33 ± 0.02 ^{b}	0.40 ± 0.01 ^{ab}	0.76 ± 0.03	0.75 ± 0.02	0.74 ± 0.03
∑MUFA	49.89 ± 0.71	48.34 ± 0.15	50.22 ± 2.72	42.96 ± 1.22	42.57 ± 0.60	43.72 ± 0.05	29.86 ± 0.54	29.46 ± 0.60	30.16 ± 0.24
C18:2n6c	17.06 ± 1.91	19.26 ± 0.47	15.97 ± 4.03	16.96 ± 0.08	17.83 ± 0.26	16.84 ± 1.41	11.25 ± 0.90	10.60 ± 0.26	9.69 ± 0.04
C18:3n3	1.84 ± 0.03	2.12 ± 0.20	1.59 ± 0.36	3.10 ± 0.11	3.85 ± 0.23	3.31 ± 0.18	1.58 ± 0.10	1.61 ± 0.01	1.36 ± 0.18
C20:2n6	1.18 ± 0.03 ^{a}	0.90 ± 0.02 ^{b}	0.87 ± 0.06 ^{b}	1.07 ± 0.07	0.92 ± 0.04	0.90 ± 0.01	1.68 ± 0.05	1.62 ± 0.03	1.58 ± 0.14
C20:4n6 (ARA)	0.96 ± 0.09	0.80 ± 0.10	0.76 ± 0.09	3.38 ± 0.03	3.32 ± 0.11	3.29 ± 0.09	6.73 ± 0.17	7.02 ± 0.18	7.14 ± 0.24
C20:3n3	0.37 ± 0 ^{a}	0.30 ± 0.03 ^{b}	0.26 ± 0.01 ^{b}	0.49 ± 0.04	0.47 ± 0	0.45 ± 0.02	0.56 ± 0.02 ^{b}	0.61 ± 0.01 ^{a}	0.58 ± 0.01 ^{ab}
C20:5n3 (EPA)	1.03 ± 0.04	0.84 ± 0.14	0.72 ± 0.18	6.48 ± 0.06	6.58 ± 0.22	6.35 ± 0.07	16.50 ± 0.28	17.15 ± 0.16	16.34 ± 0.33
C22:6n3 (DHA)	0.44 ± 0.12	0.41 ± 0.08	0.36 ± 0.07	3.24 ± 0.18	3.36 ± 0.18	2.81 ± 0.09	7.62 ± 0.48	8.22 ± 0.22	7.68 ± 0.21
∑PUFA	23.60 ± 0.72	25.21 ± 0.05	21.15 ± 2.06	35.35 ± 0.28	36.86 ± 0.23	34.51 ± 0.42	46.32 ± 0.25 ^{ab}	47.23 ± 0.04 ^{a}	44.83 ± 0.13 ^{b}
∑EFA	19.18 ± 0.84	21.58 ± 0.12	17.78 ± 1.95	20.29 ± 0.08	21.86 ± 0.22	20.34 ± 0.55	13.04 ± 0.44	12.40 ± 0.12	11.31 ± 0.09
∑HUFA	5.09 ± 0.16	4.86 ± 0.29	4.09 ± 0.34	17.08 ± 0.24	17.93 ± 0.15	16.57 ± 0.23	33.19 ± 0.14 ^{b}	34.82 ± 0.06 ^{a}	33.30 ± 0.09 ^{b}
∑n-3 PUFA	3.67 ± 0.13	3.66 ± 0.32	2.93 ± 0.43	13.30 ± 0.27	14.26 ± 0.13	12.91 ± 0.24	26.25 ± 0.08 ^{b}	27.58 ± 0.05 ^{a}	25.95 ± 0.03 ^{b}
∑n-6 PUFA	19.93 ± 1.04	21.55 ± 0.19	18.22 ± 2.31	22.05 ± 0.14	22.60 ± 0.20	21.60 ± 0.74	20.80 ± 0.40	19.65 ± 0.02	18.89 ± 0.19
n-3/n-6 PUFA	0.19 ± 0.03	0.17 ± 0.02	0.16 ± 0	0.60 ± 0.01	0.63 ± 0	0.60 ± 0.05	1.31 ± 0.05	1.40 ± 0	1.37 ± 0.03
∑DHA + EPA	1.46 ± 0.15	1.25 ± 0.22	1.08 ± 0.25	9.72 ± 0.23 ^{ab}	9.94 ± 0.04 ^{a}	9.16 ± 0.16 ^{b}	24.11 ± 0.20 ^{b}	25.37 ± 0.05 ^{a}	24.02 ± 0.13 ^{b}
DHA/EPA	0.42 ± 0.10	0.48 ± 0.02	0.51 ± 0.03	0.50 ± 0.02	0.51 ± 0.04	0.44 ± 0.01	0.46 ± 0.04	0.48 ± 0.02	0.47 ± 0.02

Table 2. *Cont.*

Fatty Acid	Hepatopancreas			Gonad			Muscle		
	FW	SW	AW	FW	SW	AW	FW	SW	AW
Male									
C14:0	1.56 ± 0.07	1.37 ± 0.13	1.62 ± 0.11	0.52 ± 0.07	0.41 ± 0.05	0.49 ± 0.01	—	—	—
C15:0	0.83 ± 0.04 ^{a}	0.64 ± 0.01 ^{b}	0.73 ± 0.05 ^{ab}	—	—	—	—	—	—
C16:0	19.92 ± 0.83 ^{a}	17.15 ± 0.07 ^{b}	17.78 ± 0.24 ^{ab}	10.25 ± 0.39	9.63 ± 0.44	10.21 ± 0.02	13.31 ± 0.45	12.18 ± 0.04	12.60 ± 0.05
C17:0	0.91 ± 0.03 ^{a}	0.69 ± 0.01 ^{b}	0.87 ± 0.06 ^{a}	0.90 ± 0.05	0.75 ± 0.05	0.84 ± 0.01	1.10 ± 0.02	0.93 ± 0.04	1.11 ± 0.09
C18:0	3.63 ± 0.06 ^{a}	3.13 ± 0.12 ^{b}	3.40 ± 0.02 ^{ab}	7.79 ± 0.50	7.48 ± 0.04	7.62 ± 0.07	10.27 ± 0.09	9.49 ± 0.07	10.21 ± 0.30
∑SFA	27.90 ± 0.90 ^{a}	23.96 ± 0 ^{b}	25.42 ± 0.42 ^{ab}	19.90 ± 0.09	18.80 ± 0.58	19.63 ± 0.11	25.33 ± 0.48 ^{a}	23.07 ± 0.12 ^{b}	24.50 ± 0.35 ^{ab}
C15:1n5	0.39 ± 0.02 ^{a}	0.28 ± 0.02 ^{b}	0.32 ± 0.02 ^{ab}	0.78 ± 0.04	0.83 ± 0.02	0.80 ± 0.03	0.92 ± 0.02	0.83 ± 0.03	0.93 ± 0.10
C16:1n7	8.56 ± 0.29	8.04 ± 1.14	9.40 ± 0.82	2.49 ± 0.39	2.38 ± 0.13	3.00 ± 0.13	2.87 ± 0.44	2.36 ± 0.09	2.29 ± 0.11
C18:1n9c	34.99 ± 1.21	32.33 ± 1.28	31.71 ± 0.44	24.23 ± 0.38 ^{a}	22.65 ± 0.36 ^{b}	24.39 ± 0.17 ^{a}	25.43 ± 0.21 ^{a}	23.41 ± 0.04 ^{c}	24.43 ± 0.12 ^{b}
C20:1n9	1.43 ± 0.05 ^{a}	1.07 ± 0.11 ^{b}	0.97 ± 0.02 ^{b}	1.22 ± 0.03	1.09 ± 0.06	1.13 ± 0.01	0.79 ± 0.04	0.75 ± 0.01	0.76 ± 0.02
∑MUFA	46.65 ± 1.61	42.76 ± 0.24	43.37 ± 0.46	29.21 ± 0.77	27.39 ± 0.49	29.80 ± 0.32	30.47 ± 0.77 ^{a}	27.72 ± 0.09 ^{b}	28.79 ± 0.12 ^{ab}
C18:2n6c	16.18 ± 2.20	22.52 ± 0.38	19.53 ± 1.37	8.74 ± 1.14	10.71 ± 0.17	9.83 ± 0.18	10.11 ± 0.58 ^{b}	12.18 ± 0.33 ^{a}	9.90 ± 0.20 ^{b}
C18:3n3	2.56 ± 0.30	3.17 ± 0.44	3.45 ± 0.22	1.13 ± 0.22	1.07 ± 0.04	1.16 ± 0.16	1.60 ± 0.16	1.56 ± 0.09	1.37 ± 0.06
C20:2n6	1.01 ± 0.05	0.95 ± 0.18	0.83 ± 0.01	2.54 ± 0.13	2.63 ± 0.11	2.55 ± 0.18	1.50 ± 0.08 ^{b}	1.72 ± 0.03 ^{a}	1.64 ± 0.02 ^{ab}
C20:4n6 (ARA)	1.61 ± 0	1.65 ± 0.05	1.92 ± 0.13	15.21 ± 0.89	13.97 ± 0.40	14.54 ± 0.53	7.63 ± 0.21	7.36 ± 0.23	8.03 ± 0.26
C20:3n3	0.37 ± 0.01	0.38 ± 0.02	0.38 ± 0.01	0.66 ± 0.01 ^{a}	0.55 ± 0.01 ^{b}	0.63 ± 0.02 ^{a}	0.61 ± 0.03	0.59 ± 0.01	0.62 ± 0.02
C20:5n3 (EPA)	1.36 ± 0.09	1.85 ± 0.21	2.09 ± 0.16	10.22 ± 0.97	10.79 ± 0.20	10.40 ± 0.30	13.87 ± 0.19 ^{b}	15.48 ± 0.20 ^{a}	15.68 ± 0.26 ^{a}
C22:6n3 (DHA)	1.40 ± 0.01 ^{b}	1.96 ± 0.14 ^{a}	2.17 ± 0.08 ^{a}	6.53 ± 0.38	6.82 ± 0.32	6.00 ± 0.07	8.64 ± 0.17	9.77 ± 0.26	9.35 ± 0.31
∑PUFA	25.45 ± 1.13 ^{b}	33.32 ± 0.10 ^{a}	31.23 ± 0.40 ^{ab}	45.37 ± 0.47	46.81 ± 0.36	45.49 ± 0.14	44.20 ± 0.13 ^{c}	48.84 ± 0.10 ^{a}	46.82 ± 0.06 ^{b}
∑EFA	19.00 ± 1.10	25.90 ± 0.02	23.19 ± 0.51	10.22 ± 0.60	12.05 ± 0.10	11.39 ± 0	11.96 ± 0.17 ^{b}	13.93 ± 0.19 ^{a}	11.51 ± 0.14 ^{b}
∑HUFA	8.01 ± 0.20 ^{b}	9.65 ± 0.40 ^{ab}	10.67 ± 0.24 ^{a}	33.75 ± 1.01	33.20 ± 0.44	32.72 ± 0.01	32.34 ± 0.16 ^{b}	34.75 ± 0.30 ^{a}	35.04 ± 0.07 ^{a}
∑n-3 PUFA	5.69 ± 0.28 ^{b}	7.35 ± 0.54 ^{ab}	8.08 ± 0.21 ^{a}	18.54 ± 0.80	19.23 ± 0.34	18.19 ± 0.39	24.71 ± 0.08 ^{b}	27.39 ± 0.25 ^{a}	27.01 ± 0.09 ^{a}
∑n-6 PUFA	19.76 ± 1.23	25.97 ± 0.32	23.15 ± 0.68	26.84 ± 0.05	27.58 ± 0.19	27.31 ± 0.49	19.49 ± 0.23 ^{b}	21.45 ± 0.08 ^{a}	19.81 ± 0 ^{b}
n-3/n-6 PUFA	0.29 ± 0.01	0.28 ± 0.04	0.35 ± 0.03	0.69 ± 0.04	0.70 ± 0.01	0.67 ± 0.04	1.27 ± 0.03	1.28 ± 0.03	1.36 ± 0.01
∑DHA + EPA	2.76 ± 0.10 ^{b}	3.80 ± 0.35 ^{a}	4.26 ± 0.08 ^{a}	16.75 ± 1.35	17.61 ± 0.52	16.40 ± 0.36	22.51 ± 0.02 ^{b}	25.25 ± 0.46 ^{a}	25.03 ± 0.05 ^{a}
DHA/EPA	1.03 ± 0.06	1.07 ± 0.05	1.05 ± 0.12	0.64 ± 0.02	0.63 ± 0.02	0.58 ± 0.01	0.62 ± 0.02	0.63 ± 0.01	0.60 ± 0.03

Notes: The data presented as means ± standard error (SE) (n = 20). Fatty acids less than 0.3% were not listed in the table or instead of —. Values in the same row with different superscripts are significantly different ($p < 0.05$). Thirty mature individuals were collected from each culture environment. Abbreviations: FW, freshwater; SW, saline water; AW, alkaline water; ∑SFA, sum of saturated fatty acids; ∑MUFA, sum of monounsaturated fatty acids; ∑PUFA, sum of polyunsaturated fatty acids; ∑EFA, sum of essential fatty acids; ∑HUFA, sum of highly unsaturated fatty acids; ∑n-3 PUFA, sum of ω-3 polyunsaturated fatty acids; ∑n-6 PUFA, sum of ω-6 polyunsaturated fatty acids.

In the gonad (Table 2), the ovaries and testes had apparent differences in their fatty acid composition and concentration. Females have higher concentrations of ∑SFA, ∑MUFA and ∑EFA, but lower percentages of ∑PUFA, ∑HUFA, ∑n-3 PUFA, ∑n-6 PUFA and ∑DHA + EPA compared with males. A significantly decreasing trend was found in 18:1n9c from FW to SW ($p < 0.05$). The ∑EFA content of the testis in SW and AW was slightly higher than that of FW, but no significant difference was detected ($p > 0.05$).

The lower crude lipid concentration of the testis and muscle was recorded. Therefore, the fatty acid compositions of the muscle were similar to those of the testis, but obviously different from those of the ovary and hepatopancreas (Table 2). Regardless of gender, concentrations of ∑SFA in muscle significantly decreased from FW to SW ($p < 0.05$). This trend was mainly driven by a reduction in 16:0 and 18:0. Whereas the female muscle content of ∑SFA in AW was significantly higher than that of FW, male muscle content was lower compared with that of FW ($p < 0.05$). A significantly decreasing trend was detected of ∑MUFA in the male muscle from FW to SW and AW ($p < 0.05$), which was mainly driven by a reduction in 18:1n9c. The content of ∑PUFA in male muscle significantly increased from FW to SW and AW ($p < 0.05$). This trend was mainly driven by an increase in 18:2n6c, DHA, and EPA. As for the indices describing combinations of PUFA, the contents of ∑HUFA, ∑n-3 PUFA, ∑DHA + EPA showed a significantly increasing trend from FW to SW and AW ($p < 0.05$).

Seventeen FAAs including seven essential free amino acids (EFAAs) for adult human beings, and two tastes including pleasant taste (umami and sweetness) and unpleasant taste (bitterness) are detected, respectively (Tables 3 and 4).

With respect to total essential free amino acids (∑EFAA) and total free amino acids (∑FAA) of *E. sinensis* hepatopancreas, the concentrations in females significantly increased from FW to SW and AW ($p < 0.05$) (Table 3). This trend was mainly driven by a significant increase in aspartic acid (Asp), cysteine (Cys), histidine (His), proline (Pro), tyrosine (Tyr), isoleucine (Ile), leucine (Leu), lysine (Lys), methionine (Met), phenylalanine (Phe), and threonine (Thr), which ∑EFAA increased from 551.55 ± 18.37 mg/100 g in FW to 696.63 ± 5.77 mg/100 g in SW, and ∑FAA increased from 1476.65 ± 4.09 mg/100 g in FW to 1721.86 ± 46.53 mg/100 g in AW. A similar trend was also observed in males, but no significant difference existed ($p > 0.05$). Regardless of gender, the total umami values (∑TUV) containing Asp and glutamic acid (Glu) in SW and AW were obviously higher than those of FW, which increased from 3.89 to 4.47 in females and from 3.40 to 3.89 in males with the domain umami compound Glu (Table 4). Nevertheless, the total sweetness values (∑TSV) containing alanine (Ala), glycine (Gly), serine (Ser), Thr, and Pro in females and total bitterness values (∑TBV) including arginine (Arg), Lys, valine (Val), Met, histidine (His), Ile, Leu, and Phe in males, were slightly decreased from FW to SW and AW. A slightly increasing trend of ∑TSV in males and ∑TBV in females was detected from FW to SW and AW. The domain sweetness compound was Ala, and the bitterness compounds were Arg, Lys, Val, Met, and His, respectively.

Except for the female lower ∑EFAA in AW, the contents of ∑EFAA increased from FW to SW and AW in gonad (Table 3). Due to the low content of female ∑FAA in FW, the percentage of ∑EFAA to ∑FAA (PETFAA) was higher than that of AW and SW. The concentrations of His, Pro, and Lys in females significantly increased from FW to SW and AW ($p < 0.05$), while a similar trend was also observed in males with higher contents of His, Ile, Leu, and Thr between SW and AW ($p < 0.05$). For the ovary, the ∑TUV and ∑TSV fattening in SW and AW were higher than that of FW (Table 4), indicating an obvious improvement in umami and sweetness. An apparent increase of ∑TBV was also found in SW compared with that of FW. The domain umami compound was Glu, while Ala was the main sweetness compound. Arg, Lys, and His presented as an unpleasant taste. For the testis, the ∑TUV, ∑TSV and ∑TBV fattening in FW were slightly higher than that in SW, but lower than that in AW. The testis possessed similar umami and sweetness compounds but a lower bitterness compound content compared with the ovary.

Table 3. Free amino acid composition in hepatopancreas, gonad, and muscle of adult *Eriocheir sinensis* fattening in different culture environments (mg/100 g, wet weight).

Free Amino Acids	Hepatopancreas			Gonad			Muscle		
	FW	SW	AW	FW	SW	AW	FW	SW	AW
Female									
Aspartic acid	41.33 ± 3.44 [b]	58.10 ± 2.80 [a]	61.89 ± 3.31 [a]	10.23 ± 2.87	7.04 ± 1.28	8.21 ± 2.04	3.72 ± 0.58	2.98 ± 0.26	3.59 ± 0.22
Arginine	299.51 ± 28.50	275.89 ± 1.46	286.47 ± 13.84	223.54 ± 18.42	291.75 ± 29.03	227.79 ± 18.41	487.49 ± 1.84 [b]	452.77 ± 17.15 [b]	544.70 ± 10.59 [a]
Alanine	164.85 ± 11.12 [a]	121.78 ± 7.78 [b]	135.10 ± 4.74 [ab]	75.95 ± 12.48	82.20 ± 24.79	66.25 ± 2.35	303.43 ± 24.60	317.08 ± 18.38	354.05 ± 32.29
Cysteine	11.55 ± 0.19 [b]	16.21 ± 0.75 [a]	15.54 ± 0.20 [a]	0.97 ± 0.65	0.41 ± 0.16	1.02 ± 0.08	3.14 ± 0.96	3.30 ± 0.46	3.41 ± 0.93
Glutamic acid	104.54 ± 8.34	111.19 ± 4.96	115.50 ± 3.16	104.86 ± 8.72	129.77 ± 34.62	120.67 ± 12.59	38.13 ± 8.16	51.94 ± 3.41	61.20 ± 17.91
Glycine	80.13 ± 0.10	80.09 ± 9.62	81.19 ± 3.07	43.89 ± 9.91	51.84 ± 14.00	43.84 ± 3.18	450.43 ± 33.41 [b]	595.76 ± 7.73 [a]	461.48 ± 32.33 [b]
Histidine	32.51 ± 4.82 [b]	49.74 ± 3.11 [a]	47.94 ± 2.46 [ab]	25.62 ± 1.86 [b]	33.12 ± 0.52 [a]	27.95 ± 0.57 [ab]	16.56 ± 0.72	22.12 ± 1.52	22.44 ± 2.56
Proline	96.27 ± 14.42 [b]	203.41 ± 17.57 [a]	202.96 ± 8.17 [a]	131.94 ± 17.98 [b]	272.57 ± 32.06 [ab]	315.82 ± 47.81 [a]	238.99 ± 40.97 [b]	367.13 ± 51.10 [ab]	491.15 ± 44.92 [a]
Serine	18.09 ± 2.18	13.58 ± 0.90	14.40 ± 0.37	8.48 ± 2.01	6.86 ± 0.65	4.92 ± 0.01	10.08 ± 1.21	9.51 ± 1.25	6.68 ± 0.15
Tyrosine	76.31 ± 1.43 [b]	86.43 ± 2.69 [a]	92.36 ± 0.44 [a]	20.80 ± 8.10	18.64 ± 2.57	13.30 ± 0.86	15.11 ± 2.14	15.69 ± 2.24	21.44 ± 3.70
Isoleucine ▲	39.21 ± 2.26 [b]	75.79 ± 1.00 [a]	49.94 ± 3.33 [b]	10.49 ± 2.94	10.34 ± 0.90	7.87 ± 0.19	10.86 ± 1.24	13.99 ± 2.29	18.23 ± 2.74
Leucine ▲	112.38 ± 6.02 [b]	167.73 ± 0.95 [a]	151.50 ± 1.42 [a]	21.59 ± 9.25	22.55 ± 0.58	15.32 ± 1.56	23.33 ± 2.59	32.57 ± 6.10	40.77 ± 5.96
Lysine ▲	158.23 ± 6.43 [b]	164.07 ± 4.15 [ab]	183.05 ± 0.78 [a]	51.43 ± 6.59 [b]	93.31 ± 6.33 [a]	71.28 ± 1.37 [ab]	33.16 ± 2.43 [b]	39.17 ± 5.33 [ab]	72.64 ± 11.85 [a]
Methionine ▲	41.07 ± 1.63 [ab]	45.69 ± 2.36 [a]	34.73 ± 2.61 [b]	16.94 ± 5.20	10.69 ± 1.67	9.11 ± 0.02	21.84 ± 0.31	19.75 ± 1.11	27.15 ± 3.86
Phenylalanine ▲	71.82 ± 1.68 [b]	94.11 ± 0.05 [a]	94.11 ± 0.41 [a]	19.83 ± 6.62	19.57 ± 1.26	11.83 ± 0.98	13.40 ± 2.08	15.13 ± 2.95	20.63 ± 3.27
Threonine ▲	61.61 ± 1.55 [b]	79.43 ± 1.42 [a]	83.23 ± 4.56 [a]	56.60 ± 0.62	79.45 ± 11.39	59.13 ± 6.06	23.66 ± 1.08	31.92 ± 7.65	38.70 ± 7.42
Valine ▲	67.23 ± 1.89	69.82 ± 0.68	71.94 ± 4.20	26.69 ± 4.85	21.57 ± 0.62	15.48 ± 1.56	30.23 ± 1.87	38.04 ± 5.00	39.45 ± 7.44
∑EFAA	551.55 ± 18.37 [b]	696.63 ± 5.77 [a]	668.50 ± 12.09 [a]	203.56 ± 34.83	257.49 ± 22.75	190.03 ± 11.74	156.47 ± 11.60	190.56 ± 30.43	257.58 ± 42.54
∑FAA	1476.65 ± 4.09 [b]	1713.04 ± 10.57 [a]	1721.86 ± 46.53 [a]	849.86 ± 80.54	1151.68 ± 95.42	1019.80 ± 94.39	1723.54 ± 40.57 [b]	2028.83 ± 84.18 [a]	2227.72 ± 12.45 [a]
PETFAA	37.35 ± 1.14	40.67 ± 0.59	38.83 ± 0.35	23.78 ± 1.85 [a]	22.35 ± 0.12 [ab]	18.69 ± 0.58 [b]	9.07 ± 0.46	9.35 ± 1.11	11.55 ± 1.84

Table 3. *Cont.*

Free Amino Acids	Hepatopancreas			Gonad			Muscle		
	FW	SW	AW	FW	SW	AW	FW	SW	AW
Male									
Aspartic acid	45.09 ± 1.29	53.68 ± 7.18	55.49 ± 5.88	49.91 ± 3.14	61.92 ± 5.04	66.65 ± 2.83	2.29 ± 0.36	3.20 ± 0.51	2.98 ± 0.63
Arginine	255.53 ± 29.47	247.79 ± 22.87	223.99 ± 6.52	47.87 ± 1.16	45.28 ± 2.40	51.45 ± 2.66	489.02 ± 32.00	537.25 ± 16.27	490.85 ± 65.55
Alanine	141.21 ± 7.13	108.50 ± 11.38	126.99 ± 6.63	71.09 ± 7.57	59.15 ± 3.12	78.11 ± 3.14	300.88 ± 20.95	314.91 ± 49.62	308.30 ± 36.41
Cysteine	13.94 ± 0.18	15.88 ± 1.73	16.23 ± 0.67	3.03 ± 0.09 [a]	0.12 ± 0.01 [c]	2.15 ± 0.16 [b]	0.95 ± 0.48	2.84 ± 0.65	2.49 ± 0.24
Glutamic acid	88.52 ± 13.75	98.54 ± 1522	99.91 ± 14.26	58.79 ± 7.22	50.39 ± 1.58	61.69 ± 3.12	54.32 ± 9.05	74.84 ± 10.81	69.55 ± 9.58
Glycine	66.90 ± 0.94	78.89 ± 0.58	77.14 ± 8.51	27.57 ± 2.43	26.67 ± 0.91	27.77 ± 0.30	437.69 ± 28.26	561.13 ± 33.12	546.93 ± 33.55
Histidine	35.56 ± 5.40	44.41 ± 4.66	45.11 ± 4.08	8.75 ± 0.61 [c]	11.42 ± 0.16 [b]	13.72 ± 0.17 [a]	13.17 ± 1.84 [b]	21.18 ± 1.86 [a]	15.31 ± 0.77 [ab]
Proline	109.18 ± 0.87	253.22 ± 19.02	262.20 ± 83.68	73.92 ± 12.06	108.29 ± 13.53	137.26 ± 35.29	208.61 ± 21.22 [b]	473.18 ± 77.05 [a]	476.86 ± 51.06 [a]
Serine	16.03 ± 1.43	13.70 ± 1.22	18.50 ± 0.16	3.25 ± 0.03	3.51 ± 0.57	4.10 ± 0.62	8.19 ± 1.47	8.31 ± 0.10	5.51 ± 1.25
Tyrosine	86.85 ± 2.53	80.89 ± 8.96	80.48 ± 5.16	19.99 ± 1.37	19.55 ± 0.41	19.56 ± 0.22	13.24 ± 1.10 [b]	21.22 ± 0.65 [a]	17.37 ± 2.43 [ab]
Isoleucine ▲	44.24 ± 5.23	50.56 ± 5.39	54.31 ± 5.51	10.55 ± 0.68 [b]	12.49 ± 0.37 [b]	15.74 ± 0.82 [a]	9.60 ± 2.17	13.45 ± 1.08	12.32 ± 2.79
Leucine ▲	126.43 ± 2.35	157.29 ± 10.77	140.48 ± 11.37	15.40 ± 2.35 [b]	17.07 ± 0.53 [ab]	21.96 ± 0.62 [a]	22.63 ± 3.96	28.93 ± 1.39	26.94 ± 5.53
Lysine ▲	174.21 ± 9.13	143.66 ± 16.03	141.24 ± 7.96	26.24 ± 2.88	24.41 ± 0.97	27.66 ± 0.97	44.22 ± 2.36	48.05 ± 2.59	55.60 ± 6.77
Methionine ▲	41.06 ± 0.05	43.75 ± 6.20	38.58 ± 2.94	5.87 ± 0.90	4.70 ± 0.70	5.86 ± 0.14	14.99 ± 0.35	22.58 ± 4.11	17.47 ± 0.19
Phenylalanine ▲	82.88 ± 2.13	88.34 ± 10.24	86.42 ± 8.34	8.15 ± 0.79	8.93 ± 0.46	10.51 ± 0.10	10.21 ± 1.64	15.42 ± 0.46	12.63 ± 1.75
Threonine ▲	70.37 ± 12.98	73.90 ± 8.24	80.67 ± 4.43	9.53 ± 0.15 [c]	13.38 ± 0.27 [b]	16.66 ± 0.98 [a]	21.79 ± 4.77	30.70 ± 0.71	35.17 ± 5.18
Valine ▲	67.41 ± 3.90	70.02 ± 6.31	72.26 ± 4.05	14.90 ± 0.07 [a]	9.92 ± 0.82 [b]	11.86 ± 0.37 [b]	29.14 ± 6.17	38.67 ± 5.90	34.42 ± 6.19
∑EFAA	606.61 ± 35.67	627.53 ± 18.24	613.95 ± 44.59	90.65 ± 7.52	90.89 ± 3.38	110.27 ± 3.99	152.58 ± 21.42	197.80 ± 9.65	194.54 ± 28.02
∑FAA	1465.42 ± 54.30	1623.03 ± 111.06	1620.00 ± 180.13	454.82 ± 23.95	477.20 ± 10.18	572.73 ± 50.49	1680.96 ± 73.19	2215.86 ± 200.09	2130.69 ± 229.01
PETFAA	41.36 ± 0.90	38.77 ± 1.53	38.06 ± 1.48	20.07 ± 2.71	19.04 ± 0.30	19.34 ± 1.01	9.04 ± 0.88	8.96 ± 0.37	9.09 ± 0.34

Notes: The data presented as means ± standard error (SE) (n = 20). ▲, Essential amino acid. Values in the same row with different superscripts are significantly different ($p < 0.05$). Thirty mature individuals were collected from each culture environment. Abbreviations: FW, freshwater; SW, saline water; AW, alkaline water; ∑EFAA, total essential free amino acids; ∑FAA, total free amino acids; PETFAA, percentage of ∑EFAA to ∑FAA.

Table 4. The threshold and taste activity value of free amino acid composition in hepatopancreas, gonad, and muscle of adult *Eriocheir sinensis* fattening in different culture environments.

Free Amino Acids	Flavor Characteristics	Threshold (mg/100 mL)	Hepatopancreas			Gonad			Muscle		
			FW	SW	AW	FW	SW	AW	FW	SW	AW
Female											
Aspartic acid	umami (+)	100	0.41	0.58	0.62	0.10	0.07	0.08	0.04	0.03	0.04
Glutamic acid	umami (+)	30	3.48	3.71	3.85	3.50	4.33	4.02	1.27	1.73	2.04
∑TUV			3.89	4.29	4.47	3.60	4.40	4.10	1.31	1.76	2.08
Alanine	sweetness (+)	60	2.75	2.03	2.25	1.27	1.37	1.10	5.06	5.28	5.90
Glycine	sweetness (+)	130	0.62	0.62	0.62	0.34	0.40	0.34	3.46	4.58	3.55
Serine	sweetness (+)	150	0.12	0.09	0.10	0.06	0.05	0.03	0.07	0.06	0.04
Threonine	sweetness (+)	260	0.24	0.31	0.32	0.22	0.31	0.23	0.09	0.12	0.15
Proline	sweetness/bitterness (+)	300	0.32	0.68	0.68	0.44	0.91	1.05	0.80	1.22	1.64
∑TSV			4.05	3.73	3.97	2.33	3.04	2.75	9.48	11.26	11.28
Arginine	sweetness/bitterness (−)	50	5.99	5.52	5.73	4.47	5.83	4.56	9.75	9.06	10.89
Lysine	sweetness/bitterness (−)	50	3.16	3.28	3.66	1.03	1.87	1.43	0.66	0.78	1.45
Valine	sweetness/bitterness (−)	40	1.68	1.75	1.80	0.67	0.54	0.39	0.76	0.95	0.99
Methionine	bitterness/sweetness/sulphur (−)	30	1.37	1.52	1.16	0.56	0.36	0.30	0.73	0.66	0.91
Histidine	bitterness (−)	20	1.63	2.49	2.40	1.28	1.66	1.40	0.15	0.17	0.23
Isoleucine	bitterness (−)	90	0.44	0.84	0.55	0.12	0.11	0.09	0.12	0.16	0.20
Leucine	bitterness (−)	190	0.59	0.88	0.80	0.11	0.12	0.08	0.12	0.17	0.21
Phenylalanine	bitterness (−)	90	0.80	1.05	1.05	0.22	0.22	0.13	0.83	1.11	1.12
∑TBV			15.66	17.33	17.15	8.46	10.71	8.38	13.12	13.06	16.00
Male											
Aspartic acid	umami (+)	100	0.45	0.54	0.55	0.50	0.62	0.67	0.02	0.03	0.03
Glutamic acid	umami (+)	30	2.95	3.28	3.33	1.96	1.68	2.06	1.81	2.49	2.32
∑TUV			3.40	3.82	3.88	2.46	2.30	2.73	1.83	2.52	2.35
Alanine	sweetness (+)	60	2.35	1.81	2.12	1.18	0.99	1.30	5.01	5.25	5.14
Glycine	sweetness (+)	130	0.51	0.61	0.59	0.21	0.21	0.21	3.37	4.32	4.21
Serine	sweetness (+)	150	0.11	0.09	0.12	0.02	0.02	0.03	0.05	0.06	0.04
Threonine	sweetness (+)	260	0.27	0.28	0.31	0.04	0.05	0.06	0.08	0.12	0.14
Proline	sweetness/bitterness (+)	300	0.36	0.84	0.87	0.25	0.36	0.46	0.70	1.58	1.59
∑TSV			3.60	3.63	4.01	1.70	1.63	2.06	9.21	11.33	11.12
Arginine	sweetness/bitterness (−)	50	5.11	4.96	4.48	0.96	0.91	1.03	9.78	10.75	9.82
Lysine	sweetness/bitterness (−)	50	3.48	2.87	2.82	0.52	0.49	0.55	0.88	0.96	1.11
Valine	sweetness/bitterness (−)	40	1.69	1.75	1.81	0.37	0.25	0.30	0.73	0.97	0.86
Methionine	bitterness/sweetness/sulphur (−)	30	1.37	1.46	1.29	0.20	0.16	0.20	0.50	0.75	0.58
Histidine	bitterness (−)	20	1.78	2.22	2.26	0.44	0.57	0.69	0.11	0.17	0.14
Isoleucine	bitterness (−)	90	0.49	0.56	0.60	0.12	0.14	0.17	0.11	0.15	0.14
Leucine	bitterness (−)	190	0.67	0.83	0.74	0.08	0.09	0.12	0.12	0.15	0.14
Phenylalanine	bitterness (−)	90	0.92	0.98	0.96	0.09	0.10	0.12	0.66	1.06	0.77
∑TBV			15.51	15.63	14.96	2.78	2.71	3.18	12.89	14.96	13.56

Notes: (+). pleasant taste; (−). unpleasant taste. ND, taste threshold not detected. Thirty mature individuals were collected from each culture environment. Abbreviations: FW, freshwater; SW, saline water; AW, alkaline water; ∑TUV, total umami values; ∑TSV, total sweetness values; ∑TBV, total bitterness values.

With respect to ∑EFAA and ∑FAA of the *E. sinensis* muscle, an apparent improvement was detected between SW and AW compared with that of FW (Table 3). This trend was mainly driven by an obvious increase in Gly, and Pro. Significant change was observed among the three treatments including Arg, Gly, Pro, Lys, and ∑FAA in females, as well as His, Pro, and Tyr in males ($p < 0.05$). Regardless of gender, the ∑TUV in SW and AW was obviously higher than that of FW, which increased from 1.31 to 2.08 in females and from 1.83 to 2.53 in males with the domain umami compound Glu (Table 4). A similar increasing trend was also observed for ∑TSV between SW and AW, with the values ranging from 9.48 to 11.28 in females and from 9.21 to 11.33 in males, accompanied by domain sweetness compounds Ala, Gly, and Pro. For ∑TBV, Arg contributed the largest amount of flavor compound with the TAV > 9.

4. Discussion

4.1. Total Edible Yield

Fattening is a highly important aspect in the farming of *E. sinensis*, where fattening performance can generally be evaluated by gonadal development status and edible yield [25]. *E. sinensis* with well-developed gonadal systems are usually sold at a higher price than less developed *E. sinensis*; therefore, the status of gonadal development directly affects the nutritional value and price of the market [11]. The present study showed that *E. sinensis* GSI parament fattening in SW and AW was not significantly different from that of fattening in FW ($p > 0.05$), which illustrated that low salinity and alkalinity could not significantly affect the gonadal development of *E. sinensis*. Nevertheless, a downward trend in GSI was detected in females, which is likely caused by the lower body weight of *E. sinensis* in SW and AW. In the same polyculture pond, the smaller the body weight of adult individuals, the earlier the puberty molt is completed. That is, the gonadal development of small body weight starts earlier than that of big body weight. Previous studies have argued that no significant GSI change was detected, fattening below 6 ppt salinity [14], and this result was consistent with our study.

4.2. Color Parameters and Biochemical Composition

The market value of *E. sinensis* is predominately driven by its visual appearance. Generally, the reddish color of crustaceans means higher market prices [26]. The color of *E. sinensis* is attributable to the deposition of carotenoids, especially astaxanthin [21]. In the present study, higher a^* values of freeze-dried carapace were observed fattening between SW and AW compared with that of FW, illustrating higher astaxanthin and canthaxanthin contents [21,27]. A similar trend was also observed with higher b^* values of freeze-dried carapace fattening between SW and AW, suggesting strongly zeaxanthin and β-carotene contents [27]. These results suggest that fattening in saline-alkaline water was helpful to carotenoid accumulation in *E. sinensis* carapace. A similar conclusion was also detected that salinity stress induced an increase in carotenoid content, such as *Synechocystis* [28], and *Golenkinia* [29].

The biochemical composition of edible tissues is an important indicator for evaluating the nutritional value of aquatic animals, and its composition is influenced by many factors, such as germplasm, culture environment, fattening stage, and diet [1,25]. This study showed that male *E. sinensis* fattening in SW and AW had lower moisture contents in muscle and hepatopancreas, but had higher protein contents in muscle. The possible explanation is that salinity or alkalinity affects osmotic pressure regulation, resulting in a decrease in the moisture of *E. sinensis* edible tissues, while the higher protein content may be due to the self-protection strategy adopted to resist the environment. Similar results were also observed in *Scylla paramamosain* [30]. Female fattening in SW and AW had lower crude protein compared with that of FW, implying that *E. sinensis* muscle tissue of different genders was reflected differently under SW and AW.

4.3. Fatty Acid Composition

Fatty acid composition is an important indicator for the evaluation of the quality of edible aquatic species, especially essential fatty acids, and unsaturated fatty acid contents [25]. In this study, the concentrations of ∑SFA significantly decreased, but the content of ∑PUFA significantly increased in male hepatopancreas from FW to SW and AW ($p < 0.05$). This result was consistent with previous studies [15,17], which may explain why it is necessary to improve membrane permeability to enhance the absorption of ions and maintain the intracellular ionic balance. Hence, increasing ∑PUFA levels might be beneficial to the intracellular and extracellular osmotic and ionic balance [15].

Balanced concentrations of fatty acids are essential for human health, which mainly refer to a higher proportion of essential fatty acids (EFA) and LC-PUFAs [31]. In this study, except for slightly low ∑EFA in SW female muscle, the ∑EFA contents of *E. sinensis* other edible tissues fattening in SW were higher than those of fattening in FW, illustrating a higher quality. DHA, EPA, and ARA are three important LC-PUFAs for human health, especially fetuses, infants, adolescents, and pregnant or lactating women [32]. DHA and EPA can inhibit the proliferation of tumor cells [33]. This indoor experiment demonstrated that fattening in SW and AW could improve the DHA and EPA contents in male hepatopancreas and muscle, which illustrated that salinity and alkalinity could regulate the accumulation of DHA and EPA and promote endogenous biosynthesis. The internal mechanism may be to increase the protein expression of the elongase of very long-chain fatty acid (Elovl) and fatty acid desaturase (Fad) in the process of LC-PUFA synthesis [34,35]. However, through the comparison between indoor and outdoor earthen pond cultures [4], our results support that the *E. sinensis* LC-PUFA accumulation mainly comes from exogenous food sources, followed by endogenous biosynthesis.

Long-chain n-3 and n-6 PUFAs and their ratios (n-3/n-6) are also considered to be significantly important for human health. The FAO/WHO [31] recommended that the appropriate dietary n-3/n-6 PUFA ratio was 0.1~0.2. If the ratio was >0.2, it would be better for human health [1]. The results in this study illustrated n-3/n-6 PUFA ratios of all the *E. sinensis* edible tissues were >0.1, suggesting that fattening in SW and AW would not affect the nutritional quality. However, compared with previous studies [1,4,11], a lower n-3/n-6 PUFA ratio in hepatopancreas was observed in this study. This is likely attributable to the black color in hepatopancreas of *E. sinensis* rearing in an indoor circulating aquaculture system because above 40% fatty acid parameters were significantly changed and a lower n-3/n-6 PUFA ratio was found between normal color and black color in *E. sinensis* hepatopancreas [36].

4.4. FAA Composition and TAV Analysis

It is widely known that FAA composition and concentration play a marked role in nutritional and non-volatile flavor quality (taste) [5,25]. Generally, Asp and Glu contribute to umami, and Ser, Gly, Thr, Pro, and Ala contribute to sweetness. His, Phe, Ile, and Leu contribute to bitterness [4,16,25], while the taste of Arg relies on its concentration. Regardless of gender, the present study showed that the ∑EFAA and ∑FAA contents of *E. sinensis* edible tissues in hepatopancreas and muscle increased fattening in SW and AW, which explained the higher nutritional quality. Similar results were also detected by previous findings [17,37].

TAV is generally used as the most classical and objective method to determine the taste intensity of a single compound in food and to evaluate its contribution to overall flavor quality [37]. Compounds with a TAV > 1 were considered to significantly contribute toward *E. sinensis* taste, while compounds with a TAV < 1 were considered to contribute less [16]. This study showed that the compositions of main flavor amino acids are consistent regardless of fattening in FW or saline-alkaline water, but the contents of the main flavor amino acids are different. Similar changes have been confirmed by previous studies [16,17]. Glu, Ala, and Arg are likely the main compounds for the strong umami and sweet taste of *E. sinensis* edible tissues. This result is similar to previous studies [4,5,16,17,25]. Although Arg has a bitter taste, abundant Arg can enhance the persistence, complexity, and strong sense of umami [38]. Even in Wang and Zhang's research, Arg was directly listed as a pleasant taste amino acid [16,37]. Studies have confirmed Arg's great contribution to the overall taste of aquatic products, and showed a positive correlation [16,39]. Previous studies have argued that Pro, Ala, Gly, and Arg may be used as osmotic regulators for crustacean exposed to salinity for extensive time periods [14], and this may be an important reason for the above FAA changes in the process of fattening in saline-alkaline water. Further, some obvious changes in FAAs were also detected in fattening between SW and AW. Although Na^+ was the common cation, different anions (salinity, Cl^-; alkalinity, HCO_3^-) may have led to the above differences. However, how salinity and alkalinity regulate the difference in FAA content still needs further research.

5. Conclusions

In the present study, no significantly increasing or decreasing trend was observed in GSI or TEY among FW, SW, and AW during the 55-day fattening period. Higher a^* and b^* values of freeze-dried carapace were observed in the fattening between SW and AW. The crude protein in gonad and male muscle, moisture in female muscle, and crude lipid in male muscle increased significantly from FW to SW and AW. Better nutritional and flavor values were also detected in male hepatopancreas and muscle. In summary, numerous advantages of fattening in SW and AW were observed, including the improvement of carotenoid accumulation in freeze-dried carapace, DHA, EPA, ∑EFAA, ∑FAA, and ∑TUV contents in male hepatopancreas and muscle. These results will be helpful in improving the quality of *E. sinensis*. However, the synthetic mechanism inside organisms needs to be further studied.

Author Contributions: Conceptualization, S.W., W.X. and Z.Z.; methodology, S.W., K.G. and Z.Z.; software, S.W.; validation, L.L., R.Z. and Y.S.; formal analysis, K.G. and Y.S.; investigation, S.W. and W.X.; resources, Z.Z.; data curation, S.W., L.L. and R.Z.; writing—original draft preparation, S.W.; writing—review and editing, S.W. and Z.Z.; visualization, R.Z.; supervision, W.X. and Z.Z.; project administration, W.X.; funding acquisition, S.W., W.X. and Z.Z. All authors have read and agreed to the published version of the manuscript.

Funding: This research was funded by the Central Public-interest Scientific Institution Basal Research Fund, CAFS (No. 2021XT05; 2020TD56) and the Central Public-interest Scientific Institution Basal Research Fund, HRFRI (No. HSY202008Q).

Institutional Review Board Statement: Not applicable.

Informed Consent Statement: Not applicable.

Data Availability Statement: The data presented in this study are available on request from the corresponding author.

Acknowledgments: We would like to thank all the students in our team for their help with sample collection.

Conflicts of Interest: The authors declare no conflict of interest.

References

1. Long, X.W.; Guo, Q.; Wang, X.C.; Francis, D.S.; Cheng, Y.X.; Wu, X.G. Effects of fattening period on ovarian development and nutritional quality of adult female Chinese mitten crab *Eriocheir sinensis*. *Aquaculture* **2020**, *519*, 734–748. [CrossRef]
2. Bureau of Fisheries and Fishery Management, Ministry of Agriculture and Rural affairs of China. *2021 China Fisheries Statistical Yearbook*; Chinese Agricultural Press: Beijing, China, 2022; p. 34.
3. Cheng, Y.X.; Wu, X.G.; Yang, X.Z.; Hines, A.H. Current trends in hatchery techniques and stock enhancement for Chinese mitten crab, *Eriocheir japonica sinensis*. *Rev. Fish. Sci. Aquac.* **2008**, *16*, 377–384. [CrossRef]
4. Wang, S.H.; Wang, Y.Z.; Wu, X.G.; Zhang, X.B.; Zhao, J.S.; Yang, J.M.; Cheng, Y.X. Gonadal development and biochemical composition of Chinese mitten crabs (*Eriocheir sinensis*) from four sources. *J. Food Sci.* **2021**, *86*, 1066–1080. [CrossRef] [PubMed]
5. Wang, Y.Z.; Li, J.H.; Wang, S.H.; Sun, B.H.; Xiong, L.M.; Cheng, Y.X. Composition and nutritional qualities of edible tissues of Chinese mitten crab (*Eriocheir sinensis*) from Ya Lake over different months. *J. Food Compos. Anal.* **2022**, *105*, 104199. [CrossRef]
6. Sumner, M.E.; Naidu, R. *Sodic Soils: Distribution, Properties, Management and Environmental Consequences*; Oxford University Press: Oxford, UK, 1998; pp. 1069–1093.
7. Lin, T.T.; Lai, Q.F.; Yao, Z.L.; Lu, J.X.; Zhou, K.; Wang, H. Combined effects of carbonate alkalinity and pH on survival, growth and haemocyte parameters of the Venus clam *Cyclina sinensis*. *Fish Shellfish Immun.* **2013**, *35*, 525–531. [CrossRef]
8. Wang, S.H.; Zhao, J.S.; Wu, X.G.; Cheng, Y.X. Growth performance, gonadal development and nutritional composition of adult *Eriocheir sinensis* from Dongying. *J. Shanghai Ocean Univ.* **2020**, *29*, 17–26.
9. Dong, Z.G.; Shen, S.Y.; LI, X.Y.; Yan, B.L.; SUN, X.W. Fatty acid composition variation and fingerprint of the swimming crab *Portunus trituberculatus* from China Sea based on multivariate analysis method. *J. Fish. China* **2013**, *37*, 192–200. [CrossRef]
10. Tang, L.; Wang, H.; Wang, C.L.; Mu, C.K.; Wei, H.L.; Yao, H.Z.; Ye, C.Y.; Chen, L.Z.; Shi, C. Temperature potentially induced distinctive flavor of mud crab *Scylla paramamosain* mediated by gut microbiota. *Sci. Rep.* **2020**, *10*, 3720. [CrossRef]
11. Wu, X.G.; Zhu, S.C.; Zhang, H.C.; Liu, M.M.; Wu, N.; Pan, J.; Luo, M.; Wang, X.C.; Cheng, Y.X. Fattening culture improves the gonadal development and nutritional quality of male Chinese mitten crab *Eriocheir sinensis*. *Aquaculture* **2019**, *518*, 734865. [CrossRef]
12. Mu, Y.C.; Wang, F.; Dong, S.L.; Huang, G.Q.; Dong, S.S. Effects of salinity fluctuation pattern on growth and energy budget of juvenile shrimp *Fenneropenaeus chinensis*. *J. Shellfish Res.* **2005**, *24*, 1217–1221.
13. Boyd, C.E.; Tucker, C.S.; Somridhivej, B. Alkalinity and hardness: Critical but elusive concepts in aquaculture. *J. World Aquacult. Soc.* **2016**, *47*, 6–41. [CrossRef]
14. Long, X.W.; Wu, X.G.; Zhao, L.; Ye, H.H.; Cheng, Y.X.; Zeng, C.S. Physiological responses and ovarian development of female Chinese mitten crab *Eriocheir sinensis* subjected to different salinity conditions. *Front. Physiol.* **2018**, *8*, 1072. [CrossRef]
15. Long, X.W.; Wu, X.G.; Zhu, S.C.; Ye, H.H.; Cheng, Y.X.; Zeng, C.S. Salinity can change the lipid composition of adult Chinese mitten crab after long-term salinity adaptation. *PLoS ONE* **2019**, *14*, e0219260. [CrossRef]
16. Zhang, L.; Yin, M.Y.; Zheng, Y.; Xu, C.H.; Tao, N.P.; Wu, X.G.; Wang, X.C. Brackish water improves the taste quality in meat of adult male *Eriocheir sinensis* during the postharvest temporary rearing. *Food Chem.* **2021**, *343*, 128409. [CrossRef]

17. Qin, K.X.; Ruan, T.S.; Chen, Y.H.; Liang, G.L.; Wang, H.; Mu, C.K.; Wang, C.L. Effects of temporary rearing time under salinity 7 on the non-volatile flavorings and fatty acids of *Eriocheir sinensis*. *J. Food Compos. Anal.* **2022**, *107*, 104366. [CrossRef]
18. Chi, B.J.; Liang, L.Q.; Liu, C.L.; Chang, Y.M.; Wang, S.; Han, Q.X.; Gao, G.Q. Adaptability of *Tribolodon brandti* (Dybowski) to NaCl concentration and alkalinity. *J. Fish. Sci. China* **2011**, *18*, 689–694. [CrossRef]
19. Ren, S.S.; Sun, B.; Luo, L.; Zhang, L.M.; Chang, Y.M.; Liang, L.Q. Tolerance of freshwater shrimp (*Macrobrachium nipponense*) to alkalinity and low temperature in northeast China. *Chin. J. Fish.* **2020**, *33*, 24–28.
20. Yang, Y.H.; Li, M.S.; Luo, L.; Wang, S.H.; Zhang, R.; Guo, K.; Liu, J.Y.; Li, H.T.; Zhao, Z.G. Study on toxicity of salinity and alkalinity on *Eriocheir sinensis*. *J. Northeast Agric. Univ. Engl. Ed.* **2022**, *53*, 36–41.
21. Long, X.W.; Wu, X.G.; Zhao, L.; Liu, J.G.; Cheng, Y.X. Effects of dietary supplementation with *Haematococcus pluvialis*, cell powder on coloration, ovarian development and antioxidation capacity of adult female Chinese mitten crab, *Eriocheir sinensis*. *Aquaculture* **2017**, *473*, 545–553. [CrossRef]
22. AOAC. *Official Methods of Analysis of Association of Official Analytical Chemists*, 16th ed.; Association of Official Analytical Chemists: Arlington, VA, USA, 1995; p. 13.
23. National Health Commission of the People's Republic of China; State Administration for Market Regulation. *GB 5009.6-2016*; National Food Safety Standard-Determination of Fat in Food; Standards Press of China: Beijing, China, 2016; pp. 1–2.
24. National Health Commission of the People's Republic of China; State Administration for Market Regulation. *GB 5009.168-2016*; National Food Safety Standard-Determination of Fatty Acids in Food; Standards Press of China: Beijing, China, 2016; pp. 10–11.
25. Shao, L.C.; Wang, C.; He, J.; Wu, X.G.; Cheng, X.G. Meat quality of Chinese mitten crabs fattened with natural and formulated diets. *J. Aquat. Food Prod. Technol.* **2014**, *23*, 59–72. [CrossRef]
26. Tume, R.K.; Sikes, A.L.; Tabrett, S.; Smith, D.M. Effect of background color on the distribution of astaxanthin in black tiger prawn (*Penaeus monodon*): Effective method for improvement of cooked color. *Aquaculture* **2009**, *296*, 129–135. [CrossRef]
27. Li, Q.Q.; Sun, Q.F.; Liu, Q.; Cheng, Y.X.; Wu, X.G. Estimation of genetic parameters for carotenoid traits in Chinese mitten crab, *Eriocheir sinensis*, females. *Aquaculture* **2021**, *532*, 735990. [CrossRef]
28. Paliwal, C.; Pancha, I.; Ghosh, T.; Maurya, R.; Chokshi, K.; Vamsi-Bharadwaj, S.V.; Ram, S.; Mishra, S. Selective carotenoid accumulation by varying nutrient media and salinity in *Synechocystis* sp. CCNM 2501. *Bioresour. Technol.* **2015**, *197*, 363–368. [CrossRef]
29. Rearte, T.A.; Vélez, C.G.; Beligni, M.V.; Figueroa, F.L.; Gómez, P.I.; Flaig, D.; De-Iorio, A.F. Biological characterization of a strain of *Golenkinia* (Chlorophyceae) with high oil and carotenoid content induced by increased salinity. *Algal Res.* **2018**, *33*, 218–230. [CrossRef]
30. Wu, Q.Y.; Shi, X.; Fang, S.B.; Xie, Z.F.; Guang, M.Y.; Li, S.K.; Zheng, H.P.; Zhang, Y.L.; Ikhwanuddin, M.; Ma, H.Y. Different biochemical composition and nutritional value attribute to salinity and rearing period in male and female mud crab *Scylla paramamosain*. *Aquaculture* **2019**, *513*, 734417. [CrossRef]
31. FAO; WHO. Fats and oils in human nutrition. In *Report of a Joint FAO/WHO Expert Consultation*; Food and Agriculture Organization of the United Nations: Rome, Italy, 1994; pp. 202–205.
32. Muskieta, F.A.J.; Kemperman, R.F.J. Folate and long-chain polyunsaturated fatty acids in psychiatric disease. *J. Nutr. Biochem.* **2006**, *17*, 717–727. [CrossRef]
33. Roynette, C.E.; Calder, P.C.; Dupertuis, Y.M.; Pichard, C. n–3 polyunsaturated fatty acids and colon cancer prevention. *Clin. Nutr.* **2004**, *23*, 139–151. [CrossRef]
34. Izquierdo, M.S.; Robaina, L.; Juárez-Carrillo, E.; Oliva, V.; Hernández-Cruz, C.M.; Afonso, J.M. Regulation of growth, fatty acid composition and delta-6 desaturase expression by dietary lipids in gilthead seabream larvae (*Sparus aurata*). *Fish Physiol. Biochem.* **2008**, *34*, 117–127. [CrossRef]
35. Xie, D.Z.; Chen, C.Y.; Dong, Y.W.; You, C.H.; Wang, S.Q.; Monroig, Ó.; Tocher, D.R.; Li, Y.Y. Regulation of long-chain polyunsaturated fatty acid biosynthesis in teleost fish. *Prog. Lipid Res.* **2021**, *82*, 101095. [CrossRef]
36. Wang, S.H.; Cheng, Y.X.; Jia, Z.Y.; Li, C.T.; Zhang, X.B.; Shi, L.Y. Comparison of nutritional quality of adult Chinese mitten handed crab (*Eriocheir sinensis*) with different hepatic color. *Chin. J. Fish.* **2021**, *34*, 27–34.
37. Wang, D.Q.; Zhang, L.; Wu, X.G.; Wang, X.C.; Li, Y.Y. Effect of 8‰ low salinity seawater on taste compounds and sensory evaluation of female Chinese mitten crab during temporary culture. *Sci. Technol. Food Ind.* **2018**, *39*, 1–12.
38. Zhao, L.; Wu, N.; Wang, X.C.; Wu, X.G.; Wang, Y.H. Comparison of the flavor components of Chinese mitten crab at different growth stages. *Mod. Food Sci. Technol.* **2016**, *32*, 261–269.
39. Dermiki, M.; Phanphensophon, N.; Mottram, D.S.; Methven, L. Contributions of non-volatile and volatile compounds to the umami taste and overall flavour of shiitake mushroom extracts and their application as flavour enhancers in cooked minced meat. *Food Chem.* **2013**, *141*, 77–83. [CrossRef] [PubMed]

foods

MDPI

Article

Effect of Respite Time before Live Transportation on Muscle Quality of Blunt Snout (Wuchang) Bream

Ling Peng [1], Juan You [1], Lan Wang [2], Shanbai Xiong [1], Qilin Huang [1] and Tao Yin [1,*]

[1] College of Food Science and Technology, National R&D Branch Center for Conventional Freshwater Fish Processing (Wuhan), Huazhong Agricultural University, Wuhan 430070, China; pengling@webmail.hzau.edu.cn (L.P.); juanyou@mail.hzau.edu.cn (J.Y.); xiongsb@mail.hzau.edu.cn (S.X.); hql@mail.hzau.edu.cn (Q.H.)

[2] Institute of Agricultural Products Processing and Nuclear-Agricultural Technology, Hubei Academy of Agricultural Sciences, Wuhan 430064, China; 2005lily@gmail.com

* Correspondence: yintao@mail.hzau.edu.cn

Abstract: To provide scientific support for improving the muscle quality of blunt snout bream, ultrahigh performance liquid chromatography, texture analyzer, and optical electron microscopy were applied to explore the effects of respite time (0, 24, 48, and 72 h) on the muscle quality of blunt snout bream before live transportation. The energy compounds (ATP and glycogen) of muscle significantly decreased with the respite time ($p < 0.05$). Lactic acid content declined and then increased, leading to a rise and then a fall in pH ($p < 0.05$). Water-holding capacity of fish muscle increased progressively when the respite time was increased to 48 h and then dramatically decreased when the respite time was further increased to 72 h ($p < 0.05$). Shear force gradually increased ($p < 0.05$), while the whiteness and lightness values remained stable ($p > 0.05$). Both the content of umami compounds (IMP) and bitter compounds (HxR, Hx), and the calculated K value decreased steadily with the respite time ($p < 0.05$). The extracellular gap of the fish muscles gradually decreased with the respite time. The most uniform and intact cellular structure occurred at 48 h. However, when the respite time was extended to 72 h, the extracellular gap and muscle fragmentation rate of the muscle increased considerably. The findings indicated that a 48 h respite time was suitable to improve the muscle quality of blunt snout bream after live transportation.

Keywords: blunt snout bream; respite time; live transportation; texture; cell structure

Citation: Peng, L.; You, J.; Wang, L.; Xiong, S.; Huang, Q.; Yin, T. Effect of Respite Time before Live Transportation on Muscle Quality of Blunt Snout (Wuchang) Bream. *Foods* **2022**, *11*, 2254. https://doi.org/10.3390/foods11152254

Academic Editor: Concetta Messina

Received: 28 June 2022
Accepted: 25 July 2022
Published: 28 July 2022

1. Introduction

At present, cross-regional logistics of freshwater fish is mostly undertaken using live fish transportation vehicles equipped with oxygen and water tanks, to resolve the issue of unbalanced supply and demand for fishing resources [1]. The live freshwater fish is transported in closed containers with a restricted amount of water; thus, mucus generated by fish and suspended particles in feces quickly contaminates water bodies [2]. Additionally, ammonia (a compound of fish metabolism) is dissolved after being expelled via the fish's gills. When the quantity of ammonia nitrogen exceeds 2 mg/L in the water, fish may die [3]. Respite before live transportation of fish is one of the most practical and effective methods for reducing transportation stress and increasing survival rates [4]. During the respite, the fish expel their excreta, thereby minimizing ammonia nitrogen and odorous chemical buildup in the culture environment [5,6]. This will also reduce metabolism rate and oxygen consumption during live transportation, lessen stress, and ultimately boost survival time and rate [7]. Domestic and international research on respite treatment has focused on its influence on ammonia nitrogen metabolism and survival time of the fish during the live transportation [8–10].

The fish's body is divided into four parts: the head, torso, tail, and fins, with the muscle of the torso being the major part used as food [11]. Muscle quality affects customer's acceptance, which ultimately determines economic value. Muscle quality (shear force, water-holding capacity, color, pH of the muscle, etc.) is affected by the fish stress [12–14]. Yu et al. [15] discovered that feeding grass carp (*Ctenopharyngodon idella*) only by faba bean (*Vicia faba L.*) enhanced hardness and elasticity of muscle greatly, which was due to the diet stress. Acerete et al. [16] demonstrated that pre-slaughter anesthetics and exposure to low temperatures reduced stress levels, which decreased lactic acid buildup in fish muscle, and enhanced sensory (odor, gill, skin, and eye) scores in European perch (*Dicentrarchus labrax*). Lv et al. [17] reported that purification treatment could improve muscle hardness, flexibility, and water-holding capacity (WHC) of the grass carp muscle, along with decrease the odoriferous volatile chemicals (nonanal and hexanal). Respite treatment has been reported to reduce transportation stress. However, no studies have been conducted on the impact of respite treatment on the muscle quality of fish after live transportation.

Blunt snout bream (*Megalobrama amblycephala*) is a type of bream widely farmed in China. It is popular with customers due to its high protein and vitamin content and low-fat content. Therefore, this study simulated the commercial transportation of live blunt snout bream (short-time respite, capture, and live transportation). Additionally, ultrahigh performance liquid chromatography, texture analyzer, and optical electron microscopy were applied to investigate the effects of respite time on the muscle quality of blunt snout bream after live transportation, in order to provide scientific support for improving the muscle quality of blunt snout bream.

2. Materials and Methods

2.1. Materials

Blunt snout bream with an average length of 35.1 ± 1.39 cm and weight of 641.25 ± 56.58 g was obtained from a pond in Ezhou National Original Breeding Farm (Ezhou, China). The water temperature in this pond was 16.68 °C, the dissolved oxygen was 12.49 mg/L, and the pH was 8.14.

Anhydrous ethanol, sodium hydroxide, phosphoric acid, potassium dihydrogen phosphate, and HE dye suit were purchased from Sinopharm Chemical Reagent Co., Ltd. (Shanghai, China). Acetonitrile and methanol were purchased from Merck (Darmstadt, Germany).

2.2. Sample Preparation

All animal standard operation procedures were approved by the Animal Care and Use Committee of Huazhong Agricultural University and performed in accordance with the Guidelines for Care and Use of Laboratory Animals of Huazhong Agricultural University. Blunt snout bream of uniform size and health status were respited (0, 24, 48, 72 h) in a net box (6 m × 1 m × 1 m) at 3 m offshore, with the bottom of the net box 0.5 m from the lake bottom. At the end of each time period of respite, blunt snout bream (10 samples) was captured from the net box and transferred to a plastic box (54.5 cm × 37 cm × 34.5 cm) with a doubled amount of water (water from the same pond) according to their weight, and ice bags were added. The live blunt snout bream were transported to the laboratory within 2 h, and then the boxes were transferred to a simulated transport platform (DK-5024, Starshow Intelligent Equipment Co., Ltd., Xiamen, China) for further transportation under a vibration frequency of 120 r/min. During the whole transportation process, the water temperature was maintained at 10 ± 3 °C and dissolved oxygen at >10 mg/L. The transportation procedure and conditions adopted in the experiment were referenced to the commercialization model of transporting blunt snout bream in China [1].

2.3. Lactic Acid and Muscle Glycogen

Lactic acid and muscle glycogen content were determined by kits (A019-2-1/A043-1-1, Nanjing Jiancheng Institute of Biological Engineering, Nanjing, China). At 620 nm, glycogen was detected, whereas lactic acid was detected at 530 nm. Three samples were used in the determination for each treatment.

2.4. pH

The electrode of an insertion pH meter (TESTO 205, DETO Instruments International Trading Co., Ltd., Shanghai, China) was inserted into the fish muscle, and the pH value of the dorsal muscle of the blunt snout bream was shown on the display. Six samples were used in the determination for each treatment.

2.5. Water-Holding Capacity

Water-holding capacity was measured according to the method of Subbaiah [18] with slight changes. About 3 g of fish muscle was weighed and wrapped in a double layer of qualitative filter paper and centrifuged at 4000 rpm for 15 min. The water-holding capacity was expressed as the ratio of the sample mass before and after centrifugation. Six samples were used in the determination for each treatment.

2.6. Color

Color was measured according to the method of Shi et al. [19] with slight changes. The dorsal muscle of blunt snout bream was cut into cubes (20 mm × 20 mm × 10 mm), and L^*, a^*, and b^* values were recorded using a portable colorimeter (CR-400, Konica Minolta, Tokyo, Japan). Muscle samples without respite (0 h) were used as a reference for ΔE calculation. Six samples were used in the determination for each treatment.

$$W = 100 - \sqrt{(100 - L*)^2 + a*^2 + b*^2} \tag{1}$$

$$\triangle E = \sqrt{\triangle L*^2 + \triangle a*^2 + \triangle b*^2} \tag{2}$$

where W indicates the whiteness of the sample; ΔE indicates the total color difference; L^* value indicates the lightness of the sample: positive a^* represents red, negative a^* represents green, positive b^* represents yellow, and negative b^* represents blue (based on AMSA Meat Color Measurement Guidelines, 2012).

2.7. Shear Force

Shear force was measured according to the method of Shi et al. [19] with slight changes. The dorsal muscle of blunt snout bream was cut into 20 mm × 20 mm × 10 mm cubes. Cubes were cut perpendicular to the direction of muscle fibers at a speed of 60 mm/min using a texture analyzer (SD-700, Akiyama Technology Co., Ltd., Dongguan, China) equipped with a blade (15 mm in diameter). After cutting, the shear force was recorded as the maximum force (g). Ten samples were used in the determination for each treatment.

2.8. ATP-Related Compounds

Separation and identification of samples were performed using ultrahigh performance liquid chromatography (Acquity UPLC-H Class, Waters Corporation, Milford, MA, USA), according to Liu [20]. The separation was performed on an ACQUITY UPLC BEH Amide (1.7 μm, 2.1 mm × 100 mm, Waters) column. The mobile phases were acetonitrile, 10 mmol/L sodium dihydrogen phosphate, and 0.1% (v/v) aqueous phosphate solution; the gradient system consisted of varying concentrations of mobile phases A, B, and C (0–6 min, 88–80% A, 7–17.5% B; 6–8 min, 80–77% A, 17.50–22% B; 8–9 min, 77–65% A, 22–35% B; 9–10.7 min, 65–55% A, 35–45% B; 10.70–10.80 min, 55–88% A, 45–7% B; 10.80–23.00 min, 88% A, 7% B). The chromatographic separation conditions were set as follows: column temperature 50 °C; flow rate 0.5 mL/min; injection volume 5 μL. Based on ATP-related compounds, the K value is usually calculated as the percentage rate of HxR and Hx to the sum of ATP and degradation products. Three samples were used in the determination for each treatment.

$$K = \frac{\text{Hx} + \text{HxR}}{\text{ATP} + \text{ADP} + \text{AMP} + \text{IMP} + \text{Hx} + \text{HxR}} \times 100\% \tag{3}$$

2.9. Morphological Observation

Morphological observation was carried out according to the method of Shi et al. [19] with slight modification. The dorsal muscle was fixed with 4% paraformaldehyde solution overnight. The fixed sample was sliced in paraffin, dewaxed, stained with hematoxylin eosin, dehydrated, and sealed. The processed samples were scanned panoramically with an optical microscope (EclipseCi, Nikon, Tokyo, Japan) and displayed using a Pannoramic Viewer (1.15.3, 3DHISTECH Ltd., Budapest, Hungary). The magnification of scanned images was adjusted to 100 times. Three samples were used in the determination for each treatment.

2.10. Statistical Analysis

SAS software (V8, SAS Institute Inc., Cary, NC, USA) was used for statistical analysis by one-way ANOVA. The significance method was LSD (least significant difference), and the detection limit was 0.05.

3. Results

3.1. Muscle Glycogen

As shown in Table 1, the muscle glycogen content of blunt snout bream decreased considerably as respite time increased ($p < 0.05$). Without respite (0 h), the glycogen content of blunt snout bream muscle was 0.62 mg/g. The increase in the respite time to 24 h resulted in a significant decrease in muscle glycogen to 0.49 mg/g ($p < 0.05$). When the respite time was extended to 48 h, muscle glycogen decreased to 0.44 mg/g, although there was no significant difference in comparison to 24 h ($p > 0.05$). A 72 h respite time resulted in a significant decrease in muscle glycogen to 0.34 mg/g ($p > 0.05$). At 72 h, muscle glycogen dropped by 45,16% in fish muscle compared to the sample without respite (0 h).

Table 1. The effect of respite time before live transportation on muscle glycogen, lactic acid, and pH of blunt snout bream (n = 3 for each respite time).

Respite Time (h)	Muscle Glycogen (mg/g)	Lactic Acid (mg/g Protein)	pH
0	0.62 ± 0.06 [a]	3.77 ± 0.17 [a]	6.71 ± 0.10 [c]
24	0.49 ± 0.00 [b]	3.22 ± 0.19 [b]	6.87 ± 0.08 [b]
48	0.44 ± 0.00 [b]	2.90 ± 0.17 [b]	6.98 ± 0.05 [a]
72	0.34 ± 0.03 [c]	3.20 ± 0.16 [b]	6.85 ± 0.05 [b]

Notes: different lowercase letters in the same column indicate significant differences ($p < 0.05$).

3.2. Lactic Acid

The lactic acid content of blunt snout muscle decreased as respite time increased, from 3.77 mg/g protein at 0 h to 2.90 mg/g protein at 48 h, and subsequently increased to 3.20 mg/g protein at 72 h ($p < 0.05$). The content of lactic acid in muscle of blunt snout bream after respite was significantly lower than that without respite (0 h). After 48 h of respite before live transportation, their muscles had the lowest amount of lactic acid, at 2.90 mg/g protein (Table 1).

3.3. pH

The muscle pH of blunt snout bream increased significantly from 6.71 to 6.98 as the respite time extended from 0 h to 48 h ($p < 0.05$). The muscle pH of blunt snout bream significantly decreased to 6.85 ($p < 0.05$) after live transportation when the respite time continued to increase to 72 h (Table 1). Muscle pH was significantly higher in blunt snout bream after different times of respite (24, 48, and 72 h) compared to that of sample without respite (0 h) (Table 1).

3.4. Water-Holding Capacity

After live transportation, the water-holding capacity of blunt snout bream without respite (0 h) was 71.77%. After 24, 48 and 72 h of respite, the muscle water-holding capacity of blunt snout bream increased to 80.45%, 82.56%, and 79.59%, respectively ($p < 0.05$). The water-holding capacity decreased considerably ($p < 0.05$) after 72 h of respite as compared to 48 h, but was significantly greater than that without respite (0 h) (Figure 1).

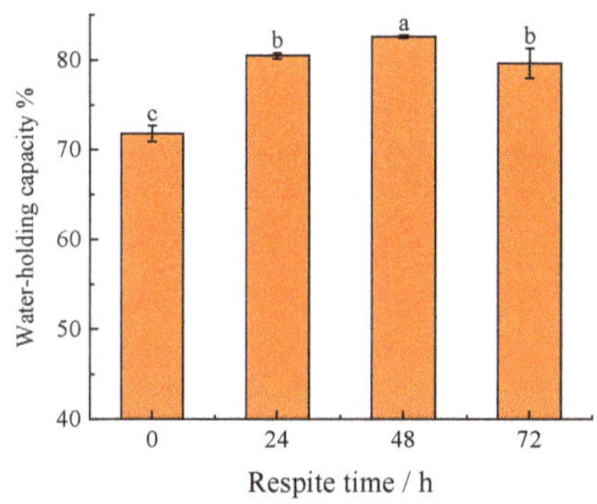

Figure 1. The effect of respite time before live transportation on water-holding capacity of blunt snout bream. Different lowercase letters indicate significant difference ($p < 0.05$).

3.5. Color

Table 2 illustrates the effect of respite time before live transportation on the muscle color of blunt snout bream. The lightness (L^*) and whiteness (W) of blunt snout bream muscles did not substantially change as respite time increased ($p > 0.05$). Without respite (0 h), the redness (a^*) and yellowness (b^*) values of blunt snout bream muscles were 0.75 and 1.82, respectively. They were significantly higher than the a^* and b^* values of blunt snout bream muscles after respite (24, 48, and 72 h) before transportation ($p < 0.05$). Furthermore, the a^* and b^* values of blunt snout bream muscle did not change significantly after 24–72 h respite time (Table 2).

Table 2. The effect of respite time before live transportation on muscle color of blunt snout bream (n = 6 for each respite time).

Respite Time (h)	L^*	a^*	b^*	W	ΔE
0	51.18 ± 1.80^{a}	0.75 ± 0.27^{a}	1.82 ± 0.88^{a}	51.13 ± 1.78^{a}	0.00
24	51.34 ± 1.83^{a}	-0.34 ± 0.10^{b}	-0.36 ± 0.15^{b}	51.33 ± 1.83^{a}	2.89 ± 0.69
48	49.47 ± 0.69^{a}	-0.37 ± 0.19^{b}	-1.06 ± 0.49^{b}	49.46 ± 0.69^{a}	3.56 ± 0.68
72	50.95 ± 1.19^{a}	-0.37 ± 0.19^{b}	-1.03 ± 0.32^{b}	50.93 ± 1.19^{a}	3.26 ± 0.32

Notes: different lowercase letters in the same column indicate significant differences ($p < 0.05$). W: whiteness; ΔE: the total color difference; L^*: lightness; a^*: redness; b^*: yellowness.

3.6. Shear Force

As shown in Figure 2, the shear force increased significantly as the respite time increased ($p < 0.05$). The muscle shear force of blunt snout bream without respite (0 h) was 486.64 g, which increased to 538.35 g when the respite time was 24 h. It increased to 595.29 and 759.50 g after 48 and 72 h of respite, respectively. Compared with 0 h, the shear force increased by 56.07% after 72 h of respite (Figure 2).

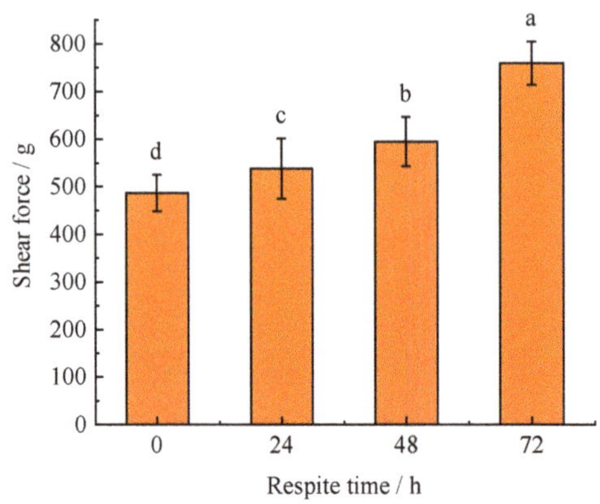

Figure 2. The effect of respite time before live transportation on shear force of blunt snout bream. Different lowercase letters indicate significant difference ($p < 0.05$).

3.7. ATP-Related Compounds

As shown in Table 3, the ATP content of muscle from blunt snout bream without respite (0 h) was found to be 25.81 mg/100 g. However, ATP was not detected in the muscles of blunt snout bream with respite (24, 48, and 72 h), showing that the muscle ATP content was low and almost totally degraded. With increasing respite time, the energy compounds ADP and ATP in muscle decreased significantly, from 41.34 and 10.09 mg/100 g to 29.96 and 0.38 mg/100 g, respectively; the umami compounds IMP decreased significantly from 435.43 to 383.01 mg/100 g, and the bitter compounds HxR and Hx decreased significantly from 25.37 and 4.01 mg/100 g to 13.07 and 2.17 mg/100 g, respectively ($p < 0.05$).

Table 3. The effect of respite time before live transportation on ATP-related compounds of blunt snout bream (n = 3 for each respite time).

Respite Time (h)	ATP	ADP	AMP	IMP	HxR	Hx	K Value
0	25.81 ± 0.41	41.34 ± 0.3 ^{a}	10.09 ± 0.37 ^{a}	435.43 ± 1.58 ^{a}	25.37 ± 0.49 ^{a}	4.01 ± 0.56 ^{a}	5.87 ± 0.13 ^{a}
24	-	38.02 ± 2.06 ^{b}	8.20 ± 1.00 ^{b}	397.76 ± 0.48 ^{b}	16.44 ± 0.78 ^{b}	3.19 ± 0.30 ^{b}	4.20 ± 0.12 ^{b}
48	-	35.08 ± 1.55 ^{c}	1.95 ± 0.95 ^{c}	389.11 ± 0.33 ^{c}	14.48 ± 2.52 ^{bc}	3.11 ± 0.38 ^{c}	3.94 ± 0.64 ^{b}
72	-	29.96 ± 0.64 ^{d}	0.38 ± 0.21 ^{d}	383.01 ± 0.18 ^{d}	13.07 ± 1.86 ^{c}	2.17 ± 0.06 ^{d}	3.51 ± 0.41 ^{b}

Notes: different lowercase letters in the same column indicate significant differences ($p < 0.05$). ATP: 5′-adenosine triphosphate; ADP: 5′-adenosine diphosphate; AMP: 5′-adenosine monophosphate; IMP: inosinic acid; HxR: inosine; Hx: hypoxanthine; K: freshness.

The K value of fish meat was less than 10% after 0, 24, 48 and 72 h of respite. Moreover, the K value of the sample without respite (0 h) was significantly higher than those after 24, 48, and 72 h of respite (Table 3).

3.8. Morphological Observation

The cross-sectional cytoarchitecture of blunt snout bream muscles with different respite time is shown in Figure 3. The dorsal muscle cells of blunt snout bream were irregularly polygonal in cross-section, with cells closely adjacent to each other. When the respite time was prolonged from 0 to 48 h, the extracellular space in the muscle of blunt snout bream progressively decreased. At 48 h, the cells were the most intact and full. However, when the time of the respite was increased to 72 h, the densely coupled cells progressively disengaged and the extracellular distance expanded substantially (Figure 3).

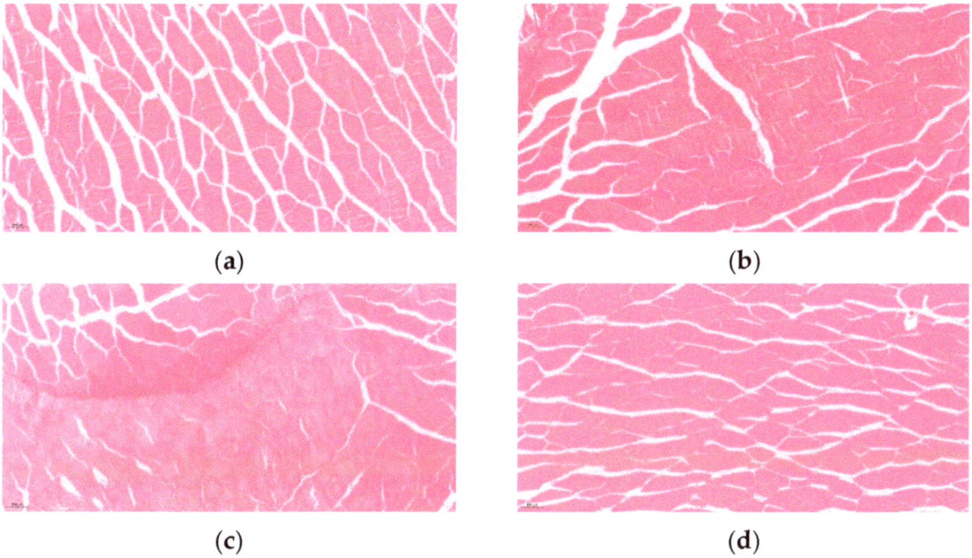

Figure 3. The effect of respite time before live transportation on cross-sectional cytoarchitecture of blunt snout bream: (a) 0 h, (b) 24 h, (c) 48 h, and (d) 72 h. Muscle fiber microstructure with 100 times magnification.

4. Discussion

During the live transportation, water is polluted due to the physiological and metabolic activities (respiration and excretion) of fish. After the deterioration of water quality, fish undergo an intense stress response, which affects their survival rate and muscle quality. Ammonia nitrogen in transportation waters comes mainly from the fish's own nitrogen excretion and degradation of nitrogenous organic matter such as feces [21]. When the ammonia nitrogen content of water is too high, nonionic ammonia competes with oxygen for hemoglobin, resulting in fish tissue hypoxia [3]. In addition, it may cause damage to the fish's liver and renal tissues, resulting in edema, congestion, and inflammation, which may even lead fish to unconsciousness and death [22]; Additionally, it may increase the quantity of reactive oxygen species (ROS), resulting in oxidative stress, impairing muscle cell function of fish, and possibly triggering death [23].

The carbohydrates in the bait provide the major energy for blunt snout bream. During the pre-stop feeding period (respite), blunt snout bream might retain a tiny quantity of undigested leftover bait in their bodies. When bait was depleted, they were unable to obtain carbohydrates. To maintain normal physiological functions, the fish would preferentially use glucose stored in the form of glycogen; when glycogen in the body was depleted, they would use muscle fat; and when severely starved, they would even break down muscle protein [23,24]. Therefore, the muscle glycogen content of blunt snout bream gradually decreased with the increase in respite time, and the sample with respite showed significantly lower glycogen content than the blunt snout bream without respite (Table 1). This result is consistent with Einen et al. [25]. Changes in the amount of ATP and its metabolic products can be indicative of changes in fish muscle energy [26]. ATP is synthesized during phosphagen (ATP-Pcr), glycolysis, aerobic oxidation, etc. [27]. ATP was detected only in the sample without respite, at 25.81 mg/100 g (Table 3). The content of ATP-related compounds, including ADP, AMP, IMP, HxR, and Hx, decreased significantly with the increase in respite time ($p < 0.05$). It was probable that the blunt snout bream could not obtain carbohydrates during the respite to replenish energy. Thereby, the glycogen content of muscles progressively diminished, resulting in an ATP breakdown rate exceeding the synthesis rate [28]. Increased ATP catabolism led to its continual breakdown to ADP, which was then broken down further by myokinase to ATP and AMP. By the action of AMP deaminase, AMP was degraded to IMP and ammonia, and IMP accumulated in muscle. The umami compounds are mainly amino acids, nucleotides, and peptides. The nucleotides of AMP and IMP both are umami compounds that contribute positively to the flavor of fish. However, further degradation of IMP produces bitter compounds such as HxR and Hx.

After being aroused by stressors such as hunger, shock, and crowding, muscular activity (swimming, wrestling, and escape behaviors) of the blunt snout bream increased, which accelerated the consumption of energy compounds in the muscles, and consequently changed the fish's physiological metabolism. In order to sustain normal physiological function, the fish might activate anaerobic glycolysis and the phosphocreatine pathway to synthesize ATP. Glycogen was degraded in muscle through glycolysis to pyruvate, which was then converted to lactate under anaerobic and lactate dehydrogenase conditions, resulting in lactate accumulation in the muscle. The muscles of the blunt snout bream that had been respited before transportation showed a low lactic acid level (Table 1). This might be because the respite slowed ammonia metabolism and emptied the fish's excretion, which contributed to alleviate blunt snout bream's stress during harvesting and transportation. As a result, it attenuated the anaerobic glycolytic response, and decreased lactic acid buildup. However, at 72 h respite time, the ATP-relatives and glycogen levels in fish muscles were low. In this case, the fish was short of energy to maintain life activities, which might weaken the capacity for adapting to the new environment [29]. During live transportation, blunt snout bream with too long a respite time (72 h) were more susceptible to ammonia [9], which exacerbated stress response. The lactic acid content increased while muscle glycogen content decreased, which might be caused by anaerobic metabolism (Table 1). The pH of muscle was closely related with the content of lactic acid, i.e., the accumulation of lactic acid led to the decrease in pH. As a result, the pH of blunt snout bream progressively increased as the respite time was prolonged to 48 h. When the respite time continuously rose to 72 h, the muscle pH decreased (Table 1). Additionally, the change in pH might be related to the creatine content. When the respite time was increased to 72 h, the activity of creatine kinase converted phosphocreatine and ADP to ATP and creatine, which may cause the decrease in pH [27].

The pH of muscle has an impact on the net surface charge of myogenic fibronectin. When the pH of muscle is low, the number of net charges on the surface of myogenic fibrous proteins is less, which generally corresponds to a lower water-holding capacity of muscles. The water-holding capacity of the fish muscle is also related to the cell structure, i.e., muscle with intact and homogeneous structure has a high water-holding capacity. After 24–72 h of respite, the cellular structure of blunt snout bream cells was obviously denser than that without respite. However, the cellular structure of blunt snout bream muscle transplanted after 72 h was more porous than that after 48 h of respite (Figure 3). The water-holding capacity of the blunt snout bream's muscle increased dramatically when the respite time was prolonged from 0 to 48 h. When the respite time was increased to 72 h, the muscle water-holding capacity fell marginally. The changes in cellular structural might be related to the muscle fiber injury and muscular atrophy mediated by the stress response [30,31] and elevated concentrations of reactive oxygen species (ROS). When fish are stressed, the concentration of reactive oxygen species (ROS) rises, prompting an oxidative stress response in the fish. Increased ROS levels may cause lipid peroxidation and the release of intracellular components (lysosomal enzymes) from muscle cells, which damage the cell membrane and cause cells to break apart [32]. Additionally, the ROS interfers with calcium binding and myogenic fibrin breakage in muscle, weakening actomyosin interactions and leading to a decrease in muscle fiber diameter [15].

The shear force of blunt snout bream muscle increased as the respite time increased, which might be mainly associated with the changes in muscle cell structure [33,34]. Excreta such as ammonia excretion and feces from blunt snout bream without respite (0 h) might severely contaminate the water, resulting in increased stress in blunt snout bream. The stressor stimulates receptors in the hypothalamus, resulting in hormonal and electrical signals that activated the hypothalamic–pituitary–adrenal axis, causing it to release glucocorticoids. The nuclear transcription factor NF-κB might be activated as the glucocorticoid production rises, which might result in cell autophagy and death, as evidenced by the ruptured cells [31]. Respite might reduce metabolism rate and oxygen consumption, which leads to lower stress response intensity and weaker cellular damage in the fish during transportation. After the blunt snout bream was subjected to respite treatment before live transportation, the fish might use the fat in the muscle to maintain normal physiological metabolism since it was short on carbohydrate sources. Therefore, significant increases in muscle shear force in blunt snout bream might be a result of muscle lipid depletion as well [35].

The redness and yellowness values of the muscles of blunt snout bream with respite (24, 48, and 72 h) were significantly lower than those without respite (0 h), which might be related to the lower level of stress during live transportation. When fish were stressed, their muscular tissues became congested [36], affecting the color of the muscles. Additionally, as the stress response was intensified, the fish muscle contractions were accelerated, affecting the capacity of astaxanthin and canthaxanthin to bind actinomycin acid [37]. This might change the redness value and the yellowness value of the muscle.

5. Conclusions

With respite time at 24 and 48 h, the muscle quality of blunt snout bream obviously improved, but it declined when the respite time was extended to 72 h. The improvement of muscle quality could be mainly attributed to the benefits of bait during the respite (< 48 h), which promoted fish evacuation of digestive tract waste. It might effectively alleviate the fish stress during live transportation. As a result, water-holding capacity and freshness (indicated by K value) increased; shear force increased significantly. However, prolonged respite led to the consumption of stored energy substances (such as glycogen). It might lead to increased stress injury, which could be confirmed by the results of structural changes in muscle cells. Consequently, the water-holding capacity decreased significantly. This research demonstrates that it is suitable to transport live blunt snout bream after 48 h of respite.

Author Contributions: Conceptualization, T.Y.; methodology, L.P.; software, L.P.; validation, J.Y., L.W.; formal analysis, T.Y.; investigation, L.P.; resources, S.X.; data curation, T.Y.; writing—original draft preparation, L.P.; writing—review and editing, T.Y.; visualization, Q.H.; supervision, T.Y.; project administration, S.X.; funding acquisition, S.X. All authors have read and agreed to the published version of the manuscript.

Funding: Major Program of Technical Innovation of Hubei Province (2020BBA048); earmarked fund for CARS (CARS-45).

Institutional Review Board Statement: The animal study protocol was approved by the Animal Care and Use Committee of Huazhong Agricultural University (protocol code HZAUFI-2022-0012 and date of approval 15 July 2022).

Informed Consent Statement: Not applicable.

Data Availability Statement: The data presented is contained within the article.

Acknowledgments: A special thank is extended to Zenan Huang, Xinyi Su, and Yiyang Lou for their valuable help in sample preparation.

Conflicts of Interest: The authors have declared no conflict interest.

References

1. Peng, L.; You, J.; Xiong, G.Q.; Wang, L.; Wu, W.J.; Zhang, T.H.; Huang, Q.L.; Du, H.Y.; Yin, T. Research progress on the effects of logistics and transportation on the quality of fish muscle. *Meat Res.* **2021**, *35*, 54–63.
2. Marçalo, A.; Mateus, L.; Correia, J.H.D.; Serra, P.; Fryer, R.; Stratoudakis, Y. Sardine (*Sardina pilchardus*) stress reactions to purse seine fishing. *Mar. Biol.* **2006**, *149*, 1509–1518.

3. Xia, W.S.; Luo, Y.K.; Xiong, S.B.; Xu, S.Y. *Storage, Transportation, Preservation and Processing Technology of Bulk Freshwater Fish*, 1st ed.; China Agriculture Press: Beijing, China, 2014; pp. 27–65.
4. Lan, Y.Y.; Zhang, Y.J.; Song, Z.F.; Wu, S.Z.; Xu, C.L.; Zhao, Z.M. Construction and application of denitrification system ofseawater temporary nutrition system at low temperature. *J. South. Agric.* **2019**, *50*, 1836–1843.
5. Davidson, J.; Schrader, K.; Ruan, E.; Swift, B.; Aalhus, J.; Juarez, M.; Wolters, W.; Burr, G.; Good, C.; Summerfelt, S.T. Evaluation of depuration procedures to mitigate the off-flavor compounds geosmin and 2-methylisoborneol from Atlantic salmon Salmo salar raised to market-size in recirculating aquaculture systems. *Aquac. Eng.* **2014**, *61*, 27–34. [CrossRef]
6. Harmon, T.S. Methods for reducing stressors and maintaining water quality associated with live fish transport in tanks: A review of the basics. *Rev. Aquac.* **2010**, *1*, 58–66. [CrossRef]
7. Burr, G.S.; Wolters, W.R.; Schrader, K.K.; Summerfelt, S.T. Impact of depuration of earthy-musty off-flavors on fillet quality of Atlantic salmon, Salmo salar, cultured in a recirculating aquaculture system. *Aquac. Eng.* **2012**, *50*, 28–36. [CrossRef]
8. Sinha, A.K.; Liew, H.J.; Diricx, M.; Blust, R.; Gudrun, D.B. The interactive effects of ammonia exposure, nutritional status and exercise on metabolic and physiological responses in gold fish (*Carassius auratus* L.). *Aquat. Toxicol.* **2012**, *109*, 33–46.
9. Diricx, M.; Sinha, A.K.; Liew, H.J.; Mauro, N.; Blust, R.; De Boeck, G. Compensatory responses in common carp (*Cyprinus carpio*) under ammonia exposure: Additional effects of feeding and exercise. *Aquat. Toxicol.* **2013**, *142–143*, 123–137. [CrossRef]
10. Tian, B.; Chen, S.R.; Yang, Y.F.; Zhang, Q.B. A primary study on the Keeping-alive technique of spams macrocephalus by waterless method. *J. Jimei Univ. (Nat. Sci.)* **2004**, 221–225. [CrossRef]
11. Periago, M.J.; Ayala, M.D.; López-Albors, O.; Abdel, I.; Martínez, C.; García-Alcázar, A.; Ros, G.; Gil, F. Muscle cellularity and flesh quality of wild and farmed sea bass, *Dicentrarchus labrax* L. *Aquaculture* **2005**, *249*, 175–188.
12. Refaey, M.M.; Tian, X.; Tang, R.; Li, D. Changes in physiological responses, muscular composition and flesh quality of channel catfish Ictalurus punctatus suffering from transport stress. *Aquaculture* **2017**, *478*, 9–15. [CrossRef]
13. Wang, W.S.; Zhang, Y.J.; Liu, Y.; Adányi, N.; Zhang, X.S. Effects of waterless live transportation on survivability, physiological responses and flesh quality in Chinese farmed sturgeon (*Acipenser schrenckii*). *Aquaculture* **2020**, *518*, 734834. [CrossRef]
14. Wu, Y.; You, X.; Sun, W.; Xiong, G.; Shi, L.; Qiao, Y.; Wu, W.; Li, X.; Wang, J.; Ding, A.; et al. Insight into acute heat stress on meat qualities of rainbow trout (*Oncorhynchus mykiss*) during short-time transportation. *Aquaculture* **2021**, *543*, 737013. [CrossRef]
15. Yu, E.; Fu, B.; Wang, G.; Li, Z.; Ye, D.; Jiang, Y.; Ji, H.; Wang, X.; Yu, D.; Ehsan, H.; et al. Proteomic and metabolomic basis for improved textural quality in crisp grass carp (*Ctenopharyngodon idellus C.et V*) fed with a natural dietary pro-oxidant. *Food Chem.* **2020**, *325*, 126906. [CrossRef]
16. Acerete, L.; Reig, L.; Alvarez, D.; Flos, R.; Tort, L. Comparison of two stunning/slaughtering methods on stress response and quality indicators of European sea bass (*Dicentrarchus labrax*). *Aquaculture* **2009**, *287*, 139–144. [CrossRef]
17. Lv, H.; Hu, W.; Xiong, S.; You, J.; Fan, Q. Depuration and starvation improves flesh quality of grass carp (*Ctenopharyngodon idella*). *Aquac. Res.* **2018**, *49*, 3196–3206. [CrossRef]
18. Subbaiah, K.; Majumdar, R.K.; Choudhury, J.; Priyadarshini, B.M.; Dhar, B.; Roy, D.; Saha, A.; Maurya, P. Protein degradation and instrumental textural changes in fresh nile tilapia (*Oreochromis niloticus*) during frozen storage. *J. Food Processing Preserv.* **2015**, *39*, 2206–2214. [CrossRef]
19. Shi, L.; Yin, T.; Wang, L.; Xiong, G.Q.; Gao, R.C.; Ding, A.Z.; Li, X.; Wu, W.J.; Qiao, Y.; Liao, L.; et al. Effect of pre-chilling time on the physicochemical properties of Channel catfish during frozen storage. *Int. J. Refrig.* **2020**, *115*, 56–62. [CrossRef]
20. Liu, J.K. Flavor Character of Sliver Carp and the Influence of Heating History on the Sliver Carp Flavor. Doctor Thesis, Huazhong Agricultural University, Wuhan, China, 2009.
21. Barbieri, E.; Bondioli, A.C.V. Acute toxicity of ammonia in Pacu fish (*Piaractus mesopotamicus*, Holmberg, 1887) at different temperatures levels. *Aquac. Res.* **2015**, *46*, 565–571. [CrossRef]
22. Xie, J.; Wang, Q. Progress in Understanding Environmental Stress and Physiological Regulation Mechanism in Aquatic Animals during Live Transportation. *Food Sci.* **2021**, *42*, 319–325.
23. Wang, T.; Hung, C.C.Y.; Randall, D.J. The comparative physiology of food deprivation: From feast to famine. *Annu. Rev. Physiol.* **2006**, *68*, 223–251. [CrossRef]
24. McCue, M.D. Starvation physiology: Reviewing the different strategies animals use to survive a common challenge. *Comp. Biochem. Physiol. A-Mol. Integr. Physiol.* **2010**, *156*, 1–18. [CrossRef]
25. Einen, O.; Waagan, B.; Thomassen, M.S. Starvation prior to slaughter in Atlantic salmon (*Salmo salar*): I. Effects on weight loss, body shape, slaughter- and fillet-yield, proximate and fatty acid composition. *Aquaculture* **1998**, *166*, 85–104. [CrossRef]
26. Speers-Roesch, B.; Sandblom, E.; Lau, G.Y.; Farrell, A.P.; Richards, J.G. Effects of environmental hypoxia on cardiac energy metabolism and performance in tilapia. *Am. J. Physiol. Regul. Integr. Comp. Physiol.* **2010**, *298*, 104–119. [CrossRef]
27. Hui, H.; Regenstein, J.M.; Luo, Y. The Importance of ATP-related Compounds for the Freshness and Flavor of Post-mortem Fish and Shellfish Muscle: A Review. *C R C Crit. Rev. Food Technol.* **2015**, *57*, 1787–1798.
28. Scheffler, T.L.; Park, S.; Gerrard, D.E. Lessons to learn about postmortem metabolism using the AMPKγ3R200Q mutation in the pig. *Meat Sci.* **2011**, *89*, 244–250. [CrossRef]
29. Shrivastava, J.; Sinha, A.K.; Cannaerts, S.; Blust, R.; De Boeck, G. Temporal assessment of metabolic rate, ammonia dynamics and ion-status in common carp during fasting: A promising approach for optimizing fasting episode prior to fish transportation. *Aquaculture* **2017**, *481*, 218–228. [CrossRef]

30. Julia, T.V.; Raúl, L.H.; Teresa, G.G.; Alejandra, G.G. Mechanisms of stress-related muscle atrophy in fish: An ex vivo approach. *Mech. Dev.* **2018**, *154*, 162–169.
31. Wan, X.; Wang, D.; Xiong, Q.; Xiang, H.; Li, H.; Wang, H.; Liu, Z.; Niu, H.; Peng, J.; Jiang, S. Elucidating a molecular mechanism that the deterioration of porcine meat quality responds to increased cortisol based on transcriptome sequencing. *Sci. Rep.* **2016**, *6*, 36589. [CrossRef]
32. Demir, S.; Yılmaz, M.; Köseoğlu, M.; Akalin, N.; Aslan, D.; Aydin, A. Role of free radicals in peptic ulcer and gastritis. *Turk. J. Gastroenterol.* **2003**, *14*, 39–43.
33. Li, X.M.; Yuan, J.M.; Fu, S.J.; Zhang, Y.G. The effect of sustained swimming exercise on the growth performance, muscle cellularity and flesh quality of juvenile qingbo (*Spinibarbus sinensis*). *Aquaculture* **2016**, *465*, 287–295. [CrossRef]
34. Kiessling, A.; Espe, M.; Ruohonen, K.; Mørkøred, T. Texture, gaping and colour of fresh and frozen Atlantic salmon flesh as affected by pre-slaughter iso-eugenol or CO_2 anaesthesia. *Aquaculture* **2004**, *236*, 645–657. [CrossRef]
35. Fuentes, A.; Fernández-Segovia, I.; Serra, J.A.; Barat, J.M. Comparison of wild and cultured sea bass (*Dicentrarchus labrax*) quality. *Food Chem.* **2010**, *119*, 1514–1518. [CrossRef]
36. Digre, H.; Erikson, U.; Misimi, E.; Standal, I.B.; Gallart-Jornet, L.; Riebroy, S.; Rustad, T. Bleeding of farmed Atlantic Cod: Residual Blood, Color, and Quality Attributes of Pre- and Postrigor Fillets as Affected by Perimortem Stress and Different Bleeding Methods. *J. Aquat. Food Prod. Technol.* **2011**, *20*, 391–411. [CrossRef]
37. Brizio, P.; Benedetto, A.; Righetti, M.; Prearo, M.; Gasco, L.; Squadrone, S.; Abete, M.C. Astaxanthin and Canthaxanthin (*Xanthophyll*) as Supplements in Rainbow Trout Diet: In Vivo Assessment of Residual Levels and Contributions to Human Health. *J. Agric. Food Chem.* **2013**, *61*, 10954–10959. [CrossRef]

foods

MDPI

Review

Cutting Techniques in the Fish Industry: A Critical Review

Wenbo Liu [1,*], Jiaqi Lyu [2], Di Wu [3], Yupeng Cao [4], Qingquan Ma [5], Yuzhen Lu [6,7] and Xin Zhang [6]

1 Coastal Research & Extension Center & Department of Agricultural and Biological Engineering, Mississippi State University, Pascagoula, MS 39567, USA
2 Department of Mechanical Engineering, Stevens Institute of Technology, Hoboken, NJ 07030, USA
3 Department of Chemical Engineering and Materials Science, Stevens Institute of Technology, Hoboken, NJ 07030, USA
4 Department of Electrical and Computer Engineering, Stevens Institute of Technology, Hoboken, NJ 07030, USA
5 Department of Civil and Environmental Engineering, New Jersey Institute of Technology, Newark, NJ 07102, USA
6 Department of Agricultural and Biological Engineering, Mississippi State University, Starkville, MS 39762, USA
7 Department of Biosystems and Agricultural Engineering, Michigan State University, East Lansing, MI 48824, USA
* Correspondence: wliu@abe.msstate.edu; Tel.: +1-(228)-762-7783

Abstract: Fish and fishery products are among the most important sources of nutritional components for human health, including high-quality proteins, essential vitamins, minerals, and healthy polyunsaturated fatty acids. Fish farming and processing technologies are continuously evolving to improve and enhance the appearance, yield, and quality of fish and fish products from farm to fork throughout the fish supply chain, including growth, postharvest, treatment, storage, transportation, and distribution. Processing of fish involves a period of food withdrawal, collection and transportation, the process of stunning, bleeding, chilling, cutting, packaging, and byproduct recycling. Cutting is a set of crucial operations in fish processing to divide the whole fish into smaller pieces for producing fish products (e.g., fish fillets, steaks, etc.). Various techniques and machinery have been introduced in the field to advance and automate cutting operations. This review aims to provide a comprehensive review of fish cutting techniques, machine vision and artificial intelligence applications, and future directions in fish industries. This paper is expected to stimulate research on enhancing fish cutting yield, product diversity, safety and quality, as well as providing advanced solutions for engineering problems encountered in the fish industry.

Keywords: fish cutting; fish processing; waterjet cutting; machine vision; artificial intelligence

Citation: Liu, W.; Lyu, J.; Wu, D.; Cao, Y.; Ma, Q.; Lu, Y.; Zhang, X. Cutting Techniques in the Fish Industry: A Critical Review. *Foods* **2022**, *11*, 3206. https://doi.org/10.3390/foods11203206

Academic Editors: Tao Yin and Liu Shi

Received: 7 September 2022
Accepted: 9 October 2022
Published: 14 October 2022

1. Introduction

Fish production is a multibillion-dollar industry worldwide since seafood is one of the major food recourses which create billions of dollars of value per year. Studies have been carried out to discuss the fishery and the related techniques in different countries [1–3]. According to the latest data from Food and Agriculture Organization, total world fisheries and aquaculture production reached 177.8 million tons in 2020, with an increase of 9.3% compared with 2010, and at an estimated sale value of USD 406 billion [4]. Due to the high prices of fish products and the varying needs of the consumer, it is critical to carry out the fish-processing tasks such that the end product is of high quality. Fish processing is the process that turns the raw fresh fish into fish products we can buy in the market or use in manufacturing other fish-related products [5]. The typical fish processing begins right after the capture of fish. It involves fish receiving (capturing), stunning, bleeding & chilling, grading, deheading (beheading), scaling, filleting, skinning, trimming & portioning, mincing, byproduct recycling, and packaging [6–8]. Briefly, once the fish is captured, the sorting of different species will be conducted. Then, the fish is transferred to the nearest processing facilities, followed by stunning and grading according to the size and quality of the fish. The stunned fish is bled and washed off by clean circulating water. After bleeding, the fish is chilled and then gutted. The bleeding and chilling process can lower bacteria and enzyme activity to extend shelf-life and quality [9,10]. The gutting process will remove the inter organs and clean the body cavity [11]. Then the fish is further scaled, beheaded, filleted, and skinned to produce fish fillets. The fish fillets need to be trimmed, portioned, sliced into pieces, or minced, depending on the requirement of the final products. The byproduct during the gutting and cutting steps will be collected for other use [12]. Among these processing works, cutting steps play major roles since they involve deheading, filleting, trimming, skinning, and portioning. Hence, the study and discussion of the cutting methods are important for the development of the fish industry.

Fish processing makes the major waste of fish come from the "butchering of fish," during which the fish are gutted and cut into fish products. This process involves the removal of non-edible portions like viscera, head, and tail that would cause up to 70% loss of fish, which is partially caused by inefficient cutting operations. In traditional fish cutting, humans played a significant role since the cutting process was completed via laborers using knives [13]. Like other food industries, fish cutting used to be labor-intensive. However, fishery operation is seasonal, which leads to difficulty in maintaining experienced laborers. Moreover, the routine and stinking environment of fish cutting somehow discourage the workers' morale. Therefore, it is desirable to increase the application of automation in the fish-cutting process for higher productivity and economic benefit [14]. The most commonly employed method for cutting fish is to use metal blades [13,15–17]. In the past three decades, research about automatic fish cutting mainly focused on how to preciously (1) detect the size of the fish [18], (2) position the fish head [19], (3) decrease the waste of raw fish [20,21], and (4) reduce the error rate of the automatic system [22]. Metal blade cutting is indeed effective in cutting operations and cost-efficient in terms of equipment maintenance. However, the force applied for cutting reduces the quality of sections. The considerable amount of cutting debris also leads to a large waste of raw fish [23]. Efforts have been invested in investigating and optimizing the cutting mechanism [24], cutting performance, processing speed, motor power, and processable fish species [25]. However, the potential heavy metal food contamination is expected under the existence of an aqueous medium [26].

Novel techniques have been invented to overcome the drawbacks of metal blade cutting. Water-jet cutting is a well-developed cutting technique which frequently applied in various industries. Recently, it has been largely applied for food cutting, including fish [27]. The rheological properties of fish meat are significantly different from common workpieces such as metal, rock, or wood [28]. With the aid of developing automatic robotic arms, waterjet cutting is believed to be faster and able to handle fish meat with bone, skin, and fiber together with fewer bone shatters and bleeding [29,30]. Compared with knife cutting, waterjet cutting exhibited advantages in cutting efficiency and quality for food processing [31]. This method mainly utilizes the high pressure of the waterjet coming through a die with a certain shape to separate the workpieces [32]. To enhance the cutting performance, certain abrasives are added to the water [33,34]. The modern waterjet cutting machine can cut intricate shapes with high precision [24]. It has the capability to operate at low temperatures, which is very important for fish cutting [35]. It is believed that waterjet cutting will become one of the major cutting methods in the food industry [8]. To further reduce the reliance on manual work, the machine vision technique has been widely utilized in modern fish automation processing systems. The machine vision technique implements optical sensors to monitor the status of fish processing. In most instances, the images of the fish's body should be acquired, stored, and analyzed. The useful information would be extracted by the machine vision algorithm for fish processing lines, like species, fish size, and body shape [36]. In the fish-cutting processes, such as de-heading and filleting, facilities require precise identification and segmentation of fish bodies. Inspired by the outstanding image process performances of artificial intelligence (AI), modern fish automation processing systems are gradually employing AI algorithms to identify the patterns of fish bodies and guide the subsequent fish-cutting processes [37].

In this review article, we considered these focused questions: what kind of advanced technologies are used at present and being considered for fish-cutting operations? How to improve the fish-cutting yield, quality, and efficiency? How to realize fish-cutting automation? Initially, several major scientific databases, including Web of Science, IEEE Xplore, ScienceDirect, Google Scholar, and Springer, were searched to collate the literature. Then we deselected some databases with limited and repeated articles since the databases with broader coverage may cover the articles coming from the other databases with limited and repeated articles. Finally, this review considered three mainstream scientific databases (IEEE Xplore, ScienceDirect, and Google Scholar) and one open-access database (arXiv). For the search strategy and inclusion criteria, we first searched the databases with individual keywords such as "waterjet fish cutting," "automatic fish cutting," "robotics for fish deheading," "automatic fish trimming," "fish cutting machine design," "machine vision in fish cutting," and "artificial intelligence in fish cutting." Then similar keywords were grouped with AND as search string combinations, like "fish-cutting yield optimization" AND "fish-cutting quality assessment," and "image processing in fish-cutting" AND "computer vision applications of fish processing." The keywords in diverse fields were also grouped with OR, such as "automation technologies for fish processing," OR "robotic trimming of fish fillets," and "deheading cuts used in fish processing" OR "fish-skinning machine design."

Although there are some review papers focusing on advanced cutting techniques for solid food [38] and fish processing automation [8], there is no paper, especially reviewing the cutting techniques for fish. The uniqueness of this paper is focusing on fish cutting steps (deheading, filleting, skinning, portioning, and trimming) and their corresponding cutting mechanism and techniques, blade design, and performance effect factors. Other related advanced technologies (waterjet cutting, machine vision and AI) are also presented with their applications in fish cutting. Finally, fish cutting machines and technologies and future perspectives are described and summarized.

2. Modern Fish Cutting

2.1. Metal Blade Cutting Machines

Metal blade cutting is the most widely used cutting method in the fish industry. In the fish processing line, the deheading of fish is usually the first cut operation. The purpose of this operation is to separate the high-value fillets from the relatively low-value head. To maximize the commercial values, the deheading machines should leave the maximum amount of meat on the fillet and ensure no part of the fish gills, and the head skeleton is included in the fillet [39–42]. The main deheading position can be classified into four classes, as shown in Figure 1, namely straight cutting, slant cutting, V-cutting, and round cutting along the gill area [43,44]. Commercial machines, like Baader 166, have been used in industry and research [39]. Buckingham et al. [40] designed a robotic solution for fish deheading called Robofish 2. The V-cutting was performed in this machine. To perform the optimum cutting, the shape of the fish changed during the deheading process. The fish was in a back-down position when it arrived. Then the head was gripped and bent to allow the metal blades to cut the flesh. An in-line weighing component was built-in this deheading machine to perform a unique cutting for each fish. Therefore, the Robofish 2 could have high cutting reliability. Ketels [41] designed a machine that could align the fish in a row in the same direction. A cutting blade driven by a motor can perform straight or slant cutting in the desired direction with the help of a designed pressure part. Rodríguez and Martínez [45] developed a method to perform round cutting for the fish silhouette. The round cutting can follow the operculum's edge to acquire the lowest meat loss. Similar to large fish cutting, deheading is the initial cutting process for small-sized fish, like sardine, horse mackerel and mackerel. However, due to the large size deviation, the deheading machines for small-sized fish need to be redesigned to increase production efficiency. The

SEAC FPM-200, which is mainly designed for cutting small-sized fish for canning, aligns several small-sized fish and performs deheading simultaneously [46]. PN-200 of Pisces Fish Machinery Inc. is designed to remove head and viscera for small fish like Sardine, Anchovy, Smelt and others under 200 mm in length.

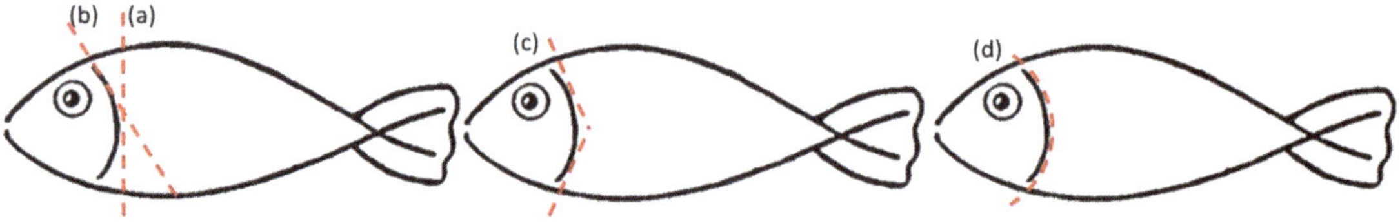

Figure 1. Schematic illustration of fish de-heading position: (**a**) Straight cutting, (**b**) Slant cutting, (**c**) V-cutting, and (**d**) Round cutting along the gill area.

For the filleting operation, as the majority of fish is symmetrical, each procedure is conducted with a pair of symmetrical knives. Modern filleting machines are fast, customized to individual fish species, and simple to adjust for various fish sizes. Currently, some machines for producing high-yielding fillets have the capability to handle short-bodied fish species, including salmon, tilapia, trout, croaker, arctic char, bar ramundi, snappers, walleye pike, and striped bass [47,48]. Therefore, these machines could achieve higher yields compared to traditional human-based filleting. The design of automatic machines is flexible. However, if the shape or size of feeding fish is out of the designed working range of filleting machines, it could result in reduced yield and even damage the fish and waste the fillets. During the filleting process, the backbone and fins are removed normally. Based on the product's specifications, if all the intermuscular bones are required to be removed, it is known as V-cut fillets, which could result in lowering yield (as much as 25%) [49]. Different machines were designed using metal blades to perform fish fillet cutting [50–54]. The main concept is to fix the fish without head and guts in a stable position. Then the blades are implemented to cut the meat from the frame bone. To minimize the meat waste during the filleting process, several blades would be designed into the system, and each of them is configured to cut a distinct segment. For example, a filleting system is designed, as shown in Figure 2 [50]. An endlessly rotating transport conveyor with saddle-shaped support bodies was designed to receive the fish and convey them along the transport direction [55]. In this design, the first two circular blades (tools 17 and 18) were designed to remove the dorsal fin. Then based on the bone structure of the fish, the angle of the following four blades (tools 9–11) could be adjusted to cut fillet meat. Specialized methods and systems designed for processing some common fish species, such as salmon, integrate and enable deheading, filleting, and gutting in a particularly reliable and efficient manner [56–58]. Besides, fillet gapping is a significant problem for the filleting process. It is different for fillets with gapping to be processed and sold. Although the causes of fillet gapping are not well understood, some actions could reduce the possibility of gapping, like rapid cooling, harvesting fish in a rest state, minimal harvest handling, season, rapid expedition to the market, etc. [49,59,60]. Production yield is a metric to evaluate the performance and effectiveness of an operation in the production lines. The fillet yield depends on the species and the structural anatomy. Fish with large heads and frames compared to the musculature would have a lower yield than the ones with smaller heads and frames. The well-chilled fish can result in a higher yield than fish at room temperature. The fish filleting machine, PASFF-110, manufactured by Peruza, is designed for small-sized fish, such as sprats, anchovies, capelin and other fish in 9–14 cm [7,61]. It could make "butterfly" type or single fillets or cut off the belly as needed. The TOYO-167 from TOYO SUISAN KIKAI is designed for fillet-cutting small-sized fish, like sardine, horse mackerel, and mackerel pike.

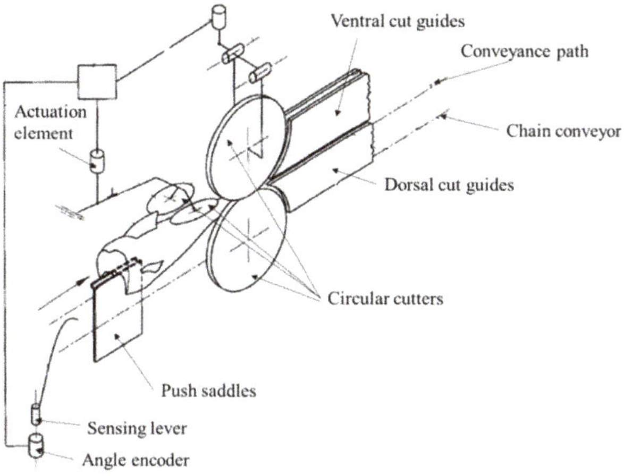

Figure 2. The representation of filleting machine for fish of Siluridae species (Braeger and Scherch, 2001).

Fish skinning machines are equipped with metal blades as well. Traditionally, the de-skinning processes could be conducted by gas flame (substation of NaOH and HCl) or steam [62]. In recent years, the mechanism of most fish skinners would press the fillets on the blades to remove the skin from the fillets [63]. A typical fish skinner consists of a feeding device, a dispatching device, a rotationally-driven separation device with a driver roller, and a main blade for removing the fish skin. The cutting gap between the main blade and the driver roller can be controlled by detecting fish fillet size [64,65]. To get a better quality of the fillets as value-added products, the deep-skinning process can remove the skin with a layer of subcutaneous fat [66,67]. Due to the relatively simple structure, some fish skinning machines are small and can be placed directly on the processing table. The well-known fish skinning machines are mainly from Baader, Cretel, Nock, and Trio [68]. Joensen and Olsen [69] compared the skinning machine Baader 51 and Trio FDS 105. The differences between these two skinning principles are the fillet feeding direction and skinning temperature. The fillet should be fed into the Baader 51 with the skin side down and Trio FDS 105 with the skin side up. In the Trio FDS 105, the skin would be frozen to a cold drum and then cut by the rotating blades. The Baader 51, like most skinners, used a drum to press the fillet to the blades. The results indicate that the fillet gapping increased after using FDS 105 compared with Baader 51. Arnþórsdóttir et al. [70] compared Baader 51 and Skaginn-skinner S3 machines. The super-chilled fillets using Skaginn S3 with Combined Blast and Contact (CBC) technique could have fewer gapping issues and fresher appearances. However, the fish skinning machines are also shown potential damage. Waterston and Holmes [71] studied the hand trauma caused by fish skinning machines. For small-sized fish, the ST600V and ST700V from STEEN can be implemented to remove small-sized fish skin.

To satisfy the specifications of a fish product, the fillets are trimmed and portioned. The trimming operations are aimed at removing the defects and unwanted regions of fish (bones, fin, belly fat area, etc.) and correcting the shape of the fillet [7]. In the automatic fish processing lines, machines are designed for trimming operations. Baader 988, a well-known high-speed auto-trimming machine in the industry, was studied by Ørnholt-Johansson et al. [72]. Based on the requirements, the trimming machine implements rotatory blades to cut the fillets according to pre-set demands. They found that the Baader 988 could achieve 88.0% ± 9.3% weight of fed salmon fillets. After that, fillets can be further portioned to create various products. Because the different parts of a fillet represent the different qualities of meat, the portioning operation is necessary for commercial benefit. The loin is thought to be the most valuable part of the fillet. Usually, the belly flap would fetch the lowest price. Commonly, the cod fillets are portioned in loins, centercuts, belly flaps, and tails. Mathiassen et al. [73] introduced machines in automatic fish processing lines. The Marel I-Cut line of products are metal knife-based portioning machines. The control system of them could change the cutting angle and speed based on the weight or dimensions of feeding non-frozen fillets. Slicing fish fillets is also included in some industrial production lines [7]. Thicker fish fillets can be further sliced with a horizontal slicing motion [74,75]. Whole fish can be vertically sliced or cut to produce fish steak with bone inside [76,77]. The measurements and estimation of weight and volume have been applied as a solution for slicing portions of whole fish and fish fillets [78,79]. Advanced slicer for producing a wide range of salmon products at high speed is designed to provide flexible multi-angle slicing motions [80,81]. The Baader 220 claims to have the capability to perform portion process on small-sized fish, like Herring, Herb herring, and Sardine in 17–30 cm.

2.2. Metal Blade Design and Operation Parameters

Based on the type of machine and process parameters, various types of planar cutting blades are used in the filleting machine. The blades were produced from different types of carbon and alloy tool steels and high-grade stainless steel. For some important components, cryogenically hardened stainless steel with a polished surface could be utilized [82]. Handling fish products is crucial as these foods are susceptible to numerous degradation factors. Hence, the utensils, materials, and equipment used in this process must be sharp and easily maintained or available. Thus, stainless steel (corrosion resistant), aluminum, approved plastic material, and galvanized steel equipment are easily sanitized and cleaned [83]. A sharp knife or blade is essential in fish cutting as it improves the cutting moment, lowers grip force, and shortens the cutting time.

Researchers studied the shape effect of the cutting edge and the sharpness of knives on the cutting forces [84]. The following knife quality indicators are available for inspection under production conditions: sharpening angle, uniformity of width, straightness of the cutting edge, depth of chipping, absence of cracks on the cutting edge and sharpness of knives. The effect of the blade sharpening angle on the parameters of this process was analyzed [85]. They evaluated resistance force with different sharpness half-angle (5°, 10°, 20°, 50°) of the inclined back edge at different cutting speeds. With increasing the sharpening angle at low cutting speed (0–0.8 mm/s) and high cutting speed (10–25 mm/s), the resistance forces both decreased. Ageev et al. [15,86] also proposed theoretical concepts and determined the optimal parameters that influence the cutting process of the blade configurations. The fish samples were processed using a wire knife with different sharpness (0.025–0.400 mm), at different temperatures (2–12 °C), at different cutting speeds (2, 4, 33, 62, 91, and 126 mm/s) and with various blade thickness (1–5 mm) to evaluate the quantitative dependences of fracture resistance and friction resistance forces on each operation parameters. The force of fracture resistance rises nonlinearly as the knife's sharpness and operational temperature increase. The force increases noticeably as the blade thickness increases. According to the measurement results, the total force of friction resistances reduces while fracture resistances increase as temperature and material elasticity increase. The reduction of undesirable energy consumption for cutting could be achieved by changing the geometry of the knife. This could be done by reducing the sharpening angle and the thickness of the knife, as well as by constructively introducing rear-inclined edges and eliminating the side edges [87]. The reduction of friction forces also could be facilitated by a decrease in the roughness of the edges, which was achieved by polishing and the use of anti-friction coatings [86,88]. Three different blades have been tested with different sliding angles and friction forces: bare blades (31.6°) and (35 μN), Ti-coated blades (20.3°) and (23.7 μN), and Z-TFMG (Zr-based thin film metallic glasses) coated blades (16.2°) and (19.2 μN). The results showed that the Teflon coating could reduce the cutting forces of an uncoated microtome blade by 80%, whereas the proposed Z-TFMG achieved a 51% reduction. Moreover, Z-TFMG was shown to protect blades during skin grafting surgery by providing a smooth surface morphology to reduce friction force and thereby improve blade sharpness. The finished fish product obtained by cutting must meet certain requirements for the accuracy of the shape, size, and smoothness of the cut. There is a compromise between cutting forces and flatness of the cutting surface that must be adopted in the design of the cutting tool. Besides the optimization of cutting conditions and geometry of the knives, the operation of cutting, and the fish product properties are also important factors contributing to reducing cutting forces and improving quality.

For a basic understanding of the cutting process as well as for modeling approaches, it is important to link operation conditions to undergoing cutting forces, including resistance and friction forces. Dowgiałło [89] proposed fundamental models for calculating the resistance forces that arise during fish cutting. The operating conditions are cutting with a flat knife with a thickness of b = 0.7 mm, which was sharpened on one side (β = 22°). And the length of the cutting line was a = 20 mm, and the speeds were applied according to the cut materials. The values of the cutting forces and pressure for sea and freshwater fish were determined for various regime parameters of processing with plate and circular knives. Ageev et al. [15,86] proposed a set of theoretical models for calculating contact pressures and resistance forces during fish cutting. A system of resistance forces that affect the knife during fish cutting was proposed, including the cutting speed (0.0–1.200 mm/s), at different rheological parameters (1.5–3.0 $\times$ 10^5 N·s^{-1}), the sharpness of the cutting tool (half-angles = 5°, 10°, 15°, 20°, 50°, 60°, 70°, 80° and 90°), the half-thickness of the knife (1.5, 2.0, 2.5, and 3.0 mm), and the elasticity of the fish muscle tissue (e= 3, 5, 7,11). The dimensional resistance force of the profile significantly depends on the rheological properties of the material. The sharpening angle and speed have very little influence on the resistance force that occurs during the cutting of the material; however, when cutting material in a low-viscous state, for example, upon defrosting and blanching, while with the increased blade thickness, the sharpening angle and speed have a noticeable influence on the value of the force in question. Jayraj et al. [90] also developed the relationship between friction and the inclination angles of fish on steel surfaces (20°–22°) and plastic surfaces (18°–20°) during feeding. By smoothing the interface between the fish's body and the surface, the slime the fish secretes decreases the frictional impact. As a result, the surface of the belt conveyors used to transport fish for unit activities in fish processing should be rough or have grips. This should prevent the fish from moving. Furthermore, they also recorded the power consumption during fish slicing, which is related to the speed of the blade and fish weight. The freshly harvested fish samples are required to accurately measure the slicing force and power requirement because the textural parameters of skin hardness, stiffness, and toughness would decay, and the force required to slice the fish would be reduced [91]. The skin hardness ranged between 86.911 and 95.656 N within five days of storage and thereafter reduced within the range from 48.714 to 65.920 N. The stiffness ranged between 3.1474 and 4.6340 N·mm^{-1} and toughness, 588.9–713.2 N·mm for five days. After five days of storage, the stiffness and toughness reduced in the range of 2.0030–2.8111 N·mm^{-1} and 415.0–526.3 N·mm, respectively.

2.3. Drawbacks and Limitations

We have visited several fish-cutting industries to learn about the challenges and drawbacks of current fish-cutting technologies. Current metal blade cutting has too much manual and repetitive work in several cutting steps. Some fish-cutting facilities have no gutting step because of the cost and space limitations. Also, some machinery companies mention that their equipment eliminates the need for gutting the fish before filleting and saves at least one operation labor by combining header gutter and filleting machines. However, the cutting performances of the filleting machines can be easily affected by the jamming issues of the cutting blades since some fish guts, and other byproducts are left inside the machines, as shown in Figure 3. Therefore, a certain number of labors are required to keep monitoring the cleanliness of the cutting blades and do maintenance work (Figure 4). Figure 5 presents another issue with the metal blade cutting: low-level fish fillet yield and high-level waste. Even though some machines are designed to recycle the mid-rib bones for producing the mince [92], increasing the fish fillet yield can provide more income opportunities than post-processing the mid-rib bones. The metal blade cutting setup, such as the cutter gap, is designed for a specified range of fish sizes. Consequently, it will cause a lower yield for the smaller fishes, and smashing damage the bones of the larger fish. Especially the smashing damage of the bones, a small piece of the bones left inside the fish fillet may be stuck in people's throats and lead to choking [93]. Fish larger than the size settings cannot be normally processed by machines, which will increase the waste. Another example of generating waste in the deheading machine is shown in Figure 6. The fish heads of the larger fish may not be fully cut off. The fish body is still partially connected with the fish head then the whole fish is dumped. Although some blades can be adjusted to fit different fish sizes, it is time-consuming and inefficient. Therefore, the current metal blade-based cutting technologies have an urgent need for innovation to increase fillet yield and reduce labor and turnover.

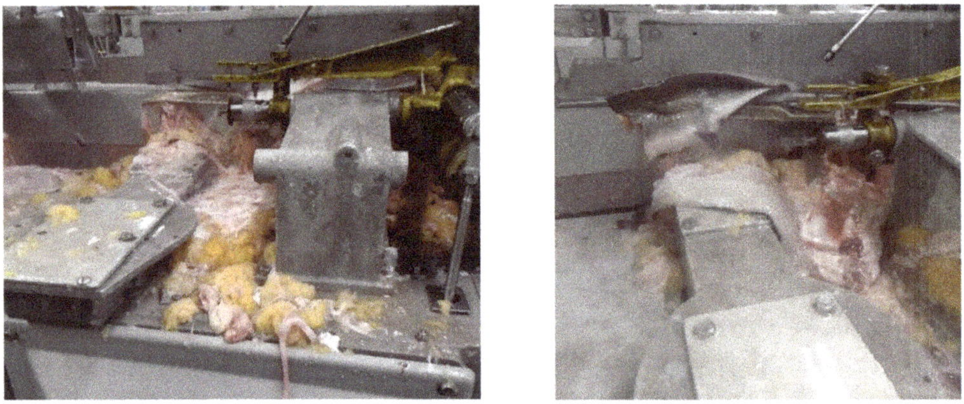

Figure 3. The filleting machine is frequently jammed because the fish guts are left in the machine. The normal filleting processing is affected by the jamming issue, which makes some fillets drop out of the machine as waste.

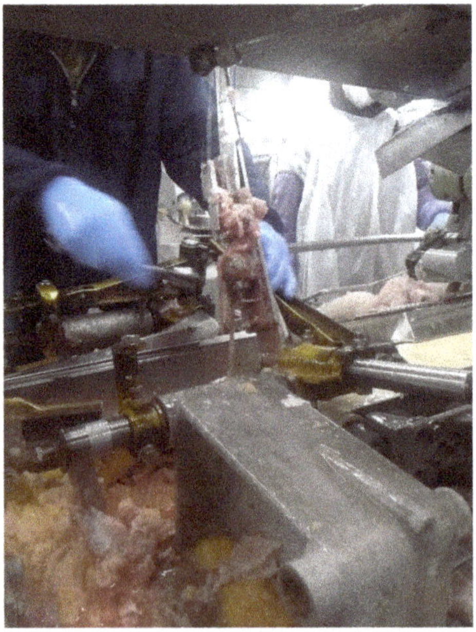

Figure 4. The manual cleaning process for the filleting machine.

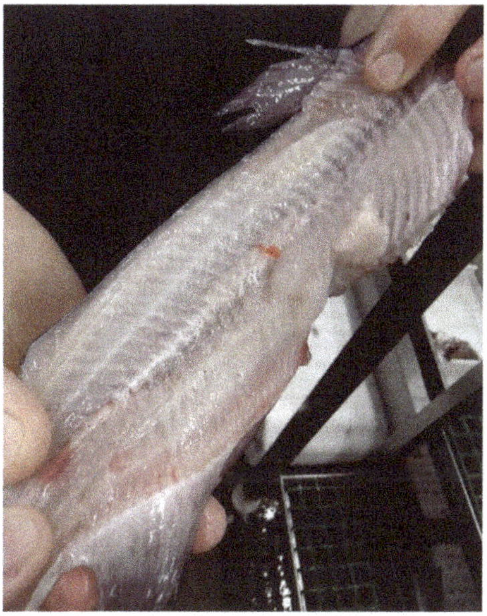

Figure 5. A fish rib bone from the filleting machine has a lot of fish meat left on it.

Figure 6. The waste due to deheading processes.

3. Waterjet Cutting

To revolutionize the catfish processing industry, it appears that a cost-effective non-conventional and non-contact cutting technology is needed with higher productivity, lower waste, and less labor-intensive requirement. The waterjet cutting technology can be an excellent alternative option to meet these expectations since it has many advantages over other traditional technologies. The excellent mobility and flexibility allow all directions cutting behaviors and complicated curved cutting lines [94]. Compared with laser cutting, there is no heat generated and no radiation zones, so lower cutting temperatures can guarantee the food's freshness and extend shelf life [95]. Besides, cross-contamination and bacterial transmission can be prevented because of no blades [96]. In addition, a large quantity of water is required in traditional fish processing factories to clean the fish product for every processing step [97]. Therefore, the waterjet cutting process can realize the integration of the processing and cleaning functions. Furthermore, the jamming issue described above will not happen, so the cutting performance would be much more constant and steadier, and the maintenance and inspection labor can be saved. The modern waterjet cutting machines typically consist of: (1) a control unit including a micro-computer, cameras, and supporting software, (2) a waterjet table that can withstand the strong impact of water flow during cutting assays, (3) a position traverse system which is controlled by the micro-computer for adjusting the related position of the nozzle and target workpieces, (4) a high-pressure pump for generating enough high pressure for the water, (5) a waterjet nozzle with certain shape and diameters of die for passing the high-pressure waterjets [98]. High-pressure waterjet has been applied to several fish-cutting steps: portioning, trimming, and scaling [27,99–101]. Due to the development of automatic technologies in the past two decades, high-performance robot arms have been largely employed in recent industries [8,102]. With the integration of robots, advanced waterjet cutting can conduct a highly precise cutting process with high speed, which favors the efficiency of fish fillet yield production [102]. However, the current waterjet cutting techniques are still facing certain issues. Unlike cutting other materials, which require precise size control, factors such as relatively large kerf width, blind cuts, and taper-long cutting lines are not considered the top concerns in fish cutting. One of the major obstacles that block the application of waterjet in fish cutting is its relative slow cutting speed and much higher starting and maintenance costs compared to blade cutting due to the system's complexity. In addition, waterjet cutting generates wastewater containing minced fish and bone. Disposal equipment and related specialized training will then furtherly increase the cost of waterjet fish cutting.

The design and development of waterjet cutting equipment for fish cutting aim to gain faster and cheaper processing, accurate workpieces identification, and automatic cutting strategy generation and operation. The performance of meat cutting can be evaluated in different grades with the fish fillet recovery rate, the damage of fish bones, the amount of saw mince, the cleanness of skin cut, the cleanness of muscle cut, and the cut-through of connective tissue [27,82,103]. To achieve the best cutting performance, the components of waterjet cutting should be well-designed. It is believed that factors majorly, including the water pressure and flow rate, transverse speed, nozzle shape and diameter, the nozzle stand-off distance, nozzle tilting angle, number of passes, and fish temperature, affect the overall waterjet cutting quality. The supplying water pressure is considered the most important parameter governing the waterjet cutting quality [104,105]. The water pressure varies according to the strength of the targeting materials and exhibits a different order magnitude [106]. By focusing on the rainbow trout, and fixing the stand-off distance and incidence angle, Kasperowicz et al. [32] have investigated the effect of the waterjet pressure, nozzle moving speed, nozzle size, and nozzle geometry on the cutting performance. It turns out that, for good cutting quality, the supply water pressure changes with the size of carcasses and cutting sites. For a full cut of the skinless lobe, the water pressure should reach 3.5 MPa or 35 atm. Within the selected nozzle size (0.175 mm–0.95 mm) and nozzle traverse speed (0–50 mm/s), no obvious effect on cutting quality was observed. The complexity of setting cutting parameters for specific fish meat is reflected in this study. This implies the fact that the fish waterjet cutting may also follow the principles summarized during the cutting assays of other workpieces. Research has also found that with a lower sample traverse speed, a smoother and deeper cutting cross-section was obtained [107]. This is important for obtaining high-quality cutting while increasing the raw fish meat yield. The quality of waterjet cutting is also influenced by the temperature of the fish meat [101]. The systematic study of cutting parameters for specific targeting materials in the food industry can be a hot topic for future research on waterjet cutting. Moreover, the fluid mechanical properties of the waterjet also affect the cutting performance and should be considered during the parameters design.

Although commercial waterjet cutting instruments have been applied to fish cutting to a certain extent, the fundamental research of waterjet cutting in different fish under different parameters still lacks and deserves more attention. The quantitative reports in the literature about the effect of these factors on fish cutting quality are very limited. One of the reasons is that the experimental data usually are trade secrets. On the other hand, it is difficult to obtain precise and universal experiment results because (1) the fish meat from the same species may have different properties according to the place and season [108]. (2) The structure and properties of fish meat may differ even within the same species [109]. Hence, reliable and universal results require extensive experiments. Nevertheless, one can still get insight into the effect of the design parameters on cutting performance by limiting the study variables.

4. Machine Vision and Artificial Intelligence for Fish Cutting

Fish cutting heavily relies on manual labor, and the working environment is usually cold and wet to ensure the freshness of the fish. In addition, many procedures involved in fish cutting are tedious, repetitive, and unsafe. Therefore, increasing automation levels of fish cutting have gained tremendous attention in recent years [110]. With the advancements in imaging and computing technologies, machine vision and AI have been applied to the fish industry for improved precision and automation levels of fish cutting.

Machine vision enables inspecting objects or scenes objectively and efficiently through image acquisition and analysis (e.g., object detection/localization, semantic/instance segmentation) to assist in automating various tasks (e.g., fish species sorting, cutting), which otherwise would have been done manually. Machine vision has been widely used for postharvest product inspection in food industries, including fish processing, such as fish morphology identification (e.g., size, volume, weight, and shape estimation) [111,112], species recognition [113,114], physical or chemical properties [8,18], and quality and damage inspection [115]. In general, machine vision relies on the acquisition, processing, and modeling of two-dimensional (2-D) and three-dimensional (3-D) images to assist in operations in fish processing.

There is considerable research on using 2D images in fish processing, where acquired color images are segmented for interested regions based on the fish/fillet color [22,116–118]. Several commercial fish cutting systems, such as Baader 988 [118] and Marel SensorX [119], used CCD cameras in the control systems of blades. For fish deheading machines (Section 2.1), the cutting position was decided based on the gill area of the fish. Accurate localization of fish gill regions would be hence critical in the cutter controller [13]. Jain et al. [22] used the machine vision system in a deheading machine. The authors fused the data from CCD cameras, optical encoders, and ultrasonic displacement to detect the collar bone position and adjust the blades accordingly. Such a machine vision system can potentially improve the accuracy of gill recognition and thus reduce protein waste during fish cutting. Sivertsen et al. [118] applied a CCD camera to scan cod fillets using the ridge detection algorithm, where the centerline of fillets can be identified with an average accuracy of 1 mm from the tail.

Since regular color cameras are limited to providing information in a 2D plane, 3D imaging systems are useful in fish processing lines by obtaining the depth information of target fish [120–122]. 3D imaging can be realized using techniques such as structured light, stereovision, or time-of-flight (ToF). The laser-scanning profilometry (line light-based 3D imaging), which only needs to capture the reflection of laser pulses and is not readily influenced by ambient light, can potentially achieve high measurement accuracy. This method can be employed in commercial fish-cutting systems, such as the Marel I-Cut series (Marel, 2022a), to empower a trimming robot with 6 degrees of freedom (DoF) for 3D cutting [122]. Bondø et al. [121] implemented a laser-based 3D vision system to obtain the point cloud data of fish with the accuracy of 1 mm, which was analyzed to detect the position of fish gill that can be used to guide a robot arm to cut along the gill arch.

In recent years, AI has been used to empower machine vision systems for enhanced fish cutting. AI through machine learning (ML), especially deep learning (DL), has demonstrated great remarkable performance in visual recognition tasks (e.g., object detection and segmentation). AI methods have been used in fish recognition to improve the blade-cutting accuracy of automated production lines [19,116,123–125]. They could be further implemented in fish deheading, trimming, and portioning operations. Gamage et al. [19] conducted pioneering work to apply ML in fish-cutting tasks. Fish head images were acquired with a CCD camera and processed to obtain edge-enhanced, 2D Gaussian-smoothed images. Features were then computed from enhanced images and fed into multiple regression algorithms to estimate the point of interest on the fish head. Odone et al. [126] proposed a machine vision-based fish grading system. This system performed fish shape measurement and used support vector machine (SVM) models to learn the relation between fish weight and shape parameters, enabling the fish grading at a rate of 3 items per second.

DL methods have been recently used in machine vision systems for fish processing. Xu and Sun [127] used convolutional neural networks (CNNs) to detect the salmon muscle gaps to reduce irregular voids or undesirable lace-like appearance in the final product. Taheri-Garavand et al. [128] constructed a CNN classifier to classify a caught fish into fresh and non-fresh with an overall accuracy of 98.21%. Laradji et al. [129] proposed a CNN-based point-level fish segmentation model achieving an accuracy of 87.9%. Diamond et al. [37] used U-Net to perform the segmentation of different parts of the fish on 2D images. The dataset of fish images used in the study was made open-source.

These studies demonstrate the promise of machine vision and AI as valuable tools for automated fish processing. Their full potential in fish cutting, however, remains to be investigated, given only a handful of relevant publications. A key question remains to be answered as to how satisfactorily machine vision and AI-driven fish cutting automated machines will perform in commercial production settings. Although more validation experiments, especially with real-time machine vision prototyping, are urgently needed, it is generally agreed that assistance from machine vision and AI will substantially automate fish-cutting operations and reduce labor costs.

Further, there is still a great deal of machine vision and AI that is worth investigating and can be potentially applied in fish cutting and whole fish industries. Studies [130,131] have shown that AI methods can be used for the optimization of cutting parameters of fish cutting machines. For cutting fault diagnosis, the precise location of fault points can be obtained by supervised pattern classifiers [132–134]. Moreover, vision- and speech-based multimodal human-machine interaction systems can be applied to fish-cutting machines to provide workers with a better operating experience [135]. Machine vision and AI algorithms can power waterjet cutting or metal blade cutting for improved fish cutting productivity. For instance, Lin et al. [136] proposed a machine vision and image recognition guided waterjet knife strawberry berry calyx removal machine. Similar approaches based on machine vision and AI can be applied to finer fish cutting operations. In addition, AI technology can also be used to create a reliable and intelligent metal blade wear detection system to mitigate the negative effects of the fish-cutting processes [137].

5. Future Perspectives

Major manufacturers of fish-cutting devices around the world have been summarized in Table 1. Potential directions for future research on fish-cutting techniques are presented as follows:

(1) Some fish species deserve more attention, such as catfish. Catfish was ranked 8th in the top 10 seafood of U.S. per capita consumption in 2019 [138]. Mississippi and Alabama were ranked first and second in catfish farming and production in 2021, respectively [139]. However, compared with 300 million kilograms in 2003, the U.S. catfish industry's yield decreased to 136.5 million kilograms in 2014 because of international market competition and increasing feed costs [140]. Furthermore, compared with other fish species, catfish have just a few rib bones, so the catfish belly flap portion can be trimmed and sold as fish nuggets [141]. Therefore, developing special cutting machines or trimming and portioning methods for catfish can keep catfish farming and fillet processing industries (with a total impact of more than one billion dollars) as an important agricultural and food industry, which their existence is vital to the success of the US rural economy.

(2) Waterjet cutting should be further developed for other fish-cutting steps. Although the waterjet cutting works well for fish fillet portioning and trimming steps, there is no waterjet cutting application for other fish cutting steps, such as filleting and deheading. Also, the small-scale fish cutting processes are all performed by a metal blade system, and no waterjet cutting machine is designed or modified for small-scale fish. The cutting parameters can be adjusted to reach a critical condition: the water stream will cut along the edge of the fish meat and touch the fish bones but cut the meat only, which means that the water cutting force is adjusted to be just good for cutting meat but not strong enough for cutting bones. Under critical conditions, waterjet cutting can realize the maximum fish fillet yield. Besides, there is no research on abrasive waterjet fish cutting. It can be a better solution for further improving the fish-cutting quality and efficiency in deheading and filleting processing steps. The abrasives, including salt, sugar, ice, and starch particles, can be added to the high-pressure waterjet, which can reduce the surface roughness, favor energy saving, and lower the overall costs of equipment maintenance [33,34,142,143].

(3) Other novel cutting techniques should be developed for the fish-cutting process. Another two common food cutting techniques, ultrasonic cutting and laser cutting, are currently not good for fish processing. Even though some companies have designed ultrasonic cutting for slicing frozen fish [144], regular raw fresh fish is too soft, which cannot promise cutting quality and efficiency. Laser cutting will have heat-affected zones, which reduces the shelf life and quality of the fish meat. As potential directions for future research, ultrasonic cutting and laser cutting techniques can be adapted into fish cutting. Other novel direct-contact and non-contact cutting techniques can also be developed for fish-cutting innovations.

(4) Fully automated cutting production lines and integrated control systems should be designed by advancing singulation, machine vision, and AI technologies. Current automation cutting machines need manual singulation and placement processes for fish supply. Novel singulation methods can help realize fully automated systems to further reduce the total cutting and processing time, labor, and turnover. Considering the uniqueness and characteristic of each fish, the machine vision system in fish cutting lines could still be improved in both hardware and software aspects. As for the hardware system, the machine vision system should be able to capture the depth information. Therefore, the 3D sensing system, such as a laser profilometer or stereo cameras, could provide more comprehensive detail to guide the fish-cutting process. Meanwhile, due to the high computation requirement of 3D information, the data processing efficiency should be improved to satisfy the high-speed requirement of fish cutting lines. As for software development, first, key-point detection is widely used in machine vision applications and has made great progress in the past few

years, where it can be applied to detect critical points in fish [145], such as gill [13] and mouth [146,147]. If the key points on the fish can be correctly located, this can facilitate the machine to cut the fish more accurately. Second, training DL models with large parameters is another major trend in machine vision and AI for visual pattern recognition tasks. Many studies have demonstrated that DL models better learn the representations of images. Therefore, visual transformer [148] and self-supervised learning [149,150] are two other techniques that can be used to extract meaning information from fish images. A better image feature extractor helps the system to analyze fish images and optimize fish-cutting operations.

(5) More research and simulation development should be devoted to quantifying and predicting fish cutting quality and efficiency. Especially for waterjet cutting, the scientific reports only define the cutting quality and efficiency for metals, woods, composite materials, and other food species. Except for the cutting grade definition, there is no specified parameter quantifying fish cutting quality and efficiency. Some parameters, such as surface roughness, need to be further investigated and adapted to evaluate the fish-cutting quality. As for simulation development, current simulation models for cutting processes and quality are mainly based on regression analysis, response surface methodology (RSM), computational fluid dynamics (CFD), and finite element method (FEM) [151–154]. Since the waterjet cutting method has great potential for fish cutting, future fish cutting simulation and theory development should pay more attention to pure and abrasive waterjet cutting processes of whole fish and fish fillets with higher accuracy and robustness.

Table 1. Major manufacturers of fish-cutting devices around the world.

Manufacturers	Cutting Operations	Fish Species	Product Advantages	Country	Reference
Arenco	Deheading, Filleting, Skinning	Pelagics	• Stainless steel fish pockets carry the fish to a precision measuring device that customizes the position of the head cut and tail cut for each individual fish, achieving a high yield; • The skinning machine works with around 300 fish per minute.	Sweden	[59,60]
BAADER	Deheading, Filleting, Portioning, Skinning, Trimming	Salmonids, Tuna, White fish, Pelagics, Aquaculture	• Committing to 100% fish utilization and zero waste; • Sophisticated automated cutting systems for quality fish products; • Smart inspection systems and software provide meaningful insights and detailed reports for improved production control.	Germany	[50,53–55,69,70]
Cabinplant	Skinning	Pelagics	• Thermal and mechanical treatments for skinning; • Only 5% product loss; • Limited amount of wastewater and no disposal of hazardous fluids.	Denmark	[7]
Cretel	Skinning	White fish, Pelagics, Aquaculture	Easy thickness control for silver, regular and deep skin.	Belgium	[71]
JBT	Portioning, Trimming	Salmonids, Tuna, White fish, Aquaculture	Computerized scanning of every piece coming into the machine, and sophisticated programming that controls waterjet portioners, horizontal slicers, 3D portioners, X-ray guided solutions, and systems with multiple cutting heads for portioning, slicing, stripping, and dicing.	USA	[73,122]
Kaj Olesen	Portioning, Skinning	Salmonids, White fish, Aquaculture	• The slicing machine has a capacity of up to 250 slices per minute; • The thickness of the slices can be adjusted from 1 mm up to 5 mm, and can slice fillets from the smallest to the largest sizes; • Without the use of any tools, the salmon slicer can be transformed into a tuna slicer—only by changing the "tower" of the machine.	Denmark	[10]
Kroma	Deheading, Filleting	Salmonids, Pelagics, Aquaculture	• The fish is turned during head cutting to obtain the U cut; • The workplaces for the operators are designed with maximum consideration for ergonomics.	Denmark	[146,147]

Table 1. *Cont.*

Manufacturers	Cutting Operations	Fish Species	Product Advantages	Country	Reference
Marel	Deheading, Filleting, Portioning, Skinning, Trimming	Salmonids, Tuna, White fish, Aquaculture	• Filleting and trimming lines streamline processing with less product handling and continuous data collection. The fish is weighed, graded, cut, and packed in a continuous flow, increasing capacity, throughput and labor efficiency; • Have both metal blade cutting and waterjet cutting technologies to create an endless variety of value-added products, from fixed-sized strips, dices, and splits to high-value portions of fixed weight and length.	Iceland	[27,51,80,101]
MARELEC Food Technologies	Portioning, Trimming	Salmonids, White fish	• Have both metal blade cutting and waterjet cutting technologies. Max Waterjet pressure 600 MPa; • High accuracy with machine vision technologies; • Flexible cutting algorithms for portioning.	Belgium	[73,78,79]
Nikko	Deheading, Filleting, Portioning	Salmonids, White fish, Pelagics, Aquaculture	• The filleting machine can do both center cutting and regular filleting without having to re-arrange the line; • The "Kamaless Header" cuts the fish head with the collar intact to increase yield.	Japan	[9]
NOCK Maschinenbau	Skinning	Salmonids, White fish, Aquaculture	CBF and SB machine models are suitable for thin, pulling-off the skin, and deep skinning conditions.	Germany	[64,65]
PERUZA	Deheading, Filleting	Pelagics	Laser fish measurement system to realize the maximum yield.	Latvia	[46,61]
Pisces Fish Machinery	Deheading, Filleting	Salmonids, White fish, Pelagics, Aquaculture	• A complete range of filleting systems can process fish from 10 g to 10 kg; • Incorporating automatic positioning of each individual fish to ensure the head cut is in the optimum position for maximum yield; • Some machines allow for head-on gutted fish to be processed into fillets with one operator.	USA	[47,48]
ROSOMA	Deheading, Portioning	Salmonids, White fish, Pelagics, Aquaculture	• On the basis of a fish typical of deheading, the head cutting angle desired is set up by adjusting the tappets among one another; • For pelagic fish, the machine can do beheading and subsequent gutting by means of a vacuum. A suction extractor consisting of a vacuum pump and a separator forms part of the vacuum nobbing.	Germany	[25]
Ryco Equipment	Deheading, Filleting, Portioning	Salmonids, White fish, Aquaculture	Collar-on head cutting with maximum recovery to process fish from 1 to 50 pounds.	USA	[56–58]

Table 1. *Cont.*

Manufacturers	Cutting Operations	Fish Species	Product Advantages	Country	Reference
Salmco	Portioning	Salmonids	The product range offers everything from simple hand slicers to semi- and fully automatic cold slicers and various fresh slicers with a variable 0° to 90° cutting angle.	Germany	[81]
SEAC AB	Deheading, Filleting	Pelagics	• Machines can process fish down to 70–100 pieces/kg at speeds up to 320 fish/min; • The accuracy of processing is up to 98%.	Sweden	[46]
Skaginn 3X	Skinning	White fish, Pelagics, Aquaculture	Skinning after SUB-CHILLING™ preserves the texture of the fillets for better quality and texture and lower yield loss.	Iceland	[70]
STEEN	Deheading, Skinning, Trimming	Salmonids, Tuna, White fish, Pelagics, Aquaculture	Skinning and de-frilling in two perfectly synchronized units.	Belgium	[82]
Tecnotrans UG	Portioning, Skinning	Salmonids, Tuna, White fish, Aquaculture	Both fresh and frozen goods can be processed to cut weight-accurate and uniform slices & cubes.	Germany	[6]
TOYO SUISAN KIKAI	Deheading, Filleting, Portioning, Skinning, Trimming	Salmonids, Tuna, White fish, Pelagics	Cover all the cutting steps and a wide range of fish species.	Japan	[11,92]
Uni-Food Technic	Deheading, Filleting, Portioning, Skinning	Salmonids, White fish, Pelagics	• Produce intelligent automatic fish machinery that reduces the need for labor costs; • Complete filleting lines with de-heading machines, de-sliming machines, de-scaling machines, trimming lines with yield control, skinning machines, and pin-bone removers.	Denmark	N/A
VARLET	Portioning, Skinning	Salmonids, White fish	• Slicing offers a constant thickness adjustment from 5 to 10mm with a dismountable spiked pusher; • Skinning has a single or double track with a removable conveyor.	France	N/A
Velfag	Deheading, Filleting, Skinning	White fish	• Deheading is designed to both head whole fish and take the collarbone off pre-headed fish; • Filleting has the capability of handling large white fish (20 kg+); • The combined skinning machines attached directly to the filleting machines allow the fillets to be skinned directly, with no extra employees needed between filleting and skinning.	Iceland	[115]

6. Conclusions

The fisheries and fish farming industries today operate in a large and global market. In such an increasingly competitive marketplace, it becomes necessary for the fisheries and fish farming industries to explore advanced technological solutions for improving productivity and profitability. As the key part of the fish processing industry, efficient cutting of fish is important given that top quality, maximum yield, and highest possible profits are to be pursued keenly. This paper evaluates and summarizes the current fish cutting techniques and highlights the research on cutting mechanisms, blade design, and performance effect factors. The applications of machine vision and AI are reviewed and discussed. With the proposed future perspectives, it could be possible to bridge the knowledge gap on cutting operations and achieve better cutting efficiency and quality.

Author Contributions: Conceptualization, W.L., Y.L., and X.Z.; methodology, W.L., J.L., and D.W.; formal analysis, W.L., J.L., and D.W.; investigation, W.L., J.L., D.W., Y.C., and Q.M.; data curation, W.L., J.L., and D.W.; writing—original draft preparation, W.L., J.L., D.W., Y.C., and Q.M.; writing—review and editing, W.L., Y.L., and X.Z.; supervision, W.L., Y.L., and X.Z. All authors have read and agreed to the published version of the manuscript.

Funding: This research received no external funding.

Institutional Review Board Statement: Not applicable.

Informed Consent Statement: Not applicable.

Data Availability Statement: The data presented in this study are available on request from the corresponding author.

Conflicts of Interest: The authors declare no conflict of interest.

References

1. Bjørndal, T.; Brasão, A.; Ramos, J.; Tusvik, A. Fish processing in Portugal: An industry in expansion. *Mar. Policy* **2016**, *72*, 94–106. [CrossRef]
2. Jayathilakan, K.; Sultana, K.; Radhakrishna, K.; Bawa, A.S. Utilization of byproducts and waste materials from meat, poultry and fish processing industries: A review. *J. Food Sci. Technol.* **2012**, *49*, 278–293. [CrossRef]
3. Jeebhay, M.F.; Robins, T.G.; Lopata, A.L. World at work: Fish processing workers. *Occup. Environ. Med.* **2004**, *61*, 471–474. [CrossRef] [PubMed]
4. FAO. *The State of World Fisheries and Aquaculture 2022*; Towards Blue Transformation; FAO: Rome, Italy, 2022.
5. Adeyeye, S.A.O. Traditional fish processing in Nigeria: A critical review. *Nutr. Food Sci.* **2016**, *46*, 321–335. [CrossRef]
6. Aldás Guerrero, R.F. Diseño de un Negocio Dedicado a la Exportación de Filete de Tilapia en Camas Frías al Mercado Canadiense Período 2014–2018. Bachelor's Thesis, UIDE, Quito, The Republic of Ecuador, 2013.
7. Buljo, J.; Gjerstad, T. Robotics and automation in seafood processing. In *Robotics and Automation in the Food Industry*; Woodhead Publishing: Sawston, UK, 2013; pp. 354–384.
8. Einarsdóttir, H.; Guðmundsson, B.; Ómarsson, V. Automation in the fish industry. *Anim. Front.* **2022**, *12*, 32–39. [CrossRef] [PubMed]
9. Nagaishi, H.; Inada, T.; Yoshioka, T.; Sato, A. Development of a compact, onboard slurry icemaker to rapidly produce optimal ice for maintaining freshness of marine products. *Synth. Engl. Ed.* **2017**, *10*, 1–10.
10. Thordarson, G.; Karlsdottir, M.; Pedersen, R.; Johannsson, M.; Hognason, A. *Sub-Chilling of Salmon*; Ísafjordur: Matis, France, 2015.
11. Tsukagoshi, T.; Uchita, T. Toyo Suisan Kikai Co Ltd, 2019. Method for Removing Guts of Fish Body and Device for Same. U.S. Patent 10,470,472, 12 November.
12. Shirai, K.; Ramirez-Ramirez, J.C. Utilization of Fish Processing By-products for Bioactive Compounds. In *Fish Processing: Sustainability and New Opportunities*; Wiley-Blackwell: Hoboken, NJ, USA, 2010; pp. 236–265. [CrossRef]
13. de Silva, C.; Wickramarachchi, N. An innovative machine for automated cutting of fish. *IEEE/ASME Trans. Mechatron.* **1997**, *2*, 86–98. [CrossRef]
14. Lang, H.; Wang, Y.; de Silva, C.W. An automated industrial fish cutting machine: Control, fault diagnosis and remote monitoring. In Proceedings of the 2008 IEEE International Conference on Automation and Logistics, Qingdao, China, 1–3 September 2018; IEEE: Piscataway, NJ, USA, 2008; pp. 775–780.
15. Ageev, O.V.; Dowgiałło, A.; Sterczyńska, M.; Piepiórka-Stepuk, J.; Giurgiulescu, L.; Janowicz, M.; Jakubowski, M. Experimental characterization and theoretical modeling of fracture and friction resistance forces during tuna cutting. *J. Food Eng.* **2021**, *307*, 110648. [CrossRef]

16. Ashwinkumar, N.; Bhuvaneshkumar, S.; Adithya, K. Development and Study of Universal Fish Cutting Apparatus. *Int. J. Res. Eng. Sci. Manag.* **2021**, *4*, 306–308.
17. Kamaruzzaman, K.A.; Mahfurdz, A.; Hashim, M.; Bidin, M.N. Design and Performance Evaluation of Semi-Automatic Fish Cutting Machine for Industry. In *IOP Conference Series: Materials Science and Engineering*; IOP Publishing: Bristol, UK, 2020; Volume 864, p. 012112.
18. Dowlati, M.; de la Guardia, M.; Mohtasebi, S.S. Application of machine-vision techniques to fish-quality assessment. *TrAC Trends Anal. Chem.* **2012**, *40*, 168–179. [CrossRef]
19. Gamage, L.B.; De Silva, C.W.; Gosine, R.G. Statistical pattern recognition for cutter positioning in automated fish processing. In Proceedings of the IEEE Pacific Rim Conference on Communications Computers and Signal Processing, Victoria, BC, Canada, 19–21 May 1993; IEEE: Piscataway, NJ, USA, 1993; Volume 2, pp. 786–789.
20. Arvanitoyannis, I.S.; Kassaveti, A. Fish industry waste: Treatments, environmental impacts, current and potential uses. *Int. J. Food Sci. Technol.* **2008**, *43*, 726–745. [CrossRef]
21. Goossens, Y.; Schmidt, T.G.; Kuntscher, M. Evaluation of Food Waste Prevention Measures—The Use of Fish Products in the Food Service Sector. *Sustainability* **2020**, *12*, 6613. [CrossRef]
22. Jain, A.; De Silva, C.W.; Wu, Q.M.J. Intelligent fusion of sensor data for product quality assessment in a fish cutting machine. In Proceedings of the Joint 9th IFSA World Congress and 20th NAFIPS International Conference (Cat. No. 01TH8569), Vancouver, BC, Canada, 5–28 July 2001; IEEE: Piscataway, NJ, USA, 2001; Volume 1, pp. 316–321.
23. Atkins, A.; Xu, X. Slicing of soft flexible solids with industrial applications. *Int. J. Mech. Sci.* **2005**, *47*, 479–492. [CrossRef]
24. Liu, S.; Wang, H.; Cai, Y. Research on Fish Slicing Method Based on Simulated Annealing Algorithm. *Appl. Sci.* **2021**, *11*, 6503. [CrossRef]
25. A Saltykov, M.; I Tkachenko, T. Multidimensional Classification for Systematization of Fish Processing Equipment. *IOP Conf. Series Earth Environ. Sci.* **2021**, *666*, 022087. [CrossRef]
26. Tomaszewska-Gras, J. Rapid quantitative determination of butter adulteration with palm oil using the DSC technique. *Food Control* **2016**, *60*, 629–635. [CrossRef]
27. Franklínsdóttir, H. Application of Water Jet Cutting in Processing of Cod and Salmon Fillets. Master's Thesis, University of Iceland, Reykjavik, Iceland, 2014.
28. Schreuders, F.K.G.; Sagis, L.M.C.; Bodnár, I.; Boom, R.M.; van der Goot, A.J. Non-linear rheology reveals the importance of elasticity in meat and meat analogues. *Sci. Rep.* **2022**, *12*, 1334. [CrossRef]
29. Bogue, R. Cutting robots: A review of technologies and applications. *Ind. Robot. Int. J. Robot. Res. Appl.* **2008**, *35*, 390–396. [CrossRef]
30. Khodabandehloo, K. Achieving robotic meat cutting. *Anim. Front.* **2022**, *12*, 7–17. [CrossRef] [PubMed]
31. Carreño-Olejua, R.; Hofacker, W.C.; Hensel, O. High-Pressure Water-Jet Technology as a Method of Improving the Quality of Post-Harvest Processing. *Food Bioprocess Technol.* **2010**, *3*, 853–860. [CrossRef]
32. Kasperowicz, M.B.; Chomka, G.P.; Bil, T. Determination of Supply Pressure during Cutting Fish Using High-Pressure Water Stream Taking into Account the Cutting Place and Diameter of the Water Nozzle. *Int. J. Food Eng.* **2019**, *16*. [CrossRef]
33. McGeough, J. Cutting of Food Products by Ice-particles in a Water-jet. *Procedia CIRP* **2016**, *42*, 863–865. [CrossRef]
34. Wang, J.; Shanmugam, D. Cutting meat with bone using an ultrahigh pressure abrasive waterjet. *Meat Sci.* **2009**, *81*, 671–677. [CrossRef] [PubMed]
35. Pogrebnyak, A.; Pogrebnyak, V. Mechanism of the High Efficiency of the Cutting Frozen Food Products Using Water-Jet with Polymer Additions. *Food Sci. Technol.* **1567**, *11*, 73–78. [CrossRef]
36. Hao, M.; Yu, H.; Li, D. The measurement of fish size by machine vision—A review. In Proceedings of the International Conference on Computer and Computing Technologies in Agriculture, Beijing, China, 27–30 September 2015; Springer: Cham, Switzerland, 2015; pp. 15–32.
37. Diamond, K.M.; Avants, B.B.; Maga, A.M. Machine learning-based segmentation and landmarking of 2D fish images. In *Integrative and Comparative Biology*; Oxford University Press Inc.: Cary, NC, USA, 2021; Volume 61, pp. E1100–E1101.
38. Xu, W.; Wang, J.; Deng, Y.; Li, J.; Yan, T.; Zhao, S.; Yang, X.; Xu, E.; Wang, W.; Liu, D. Advanced cutting techniques for solid food: Mechanisms, applications, modeling approaches, and future perspectives. *Compr. Rev. Food Sci. Food Saf.* **2022**, *21*, 1568–1597. [CrossRef]
39. Bland, J.M.; Bett-Garber, K.L.; Li, C.H.; Brashear, S.S.; Lea, J.M.; Bechtel, P.J. Comparison of sensory and instrumental methods for the analysis of texture of cooked individually quick frozen and fresh-frozen catfish fillets. *Food Sci. Nutr.* **2018**, *6*, 1692–1705. [CrossRef]
40. Buckingham, R.; Graham, A.; Arnarson, H.; Snaeland, P.; Davey, P. Robotics for de-heading fish—A case study. *Ind. Robot. Int. J.* **2001**, *28*, 302–309. [CrossRef]
41. Ketels, D. Apparatus for Positioning Fish for Heading. U.S. Patent 7,467,995, 23 December 2008.
42. Sampels, S. The effects of processing technologies and preparation on the final quality of fish products. *Trends Food Sci. Technol.* **2015**, *44*, 131–146. [CrossRef]
43. Dowgiallo, A. The effect of cutting and fish-orientation systems on the deheading yield of carp. *Int. J. Food Sci. Technol.* **2008**, *43*, 1688–1692. [CrossRef]

44. Tomczak-Wandzel, R.; Vik, E.A.; Wandzel, T. *BAT in Fish Processing Industry: Nordic Perspective*; Nordic Council of Ministers: Copenhagen, Denmark, 2015.
45. Martín Rodríguez, F.; Barral Martínez, M. Automatic turbot fish cutting using machine vision. *Instrum. Viewp.* **2015**, *18*, 24–25.
46. Sharapov, S. Compact Design of Fish Processing Equipment and Implementation of Lean Tools. Master's Thesis, KTH Royal Institute of Technology, Stockholm, Sweden, 2013.
47. Kaufman, D.; Fisher, R.A.; Wanchese Fish Company. Feasibility Study for Machine Processing Croakers into Fillets and for Forming the Fillets into Larger Portions. Fishery Resource Grant FRG 1999 - 24. Virginia Institute of Marine Science, William & Mary. Available online: https://scholarworks.wm.edu/reports/2216 (accessed on 7 September 2022).
48. Wastell, T.T. Pisces Fish Machinery Inc. Fish Filleting Machine. U.S. Patent 11,140,907, 12 October 2021.
49. Rora, A.M.B.; Mørkøre, T.; Einen, O. Primary processing (evisceration and filleting). In *Farmed Fish Quality*; Kestin, S.C., Warriss, P.D., Eds.; Blackwell Science: Oxford, UK, 2001; pp. 249–260.
50. Braeger, H.; Scherch, R.P. Baader North America Corp. Process for Fileting Fish and Machine for Performing This Process. U.S. Patent 6,200,211, 13 March 2001.
51. Jacobsen, P.H.; Jakobsen, B.K. Marel Salmon, A.S. Fish filleting machine. U.S. Patent 8,715,045, 6 May 2014.
52. Jakobsen, B.; Jacobsen, P.H. Carnitech, A.S. Fish Filleting Machine. U.S. Patent 6,994,617, 7 February 2006.
53. Kowalski, W. Nordischer Maschinenbau Rud Baader GmbH; Co, K.G. Method for Removing Blood Released during Filleting from the Backbone of Fish, and Device for Removing Such Blood. U.S. Patent 8,956,205, 17 February 2015.
54. Kowalski, W. Nordischer Maschinenbau Rud Baader GmbH; Co, K.G. Method for Mechanically Removing Pin Bones from Fillet Parts of Conveyed Fish and Device for Performing Said Method. U.S. Patent 9,357,789, 7 June 2016.
55. Jürs, M.; Schroeder, M. Nordischer Maschinenbau Rud Baader GmbH; Co, K.G. Apparatus and Method for Filleting Beheaded and Eviscerated Fish. U.S. Patent 8,814,637, 26 August 2014.
56. Ryan, R.M. RYCO EQUIPMENT Inc. Fish Processing System and Method. U.S. Patent 8,512,106, 20 August 2013.
57. Ryan, R.M. RYCO EQUIPMENT Inc. Fish Processing System and Method. U.S. Patent 8,834,238, 16 September 2014.
58. Ryan, R.M. RYCO EQUIPMENT Inc. Fish Processing Systems and Methods. U.S. Patent 9,839,223, 12 December 2017.
59. Sone, I.; Sveinsdóttir, H.I.; Stefánsson, G.; Larsson, K.; Undeland, I.; Skåra, T.; Romotowska, P.E.; Karlsdóttir, M.G. Investigating commercially relevant packaging solutions to improve storage stability of mechanically filleted Atlantic mackerel (*Scomber scombrus*) produced under industrial conditions. *Eur. Food Res. Technol.* **2020**, *246*, 693–701. [CrossRef]
60. Sveinsdóttir, H.I.; Karlsdóttir, M.G.; Arason, S.; Stefánsson, G.; Sone, I.; Skåra, T.; Rustad, T.; Larsson, K.; Undeland, I.; Gudjónsdóttir, M. Effect of antioxidants on the sensory quality and physicochemical stability of Atlantic mackerel (*Scomber scombrus*) fillets during frozen storage. *Food Chem.* **2020**, *321*, 126744. [CrossRef] [PubMed]
61. Da Mota, A.M. Optimização da Estratégia de Serviço Pós-Venda da Peruza. Ph.D. Thesis, Instituto Politecnico do Porto, Porto, Portugal, 2019.
62. Thrane, M.; Nielsen, E.H.; Christensen, P. Cleaner production in Danish fish processing–experiences, status and possible future strategies. *J. Clean. Prod.* **2009**, *17*, 380–390. [CrossRef]
63. Zieliński, B.; Kapłonek, W.; Nadolny, K. Regeneration of industrial cutting blades made from X39Cr13 steel used in skinning process of Pleuronectidae-family flatfishes. *J. Mech. Energy Eng.* **2018**, *2*, 277–284. [CrossRef]
64. Schwarz, O. Nordischer Maschinenbau Rud Baader GmbH; Co, K.G. Conveying Apparatus Comprising a Conveying Path and Designed to Supply a Plurality of Products for Consumption Having Soft Parts to a Processing Device, and Processing Machine Comprising a Conveying Apparatus and a Processing Device. U.S. Patent 9,039,498, 26 May 2015.
65. Schwarz, O. Nordischer Maschinenbau Rud Baader GmbH; Co, K.G. Device and Method for Removing a Surface Layer Including the Skin from Fish Fillets. U.S. Patent 9,872,507, 23 January 2018.
66. Arnesen, J.A.; Gildberg, A. Extraction and characterisation of gelatine from Atlantic salmon (Salmo salar) skin. *Bioresour. Technol.* **2007**, *98*, 53–57. [CrossRef]
67. Bland, J.M.; Grimm, C.C.; Bechtel, P.J.; Deb, U.; Dey, M.M. Proximate Composition and Nutritional Attributes of Ready-to-Cook Catfish Products. *Foods* **2021**, *10*, 2716. [CrossRef] [PubMed]
68. Zieliński, B.; Kapłonek, W.; Sutowska, M.; Nadolny, K. Analysis of a Feasibility Study of a Precision Grinding Process for Industrial Blades Used in the Cutting of Soft Tissues by a Prototype 5-Axis CNC Grinding Machine. *Appl. Sci.* **2019**, *9*, 3883. [CrossRef]
69. Joensen, S.; Olsen, J.V. Bløt Hyse. In *Spalting av Hysefilet Etter Skinning*; Fiskeriforskning: Tromsø, Norway, 27 August 2003.
70. Arnþórsdóttir, M.G.; Arason, S.; Margeirsson, B. *Combined Blast and Contact*; Skýrsla Matís: Reykjavík, Iceland, 2008.
71. Waterston, S.W.; Holmes, J.D. The Fish-Skinning Machine: An Unusual Source Of Hand Trauma. *Plast. Reconstr. Surg.* **2005**, *116*, 1831–1832. [CrossRef] [PubMed]
72. Ørnholt-Johansson, G.; Gudjónsdóttir, M.; Nielsen, M.E.; Skytte, J.L.; Frosch, S. Analysis of the production of salmon fillet—Prediction of production yield. *J. Food Eng.* **2017**, *204*, 80–87. [CrossRef]
73. Mathiassen, J.R.; Misimi, E.; Bondø, M.; Veliyulin, E.; Østvik, S.O. Trends in application of imaging technologies to inspection of fish and fish products. *Trends Food Sci. Technol.* **2011**, *22*, 257–275. [CrossRef]
74. Grasselli, G. Industrial Slicer. U.S. Patent Application 14/283,670, 4 December 2014.
75. Grasselli, G. Industrial Slicer. U.S. Patent 9,751,233, 5 September 2017.

76. Ross, K.; Edwards, J. Spatial Variation in the Mercury Concentration of Muscle Myomeres in Steaks of Farmed Southern Bluefin Tuna. *Foods* **2015**, *4*, 254–262. [CrossRef]
77. Singh, A.; Surasani, V.K.R. Fish processing: An entrepreneurial opportunity for livelihood and income generation. *J. Krishi Vigyan* **2020**, *9*, 144–149. [CrossRef]
78. Lorentzen, G.; Ageeva, T.N.; Heia, K. Desalting of dried salt-cured cod (*Gadus morhua* L.) without water renewal-3D imaging of volume change. *Food Control* **2021**, *121*, 107613. [CrossRef]
79. Sture, Ø.; Øye, E.R.; Skavhaug, A.; Mathiassen, J.R. A 3D machine vision system for quality grading of Atlantic salmon. *Comput. Electron. Agric.* **2016**, *123*, 142–148. [CrossRef]
80. Bro, T. Marel Salmon, A.S. D-Cut Slicer. U.S. Patent 9,180,601, 10 November 2015.
81. Manchay Aparco, L.D. Evaluación de Conservas en Base a Productos Hidrobiológicos. Bachelor's Thesis, César Vallejo University, Trujillo, Peru, 2020.
82. Kapłonek, W.; Nadolny, K.; Zieliński, B.; Plichta, J.; Pimenov, D.Y.; Sharma, S. The Role of Observation–Measurement Methods in the Surface Characterization of X39Cr13 Stainless-Steel Cutting Blades Used in the Fish Processing Industry. *Materials* **2020**, *13*, 5796. [CrossRef]
83. Faostat: Statistical Databases. Available online: http://faostat.fao.org/ (accessed on 23 December 2021).
84. Karltun, J.; Vogel, K.; Bergstrand, M.; Eklund, J. Maintaining knife sharpness in industrial meat cutting: A matter of knife or meat cutter ability. *Appl. Ergon.* **2016**, *56*, 92–100. [CrossRef] [PubMed]
85. Viatcheslavovich, A.O.; Arkadievich, N.V.; Adgamovich, F.Y. Mathematical simulation of knife profile resistance force during fish cutting. *Вестник Астраханского Государственного Технического Университета. Серия: Рыбное Хозяйство* **2019**, *3*, 150–158.
86. Ageev, O.V.; Fatykhov, Y.; Ivanova, E.E. Optimization of the knife profile for resource-saving primary fish processing. News of institutes of higher education. *Food Technol.* **2020**, *1*, 77–80.
87. Ageev, O.; Naumov, V.A.; Fatykhov, J.A. Mathematical Modeling of the Resistance Force of the Profile of a Flat-Back Knife. *J. Frict. Wear* **2019**, *40*, 580–587. [CrossRef]
88. Chu, J.P.; Diyatmika, W.; Tseng, Y.-J.; Liu, Y.-K.; Liao, W.-C.; Chang, S.-H.; Chen, M.-J.; Lee, J.-W.; Jang, J.S.C. Coating Cutting Blades with Thin-Film Metallic Glass to Enhance Sharpness. *Sci. Rep.* **2019**, *9*, 15558. [CrossRef] [PubMed]
89. Dowgiallo, A. Cutting force of fibrous materials. *J. Food Eng.* **2005**, *66*, 57–61. [CrossRef]
90. Jayraj, P.; Machavaram, R.; Sahu, G.; Paradkar, V. Measurement of Morphometric Dimensions and Mechanical Properties of Rohu Fish for Design of Processing Machines. *J. Aquat. Food Prod. Technol.* **2019**, *28*, 150–164. [CrossRef]
91. Jain, D.; Pathare, P.B.; Manikantan, M. Evaluation of texture parameters of Rohu fish (Labeo rohita) during iced storage. *J. Food Eng.* **2007**, *81*, 336–340. [CrossRef]
92. Yamase, S.; Tsukagoshi, T.; Morita, K.; Takeuchi, K.; Obara, T.; Maloney, P.J. Toyo Suisan Kikai Co Ltd; Nippon Suisan, K.K.; UniSea Inc. Method of Separation of Backbone Part of Fish and Device Therefor. U.S. Patent 7,988,542, 2 August 2011.
93. Vallamkondu, V.; Carlile, S.; Shakeel, M.; Ah-See, K.W. Neck abscess and vocal cord paresis: Delayed complications of a self-extruded long fishbone stuck in throat. *BMJ Case Rep.* **2013**, *2013*, bcr2013201832. [CrossRef] [PubMed]
94. Liu, X.; Liang, Z.; Wen, G.; Yuan, X. Waterjet machining and research developments: A review. *Int. J. Adv. Manuf. Technol.* **2019**, *102*, 1257–1335. [CrossRef]
95. Krajcarz, D. Comparison Metal Water Jet Cutting with Laser and Plasma Cutting. *Procedia Eng.* **2014**, *69*, 838–843. [CrossRef]
96. Wulfkuehler, S.; Stark, S.; Dietz, J.; Schmidt, H.; Weiss, A.; Carle, R. Effect of Water Jet Cutting and Moderate Heat Treatment on Quality of Fresh-Cut Red Oak Leaf Lettuce (*Lactuca sativa* L. var. crispa). *Food Bioprocess Technol.* **2014**, *7*, 3478–3492. [CrossRef]
97. Muthukumaran, S.; Baskaran, K. Organic and nutrient reduction in a fish processing facility—A case study. *Int. Biodeterior. Biodegrad.* **2013**, *85*, 563–570. [CrossRef]
98. Hace, A.; Jezernik, K. Control system for the waterjet cutting Machine. *IEEE/ASME Trans. Mechatron.* **2004**, *9*, 627–635. [CrossRef]
99. Huang, S.-W.; Chou, J.-H.; Tsai, J.-T. Uniform Design and Regression Analysis Methods for Optimal Operational Parameter Design of High-pressure Waterjet Machine. *Int. J. Autom. Smart Technol.* **2018**, *8*, 85–88. [CrossRef]
100. Omar, F.K.; de Silva, C.W. Optimal portion control of natural objects with application in automated cannery processing of fish. *J. Food Eng.* **2000**, *46*, 31–41. [CrossRef]
101. Thorarinsdottir, K.A. *APRICOT-Automated Pinbone Removal in Cod and Whitefish*; Nordic Innovation: Oslo, Norway, 2015.
102. Barbut, S. Meat industry 4.0: A distant future? *Anim. Front.* **2020**, *10*, 38–47. [CrossRef] [PubMed]
103. Barbut, S. Review: Automation and meat quality-global challenges. *Meat Sci.* **2014**, *96*, 335–345. [CrossRef]
104. Folkes, J. Waterjet—An innovative tool for manufacturing. *J. Mater. Process. Technol.* **2009**, *209*, 6181–6189. [CrossRef]
105. Irwansyah, I.; Ibrahim, M.; Ferdiansyah, H. Influence of water-jet nozzle geometry on cutting ability of soft material. *J. Rekayasa Kim. Lingkung.* **2012**, *9*, 6–11.
106. Kasperowicz, M.; Chudy, J.; Chomka, G. Determining the supply pressure depending on the feed speed and the diameter of the nozzle. *Carpathian J. Food Sci. Technol.* **2018**, *10*, 17–23.
107. Pogrebnyak, A.; Pogrebnyak, V.; Perkun, I.; Vasyliv, N. Influence of geometric and dynamic parameters of a water-polymer jet on characteristics of food products hydro-cutting process. *Ukr. Food J.* **2020**, *9*, 197–208. [CrossRef]
108. Sandor, Z.; Papp, Z.G.; Csengeri, I.; Jeney, Z. Fish meat quality and safety. *Sci. J. Meat Technol.* **2011**, *52*, 97–105.
109. Hyldig, G.; Nielsen, D. A review of sensory and instrumental methods used to evaluate the texture of fish muscle. *J. Texture Stud.* **2001**, *32*, 219–242. [CrossRef]

110. Komlatsky, V.I.; Podoinitsyna, T.A.; Verkhoturov, V.V.; Kozub, Y.A. Automation technologies for fish processing and production of fish products. *J. Phys. Conf. Ser.* **2019**, *1399*, 044050. [CrossRef]
111. Kong, F.; Tang, J.; Rasco, B.; Crapo, C.; Smiley, S. Quality Changes of Salmon (Oncorhynchus gorbuscha) Muscle during Thermal Processing. *J. Food Sci.* **2007**, *72*, S103–S111. [CrossRef]
112. Mohd, R.M.S.; Amjad, R.; Rosely, K.; Norhaida, A.; Tanzila, S. FiLeDI framework for measuring fish length from digital images. *Int. J. Phys. Sci.* **2012**, *7*, 607–618.
113. Sharmin, I.; Islam, N.F.; Jahan, I.; Joye, T.A.; Rahman, R.; Habib, T. Machine vision based local fish recognition. *SN Appl. Sci.* **2019**, *1*, 1529. [CrossRef]
114. Storbeck, F.; Daan, B. Fish species recognition using computer vision and a neural network. *Fish. Res.* **2001**, *51*, 11–15. [CrossRef]
115. Tveit, G.M.; Sistiaga, M.B.; Øye, E.R.; Schei, M. *Kvalitetsvurdering av Fisk Fanget Med to-og Fire-Panels Seleksjonsinnretninger: Bidrar 4-Panelkonstruksjoner og Knuteløst lin til Økt Kvalitet? Tokt Ombord F/Tr Havtind 28.06. 16–11.07. 16*; SINTEF: Trondheim, Norway, 2017.
116. Azarmdel, H.; Mohtasebi, S.S.; Jafari, A.; Muñoz, A.R. Developing an orientation and cutting point determination algorithm for a trout fish processing system using machine vision. *Comput. Electron. Agric.* **2019**, *162*, 613–629. [CrossRef]
117. Misimi, E.; Erikson, U.; Skavhaug, A. Quality Grading of Atlantic Salmon (Salmo salar) by Computer Vision. *J. Food Sci.* **2008**, *73*, E211–E217. [CrossRef]
118. Sivertsen, A.H.; Chu, C.-K.; Wang, L.-C.; Godtliebsen, F.; Heia, K.; Nilsen, H. Ridge detection with application to automatic fish fillet inspection. *J. Food Eng.* **2009**, *90*, 317–324. [CrossRef]
119. Andersen, K. Processing Quality Seafood. In *International Seafood Trade: Challenges and Opportunities*; FAO: Rome, Italy, 2009; pp. 93–96.
120. Bar, E.; Mathiassen, J.R.; Eilertsen, A.; Mugaas, T.; Misimi, E.; Linnerud, S.; Salomonsen, C.; Westavik, H. Towards robotic post-trimming of salmon fillets. *Ind. Robot. Int. J. Robot. Res. Appl.* **2016**, *43*, 421–428. [CrossRef]
121. Bondø, M.S.; Mathiassen, J.R.; Vebenstad, P.A.; Misimi, E.; Bar, E.M.S.; Toldnes, B.; Østvik, S.O. An automated salmonid slaughter line using machine vision. *Ind. Robot. Int. J. Robot. Res. Appl.* **2011**, *38*, 399–405. [CrossRef]
122. Mathiassen, J.R.; Misimi, E.; Østvik, S.O.; Aursand, I.G. Computer vision in the fish industry. In *Computer Vision Technology in the Food and Beverage Industries*; Woodhead Publishing: Sawston, UK, 2012; pp. 352–378.
123. Sun, M.; Yang, X.; Xie, Y. Deep learning in aquaculture: A review. *J. Comput.* **2020**, *31*, 294–319.
124. Xu, J.-L.; Sun, D.-W. Identification of freezer burn on frozen salmon surface using hyperspectral imaging and computer vision combined with machine learning algorithm. *Int. J. Refrig.* **2017**, *74*, 151–164. [CrossRef]
125. Zhao, S.; Zhang, S.; Liu, J.; Wang, H.; Zhu, J.; Li, D.; Zhao, R. Application of machine learning in intelligent fish aquaculture: A review. *Aquaculture* **2021**, *540*, 736724. [CrossRef]
126. Odone, F.; Trucco, E.; Verri, A. A trainable system for grading fish from images. *Appl. Artif. Intell.* **2001**, *15*, 735–745. [CrossRef]
127. Xu, J.; Sun, D.-W. Computer Vision Detection of Salmon Muscle Gaping Using Convolutional Neural Network Features. *Food Anal. Methods* **2017**, *11*, 34–47. [CrossRef]
128. Taheri-Garavand, A.; Nasiri, A.; Banan, A.; Zhang, Y.-D. Smart deep learning-based approach for non-destructive freshness diagnosis of common carp fish. *J. Food Eng.* **2020**, *278*, 109930. [CrossRef]
129. Laradji, I.; Saleh, A.; Rodriguez, P.; Nowrouzezahrai, D.; Azghadi, M.R.; Vazquez, D. Affinity lcfcn: Learning to segment fish with weak supervision. *arXiv* **2020**, arXiv:2011.03149.
130. Savkovic, B.; Kovac, P.; Rodic, D.; Strbac, B.; Klancnik, S. Comparison of artificial neural network, fuzzy logic and genetic algorithm for cutting temperature and surface roughness prediction during the face milling process. *Adv. Prod. Eng. Manag.* **2020**, *15*, 137–150. [CrossRef]
131. Tanikić, D. Computationally intelligent optimization of metal cutting regimes. *Measurement* **2020**, *152*, 107358. [CrossRef]
132. Choudhary, A.; Mian, T.; Fatima, S. Convolutional neural network based bearing fault diagnosis of rotating machine using thermal images. *Measurement* **2021**, *176*, 109196. [CrossRef]
133. He, M.; He, D. Deep Learning Based Approach for Bearing Fault Diagnosis. *IEEE Trans. Ind. Appl.* **2017**, *53*, 3057–3065. [CrossRef]
134. Jia, F.; Lei, Y.; Guo, L.; Lin, J.; Xing, S. A neural network constructed by deep learning technique and its application to intelligent fault diagnosis of machines. *Neurocomputing* **2018**, *272*, 619–628. [CrossRef]
135. Abioye, A.O.; Prior, S.D.; Thomas, G.T.; Saddington, P.; Ramchurn, S.D. The multimodal speech and visual gesture (mSVG) control model for a practical patrol, search, and rescue aerobot. In *Annual Conference Towards Autonomous Robotic Systems*; Springer: Cham, Switzerland, 2018; pp. 423–437.
136. Lin, J.; Holmes, M.; Vinson, R.; Ge, C.; Pogoda, F.C.; Mahon, L.; Gentry, R.; Seibel, G.E.; Chen, X.; Tao, Y. Design and testing of an automated high-throughput computer vision guided waterjet knife strawberry calyx removal machine. *J. Food Eng.* **2017**, *211*, 30–38. [CrossRef]
137. Marani, M.; Zeinali, M.; Kouam, J.; Songmene, V.; Mechefske, C.K. Prediction of cutting tool wear during a turning process using artificial intelligence techniques. *Int. J. Adv. Manuf. Technol.* **2020**, *111*, 505–515. [CrossRef]
138. National Marine Fisheries Service (2021) Fisheries of the United States, 2019. U.S. Department of Commerce, NOAA Current Fishery Statistics No. 2019. Available online: https://www.fisheries.noaa.gov/national/sustainable-fisheries/fisheries-united-states (accessed on 7 September 2022).

139. USDA United States Department of Agriculture. *Catfish Production Reports from National Agricultural Statistics Service (NASS)*; USDA: Washington, DC, USA, 2022.
140. Tan, Y.; Gao, H.; Chang, S.K.; Bechtel, P.J.; Mahmoud, B.S. Comparative studies on the yield and characteristics of myofibrillar proteins from catfish heads and frames extracted by two methods for making surimi-like protein gel products. *Food Chem.* **2018**, *272*, 133–140. [CrossRef] [PubMed]
141. Hill, J.I.; Nelson, R.G.; Woods, K.L.; Weese, J.O.; Whitis, G.N. Consumer preferences for attributes of catfish nuggets: Price, breading color, cooking method, and country of origin. *Aquac. Econ. Manag.* **2013**, *17*, 123–147. [CrossRef]
142. Ashrafi, N. Viscoelastic abrasive waterjet. In Proceedings of the ASME International Mechanical Engineering Congress and Exposition, Denver, CO, USA, 11–17 November 2011; Volume 54891, pp. 677–681.
143. Shakouri, E.; Abbasi, M. Investigation of cutting quality and surface roughness in abrasive water jet machining of bone. *Proc. Inst. Mech. Eng. Part H J. Eng. Med.* **2018**, *232*, 850–861. [CrossRef] [PubMed]
144. Sonikel Ultrasonics. Frozen Fish Slicing with Ultrasonic. Available online: https://www.youtube.com/watch?v=nAe071BoMFY (accessed on 1 August 2022).
145. Wang, N.; Gao, X.; Tao, D.; Yang, H.; Li, X. Facial feature point detection: A comprehensive survey. *Neurocomputing* **2018**, *275*, 50–65. [CrossRef]
146. Kristensen, I.; Jorgensen, D.B.; Kroma, A.S. Fish Processing Machine and a Method for Processing Fish. U.S. Patent Application 15/030,078, 8 September 2016.
147. Kristensen, I.; Jorgensen, D.B.; Kroma, A.S. Fish Processing Machine and a Method Enabling That Fish Can Be Processed through the Mouth. U.S. Patent 9,622,493, 18 April 2017.
148. Dosovitskiy, A.; Beyer, L.; Kolesnikov, A.; Weissenborn, D.; Zhai, X.; Unterthiner, T.; Dehghani, M.; Minderer, M.; Heigold, G.; Gelly, S.; et al. An image is worth 16x16 words: Transformers for image recognition at scale. *arXiv* arXiv:2010.11929, 2020.
149. Chen, T.; Kornblith, S.; Norouzi, M.; Hinton, G. A simple framework for contrastive learning of visual representations. In International Conference on Machine Learning. In Proceedings of the International Conference on Machine Learning, Vienna, Austria, 21 November 2020.
150. He, K.; Fan, H.; Wu, Y.; Xie, S.; Girshick, R. Momentum contrast for unsupervised visual representation learning. In Proceedings of the IEEE/CVF Conference on Computer Vision and Pattern Recognition, Seattle, WA, USA, 13–19 June 2020; IEEE: Piscataway, NJ, USA, 2020; pp. 9729–9738.
151. Annoni, M.; Arleo, F.; Malmassari, C. CFD aided design and experimental validation of an innovative Air Assisted Pure Water Jet cutting system. *J. Mater. Process. Technol.* **2014**, *214*, 1647–1657. [CrossRef]
152. Gzaiel, M.; Triki, E.; Barkaoui, A. Finite element modeling of the puncture-cutting response of soft material by a pointed blade. *Mech. Mater.* **2019**, *136*, 103082. [CrossRef]
153. Hu, H.; Li, H.; Wang, Q.; He, J.; Lu, C.; Wang, Y.; Liu, P. Anti-blocking performance of ultrahigh-pressure waterjet assisted furrow opener for no-till seeder. *Int. J. Agric. Biol. Eng.* **2020**, *13*, 64–70. [CrossRef]
154. Polyakov, A.; Zhabin, A.; Averin, E.; Polyakov, A. Generalized equation for calculating rock cutting efficiency by pulsed water jets. *J. Rock Mech. Geotech. Eng.* **2019**, *11*, 867–873. [CrossRef]

foods

MDPI

Article

Effect of Washing Times on the Quality Characteristics and Protein Oxidation of Silver Carp Surimi

Xiaodi Zhang [1], Yiqi Zhang [1], Haochen Ding [2], Wenhai Zhang [2] and Zhiyuan Dai [1,*]

1 Collaborative Innovation Center of Seafood Deep Processing, Zhejiang Province Joint Key Laboratory of Aquatic Products Processing, Institute of Seafood, Zhejiang Gongshang University, Hangzhou 310035, China
2 Anjing Foods Group Co., Ltd., Xiamen 361022, China
* Correspondence: dzy@zjsu.edu.cn; Tel.: +86-150-7617-0311

Abstract: The aim of this work is to evaluate the effects of different washing times (zero (W0), one (W1), two (W2), and three (W3) times) on the physicochemical characteristics, gel property, and protein oxidation of silver carp surimi during 4 °C refrigeration. The results showed that the yield, types of fatty acids, redness (a^*), total volatile basic nitrogen, and thiobarbituric acid reactive substances of the surimi tended to decrease, and the whiteness, pH, gel strength, and water retention tended to increase with the increase of washing times. Meanwhile, washing removed some fatty acids and the fatty acid species showed a decreasing trend. The FTIR spectra showed that washing did not change the functional group composition but changed the content of each group of the functional groups, while decreasing the proportion of β-sheet structures. Compared with the unwashed surimi, washing caused some of the immobilized water in the minced fish to be transferred to free water, and the water fluidity was enhanced. The washing enhanced the water holding capacity in the surimi gels, and the microstructure of the surimi gels was denser and delayed the protein oxidation during refrigeration. However, the difference between W2 and W3 surimi was not significant ($p > 0.05$). In practice, W2 can be used to produce surimi to improve its yield and reduce water consumption.

Keywords: washing times; silver carp surimi; quality characteristics; protein oxidation

Citation: Zhang, X.; Zhang, Y.; Ding, H.; Zhang, W.; Dai, Z. Effect of Washing Times on the Quality Characteristics and Protein Oxidation of Silver Carp Surimi. *Foods* **2022**, *11*, 2397. https://doi.org/10.3390/foods11162397

Academic Editors: Tao Yin and Liu Shi

Received: 2 July 2022
Accepted: 6 August 2022
Published: 10 August 2022

1. Introduction

Surimi has the advantages of high protein, low fat, and delicate taste, which is popular among consumers [1]. At present, the raw materials for the production of surimi are mainly seawater fish, but due to human overfishing and changes in the natural environment, marine economic fish resources cannot meet the human consumption demand for surimi products [2]. In recent years, the production of freshwater fish in China has increased significantly, and the development of freshwater fish for the production of surimi has become important research in the field of surimi processing.

Silver carp (*Hypophthalmichthys molitrix*), one of the four major fish species in China, is the main low-value freshwater fish in China, with a production of about 3.81 million tons in 2020 [3]. Silver carp is the best freshwater fish for producing surimi, with its short growth cycle, low price, and tender meat. The excess fat and protein in fish react and reduce the nutritional and organoleptic properties of the product [4]. In addition, protein oxidation can induce cross-linking reactions to denature proteins and reduce their functionality [5]. Therefore, washing is an essential process in the production of surimi, which removes impurities, fat, and water-soluble proteins [6] and thus improves the color, texture, and odor of surimi products [7]. The amount of washing times depends on the type of fish and washing solution. Granata, Flick Jr, and Martin [8] showed that producing 1 kg of surimi requires nearly 15 kg of water. A higher amount of washing solution can produce surimi with better quality, but it increases the waste-water discharge of the factory, decreases surimi yield, and also has an impact on the environment [6]. Therefore, moderate washing of surimi can maximize surimi yield and reduce water consumption.

In order to achieve sustainable and efficient development of the surimi processing industry, it is necessary to understand the effect of washing times on surimi [9]. However, there are few studies that are related to the effect of washing times on the quality of silver carp surimi. In view of this, we investigated the effect of washing times on the quality of surimi and evaluated the effect of different washing times on the physicochemical characteristics, gel property, and protein oxidation of silver carp surimi to provide a theoretical reference for the production of high quality surimi by loss reduction treatment.

2. Materials and Methods

2.1. Materials

Silver carp surimi was provided by Anjoyfood Co., Ltd. (Liaoning, China). A 10× phosphate buffered solution (PBS) and color Mixed Protein Marker (11–245 KDa) were purchased from Beijing Solarbio Science & Technology Co., Ltd. (Beijing, China). The 0.1 mol/L HCl standard titration solution, 10× electrode buffer, and bicarbonate reagent were purchased from Fuzhou Feixing Biotechnology Co. (Fuzhou, China). All other reagents were analytical grade and were purchased from Sinopharm Chemical Reagent Co., Ltd. (Shanghai, China).

2.2. Preparation of Surimi

The silver carp was divided into four groups for washing after harvesting, the first group of minced fish was not washed (W0), the second group was washed one time (W1), the third group was washed two times (W2), and the fourth group was washed three times (W3). The fish that were washed with five times their weight in tap water for 15 min, and the washing water was agitated and kept below 4 °C during the washing process. After washing, the surimi was centrifuged and plate shaped. The above operations were carried out in the surimi production workshop of Liaoning Anjing Food Co., LTD (Anshan, China). The four groups of surimi were placed at 4 °C, and each index was measured.

2.3. Proximate Analysis

In brief, the moisture content was measured using the oven-drying method. The protein content was estimated from nitrogen (N × 6.25) using constant Kjeldahl method. The fat content of the sample was extracted from the sample by Soxhlet using ether as solvent. The ash content of the sample was determined using the method Muffle furnace ashing [10]. All the values that were analyzed were expressed as a percentage on a wet weight basis.

2.4. Fatty Acid Composition

The fat of surimi was extracted and methylated according to the method of Li et al. [4]. The fatty acid composition of surimi was determined by gas chromatography coupled with a flame ionization detector (FID) (7890B, Agilent, Palo Alto, CA, USA). The injection port and detector temperatures were 260 °C. A HP-88 column (100 m × 0.25 mm) was used for determination. The column temperature was set at 180 °C for the first 20 min, and then increased to 200 °C at a rate of 0.5 °C/min. N_2 was used as the carrier gas (flow rate: 1.5 mL/min). The air flow and H_2 flow were 400 and 30 mL/min, respectively.

2.5. Determination of Whiteness

The whiteness was determined with a HunterLab XE Chroma Meter (Hunter Engineering Company, Bridgeton, MO, USA) according to the method of Xiong et al. [11], by measuring the L^* (lightness), a^* (redness), and b^* (yellowness) values. A total of nine readings were made from the surface of the samples. The *whiteness* was calculated by the following:

$$Whiteness = 100 - [(100 - L^{*2}) + a^{*2} + b^{*2}]^{1/2} \quad (1)$$

2.6. Determination of pH

The surimi was weighed for 5.0 g, then 45 mL of distilled water was added, and the sample was homogenized for 30 s at 8000 r/min. The pH was measured using a hand-held pH meter (Seven 2Go-S2, Mettler Toledo, Greenville, SC, USA) [12].

2.7. Chemical Analysis of Surimi

The total volatile base nitrogen (TVB-N) was estimated by the micro diffusion method. The 2-thiobarbituric acid-reactive substances (TBARS) of surimi was estimated spectrophotometric method. In brief, the surimi samples (5.0 g) were treated according to the method of Priyadarshini et al. [13] and reacted with 0.02 mol/L 2-thiobarbituric acid (TBA) by heating in a boiling water bath for 40 min. The absorbance of the supernatant was measured at 532 nm.

2.8. Preparation of Surimi Gels

The moisture content of the surimi samples was measured using a halogen moisture meter (SN-DHS-20A, Shampo Instruments Co., Shanghai, China), and the moisture content of the surimi was calculated and adjusted. The moisture of the surimi was adjusted to 80% and 2.5% sodium chloride was added. The surimi was placed in a mixer with a paddle attachment for 3 min. The surimi paste was stuffed into polyvinylidine casing (21 mm in diameter) and both ends of the casing were sealed tightly. Subsequently, the surimi pastes were incubated at 40 °C for 30 min and then heated at 90 °C for 20 min [14]. The gels were cooled in iced water and stored overnight at 4 °C prior to analysis. The gel casings were peeled prior to the test.

2.9. Determination of Gel Strength

The gel strength was determined according to the method that was described by Zhang et al. [15]. The gels were cut into cylinders (2.5 cm in height) prior to the test. A spherical probe (5 mm in diameter) was used to penetrate 15 mm into the samples at a speed of 1 mm/s by a gel strength analyzer (SD-700, Sun Scientific Co., Ltd., Tokyo, Japan). The breaking force (maximum breaking force, g) and deformation (cm) were determined on at least nine specimens per treatment. The gel strength (g·cm) is the product of breaking force and deformation.

2.10. Determination of Water Holding Capacity (WHC)

The surimi gels (5.0 g) were weighed into centrifugation tubes. The WHC was performed by the sample as described by Jin et al. [16]. The WHC was expressed as percentage of water that was retained with respect to the water that was present in the gel prior to centrifugation.

2.11. Determination of Low Field Nuclear Magnetic Resonance (LF-NMR)

The LF-NMR of samples was measured according to the method of Yin et al. [17] using an NMR analyzer (MesoMR23-060H-I, Niumag analytical instrument Co., Suzhou, China). Approximately 5.0 g of cylindrical surimi gel was placed into an NMR tube (size: 2 cm in diameter × 4 cm in height). The Carr Purcell Meiboom Gill (CPMG) pulse sequence was used to show the transverse relaxation time (T_2).

2.12. Fourier Transform Infrared (FTIR) Spectroscopy

We performed FTIR according to the method of Zhang, Dai, Zhang, Dong, and Hu [18]. The surimi gels were freeze-dried and mixed with potassium bromide (KBr). The mixture was ground uniformly and then compressed into a tablet. FTIR spectroscopy was carried out in a spectrometer (IN10, Thermo Fisher Scientific, Waltham, MA, USA).

2.13. Scanning Electron Microscopy (SEM)

We performed microstructure observation according to the method of Zhou et al. [1] and Shi et al. [19]. The freeze-dried surimi gel samples were sublimated and gilded. The microstructures were observed by SEM (Sigma 300, Zeiss, Oberkochen, Germany).

2.14. Preparation of Myofibrillar Protein (MP)

MP was prepared from according to the method of Liu, Chen, Kong, Han, and He [20].

2.15. Determination of Carbonyl Content

The carbonyl group reacts with 2,4-dinitrophenylhydrazine to form the red 2,4-dinitrophenylhydrazone, with a characteristic absorption peak at 370 nm. The carbonyl content was analyzed by an assay kit (Nanjing Jiancheng Technology Co. Ltd., Nanjing, China).

2.16. Determination of Total Sulphydryl Group Content

Total sulphydryl group content was determined according to the method of Balange, and Benjakul [21] with slight modifications. To 0.5 mL sample solution (5.0 mg/mL), 4.5 mL of buffer A (pH 6.8, containing 10 mmol/L EDTA, 8 mol/L urea, 1% SDS, and 0.2 mol/L Tris-HCI buffer) were added. To 4 mL of the mixture, 0.5 mL of buffer B (pH 8.0, containing 0.1% DTNB and 0.2 mol/L Tris-HCI buffer) was added and incubated at 40 °C for 30 min. A blank was conducted by replacing the sample with 0.6 mol/L NaCl. The absorbance was measured at 412 nm (Spectra MAX 190, Molecular Devices Co., Sunnyvale, TX, USA) and sulfhydryl group content was calculated using the extinction coefficient of 13,600 $mol/L^{-1}cm^{-1}$.

2.17. SDS-Polyacrylamide Gel Electrophoresis (SDS-PAGE)

Protein patterns of the surimi were analyzed by SDS-PAGE according to the method of Zhang et al. [22]. In brief, the surimi samples (3.0 g) were mixed with 5% SDS solution (30 mL), homogenized, and held at 85 °C for 1 h. After centrifugation (3500 r/min, 15 min) the supernatant was added to an equal volume of 1× loading buffer, dispersed and then boiled in a water bath for 3 min. Protein patterns of surimi samples (15 μL) were loaded onto polyacrylamide gels comprised of a 10% running gel and a 4% stacking gel and subjected to electrophoresis using an electrophoresis apparatus (BIO-RAD Co., Hercules, CA, USA). The gels were stained with 0.125% Coomassie brilliant blue R-250 and decolorized in a solution containing 65% acetic acid and 25% ethanol until the bands were clear.

2.18. Statistical Analysis

The colors and gel strengths of the samples were run in 9 replicates, and the rest of experiments were run in triplicate. The data are presented as the means ± standard deviations of duplicate measurements and subjected to analysis of variance (ANOVA) using SPSS 22.0 Statistics (SPSS Inc., Chicago, IL, USA). The means were compared by the Duncan's multiple range test at the 95% significant level ($p < 0.05$).

3. Results and Discussion

3.1. Basic Composition of Surimi

The basic composition of silver carp surimi is shown in Table 1. The results showed that with the increase of washing times, the moisture content of the surimi increased from 73.14% to 76.30%. Washing caused the hydration of myofibrillar protein of surimi, and the moisture content of the surimi increased. The moisture content of the surimi after washing was significantly higher than that of the unwashed surimi ($p < 0.05$). After repeated washing, the protein content of the surimi decreased. The protein content of the surimi in W2 (22.87%) and W3 (22.21%) was significantly lower than that in W0 (24.71%) and W1 (24.19%). This result is consistent with a previous study that was conducted by Li et al. [4], in which it was found that in the washing process, a large number of water-soluble proteins (myosinogen, hemoglobin, myoglobin, and so on) are removed, as well as part of the volatile amino compounds. The fat content of the surimi decreased significantly from 1.52% (W0) to 1.10% (W3) in the washing process, but there was no significant difference between the fat content of W2 and W3 ($p > 0.05$). Silver carp belong to low-fat fish, the overall fat content is low, but part of the fat is removed in the washing process. The fats in surimi are oxidized and interact with proteins to deteriorate the quality of the surimi [16]. After washing, most of water-soluble protein and part of the fat could be removed from surimi, but there was no significant difference in the ash content ($p > 0.05$). Producing high-quality surimi requires high protein, high myofibrillar protein, low fat, and adequate moisture [16]. The appropriate number of washing times retains some of the sarcoplasmic proteins and endogenous transglutaminases in the surimi, which can improve the gel strength of surimi gels.

Table 1. Basic composition of surimi with different washing times.

	Moisture Content (%)	Protein Content (%)	Lipid Content (%)	Ash Content (%)
W0	73.14 ± 0.98 ^{a}	24.71 ± 0.65 ^{b}	1.52 ± 0.06 ^{c}	0.53 ± 0.03 ^{a}
W1	74.00 ± 0.91 ^{ab}	24.19 ± 0.53 ^{b}	1.24 ± 0.04 ^{b}	0.53 ± 0.04 ^{a}
W2	75.44 ± 0.32 ^{bc}	22.87 ± 0.39 ^{a}	1.19 ± 0.06 ^{ab}	0.53 ± 0.02 ^{a}
W3	76.30 ± 0.46 ^{c}	22.21 ± 0.44 ^{a}	1.10 ± 0.04 ^{a}	0.52 ± 0.02 ^{a}

Note: W0, W1, W2, and W3 represent the surimi with 0 washing times, 1 washing time, 2 washing times, and 3 washing times, respectively. Different letters in the same line indicate significant differences ($p < 0.05$).

3.2. Fatty Acid Composition of Surimi

As can be seen from Table 2, the types of fatty acids in the surimi decreased with the increase of washing times, and the washing removed some saturated fatty acids. The fatty acid species of surimi decreased from 20 to 15 after three washing times, and the washing reduced the fatty acid species of surimi. Due to the low density of fat, in the upper layer of the washing water, the fatty acid species decreased gradually as the washing times increased fat removal. A similar result was reported by Li et al. [4], who likewise found that washing reduced the types of fatty acids in surimi. Surimi contains a large amount of C18:1 monounsaturated fatty acids, the most abundant polyunsaturated fatty acids are C20:4, followed by EPA and DHA, which are very important fatty acids for humans. With the increase of washing times, the change of EPA and DHA content is small, and the surimi with different washing times have s high nutritional value.

3.3. Physicochemical Characteristics of Surimi

The influence of different washing times on the physicochemical characteristics of surimi is shown in Table 3. The yield of surimi decreased with the increase of washing times, from 88.58% to 76.73%. Among them, the yield of twice-rinsing surimi was increased by 5.92% compared with three-rinsing surimi. In the traditional surimi production process, three times of cold-water washing is usually carried out to remove impurities (including sarcoplasmic protein, blood, pigment, and fat) in fish mince, thereby improving the whiteness and quality of the surimi. The redness (a^*) and whiteness of unwashed surimi (W0) were 1.44 and 65.23, respectively. The a^* and whiteness of W3 were −1.08 and 71.96, respectively. With the increase of washing times, the a^* of surimi showed a decreasing trend, while the whiteness showed an increasing trend. Myoglobin is a heme protein that is soluble in water and has an important effect on the formation of fish minced color and rancidity [7]. During the washing process of surimi, most of the myosinogen, myoglobin, hemoglobin, and other compounds are removed, thereby reducing the a^* of surimi and improving the whiteness of surimi [23]. Although increasing the washing times improved the whiteness of the surimi, there was no significant difference between the W2 and W3 groups ($p > 0.05$). At the same time, washing can cause a waste of water and a loss of nutrients in the surimi, so it is important to choose the appropriate number of surimi washing times in production.

Table 2. Fatty acid composition of surimi with different washing times.

Fatty Acid Composition (%)	W0	W1	W2	W3
C12:0	0.16 ± 0.00 ^{c}	0.71 ± 0.01 ^{d}	0.11 ± 0.00 ^{b}	nd ^{a}
C13:0	20.10 ± 0.09 ^{c}	20.74 ± 0.21 ^{d}	18.42 ± 0.23 ^{b}	15.26 ± 0.17 ^{a}
C14:0	3.08 ± 0.01 ^{b}	2.99 ± 0.03 ^{a}	7.76 ± 0.00 ^{c}	9.33 ± 0.01 ^{d}
C16:0	14.71 ± 0.00 ^{b}	14.50 ± 0.02 ^{a}	14.52 ± 0.09 ^{a}	14.66 ± 0.10 ^{ab}
C17:0	1.07 ± 0.00 ^{b}	1.22 ± 0.00 ^{c}	nd ^{a}	nd ^{a}
C18:0	2.22 ± 0.00 ^{b}	3.71 ± 0.01 ^{d}	2.78 ± 0.00 ^{c}	2.14 ± 0.05 ^{a}
C20:0	4.42 ± 0.00 ^{b}	nd ^{a}	nd ^{a}	nd ^{a}
C22:0	0.24 ± 0.00 ^{b}	0.29 ± 0.00 ^{c}	nd ^{a}	nd ^{a}
∑Saturated	**46.00 ± 0.09 ^{d}**	**44.15 ± 0.16 ^{c}**	**43.60 ± 0.13 ^{b}**	**41.40 ± 0.12 ^{a}**
C14:1	1.34 ± 0.00 ^{d}	0.87 ± 0.00 ^{c}	0.71 ± 0.01 ^{b}	0.51 ± 0.01 ^{a}
C16:1	9.55 ± 0.00 ^{b}	9.42 ± 0.01 ^{b}	9.02 ± 0.06 ^{a}	11.55 ± 0.17 ^{c}
C18:1	17.56 ± 0.21 ^{a}	18.84 ± 0.15 ^{b}	19.66 ± 0.20 ^{c}	19.55 ± 0.29 ^{c}
C20:1	2.54 ± 0.01 ^{a}	3.07 ± 0.02 ^{d}	2.94 ± 0.04 ^{c}	2.72 ± 0.04 ^{b}
C24:1	1.62 ± 0.00 ^{d}	1.17 ± 0.00 ^{a}	1.29 ± 0.00 ^{c}	1.20 ± 0.01 ^{b}
∑Monounsaturated	**32.62 ± 0.21 ^{a}**	**33.37 ± 0.18 ^{b}**	**33.62 ± 0.22 ^{b}**	**35.53 ± 0.08 ^{c}**
C18:2	2.61 ± 0.02 ^{a}	3.22 ± 0.01 ^{c}	3.22 ± 0.01 ^{c}	2.98 ± 0.01 ^{b}
C18:3n6	4.22 ± 0.08 ^{a}	4.83 ± 0.01 ^{b}	4.74 ± 0.03 ^{b}	4.73 ± 0.05 ^{b}
C20:2	0.35 ± 0.00 ^{b}	0.43 ± 0.00 ^{c}	nd ^{a}	nd ^{a}
C20:3n3	1.62 ± 0.00 ^{a}	1.82 ± 0.01 ^{b}	1.94 ± 0.01 ^{c}	2.14 ± 0.02 ^{d}
C20:4	5.25 ± 0.00 ^{a}	5.60 ± 0.01 ^{b}	6.26 ± 0.10 ^{c}	6.86 ± 0.05 ^{d}
C20:5 (EPA)	4.23 ± 0.02 ^{d}	4.07 ± 0.00 ^{c}	3.85 ± 0.13 ^{b}	3.52 ± 0.06 ^{a}
C22:6 (DHA)	3.09 ± 0.05 ^{c}	2.53 ± 0.06 ^{a}	2.77 ± 0.07 ^{b}	2.85 ± 0.01 ^{b}
∑polyunsaturated	**21.23 ± 0.12 ^{a}**	**22.49 ± 0.09 ^{b}**	**22.78 ± 0.36 ^{bc}**	**23.08 ± 0.20 ^{c}**

Note: Nd indicates that the substance that was tested was not detected. Different letters in the same line indicate significant differences ($p < 0.05$).

Table 3. Physicochemical characteristics of surimi with different washing times.

	W0	W1	W2	W3
Appearance				
Yield (%)	100.00 ± 0.00 ^{d}	88.58 ± 1.26 ^{c}	82.65 ± 0.89 ^{b}	76.73 ± 0.96 ^{a}
Color (a^*)	1.44 ± 0.44 ^{a}	0.29 ± 0.49 ^{b}	−1.31 ± 0.37 ^{c}	−1.08 ± 0.39 ^{c}
Whiteness	65.23 ± 1.03 ^{a}	71.97 ± 1.10 ^{b}	72.01 ± 0.89 ^{b}	71.96 ± 0.80 ^{b}
pH	6.32 ± 0.02 ^{a}	6.35 ± 0.01 ^{a}	6.66 ± 0.02 ^{b}	6.83 ± 0.02 ^{c}
TVB-N (mg N/100 g)	6.57 ± 0.24^{c}	5.86 ± 0.46 ^{bc}	5.27 ± 0.39 ^{ab}	4.62 ± 0.35 ^{a}
TBARS (mg/kg)	0.18 ± 0.00 ^{d}	0.13 ± 0.00 ^{c}	0.12 ± 0.00 ^{b}	0.09 ± 0.00 ^{a}
Breaking force (g)	369.67 ± 31.48 ^{a}	403.29 ± 25.94 ^{a}	419.33 ± 19.94 ^{a}	413.67 ± 35.20 ^{a}
Deformation (cm)	0.86 ± 0.04 ^{a}	0.92 ± 0.03 ^{ab}	0.97 ± 0.04 ^{bc}	1.00 ± 0.01 ^{c}
Gel strength (g·cm)	320.20 ± 38.60 ^{a}	369.89 ± 32.94 ^{b}	408.51 ± 36.03 ^{c}	415.18 ± 35.68 ^{c}
WHC (%)	88.46 ± 1.49 ^{a}	90.90 ± 1.39 ^{b}	92.53 ± 2.02 ^{b}	92.87 ± 1.75 ^{b}

Note: Different letters in the same line indicate significant differences ($p < 0.05$).

The pH of the surimi increased from 6.32 (W0) to 6.83 (W3) with an increasing number of washing times. Water-soluble acidic compounds (such as free acidic amino acids and lactic acid) are removed from the surimi [24], so the pH of surimi increased significantly ($p < 0.05$). Jin et al. [16] noted that the pH has a profound effect on the physical characteristics of meat, such as moisture content, tenderness, and color. An increase in the pH may lead to a decrease in the fat content of surimi. When the pH is not at the isoelectric point of myofibrillar protein, neutral and polar fats in minced meat can be removed and fat in minced meat can be released [25]. However, during the washing process, both alkaline and acidic substances in the surimi can be washed away, so that the surimi reaches a new pH balance after washing [23].

TVB-N is mainly composed of trimethylamine, dimethylamine, and ammonia. Protein oxidation that is caused by bacterial metabolism in fish leads to an increase in TVB-N, which is associated with fish freshness [26]. Generally speaking, the acceptable limit of TVB-N for consumers is 25–35 mg N/100 g, higher than the limit indicates that the minced meat has deteriorated [27]. The TVB-N of W0 was 6.57 mg N/100 g, indicating that the quality of fresh fish was better. With the increase of washing times, the TVB-N of surimi gradually decreased ($p < 0.05$). Therefore, the washing times of surimi can be appropriately increased in industrial production to improve the quality of surimi [28].

Fat oxidation can reduce the color, flavor, texture, and nutritional value of surimi. W0 contained more fat and endogenous pro-oxidants and, therefore, had the highest TBARS (0.183 mg/kg) [7]. With the increase of washing times, the TBARS of surimi decreased significantly ($p < 0.05$). In general, the oxidation of fat and myoglobin in minced meat occurs simultaneously and synergistically [29]. W3 had the lowest TBARS in this study (0.086 mg/kg, $p < 0.05$), which might be related to the removal of some fat and oxidized components in fish mince during washing.

Gel strength is an important index to evaluate the gel quality of surimi. As shown in Table 2, the gel strength of W0 was the lowest (330.20 g·cm, $p < 0.05$), which may be because the heat-induced proteases and fats interfered with the cross-linking of myosin in the gel matrix of surimi [6]. With the increase of washing times, some substances that interfere with myosin cross-linking were removed, so the gel strength of surimi significantly improved ($p < 0.05$). Kong, Geng, Gao, and Li [30] observed that washing could improve the gel properties of surimi, but too many washing times would also reduce the gel properties of surimi, which should be controlled at two to three times. In this study, it was found that gel strength of surimi of W2 and W3 (>400 g·cm) was higher than the Chinese standard of AAA surimi (≥400 g·cm), which indicated that the surimi washing two times still met the standard.

The WHC of surimi gel is related to myofibrillar protein molecular structure and charge change [24]. When the pH of minced meat is close to the isoelectric point of myofibrillar protein (pH = 5.5), the protein agglutinates, leading to the release of water molecules and thus affecting the formation of surimi gel [23]. With the increase of washing times, the WHC of the surimi gel increased significantly ($p < 0.05$, 90.90–92.87%) compared to W0 (88.46%). After washing, the pH of the surimi deviated from the isoelectric point of myofibrillar protein, thus improving the WHC of surimi gel, which was consistent with the results of Baxter and Skonberg [31]. Karthikeyan Dileep, and Shamasundar, [32] showed that compared with unwashed *Nemipterus Japonicus* surimi, the WHC of the washed three times surimi gel was 10% higher than that of the control. The difference in WHC depends on the stability of the protein network, and components (such as fat) that interfere with the stability of the protein network may be removed in washing [31].

In summary, washing treatment could improve the physicochemical characteristics of surimi, but there was no significant difference between W2 and W3 in color, TVB-N, TBARS, gel strength, and water holding capacity ($p < 0.05$). It can be estimated that by reducing the number of surimi rinsing steps from three times to two times, the discharge of rinsing wastewater can be reduced by at least 30%, which can effectively reduce the cost of surimi wastewater treatment and improve the utilization rate of fish resources.

3.4. FTIR Spectra

The FTIR spectra of surimi gel with different washing times are shown in Figure 1. The peak at 2980–2850 cm^{-1} represented the asymmetric stretching of CH_2 with a characteristic absorption peak of 2925.47 cm^{-1}, usually occurring in the aliphatic side chain of proteins. The peak at 1700–1600 cm^{-1} was the amide I band, representing the stretching vibration of C=O. And the peak at 1350–1200 cm^{-1} was associated with vibration of C-O and C-O-O functional groups [33]. The absorption spectra of the four groups of surimi in the figure were similar, indicating that the washing times had no effect on the composition of the functional groups in the surimi gels [7]. However, the intensity of the absorption peaks of the surimi samples increased with the increase in washing times, which indicated that the washing process affected the content of the functional groups in the surimi gels.

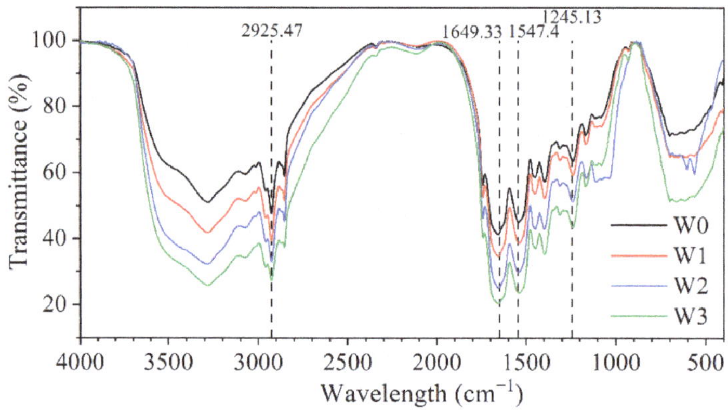

Figure 1. The FTIR spectra of surimi gel with different washing times.

The most characteristic spectral band in myofibrillar protein was the amide I band, which reflected hydrogen bonding patterns, dipole interactions, and the geometry of the protein polypeptide backbone [33]. In this study, we compared the Gaussian fitting curves of the amide I band of surimi gels with different washing times (Figure S1) to obtain the secondary structure content of surimi gel proteins, and the results are shown in Table 4. The α-helix, β-sheet, β-turn, and random coil structure contents of the unwashed surimi gels (W0) were 30.87%, 44.06%, 12.18%, and 12.88%, respectively. Washing increased the α-helix content and significantly decreased the β-sheet content ($p < 0.05$) of surimi gels, indicating that the surimi gels were more compact after washing [13]. Liu et al. [34] also found that compared to unwashed minced, the β-sheet content of washed surimi gel decreased by 13% and the β-turn content increased by 39%, while the changes in α-helical and random coil content were not significantly different ($p > 0.05$). A similar result was reported by Yuan et al. [35], who likewise found that β-sheet content decreased with the increase of washing times.

Table 4. Secondary structure of surimi with different washing times.

Secondary Structure	W0	W1	W2	W3
α-helical (%)	30.87 ± 0.24 [a]	31.30 ± 0.34 [a]	31.51 ± 0.26 [a]	31.91 ± 0.29 [a]
β-sheet (%)	44.06 ± 0.37 [b]	42.88 ± 0.31 [a]	41.85 ± 0.28 [a]	42.62 ± 0.29 [a]
β-turn (%)	12.18 ± 0.17 [a]	12.62 ± 0.19 [a]	13.52 ± 0.14 [a]	12.69 ± 0.18 [a]
Random coil (%)	12.88 ± 0.20 [a]	13.20 ± 0.22 [a]	13.12 ± 0.16 [a]	12.78 ± 0.16 [a]

Note: Different letters in the same line indicate significant differences ($p < 0.05$).

3.5. Low Field NMR Relaxation

LF-NMR analysis results could characterize changes in water mobility and distribution in surimi gels, which correlated with protein denaturation and aggregation [36]. The T_2 relaxation time distribution results of surimi gel with different washing times are shown in Figure 2 and Table 5. As relaxation components, T_{2b} represented the protein binding water that was tightly bound with macromolecules (bound water), T_{21} represented the water embedded in the protein network structure (immobilized water), and T_{22} represented the free water outside the spatial structure of surimi gel (free water) [37]. The results showed that there was more immobilized water and less bound water and free water in the surimi gel, and most of the water in the surimi gel was bound within the spatial structure, which was consistent with the findings of Wang et al. [38] and Pan et al. [39]. Compared to the unwashed mince, washing increased the percentage of T_{2b} and T_{22} and decreased the percentage of T_{21} in the surimi gel, indicating that the interaction between water and protein molecules was weaker, some immobilized water was transferred to free water, and the mobility of water was enhanced [38].

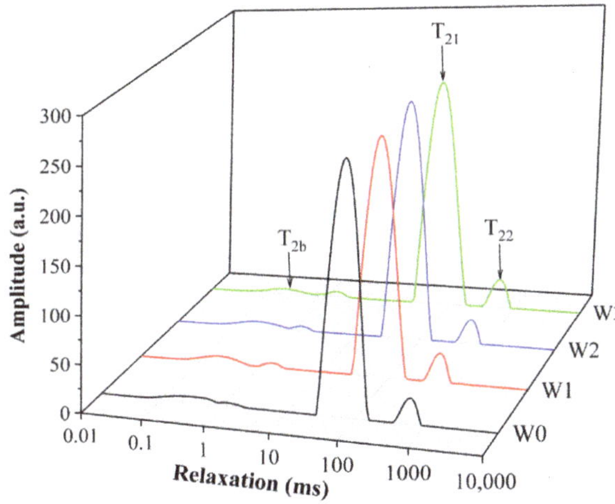

Figure 2. T_2 relaxation times of surimi gel with different washing times.

Table 5. Changes of T2 relaxation times of surimi gel with different washing times.

	W0	W1	W2	W3
T_{2b} (ms)	0.19 ± 0.00 [a]	0.21 ± 0.01 [b]	0.25 ± 0.01 [c]	0.40 ± 0.01 [d]
T_{21} (ms)	60.80 ± 0.00 [a]	60.80 ± 0.00 [a]	60.80 ± 0.00 [a]	60.80 ± 0.00 [a]
T_{22} (ms)	587.28 ± 0.00 [a]	587.28 ± 0.00 [a]	587.28 ± 0.00 [a]	587.28 ± 0.00 [a]
Peak areas T_{2b} (%)	2.78 ± 0.06 [a]	3.66 ± 0.03 [b]	3.74 ± 0.01 [b]	3.86 ± 0.01 [c]
Peak areas T_{21} (%)	93.18 ± 0.13 [b]	90.85 ± 0.15 [a]	90.95 ± 0.21 [a]	90.56 ± 0.17 [a]
Peak areas T_{22} (%)	4.05 ± 0.00 [a]	5.49 ± 0.007 [c]	5.31 ± 0.01 [b]	5.57 ± 0.00 [d]

Note: Different letters in the same line indicate significant differences ($p < 0.05$). T_{2b}: relaxation time of bound water; T_{21}: relaxation time of immobilized water; T_{22}: relaxation time of free water. Peak areas T_{2b}: percentage of T_{2b}; Peak areas T_{21}: percentage of T_{21}; Peak areas T_{22}: percentage of T_{22}.

3.6. Microstructures of Surimi Gel

The microstructure of surimi gels with different washing times are shown in Figure 3. All the surimi gels had a reticular structure and the cross-linking of myofibrillar protein gave the surimi gels a spatial structure that conferred some elasticity [7]. Bertram, Meyer, Wu, Zhou, and Andersen [40] demonstrated that the water in the protein structure of the surimi gels was locked and created pores. The W0 had a loose and rough surface with varying pore sizes, which was also the reason why the W0 had the lowest gel strength and WHC [21]. Compared with W0, the pores of surimi gels gradually became smaller with increasing washing, indicating that washing improved the spatial structure of surimi gels to some extent. Since washing removed substances that interfered with the gel structure (such as fat), it allowed myofibrillar protein to aggregate more efficiently, thereby inducing protein cross-linking and forming a denser gel network [21]. The W2 and W3 gels were denser and had similar microstructures. The smaller pores in W2 and W3 were uniformly distributed in the surimi gels, which had a finer and more uniform spatial structure that bound more water, giving the W2 and W3 groups of surimi a greater water retention capacity, which was consistent with their higher WHC, gel strength, water content, and peak area of T_{22} [39]. Kang et al. [41] showed that differences in the gel microstructure might be related to the pH of the gel and that interactions such as binding and aggregation of proteins in the gel network were dependent on the temperature and pH.

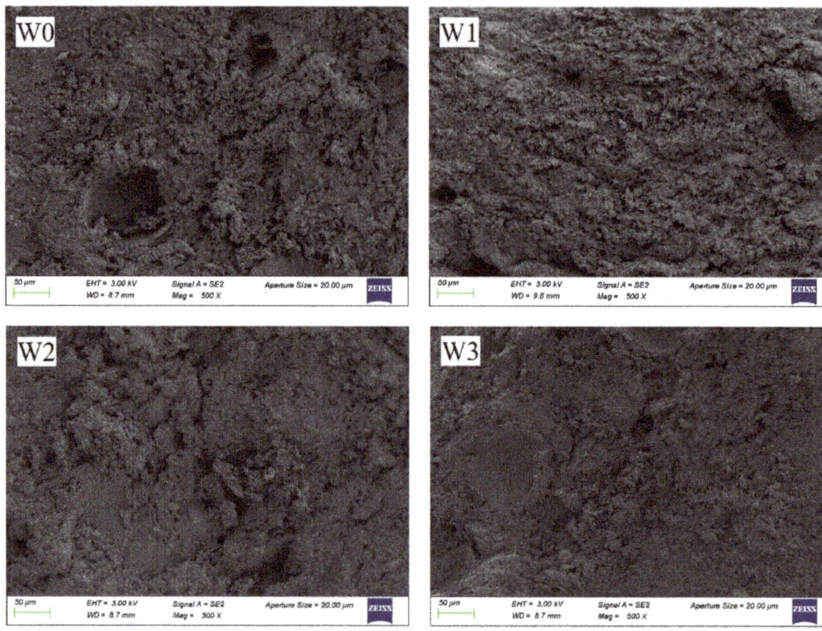

Figure 3. Microstructures (500× magnifications) of surimi gel with different washing times.

3.7. Protein Oxidation of Surimi during 4 °C Refrigeration

The extent of protein oxidation could be characterized by the carbonyl content. During storage, carbonyl compounds were produced following the breakage of amino acid side chains and peptide chains, or they could be produced by the covalent binding of proteins to oxidation products of sugars or fats in foods, making the carbonyl content an important indicator of the level of muscle protein oxidation [35]. Changes of the surimi with different washing times on the carbonyl group of the surimi during 4 °C refrigeration are shown in Figure 4a. At 0 d, there was no significant difference in the carbonyl content of the four groups of surimi (1.79–2.10 mmol/mg protein, $p > 0.05$). However, protein oxidation was very rapid during refrigeration (4 °C), and the highest values of carbonyl content (5.69–8.40 mmol/mg protein) were reached in all groups of surimi at 5 d. Among them, the carbonyl content of W0 increased rapidly, and the carbonyl content of W1, W2, and W3 increased more slowly. Li et al. [4] showed that the unwashed surimi had rapid protein oxidation during storage because of the high iron hemoglobin that was contained in the unwashed minced, whereas the washed surimi had more stable protein oxidation during storage. Some of the fat, myoglobin, pro-oxidant enzymes, and reducing sugars, which induce protein oxidation, were removed in the washing, which was able to inhibit protein carbonylation of surimi to some extent [5]. It is undeniable that lipid oxidation is critical as well as protein oxidation, which is closely related to the flavor and color of surimi samples. In addition, lipid oxidation and protein oxidation also influence each other. In the presence of carbonyl and reducing sugars, the Maillard reaction can also occur in the surimi samples during subsequent processing, resulting in color changes [42,43]. Therefore, the changes of protein oxidation, lipid oxidation, and gel quality during storage of different surimi samples still need further study.

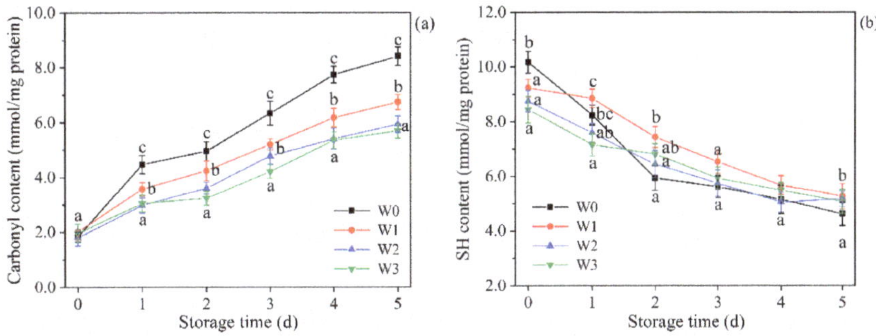

Figure 4. Changes of surimi with different washing times on the carbonyl group (**a**) and SH content (**b**) during 4 °C refrigeration. Different letters denote the significant difference ($p < 0.05$).

The combination of free radicals with hydrogen ions that are ionized by sulfhydryl groups causes the formation of disulfide bonds between (or within) protein molecules, resulting in a decrease in the total sulfhydryl content, and the change in the total sulfhydryl content also allows the evaluation of protein oxidation [44]. Figure 4b shows that the total sulfhydryl content of all four surimi groups showed a decreasing trend during storage, indicating that some sulfhydryl groups were converted to form disulfide bonds and protein oxidation occurred. During storage, oxygen continuously combined with protein, which in turn promoted the oxidation of protein. The sulfhydryl content decreased faster at 0–2 d and more slowly thereafter, which is consistent with the study of Eymard, Baron, and Jacobsen [5]. The sulfhydryl content of surimi tended to decrease as the washing times increased and washing significantly reduced the sulfhydryl content compared to W0 ($p < 0.05$). The head and tail of myosin include several reactive sulfhydryl groups. When the conformation of myosin changes, its reactive sulfhydryl groups might be exposed. Washing promotes the unfolding of the protein spatial structure, allowing buried sulfhydryl groups to be exposed [7].

3.8. SDS-PAGE Profiles of Surimi

Myosin, actin, and actomyosin are the most important proteins in myofibrillar protein [6]. Figure 5 shows the SDS-PAGE of surimi with different washing times during 4 °C refrigeration (0 and 5 d). The lanes of the electropherogram correspond to Marker, W0 (0 d of storage), W1 (0 d of storage), W2 (0 d of storage), W3 (0 d of storage), W0 (5 d of storage), W1 (5 d of storage), W2 (5 d of storage), and W3 (5 d of storage), respectively. The results showed that large molecular proteins (MW > 200 kDa) accumulated at the top of the separated gel at 0 d due to hydrophobic interaction. Myosin is a salt-soluble protein and its content was not affected by washing with water, so there was no significant difference in the intensity of the MHC bands of surimi with different washing times. The protein bands appeared below 25 kDa in all the groups, indicating the presence of myosin in surimi at different washing times. However, some smaller fish protein fractions, such as troponin C, were removed during the washing process [21]. After 5 d, protein molecules were degraded, as evidenced by the blurring, weakening, expansion, and disappearance of high molecular weight protein bands and the appearance or concentration enhancement of low molecular weight bands [45]. With the extension of storage time, the cross-linking of proteins caused fading of the bands of MHC, MLC, and paramyosin. Li et al. [4] showed that oxidation of surimi protein was generated by disulfide bonds causing the cross-linking of proteins, and the degree of oxidation was related to the processing method and the time of storage. MHC and actin were the main proteins for the gelation of surimi. The concentration of this band in surimi decreased after storage, indicating that storage was not conducive to gelation of surimi.

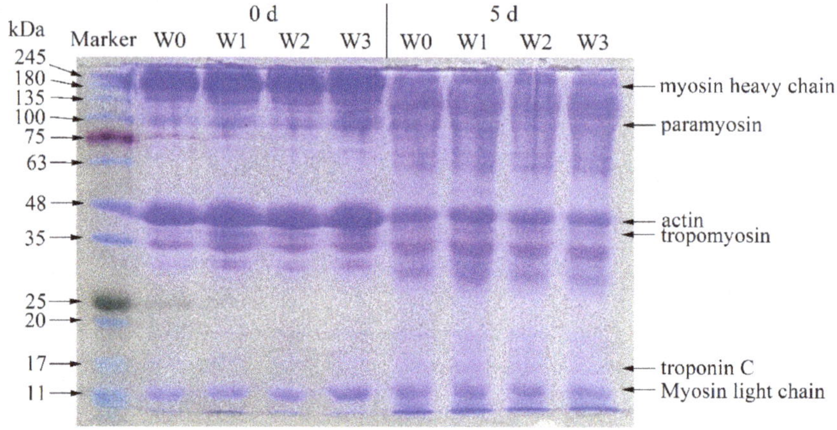

Figure 5. SDS-PAGE of surimi with different washing times during 4 °C refrigeration.

4. Conclusions

Increasing the washing times improved the physicochemical characteristics and delayed the protein oxidation of silver carp surimi, and W2 and W3 were considered to be the most suitable washes for surimi gels with the most gel strength, WHC, and best microstructure, as well as less fat oxidation and protein oxidation, and also significantly reduced the amount of water that was used for rinsing. On the other hand, the surimi yield of W2 was higher than that of W3. In practice, two washing times can be used to produce silver carp surimi to improve the quality of surimi gel. However, the effects of different rinsing treatments on protein oxidation, lipid oxidation, and gel properties during frozen storage of surimi need further study. While less washing treatment can improve the yield of surimi, the addition of antifreeze and antioxidants is still required to improve the quality of frozen surimi.

Supplementary Materials: The following supporting information can be downloaded at: https://www.mdpi.com/article/10.3390/foods11162397/s1, Figure S1: Gaussian curve fitting of amide I of surimi gel with different washing times.

Author Contributions: Conceptualization, X.Z., Y.Z. and Z.D.; methodology, X.Z. and H.D.; software, X.Z.; validation, X.Z., Y.Z. and W.Z.; formal analysis, X.Z.; investigation, X.Z., Y.Z., H.D., W.Z. and Z.D.; resources, Y.Z. and Z.D.; data curation, X.Z.; writing—original draft preparation, X.Z.; writing—review and editing, Y.Z. and Z.D.; visualization, X.Z.; supervision, Z.D.; project administration, Y.Z. and Z.D.; funding acquisition, Y.Z. and Z.D. All authors have read and agreed to the published version of the manuscript.

Funding: This research was funded by the National Key Research and Development Program of China, grant number 2019YFD0902000 and the Major Science and Technology Planed Program Projects in Xiamen City, grant number 3502Z20201032.

Institutional Review Board Statement: Not applicable.

Informed Consent Statement: Not applicable.

Data Availability Statement: Data is contained within the article or supplementary material.

Conflicts of Interest: The authors declare no conflict of interest.

References

1. Zhou, X.; Lin, H.; Zhu, S.; Xu, X.; Lyu, F.; Ding, Y. Textural, rheological and chemical properties of surimi nutritionally-enhanced with lecithin. *LWT-Food Sci. Technol.* **2020**, *122*, 108984. [CrossRef]
2. Wang, Y.F.; Li, B.F.; Zhang, Z.H. Effect of Rinsing on Gel Strength and Texture Properties of Sliver Carp Surimi. *Food Sci.* **2013**, *34*, 122–125. [CrossRef]
3. The State Statistical Bureau. *China Statistical Yearbook*; China Statistics Press: Beijing, China, 2020.
4. Li, P.; Yang, H.; Zhu, Y.; Wang, Y.; Bai, D.; Dai, R.; Ren, X.; Yang, H.; Ma, L. Influence of washing and cold storage on lipid and protein oxidation in catfish (*Clarias lazera*) surimi. *J. Aquat. Food Prod. Technol.* **2016**, *25*, 790–801. [CrossRef]
5. Eymard, S.; Baron, C.P.; Jacobsen, C. Oxidation of lipid and protein in horse mackerel (*Trachurus trachurus*) mince and washed minces during processing and storage. *Food Chem.* **2009**, *114*, 57–65. [CrossRef]
6. Priyadarshini, B.; Xavier, K.M.; Nayak, B.B.; Dhanapal, K.; Balange, A.K. Instrumental quality attributes of single washed surimi gels of tilapia: Effect of different washing media. *LWT-Food Sci. Technol.* **2017**, *86*, 385–392. [CrossRef]
7. Somjid, P.; Panpipat, W.; Cheong, L.Z.; Chaijan, M. Reduced Washing Cycle for Sustainable Mackerel (*Rastrelliger kanagurta*) Surimi Production: Evaluation of Bio-Physico-Chemical, Rheological, and Gel-Forming Properties. *Foods* **2021**, *10*, 2717. [CrossRef]
8. Granata, L.A.; Flick, G.J., Jr.; Martin, R.E. (Eds.) *The Seafood Industry: Species, Products, Processing, and Safety*; John Wiley & Sons: Hoboken, NJ, USA, 2012.
9. Zhang, L.; Li, Q.; Hong, H.; Luo, Y. Prevention of protein oxidation and enhancement of gel properties of silver carp (*Hypophthalmichthys molitrix*) surimi by addition of protein hydrolysates derived from surimi processing by-products. *Food Chem.* **2020**, *316*, 126343. [CrossRef]
10. AOAC. *Official Methods of Analysis of the Association of Analytical Chemists International*, 18th ed.; AOAC: Gaithersburg, MD, USA, 2005.
11. Xiong, G.; Cheng, W.; Ye, L.; Du, X.; Zhou, M.; Lin, R.; Geng, S.; Chen, M.; Corke, H.; Cai, Y.Z. Effects of konjac glucomannan on physicochemical properties of myofibrillar protein and surimi gels from grass carp (*Ctenopharyngodon idella*). *Food Chem.* **2009**, *116*, 413–418. [CrossRef]
12. Alipour, H.J.; Rezaei, M.; Shabanpour, B.; Tabarsa, M. Effects of sulfated polysaccharides from green alga Ulva intestinalis on physicochemical properties and microstructure of silver carp surimi. *Food Hydrocolloids* **2018**, *74*, 87–96. [CrossRef]
13. Priyadarshini, B.; Xavier, M.; Nayak, B.B.; Apang, T.; Balange, A.K. Quality characteristics of tilapia surimi: Effect of single washing cycle and different washing media. *J. Aquat. Food Prod. Technol.* **2018**, *27*, 643–655. [CrossRef]
14. Zhou, X.; Chen, T.; Lin, H.; Chen, H.; Liu, J.; Lyu, F.; Ding, Y. Physicochemical properties and microstructure of surimi treated with egg white modified by tea polyphenols. *Food Hydrocoll.* **2019**, *90*, 82–89. [CrossRef]
15. Zhang, F.; Fang, L.; Wang, C.; Shi, L.; Chang, T.; Yang, H.; Cui, M. Effects of starches on the textural, rheological, and color properties of surimi–beef gels with microbial tranglutaminase. *Meat Sci.* **2013**, *93*, 533–537. [CrossRef]
16. Jin, S.K.; Kim, I.S.; Kim, S.J.; Jeong, K.J.; Choi, Y.J.; Hur, S.J. Effect of muscle type and washing times on physico-chemical characteristics and qualities of surimi. *J. Food Eng.* **2007**, *81*, 618–623. [CrossRef]
17. Yin, T.; Yao, R.; Ullah, I.; Xiong, S.; Huang, Q.; You, J.; Hu, Y.; Shi, L. Effects of nanosized okara dietary fiber on gelation properties of silver carp surimi. *LWT-Food Sci. Technol.* **2019**, *111*, 111–116. [CrossRef]
18. Zhang, X.; Dai, Z.; Zhang, Y.; Dong, Y.; Hu, X. Structural characteristics and stability of salmon skin protein hydrolysates obtained with different proteases. *LWT-Food Sci. Technol.* **2022**, *153*, 112460. [CrossRef]
19. Shi, L.; Yin, T.; Huang, Q.; You, J.; Hu, Y.; Jia, D.; Xiong, S. Effects of filleting methods on composition, gelling properties and aroma profile of grass carp surimi. *Food Sci. Hum. Wellness* **2021**, *10*, 308–315. [CrossRef]
20. Liu, Q.; Chen, Q.; Kong, B.; Han, J.; He, X. The influence of superchilling and cryoprotectants on protein oxidation and structural changes in the myofibrillar proteins of common carp (*Cyprinus carpio*) surimi. *LWT-Food Sci. Technol.* **2014**, *57*, 603–611. [CrossRef]
21. Balange, A.K.; Benjakul, S. Effect of oxidised tannic acid on the gel properties of mackerel (*Rastrelliger kanagurta*) mince and surimi prepared by different washing processes. *Food Hydrocoll.* **2009**, *23*, 1693–1701. [CrossRef]
22. Zhang, X.; Zhang, Y.; Dong, Y.; Ding, H.; Chen, K.; Lu, T.; Dai, Z. Study on the mechanism of protein hydrolysate delaying quality deterioration of frozen surimi. *LWT-Food Sci. Technol.* **2022**, *167*, 113767. [CrossRef]
23. Somjid, P.; Panpipat, W.; Chaijan, M. Carbonated water as a novel washing medium for mackerel (*Auxis thazard*) surimi production. *J. Food Sci. Technol.* **2017**, *54*, 3979–3988. [CrossRef]
24. Chaijan, M.; Benjakul, S.; Visessanguan, W.; Faustman, C. Characteristics and gel properties of muscles from sardine (*Sardinella gibbosa*) and mackerel (*Rastrelliger kanagurta*) caught in Thailand. *Food Res. Int.* **2004**, *37*, 1021–1030. [CrossRef]
25. Chaijan, M.; Srirattanachot, K.; Panpipat, W. Biochemical property and gel-forming ability of surimi-like material from goat meat. *Int. J. Food Sci. Technol.* **2021**, *56*, 988–998. [CrossRef]
26. Fan, W.; Chi, Y.; Zhang, S. The use of a tea polyphenol dip to extend the shelf life of silver carp (*Hypophthalmicthys molitrix*) during storage in ice. *Food Chem.* **2008**, *108*, 148–153. [CrossRef]
27. Ning, H.Q.; Wang, Z.S.; Li, Y.Q.; Tian, W.L.; Sun, G.J.; Mo, H.Z. Effects of glycinin basic polypeptide on the textural and physicochemical properties of Scomberomorus niphonius surimi. *LWT-Food Sci. Technol.* **2019**, *114*, 108328. [CrossRef]
28. Phetsang, H.; Panpipat, W.; Undeland, I.; Panya, A.; Phonsatta, N.; Chaijan, M. Comparative quality and volatilomic characterisation of unwashed mince, surimi, and pH-shift-processed protein isolates from farm-raised hybrid catfish (*Clarias macrocephalus* × *Clarias gariepinus*). *Food Chem.* **2021**, *364*, 130365. [CrossRef] [PubMed]

29. Wongwichian, C.; Klomklao, S.; Panpipat, W.; Benjakul, S.; Chaijan, M. Interrelationship between myoglobin and lipid oxidations in oxeye scad (*Selar boops*) muscle during iced storage. *Food Chem.* **2015**, *174*, 279–285. [CrossRef]
30. Kong, B.; Geng, X.; Gao, X.; Li, Z. Effects of different rinsing methods on gel properties of silver carp surimi. *Food Ind.* **2000**, *1*, 41–43.
31. Baxter, S.R.; Skonberg, D.I. Gelation properties of previously cooked minced meat from Jonah crab (*Cancer borealis*) as affected by washing treatment and salt concentration. *Food Chem.* **2008**, *109*, 332–339. [CrossRef]
32. Karthikeyan, M.; Dileep, A.A.; Shamasundar, B.A. Effect of water washing on the functional and rheological properties of proteins from threadfin bream (*Nemipterus japonicus*) meat. *Int. J. Food Sci. Technol.* **2006**, *41*, 1002–1010. [CrossRef]
33. Sun, J.Y.; Qian, F.; Jiang, S.J.; Tuo, Y.F.; Mu, G.Q. Effect of heat treatments on the secondary structure of milk proteins analyzed by Fourier transform infrared spectroscopy. *Food Sci.* **2017**, *38*, 82–86. [CrossRef]
34. Liu, F.; Lin, W.; Li, L.; Wu, Y.; Yang, S.; Huang, H.; Yang, Q.; Lin, Z. Mechanism underlying protein changes during processing and gelation of sea bass surimi. *Food Sci.* **2020**, *41*, 15–22. [CrossRef]
35. Yuan, K.; Zhang, L.; Gu, D.; Fan, Z.; Wang, X.; Li, Y.; Liu, Y. Protein oxidation in minced Silver carp (*Hypophthalmichthys molitrix*) during washing process. *Food Ferment. Ind.* **2017**, *43*, 30–36. [CrossRef]
36. Li, F.; Wang, B.; Kong, B.; Shi, S.; Xia, X. Decreased gelling properties of protein in mirror carp (*Cyprinus carpio*) are due to protein aggregation and structure deterioration when subjected to freeze-thaw cycles. *Food Hydrocolloids* **2019**, *97*, 105223. [CrossRef]
37. Zhang, L.; Xue, Y.; Xu, J.; Li, Z.; Xue, C. Effects of high-temperature treatment (≥100°C) on Alaska Pollock (*Theragra chalcogramma*) surimi gels. *J. Food Eng.* **2013**, *115*, 115–120. [CrossRef]
38. Wang, R.; Gao, R.; Xiao, F.; Zhou, X.; Wang, H.; Xu, H.; Gong, C.; Huang, P.; Zhao, Y. Effect of chicken breast on the physicochemical properties of unwashed sturgeon surimi gels. *LWT-Food Sci. Technol.* **2019**, *113*, 108306. [CrossRef]
39. Pan, J.; Jia, H.; Shang, M.; Xu, C.; Lian, H.; Li, H.; Dong, X. Physiochemical properties and tastes of gels from Japanese Spanish mackerel (*Scomberomorus niphonius*) surimi by different washing processes. *J. Texture Stud.* **2018**, *49*, 578–585. [CrossRef] [PubMed]
40. Bertram, H.C.; Meyer, R.L.; Wu, Z.; Zhou, X.; Andersen, H.J. Water distribution and microstructure in enhanced pork. *J. Agric. Food Chem.* **2008**, *56*, 7201–7207. [CrossRef] [PubMed]
41. Kang, G.H.; Kim, S.H.; Kim, J.H.; Kang, H.K.; Kim, D.W.; Na, J.C.; Yu, D.J.; Suh, O.S.; Choi, Y.H. Effects of washing methods on gel properties of chicken surimi prepared from spent hen breast muscle. *Poult. Sci.* **2009**, *88*, 1438–1443. [CrossRef]
42. Wang, Z.; Tu, J.; Zhou, H.; Lu, A.; Xu, B. A comprehensive insight into the effects of microbial spoilage, myoglobin autoxidation, lipid oxidation, and protein oxidation on the discoloration of rabbit meat during retail display. *Meat Sci.* **2021**, *172*, 108359. [CrossRef]
43. Li, T.; Niu, L.; Li, X.; Wang, F.; Huang, Y.; Liu, Y. Formation of advanced glycation end-products in silver carp (*Hypophthalmichthys molitrix*) surimi products during heat treatment as affected by freezing-thawing cycles. *Food Chem.* **2022**, *395*, 133612. [CrossRef]
44. An, Y.; You, J.; Xiong, S.; Yin, T. Short-term frozen storage enhances cross-linking that was induced by transglutaminase in surimi gels from silver carp (*Hypophthalmichthys molitrix*). *Food Chem.* **2018**, *257*, 216–222. [CrossRef] [PubMed]
45. Ma, Y.; Xiong, S.; You, J.; Hu, Y.; Huang, Q.; Yin, T. Effects of vacuum chopping on physicochemical and gelation properties of myofibrillar proteins from silver carp (*Hypophthalmichthys molitrix*). *Food Chem.* **2018**, *245*, 557–563. [CrossRef] [PubMed]

foods

Article

Effect of pH-Shifting Process on the Cathepsin Activity, Muddy Off-Odor Compounds' Content and Gelling Properties of Isolated Protein from Silver Carp

Weidan Guo [1], Miao Zhan [1], Hui Liu [1], Xiangjin Fu [1,2,3,*] and Wei Wu [1]

[1] College of Food Science and Engineering, Central South University of Forestry and Technology, Changsha 410004, China
[2] Hunan Provincial Key Laboratory of Processed Food for Special Medical Purpose, Changsha 410004, China
[3] Hunan Provincial Engineering Technology Research Center of Seasonings Green Manufacturing, Changsha 410004, China
* Correspondence: drxjfu@163.com; Tel./Fax: +86-0731-85623240

Abstract: Silver carp (*Hypophthalmichthys molitrix*) is a potential source for making surimi products. However, it has the disadvantages of bony structures, high level of cathepsines and muddy off-odor which is mainly caused by geosmin (GEO) and 2-methylisoborneol (MIB). These disadvantages make the conventional water washing process of surimi inefficient (low protein recovery rate, and high residual muddy off-odor). Thus, the effect of the pH-shifting process (acid-isolating process and alkali-isolating process) on the cathepsins activity, GEO content, MIB content, and gelling properties of the isolated proteins (*IPs*) was investigated, comparing it with surimi obtained through the conventional cold water washing process (WM). The alkali-isolating process greatly boosted the protein recovery rate from 28.8% to 40.9% ($p < 0.05$). In addition, it removed 84% GEO and 90% MIB. The acid-isolating process removed about 77% GEO and 83% MIB. The acid-isolated protein (AC) displayed the lowest elastic modulus (G′), the highest TCA-peptide content (90.89 ± 4.65 mg/g) and the highest cathepsin L activity (65.43 ± 4.91 U/g). The AC modori (60 °C for 30 min) gel also demonstrated the lowest breaking force (226.2 ± 19.5 g) and breaking deformation (8.3 ± 0.4 mm), indicating that proteolysis caused by the cathepsin deteriorated the gel quality of AC. The setting (40 °C for 30 min) considerably increased the breaking force (386.4 ± 15.7 g) and breaking deformation (11.6 ± 0.2 mm) of the gel made from the alkali-isolated protein (AK) ($p < 0.05$). In AC and AK gel, a clearly visible cross-linking protein band with a molecular weight greater than MHC was seen, demonstrating the presence of endogenous trans-glutaminase (TGase) activity, that improved the gel quality of AK. In conclusion, the alkali-isolating process was an effective alternative method for making water-washed surimi from silver carp.

Keywords: silver carp (*Hypophthalmichthys molitrix*); pH-shifting process; muddy off-odor; gelling properties; cathepsin

Citation: Guo, W.; Zhan, M.; Liu, H.; Fu, X.; Wu, W. Effect of pH-Shifting Process on the Cathepsin Activity, Muddy Off-Odor Compounds' Content and Gelling Properties of Isolated Protein from Silver Carp. *Foods* **2023**, *12*, 939. https://doi.org/10.3390/foods12050939

Academic Editors: Tao Yin and Liu Shi

Received: 25 December 2022
Revised: 26 January 2023
Accepted: 13 February 2023
Published: 22 February 2023

1. Introduction

The huge annual yield and high nutritional value (rich in protein and polyunsaturated fatty acids, PUFAs) of silver carp (*Hypophthalmichthys molitrix*) are the main reasons for its global importance [1]. With the increase in the global population and the lack of protein resources, researchers have paid attention to the utilization of fish protein resources, such as the carps in fresh water in the United States [1]. Yet, due to its bony form and muddy off-odor, the commercial value of fresh silver carp is quite restricted [2]. The small bones also make the typical mechanical approach of meat recovery from carp inefficient [3,4].

Fortunately, the pH-shifting process developed by Hultin and Kelleher [5] is convenient for recovering protein from bony fish. This new method involves dissolving fish muscle protein at either low pH (2–3) or extremely high pH (10.5–12), removing small bones using high force centrifugation or filtering, and then adjusting the pH of the liquid to neutrality (pH 5–7) in order to recover the isolated protein (IP). This method has been carried out in catfish (*Silurus asotus*) [6,7], atlantic menhaden *(Brevoortia tyrannus)* [8], giant squid (*Architeuthis*) [9], tilapia (*Oreochromis mossambicus*) [10–12], Cape hake (*Merluccius capensis*) [13], yellow stripe trevally (*Selaroides leptolepis*) [14], jumbo squid (*Dosidicus gigas*) [15], kilka (*Clupeonella cultriventris*) [16,17], etc.

Additionally, the functional characteristics of silver carp protein recovered via the pH-shifting technique were revealed in the literature. This new process successfully eliminated contaminants such as bones, scales, skin, and fins from whole gutted carp [2,3], and the IPs could produce gels that were as strong as Alaska pollock surimi or stronger [18]. The principal gelling protein in muscle, myosin, was evaluated for its conformation and denaturation using differential scanning calorimetry (DSC), and the data showed that greater transition temperatures and myosin's enthalpy resulted in improved gelling qualities [2]. However, no literature has focused on the effect of the pH-shifting process on several other components such as cathepsins and trans-glutaminase (TGase) which also determine the gel properties of surimi.

The high activity of cathepsins is a major disadvantage of silver carp muscle for gel preparation [19,20], because cathepsins degrade myosin during the gelling temperature of 40–65 °C, resulting in a decrease in gel strength (called "modori"). In contrast, TGase could improve the gel strength by catalyzing formed cross-links (bonds of ε-(y-glutamyl) lysine) between proteins at 4–40 °C. Microbial TGase was found to really improve the mechanic properties of silver carp mince and surimi [4,21]. The pH-shifting process recovers a lot of sarcoplasmic proteins, including the cathepsines and TGase, but they might denature at different degrees in the extreme pH conditions of the pH-shifting process, surely affecting the gelling properties of IPs in different ways.

A muddy off-odor is another disadvantage of silver carp muscle being used as food. A muddy off-odor in silver carp is caused mainly by geosmin (GEO) and 2-methylisoborneol (MIB), with a threshold of about 0.7 μg/kg [22]. Due to their lipophilic and hemi-volatile nature, they are exceedingly difficult to be removed via traditional processes, such as washing, drying, curing and fermentation [22]. The muddy off-odor in catfish and tilapia was reportedly removed extremely effectively [6,7] using the pH-shifting technique [23]. There is no information available regarding how this treatment affects the silver carp's muddy off-odor.

Thus, it is very interesting to look at the impact of pH-shifting on the cathepsin activity, TGase activity, GEO content and MIB content in silver carp. Additionally, the gelation properties of IPs were also determined in this study.

2. Materials and Methods

2.1. Materials

Silver carp fillets were bought from a local supermarket. Fresh silver carps (1 kg average) were slaughtered, headed, gutted, filleted and washed (4 °C) manually.

All chemical reagents used were analytical grade and were purchased from Sinopharm Chemical Reagent Co., Ltd. (Shanghai, China) or Sigma Aldrich Co. (St. Louis, MO, USA).

2.2. Methods

2.2.1. Mince Collection

Skinless, boneless mince for the conventional water washing process was obtained by passing the fillet through a mini belt deboner (Hengchang Machinery Manufacture Co., Ltd., Linyi, China). The aperture of the machine was 5 mm (diameter).

The fillets for the pH-shifting process were removed from the back bone and skin manually (regardless of the small bones and connective tissue), and then cut into slices.

2.2.2. Preparation of Conventional Water-Washed Surimi

The water-washed surimi (WM) was prepared according to Luo et al. (2008) [4]. The mince was mixed three times (v/w) with cold water (4 °C), stirred for 3 min and left to stand for 15 min (4 °C). Then, we hand-extruded and filtered the washed meat with three layers of coarse cotton. Washing and filtration processes were repeated three times. The water-washed surimi was wrapped in polyethylene bags and stored on ice in a cold closet until its use on the same day.

2.2.3. The pH-Shifting Process

The procedure was performed according to Hultin [5] with slight modification. Fish slices were homogenized (DS-1 high speed homogenizer, Sample's Model Factory, Shanghai, China) 9 times (v/w) with cold water at 10,000 r/min for 30 s. Then, the meat slurry was stirred and adjusted the pH value using HCl or NaOH solution (1 mol/L). The pH values were 11.8 and 2.3 for alkali- and acid-isolating processes, respectively. After adjusting the pH to the desired (expected) value, it was continuously stirred for 30 min, followed by centrifuging (4 °C, 10,000× g for 10 min), and discarded the sediment (small bones, connective tissue and scales), adjusted the pH value of the supernatant to 5.5 using HCl or NaOH solution (1 mol/L) and then centrifuged (4 °C, 10,000× g for 30 min). The supernatant was discarded, and the isolated proteins (IPs) in the sediment, including the acid-isolated protein (AC) or alkali-isolated protein (AK), were collected and packed in polyethylene bags, stored in an ice-cold closet and used within four hours.

2.2.4. Protein Recovery Rate

After the fish fillets and IPs were recovered, the weight was recorded and the weight ratio was calculated, the ratio was the recovery rate (at the same moisture content). The recovery of the protein was calculated as follows:

$$\text{Protein recovery } (\%) = \frac{\text{Weight of recovered protein}}{\text{Weight of initial fillet}} \times 100$$

2.2.5. Cathepsin Activity

Samples were extracted with 25 mmol/L sodium acetate (v/w) buffer solution which contains 5 mmol/L cysteine and 0.3 mmol/L phenylmethylsulfonyl fluoride (PMSF), and the pH of the extraction was adjusted to 5.0, then centrifuged at 12,000× g for 2 min and repeated 4 times. The supernatant was collected as the enzyme extraction.

Enzyme activity was determined according to Liu et al. [20]. Briefly, Z-Phe-Arg-MCA and Z-Arg-Arg-MCA were used as substrates for cathepsin L and cathepsin B, respectively. The enzyme and substrate were blended, and then incubated at 40 °C for 10 min. The fluorescent intensity of aminomethylcoumarin (AMC) released via hydrolysis was determined in a fluorescence spectrophotometer (LS55, Perkin Elmer, Waltham, MA, USA) with 380 nm of excited wavelength and 460 nm of emitted wavelength. The amount of activity releasing 1 nmol of AMC per minute was recorded as one unit of enzyme activity.

2.2.6. Determination of GEO and MIB

The GEO and MIB were extracted based on the method of Phetsang [6] and Fu [24]. Microwave-mediated distillation and solid-phase microextraction (SPME) were used for extraction, followed by gas chromatography mass spectrometry (GC/MS) for analysis, using the PEG-20M capillary column (30 m × 0.25 mm × 0.25 μm). The helium carrier gas (99.99% purity) flow rate was 1.0 mL/min. The initial temperature of the GC oven was 60 °C, then it increased to 120 °C at 10 °C/min rate, then it increased to 230 °C at 20 °C/min rate and it was held for 1 min. The injector temperature was 200 °C, and the ionization energy (EI) was 70 eV; the EI ionization source was 230 °C. The detector was set in the electron impact (EI) ionization mode at 70 eV in full scan mode with a mass/charge ratio (m/z) range from 50 to 500. The characteristic ion mass/charge ratio (m/z) of MIB and GEO was 95 and 112, respectively.

Gas chromatograph grade MIB (Sigma Aldrich Co., Ltd., St. Louis, MO, USA) and GEO were used as the exterior standards.

2.2.7. Rheology Studies

The function of temperature in the gel forming procedure was determined using an AR-1000 rheometer (TA Instruments, New Castle, DE, USA). The WM and IPs were diluted with 0.60 mol/L KCl and 0.02 mol/L Tris-HCl buffer (pH 7.0) at 4 °C, until the final protein concentration was 35 mg/mL. The solution was then homogenized at 10,000 r/min for 30 s, before the air bubbles were removed via vacuuming. The conditions of the rheometer were as follows: gap 1 mm, steel cone 2°, shear strain 0.02, shear frequency 0.1 HZ and the temperature increased from 20 °C to 85 °C at 1.5 °C/min.

2.2.8. Heat-Induced Gel

The WM and IPs were chopped by a food processor (La Minerva 22A/22L, Bologna, Italy) (10 °C) for 6 min, and then sucrose (2 g/100 g), sorbitol (2 g/100 g), NaCl (3 g/100 g) and polyphosphate sodium (0.3 g/100 g) were added, with the addition of crushed ice and the moisture content being adjusted to 80 g/100 g. The paste was put into a polyvinylidine chloride casing with a diameter of 2.5 cm and was sealed tightly at both ends. The paste was incubated directly at 85 °C for 30 min (kamaboko gel), or at 40 °C (setting gel) or 60 °C (modori gel) for 30 min, and was then heated at 85 °C for 30 min in a water bath [8]. After all of the gels were heated, they were immediately cooled in ice water for 30 min, and then stored at 4 °C for analysis the next day.

2.2.9. Determination of the Gel Strength: Hardness and Elasticity

Gel hardness and elasticity were determined by a puncture test using a TA-XT2 texture analyzer (Stable Micro Systems, Godalming, UK) which was equipped with a 5 mm cylinder plunger (P/0.5). Gels were evaluated and equilibrated at ambient temperature (28–30 °C) for 2 h. Five cylinder-shaped samples (d = 2 cm) which had a length of 2.5 cm were measured. When the gel was compressed with a speed of 60 mm/min and operating depth of 2 cm, the load value (g) at the broken point indicated the hardness, while the breaking deformation (mm) represented the elasticity [25].

2.2.10. TCA-Soluble Peptide Measurement

TCA-soluble peptide was determined according to the Lowry method and the content was expressed as a mg tyrosine/g sample [25].

2.2.11. Determination of Disulfide Bond Content

The disulfide bond was assayed using 5,5′-dithiobis-(2-nitrobenzoic acid) (DTNB) according to the method of Benjakul [26]. The disulfide bond was calculated using the extinction coefficient of 13,900/mol/cm.

2.2.12. SDS-Polyacrylamide Gel Electrophoresis

Gel electrophoresis was carried out according to Laemmli [27] using 4% stacking gel and 10% acrylamide separating gel. The sample buffer was triseglycine (pH 8.8), which contained 1 g/100 mL SDS and 1 mL/100 mL L-mercaptoethanol. The sample concentration was 1 mg/mL, and the loading volume was 15 μL. Electrophoresis (Bio-RAD Mini Protein system, Bio-Rad, Hercules, CA, USA) was carried out at a constant voltage of 120 V. The film was fixed with 15% trichloroacetic acid for 30 min, dyed with 0.125% (w/w) R-250 for 12 h and then decolorized with 40% ethanol (v/v) + 10% acetic acid (v/v).

2.2.13. Statistical Analysis

All chemical analyses were carried out in triplicate. In physical analyses, e.g., structure property, at least 5 measurements for each treatment were conducted. Data were subjected to analysis of variance analysis (ANOVA) using an SPSS package (SPSS 17.0 for Windows, SPSS Inc., Chicago, IL, USA).

3. Results and Discussion

3.1. Protein Recovery, Activity of Cathepsin, Content of GEO and MIB

As shown in Table 1, the protein recovery rates of the water washing process and alkali- and acid-isolating processes were 28.8%, 38.5% and 40.9%, respectively. The pH-shifting method extended the recovery rate considerably ($p < 0.05$) and the consequences were similar to the preceding studies of other researchers [1,2] because the water soluble sarcoplasmic proteins were recovered via a new method but went away with the water via the conventional water wash process. Moreover, numerous amounts of meat sticking to the small bones were discarded in the mechanical deboning of the conventional process but were recovered in the new process, which also increased the recovery rate.

Table 1. The effect of the process on the protein recovery rate and residue activity of cathepsin B and cathepsin L.

	Protein Recovery (%)	Cathepsin B (U/g)	Cathepsin L (U/g)	GEO (ug/kg)	MIB (ug/kg)
Muscle	-	152.33 ± 8.21 [a]	92.51 ± 6.20 [a]	1.51 ± 0.08 [a]	2.06 ± 0.11 [a]
WM	28.8 ± 0.87 [b]	7.79 ± 2.13 [c]	29.14 ± 1.59 [c]	0.92 ± 0.05 [b]	1.02 ± 0.05 [b]
AC	38.5 ± 0.59 [a]	25.58 ± 4.85 [b]	65.43 ± 4.91 [b]	0.35 ± 0.03 [c]	0.33 ± 0.03 [c]
AK	40.9 ± 1.24 [a]	8.53 ± 2.60 [c]	27.67 ± 2.38 [c]	0.26 ± 0.04 [d]	0.21 ± 0.03 [d]

[a–d] Means in each column with different superscript letters are significantly different ($p < 0.05$) (n = 3). GEO indicates geosmin, MIB indicates 2-methylisoborneol. WM: water-washed surimi, AC: acid-isolated protein, AK: alkali-isolated protein.

As shown in Table 1, the cathepsins were effectively removed ($p < 0.05$). The muscle contained more cathepsin B than cathepsin L, while WM and IPs retained more cathepsin L than cathepsin B, indicating that cathepsin L bound more tightly with the muscle protein than with cathepsin B. The AC contained much higher activity of cathepsin L which might be due to the high activation of cathepsin L during acid treatment. Heidtmann et al. [28] showed that cathepsin L activity could be enhanced by acid treatment.

The muddy off-odor compounds in the silver carp muscle were effectively removed during the pH-shifting process by about 77% for GEO and 83% for MIB in the acid-isolating process, and 84% for GEO and 90% for MIB in the alkali-isolating process, respectively, while water washing removed about 39% of GEO and 50% of MIB (Table 1). This was similar with the results of Kleinholz Christina et al. [23], who found that the change process of the pH value could significantly reduce the odor of catfish ($p < 0.001$); MIB and GEO were reduced from 1.396 and 1.992 µg/kg to 0.104 µg/kg MIB and 0.258 µg/kg GEO, respectively, using phosphoric acid and to 0.0987 µg/kg MIB and 0.426 µg/kg GEO using NaOH. The removal of GEO and MIB was the result of the lipid reduction in the pH-shifting process, due to the lipophilic nature of GEO and MIB [7].

3.2. Rheology Properties

The elastic modulus (G′) of WM and IPs was shown in Figure 1. The G′ of WM included two peaks, at 47 °C and 76 °C, respectively, and a minimal peak at 57 °C, which was similar to the report of Abdollahi et al. [17]. The first G′ peak in the 40–50 °C range corresponded to the denature of the myosin head, with conformational changes and the exposure of reactive groups. This ensured a gradual sol–gel transition to form an ordered initial protein grid. Subsequently, the properties of light meromyosin changed and caused a decrease In G′, resulting in increased fluidity [29]. In addition, it has been increasingly accepted that the first increase in G′ involves the TGase-mediated covalent [26]. The second increase in G′ started at 57 °C, the peak was 76 °C and it corresponded to the myosin rod denature (at about 63 °C) and actin denature (at about 68 °C) [3]. In this step, the final kamaboko network was formed. In addition to the disulfide bonding, the formation of intermolecular hydrophobic interactions also occurs during heating [30,31].

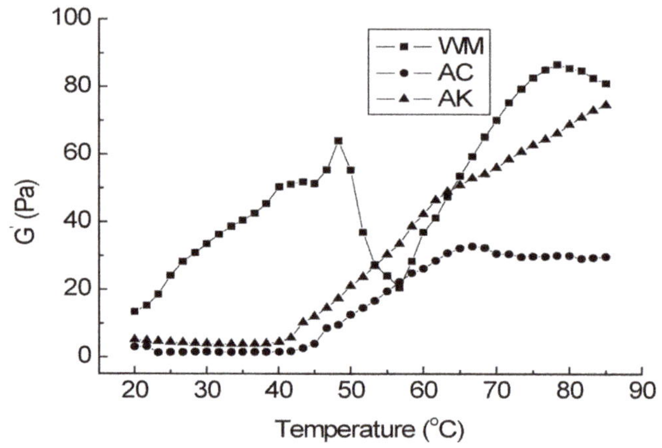

Figure 1. Rheology of surimi and isolated proteins. WM: water-washed surimi, AC: acid-isolated protein, AK: alkali-isolated protein.

The initial G′ of AC and AK was considerably lower than WM, indicating less molecule interaction in AC and AK. The G′ of AC and AK increased at about 42 °C, and the peak at 47 °C disappeared, indicating the denaturation or aggregation of the myosin head. According to Kristinsson and Hultin [32], the myosin partly unfolded in the extreme pH of the pH-shifting process, resulting in a "molten globular" state; when they then adjusted the pH to 5.5, the globulin rod refolded to the native structure, whereas the globulin head partially misfolded. Furthermore, it had to be noted that a typical 'modori' was observed for AC (the G′ of AC declined above 65 °C). The proteolytic degradation due to cathepsin enzymes was supposed to be the mechanism responsible for 'modori' [19,20], which is consistent with the result that AC retained the highest cathepsin L activity (Table 1).

3.3. The Content of TCA-Soluble Peptide

The TCA-soluble peptide is shown in Table 2. TCA-soluble peptides indicate the degree of hydrolysis of the muscle protein: the higher the content, the greater the degree of hydrolysis [25]. The TCA-soluble peptide content in WM was less than that of AC and AK ($p < 0.05$). In comparison to modori gels, the TCA-soluble peptide content of kamaboko gel was lower ($p < 0.05$). The highest TCA-soluble peptide content (90.89 mg/g) of the modori gel of AC indicates that serious hydrolysis of the proteins occurred. These results were consistent with the results of cathepsins (Table 1).

3.4. The Content of Disulfide

Numerous disulfide bonds were formed during gelling; however, there was no significant difference ($p > 0.05$) within the disulfide content of various gels. It is acknowledged that disulfide (S-S) bonds are crucial to the way that heat causes the protein to gel. SH oxidation into S-S bonds and/or the SH-induced S-S interchange could cause the covalent cross-linking of protein molecules. The disulfide bond is considered to be the main covalent bond formed in protein gel during high temperature heating (cooking at >40 °C). According to the report of Hossain et al. [33], the formation of disulfide bond polymerization in the

WM gel was almost constant, and it was proposed that disulfide bond polymerization occurred during cooking at 80 °C.

Table 2. TCA-soluble peptide content (mg/g) and disulfide content (mol/10^7 g protein) of WM, IPs and gels prepared under different conditions.

Samples	Peptide Content (mg/g)			S-S Content (mol/10^7 g Protein)		
	WM	AC	AK	WM	AC	AK
Surimi/IPs	31.71 ± 1.32 ^{c}	46.85 ± 2.50 ^{a}	36.37 ± 1.87 ^{b}	2.49 ± 0.12 ^{C}	3.97 ± 0.18 ^{B}	4.46 ± 0.21 ^{A}
Kamaboko	44.18 ± 2.41 ^{b}	61.42 ± 3.08 ^{a}	47.05 ± 1.35 ^{b}	15.26 ± 1.10 ^{A}	15.01 ± 0.81 ^{A}	15.14 ± 1.25 ^{A}
Setting gel	52.56 ± 2.64 ^{c}	77.55 ± 2.52 ^{a}	61.27 ± 3.99 ^{b}	15.35 ± 0.96 ^{A}	15.30 ± 0.64 ^{A}	15.42 ± 0.77 ^{A}
Modori gel	62.92 ± 2.86 ^{c}	90.89 ± 4.65 ^{a}	76.84 ± 2.15 ^{b}	15.18 ± 1.05 ^{A}	14.81 ± 1.32 ^{A}	15.20 ± 1.19 ^{A}

$^{a–c}$ Means in each row of peptide content with different superscript letters are significantly different ($p < 0.05$) (n = 3). $^{A–C}$ Means in each row of S-S content with different superscript letters are significantly different ($p < 0.05$) (n = 3). WM: water-washed surimi, AC: acid-isolated protein, AK: alkali-isolated protein, IPs: the isolated proteins.

3.5. Protein Pattern of WM, IPs and Gels

A decrease in the myosin heavy chain (MHC, 200 kDa) band intensity was determined in all modori gels compared with their corresponding kamaboko and setting gel (Figure 2). The lowest MHC band intensity was found in AC's modori gel. The reduction in the MHC of AC gels was synchronal with an increase in TCA-soluble peptide content (Table 2), and AC retained more cathepsins (Table 1).

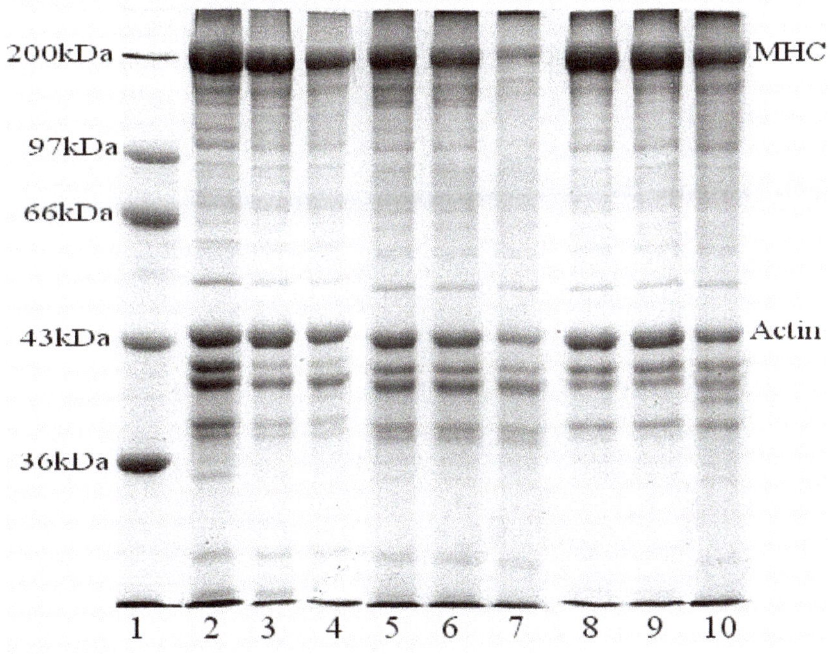

Figure 2. Protein patterns of gels prepared under different conditions: 1, standard proteins; 2, kamaboko gel of WM; 3, setting gel of WM; 4, modori gel of WM; 5, kamaboko gel of AC; 6, setting gel of AC; 7, modori gel of AC; 8, kamaboko gel of AK; 9, setting gel of AK; and 10, modori gel of AK. WM: water-washed surimi, AC: acid-isolated protein, AK: alkali-isolated protein.

The gels of AC and AK showed obvious cross-linked protein bands with higher molecular weight than MHC (Figure 2). This might be due to the endogenous TGase, which can induce protein cross-linking, because the cross-linking of water-washed surimi gel proteins mainly includes the disulfide bond and TGase-mediated bond of ε-(y-glutamyl) lysine, and the disulfide content of gels was almost the same (Table 2). Because sarcoplasmic proteins are retained throughout the pH-shifting process, IPs might retain more endogenous TGase than conventionally washed surimi, and it was indicated that the extreme pH of the pH-shifting method did not totally inactivate the TGase. In addition, the conformational changes in the myofibrillar protein during the acid and alkaline processes could expose more functional groups, which were used for TGase-induced cross-linking [32].

3.6. Gel Strength

The breaking force and deformation force of kamaboko gel prepared from WM were higher than those from AC and AK, and the breaking force of the AC gel was the lowest ($p < 0.05$) (Table 3). These results were consistent with Paker's report [3] about silver carp, finding that IP from the alkali-isolating process (solubilizing pH 11.5) showed significantly ($p < 0.05$) higher shear stress than IP from the acid-isolating process (solubilizing pH 2.5).

Table 3. Breaking force and breaking deformation of gel prepared under different processes and conditions.

Surimi/IPs		Breaking Force (g)	Deformation (mm)
WM	Kamaboko gel	467.2 ± 20.6 ^{b}	12.1 ± 0.3 ^{b}
	Setting gel	489.1 ± 16.1 ^{b}	12.3 ± 0.3 ^{b}
	Modori gel	315.9 ± 18.8 ^{e}	10.8 ± 0.5 ^{de}
AC	Kamaboko gel	347.5 ± 24.7 ^{e}	11.7 ± 0.3 ^{cd}
	Setting gel	302.6 ± 20.4 ^{e}	10.3 ± 0.4 ^{e}
	Modori gel	226.2 ± 19.5 ^{f}	8.3 ± 0.4 ^{f}
AK	Kamaboko gel	435.8± 24.6 ^{c}	11.9 ± 0.4 ^{bc}
	Setting gel	518.9 ± 21.4 ^{a}	12.8 ± 0.3 ^{a}
	Modori gel	386.4 ± 15.7 ^{d}	11.6 ± 0.2 ^{cd}

$^{a–f}$ Means in each column with different superscript letters are significantly different ($p < 0.05$) (n = 5). WM: water-washed surimi, AC: acid-isolated protein, AK: alkali-isolated protein, IPs: the isolated proteins.

Although no peak was observed at the setting temperature (about 40 °C) in the G′ curve (Figure 1) for AK, setting improved the breaking force and deformation of the AK gel considerably ($p < 0.05$), as well as the WM gel, indicating that WM and AK contain notable TGase activity, in accordance with the result of SDS-PAGE (Figure 2). It has long been assumed that the activity of endogenous Tgase plays a significant role in setting. Tsukamasa et al. [34] reported that the index of breaking strength and the index of the (γ-glutamyl) lysine cross-link concentration had a high correlation (r = 0.987). The breaking force and breaking deformation of AC gels were dramatically reduced by Modori ($p < 0.05$), which might be explained by the fact that AC retained more cathepsins (Table 1). The proteolytic degradation of myofibrillar proteins, particularly myosin, resulted in a decrease in molecular mass and a loss in structural domains, which are essential for molecular interaction [35], such as forming gel structure. The proteolytic degradation of myofibrillar has an adverse effect on the quality of water-washed surimi and significantly reduces the gel strength. Taking into account the result of rheology (Figure 1) and cathepsin activity (Table 1), it was certain that the cathepsins retained in AC were key components in causing the gel deterioration.

4. Conclusions

The pH-shifting process increased the protein recovery rate and reduced the muddy off-odor compounds in silver carp effectively.

The AC retained the highest cathepsin L activity. Furthermore, the AC showed a higher content of the TCA-soluble peptide and a lower intensity of the MHC band, and the modori gel of AC demonstrated the lowest breaking deformation and breaking force which also proved the presence of cathepsins in AC.

Setting improved the breaking force and breaking deformation of AK gel considerably ($p < 0.05$), and an apparent cross-linking protein band with molecular mass on top of MHC was determined in AC and AK gels, indicating that the Tgase might be partly retained in the pH-shifting process. In conclusion, the alkali-isolating process was an effective alternative method for making surimi from silver carp.

Author Contributions: W.G.: conceptualization, formal analysis, writing—original draft. M.Z.: conceptualization, writing—original draft. H.L.: formal analysis, writing—original draft. X.F.: supervision, project administration, writing—review and editing. W.W.: resources, the SDS-PAGE analysis. All authors have read and agreed to the published version of the manuscript.

Funding: This research was funded by Key R&D Program of Hunan Province grant number [2022NK2032].

Data Availability Statement: Anybody need the data could contact the corresponding author.

Acknowledgments: We thank Kangzi Ren for her help in writing.

Conflicts of Interest: All authors declare that they have no conflict of interest.

References

1. Taskaya, L.; Chen, Y.; Jaczynski, J. Functional properties of proteins recovered from silver carp (*Hypophthalmichthys molitrix*) by isoelectric solubilization/precipitation. *LWT-Food Sci. Technol.* **2009**, *42*, 1082–1089. [CrossRef]
2. Paker, I.; Beamer, S.; Jaczynski, J.; Matak, K.E. The effect of organic acids on gelation characteristics of protein gels made from silver carp (*Hypophthalmichthys molitrix*) protein recovered by isoelectric solubilization and precipitation. *LWT-Food Sci. Technol.* **2013**, *53*, 37–43. [CrossRef]
3. Paker, I.; Beamer, S.; Jaczynski, J.; Matak, K.E. Compositional Characteristics of Materials Recovered from Headed Gutted Silver Carp (*Hypophthalmichthys molitrix*) by Isoelectric Solubilization and Precipitation Using Organic Acids. *J. Food Sci.* **2013**, *78*, 445–451. [CrossRef] [PubMed]
4. Luo, Y.; Shen, H.; Pan, D.; Bu, G.H. Gel properties of surimi from silver carp (*Hypophthalmichthys molitrix*) as affected by heat treatment and soy protein isolates. *Food Hydrocoll.* **2008**, *22*, 1513–1519. [CrossRef]
5. Hultin, H.O.; Kelleher, S.D. Process for Isolating a Protein Composition from a Muscle Source and Protein Composition. U.S. Patent 08/797,929, 12 February 1997.
6. Phetsang, H.; Panpipat, W.; Undeland, I.; Panya, A.; Phonsatta, N.; Chaijan, M. Comparative quality and volatilomic characterisation of unwashed mince, surimi, and pH-shift-processed protein isolates from farm-raised hybrid catfish (*Clarias macrocephalus* × *Clarias gariepinus*). *Food Chem.* **2021**, *364*, 130365. [CrossRef]
7. Yarnpakdee, S.; Benjakul, S.; Penjamras, P.; Kristinsson, H.G. Chemical compositions and muddy flavour/odour of protein hydrolysate from Nile tilapia and broadhead catfish mince and protein isolate. *Food Chem.* **2014**, *142*, 210–216. [CrossRef]
8. Perez-Mateos, M.; Lanier, T.C. Comparison of Atlantic menhaden gels from surimi processed by acid or alkaline solubilization. *Food Chem.* **2006**, *101*, 1223–1229. [CrossRef]
9. Palafox, H.; Cordova-Murueta, J.H.; Toro, M.A.N.D.; Garcia-Carreno, F.L. Protein isolates from jumbo squid (*Dosidicus gigas*) by pH-shift processing. *Process Biochem.* **2009**, *44*, 584–587. [CrossRef]
10. Zhou, Y.G.; Yang, H.S. Effects of calcium ion on gel properties and gelation of tilapia (*Oreochromis niloticus*) protein isolates processed with pH shift method. *Food Chem.* **2019**, *277*, 327–335. [CrossRef]
11. Zhou, Y.G.; Liu, J.J.H.; Kang, Y.; Cui, H.J.; Yang, H.S. Effects of acid and alkaline treatments on physicochemical and rheological properties of tilapia surimi prepared by pH shift method during cold storage. *Food Res. Int.* **2021**, *145*, 110424. [CrossRef]
12. Kobayashi, Y.; Park, J.W. Biochemical and physical characterizations of fish protein isolate and surimi prepared from fresh and frozen whole fish. *LWT-Food Sci. Technol.* **2017**, *77*, 200–207. [CrossRef]
13. Tomé, A.S.; Pires, C.; Batista, I.; Sousa, I.; Raymundo, A. Protein gels and emulsions from mixtures of Cape hake and pea proteins. *J. Sci. Food Agric.* **2015**, *95*, 289–298. [CrossRef] [PubMed]
14. Arfat, Y.A.; Benjakul, S. Effect of zinc sulphate on gelling properties of phosphorylated protein isolate from yellow stripe trevally. *Food Chem.* **2013**, *141*, 2848–2857. [CrossRef] [PubMed]
15. Cortés-Ruiz, J.A.; Pacheco-Aguilar, R.; Ramírez-Suárez, C.J.; Lugo-Sánchez, M.E.; García-Orozco, K.D.; Sotelo-Mundo, R.R.; Peña-Ramos, A. Conformational changes in proteins recovered from jumbo squid (*Dosidicus gigas*) muscle through pH shift washing treatments. *Food Chem.* **2016**, *196*, 769–775. [CrossRef] [PubMed]

16. González-González, D.C.; Lugo-Sánchez, M.E.; García-Sifuentes, C.O.; Ramírez-Suárez, J.C.; Pacheco-Aguilar, R. Influence of pH, ionic strength and isoascorbic acid on the gel-forming ability of Jumbo squid muscle (*Dosidicus gigas*). *Food Chem.* **2021**, *337*, 127993. [CrossRef] [PubMed]
17. Abdollahi, M.; Rezaei, M.; Jafarpour, A.; Undeland, I. Dynamic rheological, microstructural and physicochemical properties of blend fish protein recovered from kilka (*Clupeonella cultriventris*) and silver carp (*Hypophthalmichthys molitrix*) by the pH-shift process or washing-based technology. *Food Chem.* **2017**, *229*, 695–709. [CrossRef] [PubMed]
18. Taskaya, L.; Chen, Y.C.; Jaczynski, J. Color improvement by titanium dioxide and its effect on gelation and texture of proteins recovered from whole fish using isoelectric solubilization/ precipitation. *LWT-Food Sci. Technol.* **2010**, *43*, 401–408. [CrossRef]
19. Liu, H.; Yin, L.J.; Zhang, N.; Li, S.H.; Ma, C.W. Purification and characterization of cathepsin L from the muscle of silver carp (*Hypophthalmichthys molitrix*). *J. Agric. Food Chem.* **2006**, *54*, 9584–9591. [CrossRef]
20. Liu, H.; Yin, L.J.; Zhang, N.; Li, S.H.; Ma, C.W. Isolation of cathepsin B from the muscle of silver carp (*Hypophthalmichthys molitrix*) and comparison of cathepsins B and L actions on surimi gel softening. *Food Chem.* **2008**, *110*, 310–318. [CrossRef]
21. Fang, M.X.; Luo, X.Y.; Xiong, S.B.; Yin, T.; Hu, Y.; Liu, R.; Du, H.Y.; Liu, Y.M.; You, J. In vitro trypsin digestion and identification of possible cross-linking sites induced by transglutaminase (TGase) of silver carp (*Hypophthalmichthys molitrix*) surimi gels with different degrees of cross-linking. *Food Chem.* **2021**, *364*, 130443. [CrossRef]
22. Yamprayoon, J.; Noomhorm, A. Effects of preservation methods on geosmin content and off-flavor in Nile tilapia (*Oreochromis niloticus*). *J. Aquat. Food Prod. Technol.* **2000**, *9*, 95–107. [CrossRef]
23. Kleinholz Christina, M.D.; Kleinholz, C.W.; Vann, D.G.; Bilby, C.A.; Schrader, K.K. Evaluation of Acid and Alkaline Processing to Remove Muddy Off-Flavors in Channel Catfish (*Ictalurus punctatus*). *J. Aquat. Food Prod. Technol.* **2007**, *16*, 77–90.
24. Fu, X.; Lin, Q.; Xu, S. Effect of Drying methods and Antioxidants on the Flavor and Lipid Oxidation of Silver Carp Slices. *LWT-Food Sci. Technol.* **2015**, *27*, 251–257. [CrossRef]
25. Fu, X.; Hayat, K.; Li, Z.H.; Lin, Q.L.; Xu, S.Y.; Wang, S.P. Effect of microwave heating on the low-salt gel from silver carp (*Hypophthalmichthys molitrix*) surimi. *Food Hydrocoll.* **2012**, *27*, 301–308. [CrossRef]
26. Benjakul, S.; Visessanguan, W.; Chantarasuwan, C. Effect of high-temperature setting on gelling characteristic of surimi from some tropical fish. *Int. J. Food Sci. Technol.* **2004**, *39*, 671–680. [CrossRef]
27. Laemmli, U.K. Cleavage of structural proteins during the assembly of the head of bacteriophage T4. *Nature* **1970**, *227*, 680–685. [CrossRef]
28. Heidtmann, H.H.; Salge, U.; Havemann, K.; Wiederanders, B. Secretion of a latent, acid activatable cathepsin l precursor by human non-small cell lung cancer cell lines. *Oncol. Res.* **1993**, *5*, 441–451.
29. Egelandsdal, B.; Fretheim, K.; Samejima, K. Dynamic rheological measurements on heat-induced myosin gels: Effect of ionic strength, protein concentration and addition of adenosine triphosphate or pyrophosphate. *J. Sci. Food Agr.* **1986**, *37*, 915–926. [CrossRef]
30. Fu, X.; Xu, S.; Wang, Z. Kinetics of lipid oxidation and off-odor formation in silver carp mince: The effect of lipoxygenase and hemoglobin. *Food Res. Int.* **2009**, *42*, 85–90. [CrossRef]
31. Runglerdkriangkrai, J.; Itoh, Y.; Kishi, A.; Obatake, A. Responsibility of myosin S-1 and rod for the polymerization of myosin heavy chain through disulfide bonding upon heating of actomyosin. *Fish. Sci.* **2008**, *65*, 310–314. [CrossRef]
32. Kristinsson, H.G.; Hultin, H.O. Changes in conformation and subunit assembly of cod myosin at low and high pH and after subsequent refolding. *J. Agric. Food Chem.* **2003**, *51*, 7187–7196. [CrossRef] [PubMed]
33. Hossain, M.I.; Itoh, Y.; Morioka, K.; Obatake, A. Inhibiting effect of polymerization and degradation of myosin heavy chain during preheating at 30 °C and 50 °C on the gel-forming ability of walleye pollack surimi. *Fish. Sci.* **2010**, *67*, 718–725. [CrossRef]
34. Tsukamasa, Y.; Sato, K.; Shimizu, Y.; Imai, C.; Sugiyam, M.; Minegishi, Y.; Kawabata, M. (γ-Glutamyl) lysine Cross link Formation in Sardine Myofibril Sol during Setting at 25 °C. *J. Food Sci.* **2004**, *58*, 785–787. [CrossRef]
35. Visessanguan, W.; An, H. Effects of proteolysis and mechanism of gel weakening in heat-induced gelation of fish myosin. *J. Agric. Food Chem.* **2000**, *48*, 1024–1032. [CrossRef] [PubMed]

foods MDPI

Article

A Study on the Effects of Calcium Lactate on the Gelling Properties of Large Yellow Croaker (*Pseudosciaena crocea*) Surimi by Low-Field Nuclear Magnetic Resonance and Raman Spectroscopy

Shangyuan Sang [1,2], Xiaoyun Chen [1,2], Ying Qin [1,2], Li Tong [1,2] and Changrong Ou [1,2,*]

[1] College of Food and Pharmaceutical Sciences, Ningbo University, Ningbo 315832, China
[2] Key Laboratory of Animal Protein Food Deep Processing Technology of Zhejiang Province, Ningbo University, Ningbo 315832, China
* Correspondence: ouchangrong@nbu.edu.cn

Abstract: Divalent calcium ions (Ca^{2+}) are often used in surimi gels to improve their physicochemical characteristics. The present study aimed to investigate the effect of calcium lactate on the physicochemical properties, state distribution of water, and protein structure changes of surimi gels made from large yellow croaker. The results showed that the addition of calcium lactate (0%, 0.5%, 1.5%, 2.5%, 3.5%, and 4.5% on wet surimi) significantly ($p < 0.05$) increased gel strength and whiteness, while cooking loss decreased. The water-holding capacity increased first and then decreased. When calcium lactate was added to 1.5%, the water-holding capacity reached the best value. Using low-field nuclear magnetic resonance to study the distribution of water state, the bound water content first increased and then decreased with the addition of calcium lactate, reaching the highest at 1.5%. In addition, the relaxation time of immobilized water was shortest at the addition of 1.5% calcium lactate. Analyzing the protein structural changes by Raman spectroscopy showed that there was a significant decrease ($p < 0.05$) in the α-helix accompanied by an increase in β-sheets, turns, and random coils after the addition of calcium lactate. The above changes were due to the Ca^{2+} that was bound to the negatively charged myofibrils to form a protein-Ca^{2+}-protein cross-linking. Therefore, the addition of calcium lactate had a significant positive effect on the gelling ability of surimi.

Keywords: large yellow croaker; calcium lactate; surimi; low-field nuclear magnetic resonance; Raman spectroscopy

Citation: Sang, S.; Chen, X.; Qin, Y.; Tong, L.; Ou, C. A Study on the Effects of Calcium Lactate on the Gelling Properties of Large Yellow Croaker (*Pseudosciaena crocea*) Surimi by Low-Field Nuclear Magnetic Resonance and Raman Spectroscopy. *Foods* **2022**, *11*, 3197. https://doi.org/10.3390/foods11203197

Academic Editors: Tao Yin and Liu Shi

Received: 3 September 2022
Accepted: 10 October 2022
Published: 13 October 2022

1. Introduction

Large yellow croaker (*Pseudosciaena crocea*) that is mainly limited to the coastal waters of continental East Asia is one of the three top commercial marine fishes of China [1]. It is rich in protein and fat as well as essential amino acids and polyunsaturated fatty acids. Large yellow croaker resources are very abundant in China [2]. In 2020, the seawater aquaculture production of the large yellow croakers had reached 21,353,100 tons in China. Due to the sales and transportation in the form of living life, and due to the external temperature and transportation time restrictions, its freshness cannot be effectively guaranteed. Therefore, the deep processing of the large yellow croaker into surimi can solve the problem of its decay in a short time and enrich its product form. The process of surimi involves taking meat, chopping, washing, dehydration, refining, as well as bagging. The most common types of surimi products in China's freshwater surimi market are fish balls, fish cake, fish rolls, fish intestines, etc. These products can be transported and stored in the low hand environment of minus 20 °C, suitable for urban family consumption, convenient production, and fresh and tender taste. In 2020, the production of surimi product was 1,267,727 tons in China [3]. It has a significant market consumption potential. However, the high fat content of large yellow croakers would be averse to surimi gelatin. Polyphosphates are generally used to improve the quality of surimi gels [4,5], while high doses of them can cause kidney damage and interfere with the absorption of calcium [6].

The characteristics of the physics of surimi gels is one of the important aspects of studying its quality, which is closely related to the function of myofibrillin, and further changes its microstructure. Therefore, calcium ions (Ca^{2+}) were added to improve the gelling properties of croaker surimi in this study as Ca^{2+} can increase the protein–protein interactions via the formation of a salt-bridge between negatively charged myofibrillar proteins [7,8]. In addition, different forms of calcium salts, such as lactate, citrate, sulfate, and caseinate, have been known to activate indigenous transglutaminase, which can catalyze the cross-linking reaction between γ-carboxyamide groups of glutamine and ε-amino groups of lysine to form a ε-(γ-glutamyl) lysine linkage, leading to the formation of stronger gel [8]. Calcium lactate, an excellent additive naturally present in the human body, has a higher solubility and bioavailability than calcium gluconate, calcium citrate, calcium carbonate, and calcium phosphate [9]. Previous studies have shown that calcium lactate can improve the textural properties of fish samples [10]. In addition, as one of the potential natural sources of antibacterial agents, the product of lactic acid itself has an antibacterial effect [11]. However, there is little information about the effect of calcium lactate on the characteristics of surimi of large yellow croaker.

Low-field nuclear magnetic resonance (LF-NMR) can be used to qualitatively and quantitatively analyze the water distribution and water mobility in food according to the spin–spin relaxation time (T2) of samples [12]. Raman spectroscopy can be used for rapid nondestructive detection of protein structure and qualitative and quantitative analysis of changes in protein functional groups [13]. Therefore, in this paper, LF-NMR and Raman spectroscopy were used to study the effect of calcium lactate on the protein structure and water status of large yellow croaker, as well as the relationship between calcium lactate and surimi gel characteristics parameters, to provide a theoretical basis for the quality evaluation and improvement of surimi products.

2. Materials and Methods

2.1. Materials

Dozens of 500 g large yellow croakers were purchased from local aquatic markets (Ningbo, China) and brought to the laboratory covered with ice cubes within 24 h. Calcium lactate is food grade and offered by Jindan Lactic Acid Technology Company (Zhoukou, China).

2.2. Cooking Loss

Cooking loss (CL) was measured using a method described by Yang, Wang, Wang, and Ye [14]. Before and after heating (90 °C, 20 min), the weight of surimi gel was G1 and G2, respectively. The CL was represented as the loss of liquid:

$$CL\ (\%) = (G1 - G2)/G1 \times 100 \quad (1)$$

2.3. Preparation of Surimi Gels

Fish head, skin, dark muscle, and bones were removed manually and the flesh was minced to uniformity using a mincer within about 2 min. The mince was washed at a water/mince ratio of 4:1 (w/w) and then stirred for 10 min. The homogenate was centrifuged at 4000× g for 10 min at 4 °C after washing. Washed mince was chopped for 10 min and chopped with 2% NaCl (w/w, on fish mince) for another 10 min at 4 °C. Then, the chopped mince continued to be chopped for 5 min with starch (3%, w/w, on fish mince) and calcium lactate (0%, 0.5%, 1.5%, 2.5%, 3.5%, and 4.5%, w/w, on fish mince). Finally, all the surimi samples were stuffed into a casing and heated immediately for 1 h at 40 °C and then for 30 min at 90 °C in a water bath.

2.4. Color Evaluation

Three gel samples of each treatment were subjected to color evaluation using a Hunter-Lab (CM-400d, Konica Minolta, Tokyo, Japan). Illuminant C was used as the light source of every measurement. Measurement of L* (lightness), a* (redness/greenness), and b* (yellowness/blueness) values was conducted. Whiteness was calculated using Equation (2) [15]:

$$\text{Whiteness} = 100 - [(100 - L^*)^2 + (a^*)^2 + (b^*)^2]^{1/2} \quad (2)$$

2.5. Gel Strength

Gel strength was measured according to the method of Abe, Asada, and Kajiwara using a texture analyzer (Model TA-XT2, Stable Micro Systems, Godalming, Surrey, UK) [16]. Gels were cut into cylinder-shaped samples with a diameter of 25 ± 1 mm and height of 15 ± 1 mm and equilibrated at room temperature (25 °C). A probe model of the P/0.5 s spherical probe, a maximum displacement of 15 mm, a trigger force of 5.0 g, and a test speed of 1 mm/s were set as the main parameters. Each type of sample was determined in triplicate. Gel strength is also known as elasticity as a significant indicator that represents the gel forming ability of surimi. The gel strength was calculated according to Equation (3). The value of the breaking force was read at the first peak force (g) according to the force–time deformation curve. The breaking distance was the distance traveled by the probe from the surface of the sample to the point of breakage.

$$\text{Gel strength (g} \times \text{cm)} = \text{breaking force (g)} \times \text{breaking distance (cm)} \quad (3)$$

2.6. Water-Holding Capacity of Surimi Gels

The water-holding capacity (WHC) was measured using a method described by Sánchez-González with slight modifications [17]. A 1.5 g sample of surimi packed with a piece of filter paper was introduced into centrifuge tubes and centrifuged at 4900× *g* (TDL 50C, Anting Scientific Instrument Factory, Shanghai, China) for 15 min at room temperature. The mass m1 (g) before centrifugation and the mass m2 (g) after centrifugation were measured, and the WHC was m2/m1 × 100%. Measurements were determined in triplicate.

2.7. LF-NMR Measurement of Surimi Gels

The LF-NMR relaxation measurements were performed on a Niumag Micro MR 20-025 (Niumag Electric Corporation, Suzhou, China) operating at a resonance frequency for hydrogen protons of 18.17 MHz. An approximately 5.6 g sample was placed in a 25 mm glass tube and inserted in the LF-NMR probe. A Carr–Purcell–Meiboom–Gill (CPMG) pulse sequence was employed to measure the spin–spin relaxation time, and the pulse parameters were as follows: P90 (μs) = 5.50, P180 (μs) = 11.00, SW (kHz) = 200, D3 (μs) = 20, TR (ms) = 2000, RG1 = 20, RG2 = 3, NS = 4, EchoTime (μs) = 250, and EchoCount = 10,000.

2.8. Raman Spectroscopy of Surimi Gels

The Raman spectra of surimi gels were recorded by a spectrometer equipped with a 50× lens as a microscope (Ningbo Institute of Industrial Technology, Ningbo, China). It was used to focus the excitation laser beam (532 nm exciting line of a Spectra Physics Ar-laser) on the sample spread on a glass slide and collected the Raman signal in the backscattered direction. The laser power was controlled at about 12 mW and the laser spot diameter was about 1 μm at the sample surface. The Raman spectra were recorded in the range of 200–3650 cm^{-1}. Each spectrum according to 30 scans, 20 s exposure time, and 1 cm^{-1} resolution conditions was obtained. Furthermore, Raman spectra were baseline-corrected and normalized against the phenylalanine band at 1003 cm^{-1} using the LabSpec Application.

Raman spectroscopic data were then analyzed with the PeakFit v4.12. Fourier reflexive convolution, second-derivative analysis, and band curve fitting are three methods to open and sharpen overlapping hidden peaks. In this experiment, PeakFit V4.12 was used to perform nonlinear fitting of peaks in spectral data to quantitatively calculate the area of each secondary structure, and then the second-derivative spectrum was used to locate overlapping hidden peaks. Finally, the curve fitting function was used to calculate the area of each peak, reported as a percentage [12].

2.9. Statistical Analysis

All the results were in triplicate and statistically analyzed by the SPSS 8.0 software. (International Business Machines Corporation, Chicago, IL, USA). Analysis of variance (ANOVA) was employed to determine the significance of main effects. Significant differences ($p < 0.05$) between means were identified using Duncan's multiple range test. Correlations between different indices were analyzed using Pearson type.

3. Results and Discussion

3.1. CL of Surimi

The CL of large yellow croaker surimi gels with added calcium lactate (0%, 1.5%, 2.5%, 3.5%, and 4.5%, *w*/*w*) is shown in Table 1. The CL of the control sample was higher than the sample with a calcium lactate content over 3.5% and lower than the sample with a calcium lactate content less than 2.5% ($p < 0.05$). This suggested that adding calcium lactate surpassing 3.5% could prevent the water loss during cooking and that it had a good protective effect on the CL stability compared with the control sample.

Table 1. Effect of calcium lactate on cooking loss, water-holding capacity, whiteness, L*, a*, and b*.

Calcium Lactate (%)	Cooking Loss (%)	Water-Holding Capacity (%)	Whiteness	L*	a*	b*
0	12.13 ± 0.007 c	77.86 ± 0.01 bc	74.87 ± 0.64 a	75.31 ± 0.62 a	−2.75 ± 0.09 d	3.71 ± 0.54 d
0.5	19.85 ± 0.003 e	78.72 ± 0.03 c	75.07 ± 0.43 b	75.47 ± 0.41 a	−2.55 ± 0.13 e	3.67 ± 0.39 d
1.5	20.17 ± 0.010 e	78.89 ± 0.01 c	75.92 ± 0.63 b	76.43 ± 0.66 ab	−2.67 ± 0.12 c	3.86 ± 0.46 e
2.5	16.67 ± 0.005 d	74.01 ± 0.04 b	76.84 ± 0.56 c	77.21 ± 0.55 c	−2.92 ± 0.13 b	2.82 ± 0.49 c
3.5	10.02 ± 0.003 b	71.68 ± 0.04 a	77.48 ± 0.54 d	77.91 ± 0.56 d	−3.06 ± 0.10 a	1.86 ± 0.13 a
4.5	8.47 ± 0.002 a	71.12 ± 0.02 a	77.58 ± 0.36 d	77.71 ± 0.35 d	−2.95 ± 0.15 b	1.97 ± 0.37 bc

Results are presented as the mean ± standard deviation. Different letters (a–e) in the same column represent significant differences ($p < 0.05$). L*, lightness; a*, redness/greenness; b*, yellowness/blueness.

3.2. Whiteness of Surimi Gels

The effects of calcium lactate on whiteness, L* (lightness), a* (redness/greenness), and b* (yellowness/blueness) of large yellow croaker surimi gels are shown in Table 1. The different calcium lactate contents resulted in the change in color properties of surimi gels. The whiteness of the surimi gels added with 0.5–4.5% calcium lactate was 75.07–77.58, all higher compared with the control gels of 74.87 ($p < 0.05$). Whiteness is an important parameter to determine the quality of surimi, with values greater than 75 generally considered acceptable [18,19]. In addition, with increasing calcium lactate contents in surimi, the results showed rising values of whiteness. Therefore, adding calcium lactate might improve the whiteness of surimi gels.

3.3. Strength of Surimi Gels

The effect of calcium lactate on the gel strength of surimi of large yellow croaker is shown in Figure 1. Compared with the control group, the addition of calcium lactate significantly enhanced the gel strength in the range of 0.5–4.5% and reached the maximum gel strength at 4.5% ($p < 0.05$). These results are consistent with the report of Lee and Park (1998) [9]. Calcium lactate not only has good water solubility, but divalent calcium ions may also strengthen the fixation of the tissue structure, acting equally with monovalent salt ions on surimi gels. Thus, calcium lactate obviously affected the gel strength of surimi to improve its quality [19].

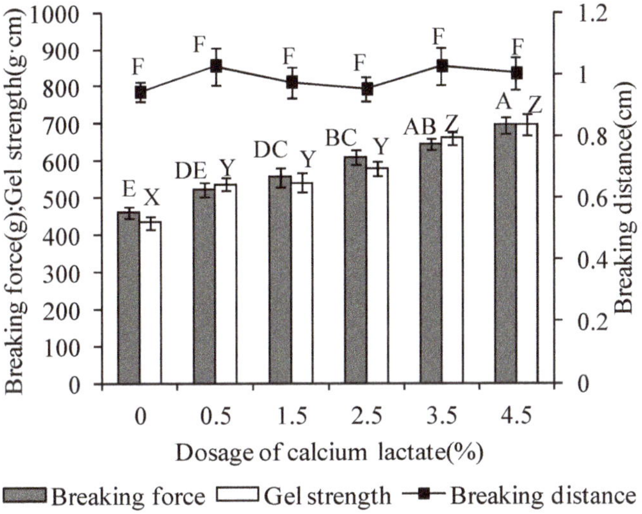

Figure 1. Effect of calcium lactate on breaking force, breaking distance, and gel strength. A–E: different letters at breaking force differ significantly ($p < 0.05$). X–Z: different letters at gel strength differ significantly ($p < 0.05$). F: indicates no significant difference ($p > 0.05$) at breaking distance.

3.4. WHC of Surimi Gels

WHC is an important parameter as the capacity of surimi gels to retain water in the microstructure. In Table 1, there is a significant difference ($p < 0.05$) in WHC between calcium-lactate-added groups and the control. The surimi gels with low addition levels (0.5–1.0%) of calcium lactate has a higher WHC than the control gels. By contrast, high addition levels (1.5–4.5%) of calcium lactate reduce the WHC of surimi gels. The WHC of surimi improved by a certain concentration of Ca^{2+} at the appropriate temperature, which could activate transglutaminase in surimi and catalyze the cross-linking between the carboxyl amide groups in glutamic acid residues and other amino acid residues [20].

3.5. Water Distribution of Surimi Gels

LF-NMR as a rapid, nondestructive monitoring method was used to investigate changes in water mobility during food processing by measuring proton relaxation [21,22]. The proton spin–spin relaxation time (T2) indicates various statuses of water and is directly related to the mobility of water in food [11]. The properties of water states and the distribution of large yellow croaker surimi gels were evaluated by the LF-NMR.

Figure 2A shows the effect of calcium lactate on the T2 distribution of large yellow croaker surimi gels. A range (T22) with a relaxation time around 30–100 ms is known as the immobilized water and a small range at about 200–400 ms (T23) is free water. The range at 1–15 ms (T21) was on behalf of the bound water with large molecules such as proteins in the gel system [15]. In Table S1, the relaxation time peaks of T21, T22, and T23 populations significantly decreased in most of the calcium lactate-added surimi ($p < 0.05$) compared with the control group, indicating that water mobility also diminished in the surimi after adding calcium lactate. The results also indicated that the addition of calcium lactate had a significant effect on the T23 ($p < 0.05$) but had no significant effect on the T21 ($p > 0.05$). When the addition level of calcium lactate was less than 1.5%, the T23 of surimi gels was shortened. Thus, a proper amount of calcium lactate can limit the flowing of water in the surimi.

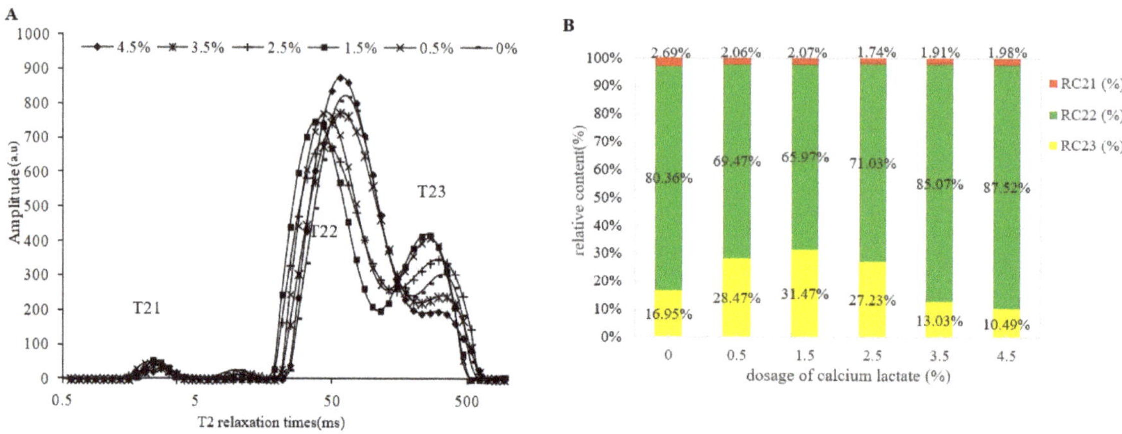

Figure 2. Effect of different dosages of calcium lactate on T2 relaxation time (**A**) and the relative content (**B**) of bound water (RC21), immobilized water (RC22), and free water (RC23) of large yellow croaker surimi gels.

Comparison of the continuous distribution profiles revealed clear differences in the distribution of water mobility between the addition of calcium lactate and control groups. The relative content of immobilized water (RC22) decreased first to the lowest at 1.5% calcium lactate and then increased with the addition of 2.5–4.5% calcium lactate (Table S1). According to the experimental results, the free water content (RC23) of surimi gels with 0.5–2.5% calcium lactate was higher than the control group, while the 3.5–4.5% calcium lactate group was lower than the control group.

3.6. Raman Spectroscopy of Proteins in Surimi Gels

Raman scattering spectroscopy was applied in this study to explain the backbone conformations of protein molecules in surimi gels. Changes in the Raman bands of protein chemical groups provided the information of changes in the secondary and tertiary structure of proteins [23–25].

In this present study, a detailed spectral analysis provided a valuable tool for the study of large yellow croaker surimi gels with calcium lactate addition. Raman spectra of large yellow croaker surimi gels with various concentrations of calcium lactate (0%, 1.5%, 2.5%, 3.5%, and 4.5% w/w) in the region are shown in Figure 3. The assignments of the corresponding bands are included in Table S2 [24,26,27].

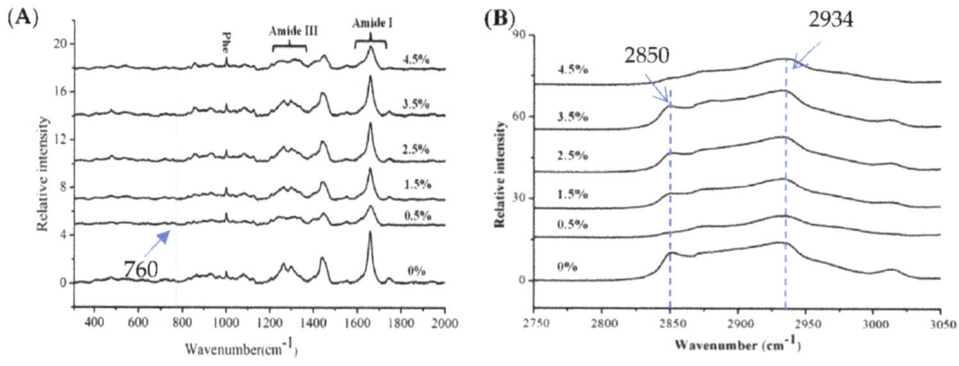

Figure 3. Original Raman spectrum (300–2000 cm^{-1} (**A**) and 2750–3050 cm^{-1} (**B**)) of large yellow croaker surimi gels with various dosages of calcium lactate.

3.6.1. Changes in the Secondary Structure of Protein

The Raman band that provides information about the secondary structure of surimi gels is amide III, which mainly involves C-N stretching, N-H in-plane bending, C_α-C stretching, and C=O in-plane bending vibrations of the peptide bond [27]. It was difficult to interpret the amide III band as the vibrational spectroscopy of surimi gel protein produced a complex pattern of bands in the 1225–1350 cm^{-1} region (Figure 3A, Table S2). The intensity of the α-helix structure representing this band around 1260–1300 cm^{-1} overlapped with the range of turns. Despite the β-sheet and random coil bands superposition in the amide III region (1250 and 1240 cm^{-1}), an increasing intensity can be observed in the range of 1225–1240 cm^{-1} in calcium-lactate-added gels. These alterations in the secondary structure content and structural properties of the proteins can be attributed to the β-sheet formed by calcium lactate addition, because it forms high-molecular-weight complexes with a compact and ordered conformation.

The Raman band around 1657 cm^{-1} was part of the amide I (1600–1700 cm^{-1}) vibrational mode (Table S2, Figure 4A), which directly provides secondary structural information about protein and involves mainly C=O stretching, C_α-C-N bending, C-N stretching, and N-H in-plane bending of peptide groups [23–25,28]. In this study, the highest intensity was at the 1657 cm^{-1} band in the spectra of large yellow croaker surimi gels with various calcium lactate concentrations. That can be put down to the high α-helix content (Figure 4A). A rational excuse for the band change may be the reduction in α-helix content resulting from calcium lactate addition.

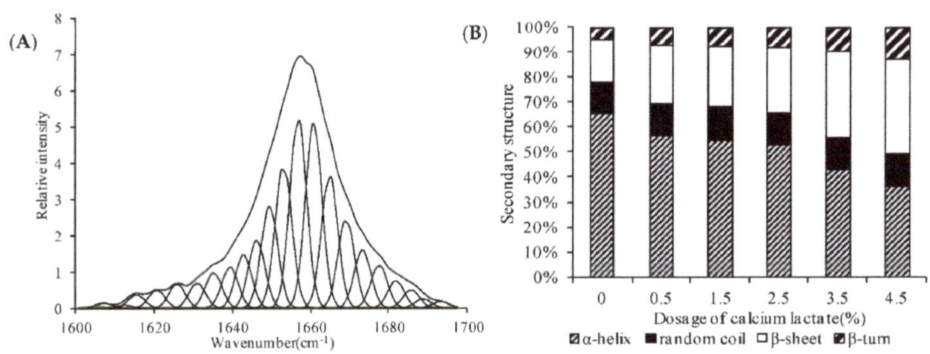

Figure 4. Original Raman spectrum (above) at the amide I band and its corresponding deconvolved spectrum (below) fitted with Gaussian components (**A**). The estimated secondary structure fractions (**B**) of proteins in large yellow croaker surimi gels with various dosages of calcium lactate.

Quantitative analysis of the protein secondary structure was performed by using Fourier reflexive convolution, second-derivative analysis, and band curve fitting. After spectral normalization, the intensity values of Raman bands from various atomic groups were determined. The visible bands were assigned to vibrational modes of amino acid side chains or peptide backbones carried out by comparing Raman spectra of model polypeptides or monographs of Raman spectra of proteins [27]. Figure 4A shows the original and deconvolved spectrum of large yellow croaker surimi gels. Then, the secondary structure fractions of the surimi gels were estimated, respectively, on the basis of the area of appropriate peaks. Similar results were also discovered in heat-induced gelation [27,29,30]. The spectral profiles and percentage of the secondary structure determined from the amide I spectral profile for large yellow croaker surimi gels with different contents of calcium lactate are shown in Figure 4B. There were remarkable ($p < 0.05$) differences in the α-helix, β-sheets, and turns structure contents after calcium lactate was added. There was a significant decrease ($p < 0.05$) in the α-helix accompanied by increased β-sheets, turns, and random coils. Upon 4.5% calcium lactate addition, the content of α-helix decreased from

65.82% to 36.19%, and β-sheets, turns, and random coils increased from 17.13%, 4.87%, and 12.18% to 37.86%, 12.97%, and 12.98% ($p < 0.05$), respectively. Higher contents of β-sheets and turns in the presence of calcium lactate indicated that the secondary structure of large yellow croaker surimi gels was likely to be calcium-lactate-dependent.

3.6.2. Local Environments of Protein Networks in the Surimi Gels

There are several Raman bands that have characteristics of the tertiary structure of surimi gels. Changes in these bands could monitor the polarity of the local environment or involvement in hydrogen bonding [26]. In this study, the normalized intensity of the Raman band near 760 cm^{-1} in the large yellow croaker surimi gels with 0.5% calcium lactate (0.20) showed a significant decrease ($p < 0.05$) compared with the control (0.36) (Figure 3A), which indicated that the hydrophobicity was involved in the surimi gels and the hydrophobic microenvironment became exposed to the polar aqueous solvent with calcium lactate addition.

The doublet bands located near 830 and 850 cm^{-1} could monitor the microenvironment around tyrosine residues and be assigned to vibrations of the para-substituted benzene ring of tyrosine residues, which were affected by the environment and the involvement of the phenolic hydroxyl group in hydrogen bonding [31]. I850/I830 is the doublet bands intensity ratio that has been proposed to determine whether the tyrosine residue is solvent-exposed or buried [28]. The ratio increased from 1.27 to 2.09 as the addition level of calcium lactate increased from 0 to 4.5%, which suggested that the tyrosine residues were exposed to the aqueous or polar microenvironment or acted as a simultaneous acceptor and donor of moderate to weak hydrogen bonds.

Proteins exhibited C-H stretching vibration (νC-H) in the 2800–3050 cm^{-1} band of the Raman spectrum [32]. There were two peaks at 2850 cm^{-1} and 2934 cm^{-1} in the νC-H band of the control sample, while one peak at 2850 cm^{-1} of the sample with 0.5% calcium lactate vanished (Figure 3B). During gelation with the addition of calcium lactate, the unfolding of the protein may result in the exposure of more methyl or methylene groups. With the increase in calcium lactate level, the strength and area of the νC-H band decreased first and then increased. These changes suggested an increased hydrophobic interaction of surimi gels with 0.5% calcium lactate, and protein cross-linking may be due to such hydrophobic contacts between protein side-chains.

3.7. Correlations Analysis

The Pearson correlation coefficient can determine whether there is a linear relationship between the two indicators. Table 2 shows the correlation between the physical properties of gels and the secondary structure of surimi protein, and it was found that both were related. The results showed that WHC was mainly negatively and significantly correlated with T23 (correlation coefficient −0.68), the peak area fraction of RC22 (correlation coefficient −0.75), and β-sheet and β-turn content (correlation coefficients −0.88 and −0.84, respectively). It was positively and significantly correlated with the peak area fraction of RC23 (correlation coefficient 0.73), RC21 (correlation coefficient 0.52), α-helix content (correlation coefficient 0.85), and CL (correlation coefficient 0.78), with a highly significant positive correlation with α-helix content and CL. GS was positively correlated with whiteness (correlation coefficient 0.94), RC22 (correlation coefficient 0.51), and β-sheet and β-turn content (correlation coefficients 0.99 and 0.96, respectively) and negatively correlated with WHC (correlation coefficient −0.87), RC21 (correlation coefficient −0.74), and α-helix (correlation coefficient −0.99), with a highly significant negative correlation with α-helix. Stangierski et al. also pointed out that the variation in the qualitative parameters was determined by the relative content of free and bound water, which can be confirmed by the relaxation time values [33]. Thus, it was seen that there was a strong correlation between structure and function, and WHC was strongly correlated with LF-NMR relaxation properties and protein secondary structure.

Table 2. Pearson correlations among different indices of large yellow croaker surimi gels.

	T	WHC	CL	GS	Whiteness	RC21	RC22	RC23	T21	T22	T23	α-Helix	Coil	β-Sheet	β-Turn
T	1	−0.9342	−0.6099	0.9661	0.9809	−0.6562	0.5863	−0.5630				−0.9644		0.9637	0.9533
WHC	−−	1	0.7836	−0.8720	−0.9397	0.5192	−0.7559	0.7325			−0.6870	0.8558		−0.8831	−0.8408
CL	−	++	1		−0.5739		−0.9883	0.9877		−0.8512	−0.9535	0.5410		−0.5770	−0.5432
GS	++	−−		1	0.9376	−0.7355	0.5088		−0.5886			−0.9896		0.9926	0.9643
Whiteness	++	−−	−	++	1	−0.705	0.5357	−0.5104	−0.5007			−0.9128		0.9212	0.8819
RC21	−	+		−	−	1			0.9114	0.5796		0.6519		−0.6492	−0.6200
RC22	+	−−	−−	+	+		1	−0.9991		0.8543	0.9335	−0.5550		0.5943	0.5529
RC23	−	+	++		−		−−	1		−0.8685	−0.9327	0.5324		−0.5705	−0.5308
T21				−	−	++			1	0.6230		0.5213		−0.5047	−0.5506
T22			−−			+	++	−−	+	1	0.9149		−0.5993		
T23		−	−−				++	−−		++	1		−0.6059		
α-helix	−−	++	+	−−	−−	+	−	+	+			1		−0.9956	−0.9859
Coil										−	−		1		
β-sheet	++	−−	−	++	++	−	+	−	−			−−		1	0.9730
β-turn	++	−−	−	++	++	−	+	−	−			−−		++	1

T, the treatment, at different addition levels of calcium lactate in the surimi gels; WHC, water-holding capacity; CL, cooking loss; GS, gel strength; Coil, random coil; RC21, RC22, RC23, the relative contents of bound water, immobilized water, and free water in the surimi gels, respectively; T21, T22, T23, the lateral relaxation times of protons from bound water, immobilized water, and free water, respectively; positive signs (+) mean positive correlations, whereas negative signs (−) mean negative correlations. Single sign indicates a correlation between two indices ($0.5 \leq |R| < 0.75$). Double signs represent a higher correlation ($0.75 \leq |R| < 1$).

4. Conclusions

Calcium lactate had significant effects on the quality of large yellow croaker surimi gels, including gel strength, water-holding capacity, and cooking loss. LF-NMR and Raman spectroscopy had also been proven useful tools for the study of the structures of large yellow croaker surimi gels fortified with calcium lactate. In our study, we thoroughly investigated changes in the relaxation time, content of bound water, immobilized water, and free water as well as the reduction in α-helix and increase in β-sheets, β-turns, and random coils. In addition, significant correlations were found between structural changes in surimi gel proteins and some other indices of surimi gels using the Pearson type of SPSS system. These results provided evidence that the addition of calcium lactate had a significant influence on the property of large yellow croaker surimi gels, and also suggested that the formation of protein-calcium-protein might occur.

Furthermore, due to the experimental cycle and objective conditions, there are still many deficiencies and imperfections in this paper. It is suggested that the following work can be carried out in the future: (1) study the effect of calcium lactate on protein aggregates (or tertiary structure) and their molecular weight distribution; (2) contrast-study the impact of anions or other cations (such as zinc and iron) on surimi gel products and their impact on human health.

Supplementary Materials: The following supporting information can be downloaded at: https://www.mdpi.com/article/10.3390/foods11203197/s1, Table S1: Effect of calcium lactate on T2 and RC2; Table S2: Tentative assignment of some bands in the Raman spectra of surimi gels.

Author Contributions: Conceptualization, methodology, supervision, project administration, and resources, S.S. and C.O; investigation and data curation, writing—original draft preparation, Y.Q., X.C. and L.T.; funding acquisition, C.O. All authors have read and agreed to the published version of the manuscript.

Funding: This work was supported by the National Key Research and Development Program under Grant 2019YFD0901705, Public Welfare Applied Research Project of Zhejiang Province (LGN18C200019), and Ningbo Agricultural Science and Technology Major Breakthrough Project (2012C10024).

Institutional Review Board Statement: Not applicable.

Informed Consent Statement: Not applicable.

Data Availability Statement: Data are contained within the article or Supplementary Material.

Acknowledgments: The authors would like to thank Lingling Jia and Weiyu Zhu for reviewing.

Conflicts of Interest: The authors declare no conflict of interest.

Glossary

CL	Cooking loss
LF-NMR	Low-field nuclear magnetic resonance
WHC	Water-holding capacity
GS	Gel strength

References

1. Liang, P.; Cheng, X.; Xu, Y.; Cheng, W.; Chen, L. Determination of fatty acid composition and phospholipid molecular species of large yellow croaker (*Pseudosciaena crocea*) roe from China. *J. Aquat. Food Prod. Technol.* **2017**, *26*, 1259–1265. [CrossRef]
2. Yuan, J.; Lin, H.-D.; Wu, L.; Zhuang, X.; Ma, J.; Kang, B.; Ding, S. Resource status and effect of long-term stock enhancement of large yellow croaker in china. *Front. Mar. Sci.* **2021**, *8*, 743836. [CrossRef]
3. China Society of Fisheries. *China Fishery Statistical Yearbook of 2020*; China Statistics Press: Beijing, China, 2021.
4. da Silva Oliveira, M.E.; Goncalves, A.A. The effect of different food grade additives on the quality of Pacific white shrimp (*Litopenaeus vannamei*) after two freeze-thaw cycles. *LWT* **2019**, *113*, 108301. [CrossRef]
5. Julavittayanukul, O.; Benjakul, S.; Visessanguan, W. Effect of phosphate compounds on gel-forming ability of surimi from bigeye snapper (*Priacanthus tayenus*). *Food Hydrocoll.* **2006**, *20*, 1153–1163. [CrossRef]

6. Panseri, S.; Arioli, F.; Biolatti, C.; Mosconi, G.; Pavlovic, R.; Chiesa, L.M. Detection of polyphosphates in seafood and its relevance toward food safety. *Food Chem.* **2020**, *332*, 127397. [CrossRef] [PubMed]
7. Ramírez, J.A.; Uresti, R.M.; Velazquez, G.; Vázquez, M. Food hydrocolloids as additives to improve the mechanical and functional properties of fish products: A review. *Food Hydrocoll.* **2011**, *25*, 1842–1852. [CrossRef]
8. Wijayanti, I.; Singh, A.; Prodpran, T.; Sookchoo, P.; Benjakul, S. Effect of asian sea bass (*Lates calcarifer*) bio-calcium in combination with different calcium salts on gel properties of threadfin bream surimi. *J. Aquat. Food Prod. Technol.* **2021**, *30*, 1173–1188. [CrossRef]
9. Lee, N.; Park, J.W. Calcium compounds to improve gel functionality of pacific whiting and alaska pollock surimi. *J. Food Sci.* **2006**, *63*, 969–974. [CrossRef]
10. Yin, T.; Park, J.W. Optimum processing conditions for slowly heated surimi seafood using protease-laden Pacific whiting surimi. *LWT-Food Sci. Technol.* **2015**, *63*, 490–496. [CrossRef]
11. Chitrakar, B.; Zhang, M.; Bhandari, B. Novel Intelligent Detection of Safer Water Activity by LF-NMR Spectra for Selected Fruits and Vegetables during Drying. *Food Bioprocess Technol.* **2019**, *12*, 1093–1101. [CrossRef]
12. Sadat, A.; Joye, I.J. Peak fitting applied to fourier transform infrared and raman spectroscopic analysis of proteins. *Appl. Sci.* **2020**, *10*, 5918. [CrossRef]
13. Wang, J.; Su, Y.; Gu, L.; Chang, C.; Xu, L.; Yang, Y.; Li, J. The inhibition of cell-free supernatants of several lactic acid bacteria on the selected psychrophilic spoilage bacteria in liquid whole egg. *Food Control* **2021**, *123*, 107753. [CrossRef]
14. Yang, Z.; Wang, W.; Wang, H.; Ye, Q. Effects of a highly resistant rice starch and pre-incubation temperatures on the physicochemical properties of surimi gel from grass carp (*Ctenopharyn Odon Idellus*). *Food Chem.* **2014**, *145*, 212–219. [CrossRef] [PubMed]
15. Cao, Y.; Zhao, L.; Huang, Q.; Xiong, S.; Yin, T.; Liu, Z. Water migration, ice crystal formation, and freeze-thaw stability of silver carp surimi as affected by inulin under different additive amounts and polymerization degrees. *Food Hydrocoll.* **2022**, *124*, 107267. [CrossRef]
16. Abe, S.; Asada, T.; Kajiwara, K. Effects of freeze-thaw cycles on Gel-Forming ability and protein denaturation in alaska pollock frozen surimi. *J. Food Qual.* **2019**, *2019*, 3760368. [CrossRef]
17. Sánchez-González, I.; Carmona, P.; Moreno, P.; Borderías, J.; Sánchez-Alonso, I.; Rodríguez-Casado, A.; Careche, M. Protein and water structural changes in fish surimi during gelation as revealed by isotopic H/D exchange and Raman spectroscopy. *Food Chem.* **2008**, *106*, 56–64. [CrossRef]
18. Priyadarshini, M.B.; Balange, A.K.; Xavier, K.A.M.; Reddy, R.; Nayak, B.B.; Sanath Kumar, H. The Effect of Lyophilized Coconut Mesocarp—Aqueous and Ethanol Phenolic Extracts on the Gel Quality of Tilapia Surimi. *J. Aquat. Food Prod. Technol.* **2021**, *30*, 1330–1343. [CrossRef]
19. Rawdkuen, S.; Jongjareonrak, A.; Benjakul, S.; Chaijan, M. Discoloration and lipid deterioration of farmed giant catfish (*Pangasianodon gigas*) muscle during refrigerated storage. *J. Food Sci.* **2008**, *73*, C179–C184. [CrossRef]
20. Yongsawatdigul, J.; Worratao, A.; Park, J.W. Effect of endogenous transglutaminase on threadfin bream surimi gelation. *J. Food Sci.* **2002**, *67*, 3258–3263. [CrossRef]
21. da Silva Carneiro, C.; Mársico, E.T.; Ribeiro, R.D.O.R.; Júnior, C.A.C.; Álvares, T.S.; de Jesus, E.F.O. Studies of the effect of sodium tripolyphosphate on frozen shrimp by physicochemical analytical methods and Low Field Nuclear Magnetic Resonance (LF ^{1}H NMR). *LWT-Food Sci. Technol.* **2013**, *50*, 401–407. [CrossRef]
22. Lan, W.; Liu, J.; Hu, X.; Xiao, L.; Sun, X.; Xie, J. Evaluation of quality changes in big-eye tuna (*Thunnus obesus*) based on near-infrared reflectance spectroscopy (NIRS) and low field nuclear magnetic resonance (LF-NMR). *J. Food Process Eng.* **2021**, *44*, e13613. [CrossRef]
23. Elkordy, A.A.; Forbes, R.T.; Barry, B.W. Study of protein conformational stability and integrity using calorimetry and FT-Raman spectroscopy correlated with enzymatic activity. *Eur. J. Pharm. Sci.* **2008**, *33*, 177–190. [CrossRef]
24. Phongpa-Ngan, P.; Aggrey, S.E.; Mulligan, J.H.; Wicker, L. Raman spectroscopy to assess water holding capacity in muscle from fast and slow growing broilers. *LWT-Food Sci. Technol.* **2014**, *57*, 696–700. [CrossRef]
25. Thawornchinsombut, S.; Park, J.W.; Meng, G.; Li-Chan, E.C.Y. Raman spectroscopy determines structural changes associated with gelation properties of fish proteins recovered at alkaline pH. *J. Agric. Food Chem.* **2006**, *54*, 2178–2187. [CrossRef] [PubMed]
26. Chen, H.; Han, M. Raman spectroscopic study of the effects of microbial transglutaminase on heat-induced gelation of pork myofibrillar proteins and its relationship with textural characteristics. *Food Res. Int.* **2011**, *44*, 1514–1520. [CrossRef]
27. Herrero, A.M.; Carmona, P.; Cofrades, S.; Jiménez-Colmenero, F. Raman spectroscopic determination of structural changes in meat batters upon soy protein addition and heat treatment. *Food Res. Int.* **2008**, *41*, 765–772. [CrossRef]
28. Sun, W.; Zhao, Q.; Zhao, M.; Yang, B.; Cui, C.; Ren, J. Structural evaluation of myofibrillar proteins during processing of cantonese sausage by raman spectroscopy. *J. Agric. Food Chem.* **2011**, *59*, 11070–11077. [CrossRef]
29. Herrero, A.M.; Cambero, M.I.; Ordóñez, J.A.; De La Hoz, L.; Carmona, P. Raman spectroscopy study of the structural effect of microbial transglutaminase on meat systems and its relationship with textural characteristics. *Food Chem.* **2008**, *109*, 25–32. [CrossRef]
30. Xu, X.-L.; Han, M.-Y.; Fei, Y.; Zhou, G.-H. Raman spectroscopic study of heat-induced gelation of pork myofibrillar proteins and its relationship with textural characteristic. *Meat Sci.* **2011**, *87*, 159–164. [CrossRef]
31. Li-Chan, E.C.Y. The applications of Raman spectroscopy in food science. *Ternds Food Sci. Technol.* **1996**, *7*, 361–370. [CrossRef]

32. Stangierski, J.; Baranowska, H.M.; Rezler, R.; Kijowski, J. Enzymatic modification of protein preparation obtained from water-washed mechanically recovered poultry meat. *Food Hydrocoll.* **2008**, *22*, 1629–1636. [CrossRef]
33. Han, M.; Wang, P.; Xu, X.; Zhou, G. Low-field NMR study of heat-induced gelation of pork myofibrillar proteins and its relationship with microstructural characteristics. *Food Res. Int.* **2014**, *62*, 1175–1182. [CrossRef]

foods

MDPI

Article

Quality Improvement of Zhayu, a Fermented Fish Product in China: Effects of Inoculated Fermentation with Three Kinds of Lactic Acid Bacteria

Yueqi An [1,2], Xiaowen Cai [1], Lin Cong [1], Yang Hu [1,2], Ru Liu [1,2], Shanbai Xiong [1,2] and Xiaobo Hu [1,*]

[1] College of Food Science and Technology/National R&D Branch Center for Conventional Freshwater Fish Processing (Wuhan), Huazhong Agricultural University, Wuhan 430070, China

[2] Engineering Research Center of Green Development for Conventional Aquatic Biological Industry in the Yangtze River Economic Belt, Ministry of Education, Wuhan 430070, China

* Correspondence: hxbhzau@126.com

Abstract: To investigate the effects of inoculation fermentation on the quality of Zhayu (a traditional fermented fish product in China), different amounts of *L. plantarum*, *P. acidilactici*, and *P. pentosaceus* were inoculated into samples, and the safety, nutritional, textural, and flavor properties of the samples were evaluated. Fermentation with lactic acid bacteria (LAB) decreased pH values and total volatile basic nitrogen content. The addition of 10^8~10^9 cfu/100 g LAB significantly increased the content of crude fat and water-soluble proteins in Zhayu. The addition of *L. plantarum* and *P. acidilactici* increased the content of soluble solids in Zhayu. Moreover, fermentation with LAB made the products tender and softer, and the samples prepared with 10^9 cfu/100 g LAB presented better overall qualities. Additionally, Zhayu fermented with *L. plantarum* and *P. acidilactici* showed the strongest sourness, while the samples prepared with *P. pentosaceus* showed the strongest umami taste, consistent with the highest contents of Asp (25.1 mg/100 g) and Glu (67.8 mg/100 g). The addition of LAB decreased the relative contents of aliphatic aldehydes, (Z)-3-hexen-1-ol, and 1-octen-3-ol, reducing the earthy and fishy notes. However, LAB enhanced the contents of terpenoids, acids, esters, and S-containing compounds, increasing the sour, pleasant, and unique odors of Zhayu.

Keywords: fermented fish product; lactic acid bacteria; inoculated fermentation; quality improvement; flavor characteristics

Citation: An, Y.; Cai, X.; Cong, L.; Hu, Y.; Liu, R.; Xiong, S.; Hu, X. Quality Improvement of Zhayu, a Fermented Fish Product in China: Effects of Inoculated Fermentation with Three Kinds of Lactic Acid Bacteria. *Foods* **2022**, *11*, 2756. https://doi.org/10.3390/foods11182756

Academic Editor: Fatih Oz

Received: 4 July 2022
Accepted: 29 August 2022
Published: 8 September 2022

1. Introduction

Fermentation is one of the main ways of the deep processing and preservation of aquatic products [1]. Fermented aquatic products have been favored by consumers in various regions because of their unique texture and flavor, such as Thai Plaa-som [2], Indian Bakasang [3], Indian Ngari [4], Chinese Suanyu [5–7], and Chinese fermented mandarin fish [8,9]. Different from other fermented fish products (such as Suanyu), Zhayu is fermented in small pieces with rice flour [10]. Zhayu is usually prepared with freshwater fish, especially grass carp (*Ctenopharyngodon idellus*). To manufacture Zhayu, washed fish fillets are cut into cubes and mixed with rice flour, salt, hot pepper, ginger, and other seasonings, and then the mixture is sealed and fermented under a solid state (Figure 1). During the solid-state fermentation, proteolysis, lipid degradation, and carbohydrate decomposition actioned by enzymes and microorganisms result in the production of organic acids, amino acids, small molecular peptides, and other substances, which enhance the nutritional value, enrich the flavor characteristics, and leads to a unique soft and loose texture of the fermented fish products [5,11]. In recent years, fermented Zhayu products have shown good market potential due to their delicious flavor and high nutritional value.

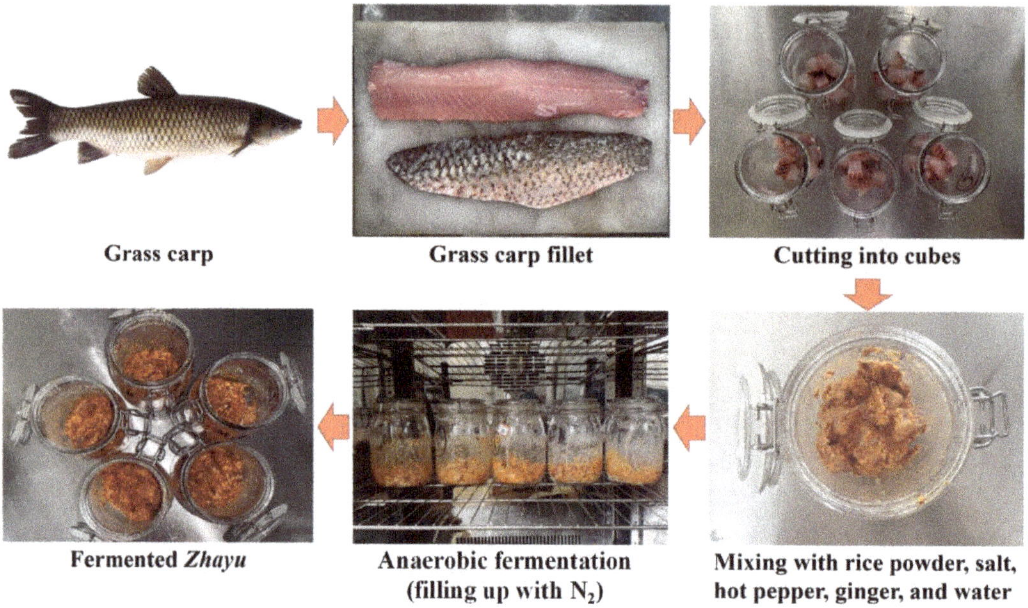

Figure 1. Schematic illustration for the process of fermented Zhayu products.

Traditional fermented fish products were usually manufactured by spontaneous fermentation without the addition of starter cultures in small-scale processing units. However, the conditions of spontaneous fermentation were difficult to control, resulting in unequal qualities of fermented fish products, which was not conducive to their industrial production. To solve this problem, yeast, mold, and bacteria (especially lactic acid bacteria and staphylococci) have been used as starter cultures for fermented fish products [12–14]. Fermentation with starter cultures not only shortens the fermentation time and prolongs the storage time of fermented products, but also improves the flavor, color, safety, and nutritional qualities of the products [15,16]. Lactic acid bacteria (LAB), especially *Lactiplantibacillus plantarum* and *Pediococcus pentosaceus*, have been isolated and identified as the dominant strains in Zhayu [17]. Moreover, *Pediococcus acidilactici* is also an important LAB in fermented fish or meat products [13,18]. LAB can reduce the pH value of the fermented product and inhibit the growth of spoilage microorganisms in the product [6,15]. LAB can also decrease the content of nitrite in fermented aquatic products and improve the safety quality of the product [19]. Moreover, LAB has been shown to make fermented sausage show better flavor properties [20]. Therefore, LAB could be used as a starter culture of Zhayu to improve its qualities. However, the effects of LAB species and inoculation amounts on the improvement of the qualities of Zhayu remain to be further studied.

In the present study, fermented Zhayu samples were prepared by inoculation with different amounts of *Lactiplantibacillus plantarum* (*L. plantarum*), *Pediococcus acidilactici* (*P. acidilactici*), and *Pediococcus pentosaceus* (*P. pentosaceus*) for starter culture fermentation, and the Zhayu sample prepared with natural fermentation was taken as the control. All Zhayu samples were fermented in the same conditions. The safety, nutritional, textural, and flavor properties of these fermented fish products were evaluated. The objectives of this research were (1) to investigate the effects of inoculation amounts and LAB species on the Zhayu products' quality and (2) to compare the effects of different kinds of LAB on the quality improvement of the fermented Zhayu products. This study will provide some theoretical basis for the production and processing of fermented aquatic products and give some new ideas for promoting industrialization and large-scale production of fermented aquatic products.

2. Materials and Methods

2.1. Materials and Chemicals

Live grass carp (*Ctenopharyngodon idellus*) (approximately 1500~2000 g) were purchased from the market of Huazhong Agricultural University (Wuhan, Hubei, China) and transported to the laboratory within 15 min in a transport case with water to keep them alive.

Rice was purchased from the market of Huazhong Agricultural University (Wuhan, Hubei, China) and smashed (<60 mesh) before use. Salt, hot pepper, and ginger were of food grade and were purchased from Zhongbai supermarket (Wuhan, Hubei, China).

The internal standard (2-octanol, >99.5%) used for gas chromatography–mass spectrometry (GC-MS) analysis was supplied by Sigma-Aldrich (St. Louis, MO, USA). The standard, reagent, and eluent solvent used for the amino acid analysis were supplied from Waters Corporation (Milford, MA, USA). All other chemicals were of analytical grade and were purchased from Sinopharm Chemical Reagent Co., Ltd. (Shanghai, China).

2.2. Preparation of Starter Culture

L. plantarum (CCTCC NO. M 2012396) was obtained from the China Center for Type Culture Collection (Peking, China). *P. acidilactici* (CICC 10344) and *P. pentosaceus* (CICC 22227) were purchased from the China Center of Industrial Culture Collection (Peking, China). These LAB were separately subcultured twice in DeMan Rogosa Sharpe (MRS) agar (Hopebio Co. Ltd., Qingdao, Shandong, China) and cultured at 30 °C for 48 h. Then, LAB was incubated in 200 mL of MRS broth (Hopebio Co. Ltd., Qingdao, Shandong, China). After the amplification of culture, the mixture of strains and MRS broth was centrifuged at 8000 r/min for 15 min at 4 °C. Cell pellets were harvested and washed with physiological saline (0.9% NaCl, *w*/*v*) twice. Then, cell pellets were resuspended with 0.9% NaCl (*w*/*v*). The growth curves of the three strains were measured by a UV-1700 ultraviolet spectrophotometer (Shimadzu, Kyoto, Japan) at 600 nm [7], and the central time point of the logarithmic phase was considered as the amplification time required for the strain's amount around 10^9 cfu/mL. Then, the strains were harvested for the exact time and the plate count agar was used to determine and verify the amount of strains. The bacterial suspension (10^9 cfu/mL) was diluted to 10^8 cfu/mL, 10^7 cfu/mL, and 10^6 cfu/mL in 0.9% NaCl (*w*/*v*) for the followed inoculation. Each bacterial suspension was stored at 4 °C and used within 24 h.

2.3. Sampling

Live grass carp were killed by a physical blow to the head, beheaded, gutted, and cleaned, following the guidance on Treating Experimental Animals developed by China's Ministry of Science and Technology in 2006 and regulations issued by the China State Council in 1988. Grass carp were cut into small pieces (approximately 1 cm × 2 cm × 2 cm). Seeds of hot pepper and ginger peals were removed, and the hot pepper and ginger were chopped. Fish pieces (100 g) were mixed in a vacuum bag (food grade) with rice flour (45 g), chopped hot pepper (20 g), bruised ginger (8 g), salt (8 g), tap water (45 mL), and different species of LAB suspension (1 mL). After mixing, the bag was vacuumed and sealed by a vacuum packaging machine. The control sample was prepared by spontaneous fermentation without the addition of starter cultures. After sealing, the mixtures were fermented at 25 °C for 60 h. Before analysis, the rice flour and seasoning were removed from the Zhayu samples.

2.4. Determination of pH, Titratable Acidity (TA), and Total Volatile Basic Nitrogen (TVB-N)

The pH values of samples were determined as reported by Zeng et al. [15] with some modifications. Ten grams of samples were homogenized with 50 mL of deionized water (discharged CO_2) at 8000 r/min for 2 min. After vacuum filtration, the pH values of filtrates were measured by a digital pH meter (Mettler Toledo FE28, Shanghai, China).

TA was determined via AOAC [21], and the results were expressed as lactic acid content (mg/g). TVB-N contents were determined by the micro-diffusion method of Conway [22].

2.5. Determination of Crude Fat, Water-Soluble Protein, and Soluble Solids

Crude fat contents were determined by an automated Soxhlet method [23] with some modifications. Two grams of homogenized and dried samples were extracted with petroleum ether (boiling point: 30~60 °C) rather than diethyl ether.

Water-soluble protein was extracted according to the method of Liao et al. [24] with some modifications. Ten grams of samples were homogenized with 0.05 mol/L phosphate buffer (pH 7.0) and diluted to 100 mL. After centrifuging at 8000 r/min for 20 min at 4 °C, the protein contents in the supernatant were measured by the Lowry method [25].

To determine the content of soluble solids, 10 g of samples was homogenized with distilled water and diluted to 100 mL. After centrifuging at 8000 r/min for 20 min at 4 °C, 10 g of the supernatant was dried at 105 ± 2 °C. The content of soluble solids was expressed as the percentage of dry matter weight.

2.6. Texture Analysis

Textural profile analysis (TPA) was carried out using a model TA-XT Plus texture analyzer (Stable Micro System, Surrey, UK) at room temperature (22 ± 2 °C). The Zhayu sample was placed on the test platform with the 2 cm × 2 cm side up. The probe type for the test was P/36R, the pre-measurement speed was 2.00 mm/s, the measurement speed was 1.00 mm/s, the post-measurement speed was 5.00 mm/s, and the compression strain was 50%. Eight parallel samples were prepared for the texture analysis.

2.7. Analysis of Flavor Characteristics of Zhayu

2.7.1. Sensory Analysis

Sensory evaluation of the taste and odor of Zhayu samples was performed by eight experienced panelists (four males and four females, aged 22 to 35) in a sensory laboratory at room temperature (22 ± 2 °C). For sensory evaluation, Zhayu samples were marked randomly with a three-digit number. Before tasting each sample, the panelists were required to rinse their mouths thoroughly with purified water, and the procedure of sensory evaluation was conducted following ISO 4120 [26] and ISO 4121 [27]. In this research, the intensities of sourness, bitterness, sweetness, umami, and saltness were evaluated as taste characteristics. Five attributes that best expressed fermented fish products' aroma characteristics were selected as "fresh", "fishy", "earthy", "oily", and "acidic". Each characteristic was rated on the following scale: 0, no taste/odor; 1, very weak; 2, weak; 3, moderate; 4, strong; and 5, very strong. The Zhayu prepared by natural fermentation was set as a reference. The standard scores of the reference were determined as sourness, 3.5; bitterness, 2; sweetness, 2; umami, 3; saltness, 3; fresh, 3.5; fishy, 1; earthy, 2; oily, 2.5; and acidic, 3.

2.7.2. Electronic Tongue (E-Tongue) Analysis

E-tongue (ALPHA MOS, Heracles, France) analysis was performed to analyze the taste characteristics of Zhayu samples. The interaction-sensitive sensor system of E-tongue includes 7 chemical sensor arrays (AHS, PKS, CTS, NMS, CPS, ANS, and SCS) and an Ag/AgCl reference electrode. Each sensor shows various sensitivities of different tastes, and AHS, CTS, NMS, ANS, and SCS can reflect the intensity of sourness, saltness, umami, sweetness, and bitterness, respectively. A total of 20 g of Zhayu samples was homogenized with 100 mL of distilled water at 8000 r/min for 0.5 min. After centrifuging at 10,000 r/min for 10 min at 4 °C, 80 mL of the supernatant was used for the E-tongue analysis. E-tongue test conditions were set as follows: sample delay time, 0 s; acquisition time, 120 s; acquisition cycle, 1.00 s; stirring speed, 60 r/min [28]. Six parallel samples were prepared for the E-tongue analysis.

2.7.3. Electronic Nose (E-nose) Analysis

An E-nose (ALPHA MOS, Heracles, France) consists of an automatic sampling device, interaction-sensitive sensor array, data acquisition system, and data analysis software. To prepare samples for E-nose analysis, samples were cut into small pieces and 2 g of each sample was added to 10 mL headspace vials. The detection conditions of E-nose were set as follows: headspace inlet temperature, 50 °C; headspace time, 120 s; stirring speed, 500 r/min; injection volume, 2.5 mL; sample collection time, 120 s; delay time, 300 s [28]. Six parallel samples were prepared for the E-nose analysis.

2.7.4. Determination of Free Amino Acids

Free amino acids in Zhayu samples (1 g) were extracted with 30 mL of hydrochloric acid (0.1 mol/L) by homogenizing at 6000 r/min for 3 min, and then shaking at ambient temperature for 15 min. After standing for 5 min, the supernatant was collected and the residue was washed twice with 20 mL hydrochloric acid (0.1 mol/L). All the obtained supernatants were pooled and brought to 100 mL with 0.1 mol/L hydrochloric acids. The free amino acid extract was filtrated through a 0.22 μm membrane, and 10 μL filtrate of each sample was taken for a derivative reaction.

For the derivative reaction, the thermostat was first heated to 55 °C, followed by adding 70 μL of borate buffer in the AccQ·Tag Ultra Derivatization kit (Waters Corporation, Milford, MA, USA) and 10 μL of each sample solution to the sample vial successively, then vortex mixing immediately, and finally adding 20 μL of reconstituted AccQ·Tag Ultra reagent. After vortex mixing for 10 s and standing at room temperature for 1 min, the mixture was heated for 10 min at 55 °C and then analyzed by UHPLC (Dionex Ultimate 3000, Thermo Scientific, Sunnyvale, CA, USA).

The amino acids were separated in an ACQUITY UPLC® BEH C18 column (1.7 μm, 2.1 × 100 mm, Waters Corporation, Massachusetts, Ireland, USA) with the column temperature and detection wavelength at 55 °C and 260 nm, respectively. Eluent A was the concentrated solution in an AccQ·Tag Ultra reagent package (diluted 20 times with high-purity water), and eluent B was 100% acetonitrile. The elution procedure was initially 0.7 mL/min (A:B = 99.9:0.1), then 0.7 mL/min for 0.54 min (A:B = 99.9:0.1), 0.7 mL/min for 5.20 min (A:B = 90.9:9.1), 0.7 mL/min for 2 min (A:B = 10:90), 0.7 mL/min for 0.3 min (A:B = 40.4:59.6), 0.7 mL/min for 0.01 min (A:B = 10:90), 0.7 mL/min for 0.59 min (A:B = 10:90), 0.7 mL/min for 0.09 min (A:B = 99.9:0.1), and 0.7 mL/min for 0.77 min (A:B = 99.9:0.1) [28]. Amino acid standards were supplied from Waters Corporation and were derivatized and separated under the same conditions as the samples. The external standard curves were built to calculate the concentrations of amino acids in the samples. The chromatogram of the standards of amino acids is shown in Figure 2.

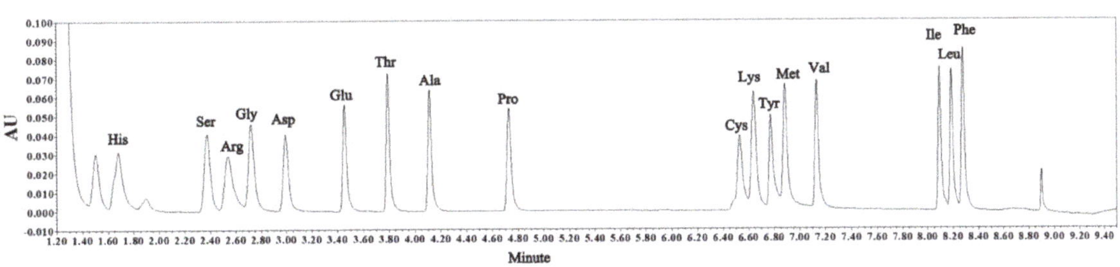

Figure 2. Chromatogram of the standards of amino acids detected by UHPLC.

2.7.5. GC-MS Analysis

Volatile aroma compounds in Zhayu were identified and semi-quantitated by headspace–solid-phase microextraction (HS-SPME) and GC-MS. The GC-MS analysis was performed on an Agilent 7890B GC equipped with an Agilent 5977B mass selective detector (Agilent Technologies, Inc., Santa Clara, CA, USA). Chopped Zhayu samples (4.0 g) were added to

6 mL of saturated sodium chloride solution in a 20 mL autosampler glass. After a volume of 10 μL of an internal standard (50 μg/mL 2-octanol) was added, the autosampler glass was capped tightly with a Teflon-faced silicone septum. The samples were equilibrated at 45 °C for 15 min, and a DVB/CAR/PDMS fiber (50/30 μm, Supelco, Inc., Bellefonte, PA, USA) was used to extract volatiles for 40 min from headspace. The stirring speed was 400 r/min. The volatiles were desorbed into the GC injection port in a splitless mode at 250 °C. The desorption time was 5 min. The separation was performed using a DB-wax column (30 m length, 0.25 mm i.d., 0.25 μm film thickness; Agilent Technologies, Inc., Santa Clara, CA, USA). The oven temperature was programmed as follows: 40 °C (initial hold for 4 min), ramp at 4 °C/min to 230 °C (hold for 5 min). The carrier gas was nitrogen with a constant flow rate of 1.0 mL/min. The electron impact (EI) energy was 70 eV, and the ion source temperature was set at 230 °C. Each sample was run in triplicate. Mass spectra of compounds were compared to those in the National Institute of Standards and Technology (NIST) library (Agilent Technologies, Inc., Santa Clara, CA, USA). A standard mixture of n-alkanes (C6~C26) was prepared and injected into GC using the same conditions as the samples. Retention indices (RIs) were calculated following a modified Kováts method [29]. Compounds were positively identified by comparing mass spectra and RIs of the standards obtained in the laboratory or tentatively identified if the RIs were from the literature. The relative contents of volatiles were expressed by the ratio of the peak area of the volatiles to that of the internal standard.

2.8. Statistical Analysis

All tests were performed in triplicate. Results of E-nose and E-tongue data were processed through Alpha Soft 12.3 software (ALPHA MOS, Heracles, France). Variance analysis and principal component analysis (PCA) were conducted using SPSS 22.0 software (SPSS Inc., Chicago, IL, USA). Differences among mean values were established using Duncan's multiple range test. $p < 0.05$ was considered statistically significant.

3. Results and Discussions

3.1. Safety Qualities

The pH values and TA contents are important indexes to evaluate the safety of fermented foods, and it was generally believed that low pH value (<4.5) and high TA content could inhibit the growth of spoilage bacteria and ensure the safety of products [30]. Figure 3 illustrates that the spontaneously fermented Zhayu samples showed the highest pH value at 5.15 and the lowest TA content at 13.41 mg/kg. The fermentation with LAB decreased pH values and increased TA content in Zhayu products significantly ($p < 0.05$). As the inoculation amount increased, the pH value decreased, and the TA content increased gradually, indicating the accumulation of LAB metabolites and the enhancement of the acidity of fermented products. When 10^9 cfu/100 g of *P. acidilactici* and *P. pentosaceus* were inoculated, or the inoculation amounts of *L. plantarum* exceeded 10^7 cfu/100 g, pH values of the fermented Zhayu were below 4.4, which is safe for consumption [31]. Additionally, samples fermented with *P. acidilactici* showed the lowest TA content compared with the other two starter cultures, indicating the differences in the acid production capacity of various LAB species. In other fermented fish samples, such as fermented silver carp sausage and plaa-som, pH values also decreased during the fermentation process with LAB, and the decrease in pH values was related to the growth of LAB [2,32].

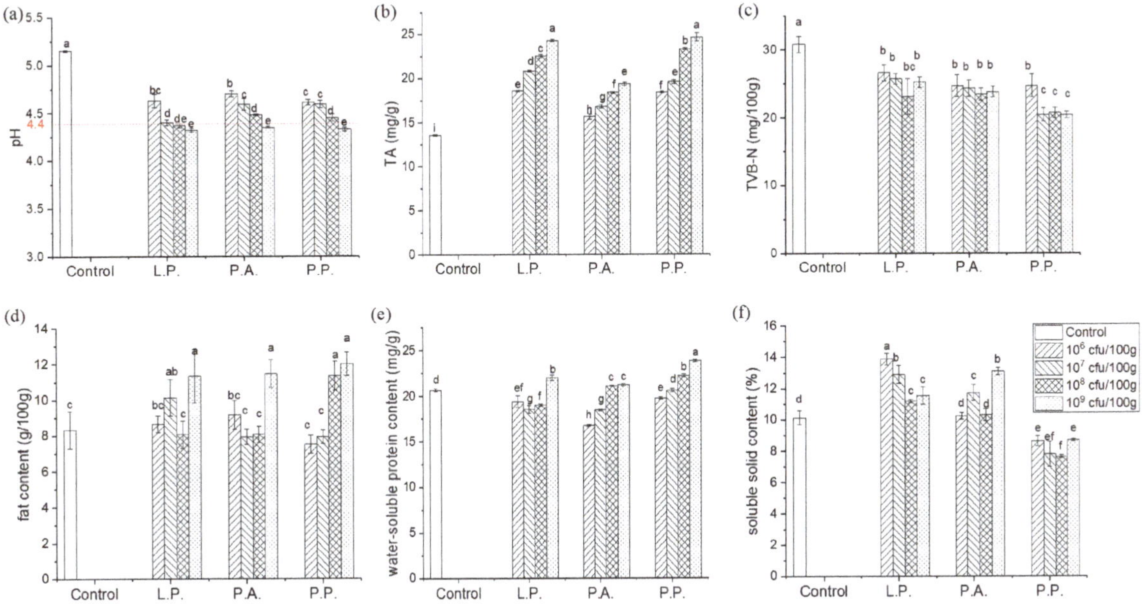

Figure 3. The pH values (**a**), TA contents (**b**), TVB-N contents (**c**), crude fat contents (**d**), water-soluble protein contents (**e**), and soluble solid contents (**f**) of Zhayu products prepared by spontaneous fermentation (control) and fermented with *L. plantarum* (L.P.), *P. acidilactici* (P.A.), and *P. pentosaceus* (P.P.) at different inoculation amounts. Different letters on the bar denote significant differences ($p < 0.05$).

The TVB-N content was closely related to the content of N-containing compounds (such as biological amines) produced by the decarboxylation of amino acids and the propagation of spoilage bacteria during the fermentation process, which could measure the spoilage of fermented fish products [33]. The contents of TVB-N of all Zhayu samples (21.0~30.8 mg/100 g) were lower than the superior limit provided by the European Union ($\leq$35 mg/100 g), indicating that the edible safety qualities of the products were guaranteed (Figure 3c). Zhayu samples prepared with LAB showed significantly lower TVB-N contents compared with the control ($p < 0.05$); however, the concentrations of *L. plantarum* and *P. acidilactici* showed no significant effect on the TVB-N contents in Zhayu samples ($p > 0.05$), and the Zhayu samples cultured with *P. pentosaceus* (10^7~10^9 cfu/100 g) presented the lowest TVB-N contents. It was reported that the addition of LAB could increase the acid-producing metabolism, resulting in the decrease in pH value and the formation of some bacteriostatic substances which could inhibit the growth of spoilage bacteria such as *Enterobacter*, and then reduce the accumulation of TVB-N [34,35]. Additionally, in Zhayu samples prepared with LAB, the increased acids could neutralize alkaline ammonia and amines, reducing the TVB-N contents. Among the three kinds of LAB, *P. acidilactici* presented a weak ability to improve the safety qualities of Zhayu samples due to its poor acid production capacity, while *P. pentosaceus* could enhance the safety qualities of Zhayu.

3.2. Nutritional Properties

The crude fat, water-soluble proteins, and soluble solids were used to evaluate the nutritional properties of Zhayu samples. The contents of these are illustrated in Figure 3d–f. It was shown that the effects of LAB-inoculated fermentation on the nutritional properties of Zhayu depended on the LAB species and concentrations. When the inoculation amounts were 10^6 cfu/100 g and 10^7 cfu/100 g, the addition of LAB did not significantly affect the contents of crude fat in the Zhayu products ($p > 0.05$). However, the contents of crude fat noticeably increased in the samples fermented with 10^9 cfu/100 g of *L. plantarum* and *P. acidilactici*, or with more than 10^8 cfu/100g of *P. pentosaceus*. LAB could produce intracellular lipase and extracellular lipase during fermentation, which could promote the degradation of lipids and increase the content of crude fat in fermented products [36]. However, the lipolysis ability of LAB was relatively weak [37]. Thus, only high amounts of LAB addition could affect the contents of crude fat in fermented fish products.

During fermentation, proteins in fish products were hydrolyzed by endogenous and microbial proteolytic enzymes to form more water-soluble proteins with low molecular weight, which were easily digested and absorbed [5]. However, due to the weak proteolytic activity of LAB, proteins in the LAB-inoculated Zhayu samples could be mainly hydrolyzed by endogenous enzymes such as cathepsins [15,37]. When the inoculation amount increased over 10^8 cfu/100 g for *P. acidilactici* and *P. pentosaceus*, and the amount increased to 10^9 cfu/100 g for *L. plantarum*, the water-soluble protein content in Zhayu products noticeably increased ($p < 0.05$) compared with the control. Additionally, the content of water-soluble proteins was the highest in the sample prepared with 10^9 cfu/100 g *P. pentosaceus*. In other fermented products, it was also found that starter culture showed a pronounced effect on protein degradation [7,32]. This suggests that the decrease in the pH value after fermentation with LAB promotes the activity of endogenous protease and then increases the content of water-soluble proteins [15].

The soluble solids in fermented fish products mainly include soluble sugar, vitamins, and minerals. Figure 3f shows that the addition of *L. plantarum* and *P. acidilactici* increased the content of soluble solids in fermented Zhayu products, while the soluble solids content of the samples fermented with *P. pentosaceus* showed a decrease compared with the control. During fermentation, the metabolism of microorganisms promotes the dissolution of soluble solids in cells, but the metabolism also consumes the soluble solids.

3.3. Texture Properties

The textural properties of fermented Zhayu products with and without starter cultures are shown in Table 1. The addition of *L. plantarum* and *P. acidilactici* significantly decreased the hardness of Zhayu samples ($p < 0.05$), but the amounts of *L. plantarum* and *P. acidilactici* did not significantly affect the hardness of the samples. When the inoculation concentration of LAB was 10^9 cfu/100 g, samples that were fermented with *P. pentosaceus* showed the lowest hardness. The reason for the decrease in hardness was the degradation of proteins caused by endogenous enzymes, resulting in a decrease in the interactions between fibrillin and the combination of musculature. The trend of the hardness of Zhayu was consistent with the water-soluble proteins' results (Figure 3e). However, it was found that fermentation with starter cultures could enhance the hardness of fermented fish sausages, Som-fug, and other fermented products [15,34,38]. This might be due to the different manufacturing processes of these fermented products. Fermentation with LAB starters also decreased the springiness and the chewiness of Zhayu products. Samples fermented with *P. pentosaceus* showed higher springiness and chewiness. It was suggested that Zhayu samples fermented with 10^9 cfu/100 g *P. pentosaceus* were tender, softer, and elastic, and presented better textural properties that were acceptable for consumers compared with other samples.

3.4. Taste Characteristics

The above results show that Zhayu samples fermented with each LAB at the inoculation amount of 10^9 cfu/100 g presented high safety qualities, contained high fats and water-soluble proteins, and showed tender and soft texture properties. Therefore, the Zhayu samples inoculated with 10^9 cfu/100 g of each LAB were chosen as the typical samples to compare the effects of LAB species on the flavor characteristics of Zhayu samples. Principal component analysis (PCA) was performed based on the responses of the E-tongue chemical sensor arrays for different Zhayu products to compare their taste profiles. Figure 4a illustrates the PCA results, and the first two principal components described 99.32% of the total taste variances, indicating that the taste characteristics of Zhayu products fermented with three kinds of LAB and the control were various. It was noticeable that the samples fermented with *L. plantarum* and *P. acidilactici* showed more similar taste characteristics.

A sensory evaluation was carried out to study the effects of LAB on the tastes of Zhayu products (Figure 4b). The inoculation of LAB dramatically increased the sourness, saltness, and umami taste of the Zhayu samples. The samples with the addition of *L. plantarum* and *P. acidilactici* presented a relatively strong sourness, while the samples fermented with *P. pentosaceus* showed the highest level of umami taste. The increase in sensory sourness was coincident with the rise in TA in Zhayu samples after starter culture fermentation (Figure 3), due to the lactic acid produced by LAB metabolism and the acid denaturation of proteins [34]. The umami taste in fish-related products mainly originates from the catabolism of proteins and nucleotides [39]. It was suggested that the positive effect of *P. pentosaceus* on the umami taste might be related to the degradation of proteins during fermentation. Meanwhile, the increase in umami tastes promoted the perception of saltiness in Zhayu samples, because of the interactions between the flavor substances [40]. Additionally, compared with the control, the inoculation with *P. pentosaceus* increased the sweetness and bitterness of Zhayu samples, while the addition of *L. plantarum* and *P. acidilactici* reduced the sweet and bitter perceptions. These differences might be due to the different protein degradation results of each LAB.

During the fermentation of aquatic products, proteins were hydrolyzed by microbes and endogenous protease, promoting the formation of free amino acids and changes in taste properties [7,41]. In Zhayu samples, the addition of LAB increased most of the contents of free amino acids (Table 2). In particular, Asp and Glu, which have an umami taste and relatively low thresholds, presented significantly higher contents and taste activity values (TAVs) in LAB-inoculated samples than those in the control. Moreover, Zhayu samples fermented with *P. pentosaceus* contained the most amounts of Asp and Glu, which was consistent with the highest water-soluble protein contents (Figure 3e) and the highest sensory umami scores (Figure 4b). In fish sauce and fish chili paste, it was also found that the inoculation fermentation significantly increased the content of free amino acids with the umami taste (Glu and Asp) [14,42]. Additionally, the contents of Thr, Lys, Tyr, and His, with a sweet or bitter taste, were higher in the LAB-inoculated Zhayu samples than in the control, suggesting that fermentation with LAB could enhance the hierarchy of taste perception of Zhayu samples.

Table 1. Textural properties of Zhayu samples prepared by spontaneous fermentation (Control) and different inoculation amounts of *L. plantarum*, *P. acidilactici*, and *P. pentosaceus*.

Inoculation Amounts (cfu/100 g)	Hardness (g)	Springiness	Chewiness (g)
Control	250.16 ± 31.54 ^a^	1.00 ± 0.01 ^a^	163.4 ± 20.3 ^a^
L. plantarum			
10^6	203.02 ± 26.84 ^bcd^	0.71 ± 0.06 ^ef^	45.53 ± 18.11 ^ef^
10^7	171.64 ± 16.33 ^de^	0.80 ± 0.19 ^ef^	42.86 ± 9.95 ^ef^
10^8	176.85 ± 21.91 ^cde^	0.88 ± 0.09 ^de^	54.19 ± 9.04 ^de^
10^9	186.78 ± 20.6 ^cd^	0.75 ± 0.10 ^f^	31.66 ± 9.43 ^f^
P. acidilactici			
10^6	178.21 ± 3.90 ^cd^	0.99 ± 0.01 ^ab^	111.4 ± 2.8 ^bc^
10^7	171.81 ± 32.46 ^de^	0.97 ± 0.02 ^b^	61.39 ± 2.65 ^de^
10^8	174.32 ± 26.46 ^cde^	0.79 ± 0.09 ^d^	67.31 ± 5.12 ^d^
10^9	162.28 ± 16.33 ^de^	0.84 ± 0.09 ^ef^	44.31 ± 6.26 ^ef^
P. pentosaceus			
10^6	246.96 ± 26.27 ^ab^	0.87 ± 0.06 ^cd^	125.1 ± 3.6 ^b^
10^7	247.29 ± 35.79 ^ab^	0.84 ± 0.03 ^d^	94.07 ± 4.92 ^c^
10^8	220.27 ± 11.41 ^abc^	0.91 ± 0.03 ^c^	98.97 ± 19.33 ^c^
10^9	130.35 ± 31.86 ^e^	0.94 ± 0.06 ^bc^	109.6 ± 18.2 ^bc^

Notes: Different letters denote significant differences in the same column ($p < 0.05$).

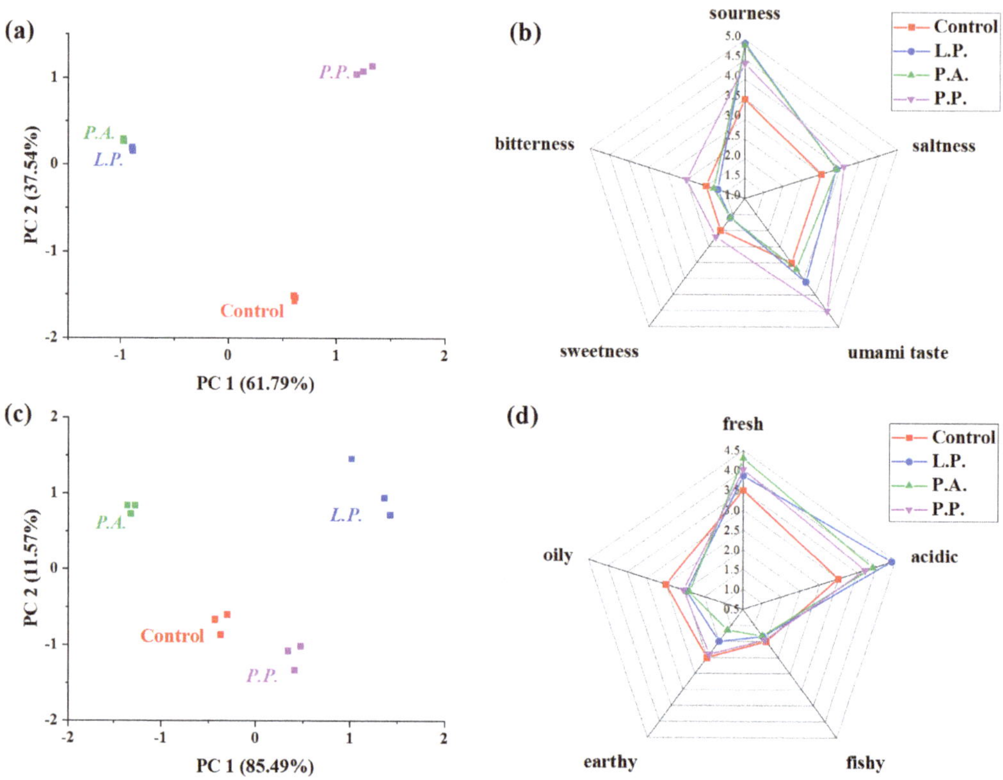

Figure 4. Principal component analysis (PCA) of Zhayu samples fermented with different starter cultures based on E-tongue results (**a**) and E-nose results (**c**), and sensory evaluation of taste characteristics (**b**) and odor profiles (**d**) of Zhayu products prepared with different kinds of LAB.

Table 2. Contents (mg/100 g) and taste activity value (TAV) of free amino acids in Zhayu samples prepared by spontaneous fermentation (SF) and fermented with 109 cfu/100 g of *L. plantarum* (L.P.), *P. acidilactici* (P.A.), and *P. pentosaceus* (P.P.).

FAA	Threshold (mg/100 g)	Taste	Concentration (mg/100 g)				TAV			
			SF	L.P.	P.A.	P.P.	SF	L.P.	P.A.	P.P.
Asp	3	umami	5.51 ± 0.32^{d}	45.6 ± 1.7^{b}	42.1 ± 1.1^{c}	52.1 ± 1.9^{a}	1.84	15.19	14.05	17.36
Glu	5	umami	49.0 ± 3.3^{d}	56.8 ± 2.7^{b}	51.8 ± 1.6^{c}	67.8 ± 2.47^{a}	9.80	14.05	17.26	22.61
Ser	150	sweet	4.61 ± 0.26^{a}	2.97 ± 0.31^{c}	2.24 ± 0.06^{d}	3.66 ± 0.13^{b}	0.03	0.02	0.01	0.02
Gly	150	sweet	23.2 ± 1.3^{a}	19.4 ± 0.8^{c}	21.1 ± 0.5^{b}	21.6 ± 0.7^{bc}	0.15	0.13	0.14	0.14
Thr	260	sweet	9.26 ± 0.53^{c}	14.2 ± 0.4^{b}	17.8 ± 0.4^{a}	18.4 ± 0.7^{a}	0.04	0.05	0.07	0.07
Ala	60	sweet	58.5 ± 3.3^{a}	44.3 ± 1.4^{c}	45.2 ± 1.2^{c}	47.5 ± 1.7^{b}	0.97	0.74	0.75	0.79
Pro	300	sweet/bitter	51.9 ± 2.9^{b}	51.8 ± 3.5^{b}	46.5 ± 1.2^{c}	56.6 ± 2.1^{a}	0.17	0.17	0.15	0.19
Lys	50	sweet/bitter	12.6 ± 0.7^{c}	45.1 ± 1.2^{ab}	44.2 ± 1.1^{b}	46.4 ± 1.6^{a}	0.25	0.90	0.88	0.93
Arg	50	bitter	1.97 ± 0.11^{b}	2.53 ± 0.36^{a}	1.11 ± 0.03^{d}	1.48 ± 0.05^{c}	0.04	0.05	0.02	0.03
Val	40	bitter	30.3 ± 1.7^{a}	25.5 ± 0.7^{c}	26.2 ± 0.7^{bc}	27.0 ± 0.9^{b}	0.76	0.64	0.65	0.68
Tyr	N.A.	bitter	20.8 ± 1.2^{c}	27.7 ± 0.7^{a}	24.6 ± 0.6^{b}	25.9 ± 0.9^{b}				
His	20	bitter	30.4 ± 1.7^{c}	76.4 ± 3.6^{a}	79.8 ± 2.1^{a}	70.8 ± 2.5^{b}	1.52	3.82	3.99	3.54
Leu	190	bitter	50.5 ± 2.9^{b}	53.4 ± 1.6^{a}	50.9 ± 1.3^{b}	53.5 ± 1.9^{a}	0.27	0.28	0.27	0.28
Ile	90	bitter	23.2 ± 1.3^{a}	18.1 ± 0.7^{b}	17.5 ± 0.5^{b}	18.4 ± 0.7^{b}	0.26	0.20	0.19	0.20
Phe	90	bitter	38.2 ± 2.1^{ab}	41.1 ± 1.0^{a}	37.7 ± 0.9^{b}	39.0 ± 1.4^{ab}	0.42	0.46	0.42	0.43
Met	30	bitter	17.8 ± 1.0^{a}	15.2 ± 0.6^{b}	14.3 ± 0.3^{c}	14.9 ± 0.5^{bc}	0.59	0.51	0.48	0.50
Cys	N.A.	salt	N.D.	N.D.	N.D.	N.D.				

Notes: N.A., not available; N.D., not detected. Different letters denote significant differences in the same row ($p < 0.05$). Thresholds for amino acids adapted from Wang et al. [38].

3.5. Odor Characteristics

Figure 4c,d illustrates the PCA results based on the responses of the E-nose sensors and the sensory results for the odor characteristics of different Zhayu samples. The first two principal components described 97.06% of the total odor variances, indicating that the responses of E-nose sensors could adequately differentiate Zhayu products fermented with various LAB. The PCA results indicate that the odor properties of each fermented Zhayu sample differed significantly. Additionally, Zhayu samples fermented with LAB showed stronger fresh and acidic notes, but weaker fishy, earthy, and oily odors compared with the control (Figure 4d), suggesting that the LAB-inoculated fermentation could enhance the aroma characteristics of Zhayu samples. Especially for the samples inoculated with *P. acidilactici*, the promotion of the pleasant fresh odor and the inhibition of earthy notes were the most pronounced. Moreover, samples prepared with *L. plantarum* presented the strongest acidic odor.

To further understand the differences in volatile compounds of Zhayu samples prepared by spontaneous and inoculated fermentation, the relative concentrations of volatile compounds in Zhayu samples were detected by HS-SPME-C-MS. The results of the relative contents of volatile compounds in the typical Zhayu samples are shown in Table 3.

Aldehydes, such as hexanal with a grassy note, octanal with a citrus odor, and (E,E)-2,4-decadienal with a fishy or fatty smell, are reported as major aroma compounds in fish-related products due to their low odor thresholds, and they are mainly derived from the oxidation of fatty acids and the degradation of proteins [8,9,38]. It was noticeable that except nonanal and benzaldehyde, the relative contents of most aldehydes were significantly lower in Zhayu samples fermented with LAB than those in the control. Especially, the amount of hexanal in the control was approximately 5.1~5.5 times higher than that in the LAB-inoculated fermented Zhayu samples. Additionally, octanal and nonanal were not detected in the Zhayu samples inoculated with *P. pentosaceus*. It was reported that in other fermented fish samples (such as *Suan zuo yu* and fermented mandarin fish), aldehydes exhibited a downward trend during fermentation [8,9]. Moreover, starter cultures showed an antioxidant capacity on unsaturated fatty acids in fermented fish products, reducing the concentrations of aliphatic aldehydes [13,42]. Additionally, it was reported that in fish paste, *L. plantarum* significantly reduced the lipolysis and inhibited the flavor generation by the other strains [14]. Thus, it was presumed that the inoculated fermentation with LAB might inhibit the lipid oxidation in Zhayu samples, decreasing the aldehydes with grassy, fishy, or fatty notes, and then increasing the aroma properties of Zhayu.

Fourteen kinds of alcohols were identified in Zhayu samples. However, most of them showed a high odor threshold and contributed little to the aroma profiles of fish-related products, except (Z)-3-hexen-1-ol, 1-octen-3-ol, and 1-heptanol [6,8]. Among them, 1-penten-3-ol, (Z)-3-hexen-1-ol, and 1-octen-3-ol, which are mainly derived from the oxidation of unsaturated fatty acids and have grassy, fishy, and mushroom-like notes [8], presented the lowest contents in Zhayu samples fermented with *P. acidilactici*, consistent with the weakest notes of "earthy" and "fishy" by the sensory analysis (Figure 4d). However, the LAB-inoculated fermentation increased the concentrations of 1-heptanol, and so did other saturated alcohols in the samples. This might be because the role of microorganisms also had an important influence on the aroma property of fermented aquatic products, which could increase the content of alcohols [16].

Zhayu samples contained many terpenoids. The source of the terpenoids in fish-related products was related to the addition of hot pepper, ginger, and other seasonings [45]. Interestingly, the relative contents of many terpenoids, contributing to lemon-like, woody, or floral notes, were noticeably higher in the samples fermented with LAB than in the samples prepared by spontaneous fermentation. It was suggested that during the fermentation of Zhayu, the metabolism of LAB could promote the formation of terpenoids. A similar phenomenon was also found in Suan yu, another kind of fermented fish product [6]. The enriched terpenoids by LAB-inoculated fermentation would enhance the overall aroma characteristics of the fermented fish products.

Table 3. Relative contents (μg/kg) of aroma compounds in Zhayu samples prepared by spontaneous fermentation (SF) and fermented with 10^9 cfu/100 g of *L. plantarum* (L.P.), *P. acidilactici* (P.A.), and *P. pentosaceus* (P.P.).

Compounds	RI	Identification	Odor	Odor Threshold (μg/kg)	Relative Concentration (μg/kg)			
					SF	L.P.	P.A.	P.P.
Aldehydes								
2/3-methylbutanal	923	MS, RIL	malty	1.5/0.5 ^{b}	0.74 ± 0.13 ^{a}	0.51 ± 0.05 ^{c}	0.61 ± 0.04 ^{b}	0.67 ± 0.07 ^{b}
hexanal	1087	MS, RI	grassy	2.4 ^{a}	2.52 ± 0.78 ^{a}	0.48 ± 0.02 ^{b}	0.49 ± 0.09 ^{b}	0.46 ± 0.17 ^{b}
octanal	1294	MS, RI	citrus	3.4 ^{a}	0.28 ± 0.13 ^{a}	0.11 ± 0.03 ^{c}	0.15 ± 0.04 ^{b}	N.D.
(E)-2-heptenal	1326	MS, RI	green	13 ^{c}	0.48 ± 0.07 ^{a}	0.41 ± 0.05 ^{b}	0.45 ± 0.06 ^{ab}	0.40 ± 0.04 ^{b}
nonanal	1399	MS, RI	citrus	2.8 ^{a}	0.09 ± 0.02 ^{a}	0.11 ± 0.02 ^{a}	0.12 ± 0.03 ^{a}	N.D.
(E)-2-octenal	1435	MS, RI	nutty	3 ^{a}	1.17 ± 0.03 ^{a}	0.59 ± 0.05 ^{b}	0.61 ± 0.09 ^{b}	0.66 ± 0.07 ^{b}
benzaldehyde	1508	MS, RI	rosy	350–3500 ^{c}	1.35 ± 0.36 ^{a}	1.07 ± 0.00 ^{a}	1.17 ± 0.03 ^{a}	1.02 ± 0.19 ^{a}
(E,E)-2,4-decadienal	1820	MS, RI	fishy, fatty	0.027 ^{a}	1.64 ± 0.11 ^{a}	0.85 ± 0.05 ^{b}	0.86 ± 0.06 ^{b}	0.67 ± 0.12 ^{c}
Alcohols								
1-propanol	1049	MS, RIL	alcoholic	7000 ^{c}	0.29 ± 0.02 ^{c}	0.38 ± 0.16 ^{b}	0.54 ± 0.00 ^{b}	0.89 ± 0.05 ^{a}
1-penten-3-ol	1164	MS, RI	grassy	400 ^{c}	0.76 ± 0.16 ^{a}	0.56 ± 0.05 ^{a}	0.42 ± 0.08 ^{b}	0.65 ± 0.12 ^{a}
1-pentanol	1260	MS, RIL	fusel-like	4000 ^{c}	3.35 ± 0.88 ^{a}	1.93 ± 0.14 ^{b}	1.75 ± 0.25 ^{b}	1.85 ± 0.15 ^{b}
2-heptanol	1322	MS, RIL	herbaceous	41–81 ^{c}	2.41 ± 0.35 ^{b}	5.06 ± 0.02 ^{a}	4.34 ± 0.21 ^{a}	4.25 ± 0.83 ^{a}
1-hexanol	1359	MS, RI	herbaceous	2500 ^{c}	7.19 ± 0.84 ^{ab}	8.14 ± 0.31 ^{a}	6.00 ± 0.42 ^{b}	8.29 ± 0.88 ^{a}
(Z)-3-hexen-1-ol	1385	MS, RI	fishy	3.9 ^{b}	1.86 ± 0.38 ^{a}	1.57 ± 0.09 ^{a}	1.27 ± 0.07 ^{b}	1.74 ± 0.13 ^{a}
3-octanol	1390	MS, RIL	oily, nutty	18–250 ^{c}	0.46 ± 0.07 ^{b}	0.67 ± 0.08 ^{a}	0.67 ± 0.04 ^{a}	0.62 ± 0.18 ^{ab}
1-octen-3-ol	1457	MS, RI	mushroom	1.2 ^{a}	1.64 ± 0.34 ^{a}	1.49 ± 0.29 ^{a}	0.51 ± 0.06 ^{b}	1.27 ± 0.48 ^{a}
1-heptanol	1468	MS, RIL	woody	3 ^{c}	0.21 ± 0.03 ^{c}	0.90 ± 0.07 ^{a}	0.65 ± 0.06 ^{b}	0.58 ± 0.07 ^{b}
2-ethylhexanol	1488	MS, RIL	green, oily	270,000 ^{c}	0.41 ± 0.07 ^{b}	1.01 ± 0.13 ^{a}	0.67 ± 0.12 ^{ab}	0.98 ± 0.27 ^{a}
1-octanol	1575	MS, RIL	soapy	110–130 ^{c}	0.19 ± 0.03 ^{c}	0.44 ± 0.01 ^{a}	0.30 ± 0.01 ^{b}	0.31 ± 0.09 ^{b}
1-nonanol	1663	MS, RIL	soapy	50 ^{c}	0.15 ± 0.00 ^{c}	0.73 ± 0.09 ^{a}	0.36 ± 0.03 ^{b}	0.43 ± 0.13 ^{b}
benzyl alcohol	1880	MS, RI	rosy	1.2–1000 ^{c}	0.73 ± 0.19 ^{b}	0.87 ± 0.11 ^{ab}	0.92 ± 0.02 ^{a}	1.05 ± 0.18 ^{a}
phenylethyl alcohol	1922	MS, RI	rosy	140 ^{b}	0.43 ± 0.01 ^{b}	0.54 ± 0.05 ^{a}	0.56 ± 0.03 ^{a}	0.67 ± 0.14 ^{a}
Terpenoids								
α-pinene	1036	MS, RIL	pine	2.2 ^{b}	3.41 ± 1.33 ^{a}	1.41 ± 0.16 ^{b}	1.93 ± 0.23 ^{a}	2.43 ± 0.34 ^{a}
camphene	1059	MS, RIL	camphoraceous	N.A.	4.06 ± 0.99 ^{b}	3.03 ± 0.21 ^{b}	4.53 ± 1.79 ^{b}	7.72 ± 1.12 ^{a}
β-myrcene	1170	MS, RIL	balsamic	1.2 ^{b}	0.56 ± 0.15 ^{a}	0.15 ± 0.01 ^{b}	0.63 ± 0.09 ^{a}	0.47 ± 0.12 ^{a}
D-limonene	1204	MS, RIL	lemon-like	10 ^{c}	0.70 ± 0.27 ^{b}	0.56 ± 0.07 ^{c}	1.14 ± 0.01 ^{a}	1.02 ± 0.23 ^{ab}
β-phellandrene	1219	MS, RIL	citrus	40–200 ^{c}	0.77 ± 0.07 ^{b}	0.87 ± 0.04 ^{b}	1.23 ± 0.70 ^{a}	1.03 ± 0.28 ^{a}
eucalyptol	1230	MS, RIL	camphoraceous	12 ^{c}	12.46 ± 2.01 ^{b}	14.26 ± 0.53 ^{a}	15.15 ± 1.68 ^{a}	15.35 ± 4.12 ^{ab}
caryophyllene	1570	MS, RIL	woody	64 ^{c}	N.D.	2.35 ± 0.58 ^{a}	1.22 ± 0.34 ^{b}	2.33 ± 0.14 ^{a}
camphor	1524	MS, RIL	minty	1000–1290 ^{c}	0.36 ± 0.07 ^{c}	0.69 ± 0.05 ^{b}	0.80 ± 0.01 ^{a}	0.83 ± 0.18 ^{ab}
linalool	1554	MS, RIL	lemon-like	0.87 ^{b}	5.23 ± 0.70 ^{b}	6.83 ± 0.26 ^{a}	6.79 ± 0.22 ^{a}	7.28 ± 1.70 ^{a}
terpinen-4-ol	1635	MS, RIL	lilac-like	N.A.	0.26 ± 0.06 ^{b}	0.58 ± 0.05 ^{a}	0.56 ± 0.04 ^{a}	0.57 ± 0.12 ^{a}
citral	1663	MS, RIL	lemon-like	32 ^{c}	0.09 ± 0.00 ^{b}	0.58 ± 0.08 ^{a}	0.68 ± 0.02 ^{a}	0.53 ± 0.21 ^{a}

Table 3. *Cont.*

Compounds	RI	Identification	Odor	Odor Threshold (μg/kg)	Relative Concentration (μg/kg)			
					SF	L.P.	P.A.	P.P.
α-terpineol	1692	MS, RIL	lilac-like	330 [c]	0.55 ± 0.05 [b]	0.88 ± 0.10 [a]	0.90 ± 0.04 [a]	1.05 ± 0.31 [a]
borneol	1696	MS, RIL	piney	140 [c]	1.62 ± 0.21 [b]	3.24 ± 0.45 [a]	3.28 ± 0.02 [a]	4.11 ± 1.24 [a]
β-bisabolene	1721	MS, RIL	balsamic	N.A.	0.25 ± 0.00 [b]	0.73 ± 0.21 [a]	0.66 ± 0.08 [a]	1.23 ± 0.41 [a]
neral	1733	MS, RIL	lemon-like	30 [c]	0.38 ± 0.01 [b]	0.59 ± 0.06 [a]	0.73 ± 0.01 [a]	0.64 ± 0.22 [a]
nerol	1808	MS, RIL	rosy	300 [c]	0.15 ± 0.03 [c]	0.32 ± 0.00 [a]	0.24 ± 0.01 [b]	0.24 ± 0.07 [b]
isogeraniol	1812	MS, RIL	rosy	N.A.	N.D.	1.23 ± 0.19 [ab]	1.40 ± 0.08 [a]	1.03 ± 0.13 [b]
geraniol	1857	MS, RIL	rosy	1.1 [b]	0.33 ± 0.04 [c]	1.91 ± 0.15 [a]	0.92 ± 0.10 [b]	1.09 ± 0.41 [b]
Ketones								
2,3-butanedione	968	MS, RI	buttery	1 [b]	2.20 ± 0.93 [a]	0.53 ± 0.02 [b]	0.52 ± 0.03 [b]	0.72 ± 0.17 [b]
3-octanone	1240	MS, RIL	fruity	21–50 [c]	0.27 ± 0.12 [ab]	0.15 ± 0.00 [b]	0.09 ± 0.01 [c]	0.31 ± 0.04 [a]
3-hydroxy-2-butanone	1286	MS, RI	yogurt	800 [c]	15.11 ± 3.08 [a]	1.87 ± 0.09 [b]	1.56 ± 0.29 [b]	1.55 ± 0.26 [b]
6-methyl-5-hepten-2-one	1341	MS, RIL	green, fatty	50 [c]	0.87 ± 0.14 [a]	0.41 ± 0.04 [b]	0.56 ± 0.10 [b]	0.67 ± 0.22 [ab]
Lactones								
butyrolactone	1634	MS, RI	sweet, buttery	20,000–50,000 [c]	0.26 ± 0.04 [a]	N.D.	N.D.	N.D.
lavender lactone	1684	MS, RIL	lavender	N.A.	0.48 ± 0.09 [a]	0.43 ± 0.03 [a]	0.44 ± 0.03 [a]	0.50 ± 0.03 [a]
γ-nonalactone	2042	MS, RI	coconut	9.7 [b]	0.10 ± 0.01 [c]	0.19 ± 0.01 [b]	0.19 ± 0.02 [b]	0.23 ± 0.01 [a]
Acids								
acetic acid	1445	MS, RIL	acid	99,000 [b]	6.93 ± 0.33 [c]	17.73 ± 1.57 [a]	13.53 ± 0.94 [b]	13.28 ± 2.07 [b]
butanoic acid	1576	MS, RI	rancid	2400 [b]	0.12 ± 0.01 [c]	0.41 ± 0.07 [a]	0.36 ± 0.05 [a]	0.26 ± 0.02 [b]
2/3-methylbutanoic acid	1627	MS, RI	rancid	2200/490 [b]	1.22 ± 0.21 [a]	1.13 ± 0.07 [a]	1.23 ± 0.24 [a]	1.09 ± 0.10 [a]
pentanoic acid	1720	MS, RIL	rancid	11,000 [b]	0.10 ± 0.01 [a]	0.08 ± 0.01 [a]	0.09 ± 0.02 [a]	0.07 ± 0.02 [a]
4-methylpentanoic acid	1817	MS, RIL	rancid	810 [c]	0.20 ± 0.02 [b]	0.42 ± 0.08 [a]	0.18 ± 0.01 [b]	0.22 ± 0.04 [b]
hexanoic acid	1839	MS, RI	rancid	890 [a]	0.53 ± 0.14 [a]	0.35 ± 0.04 [b]	0.42 ± 0.06 [a]	0.28 ± 0.09 [b]
octanoic acid	2089	MS, RI	rancid	3000 [a]	0.05 ± 0.01 [c]	0.12 ± 0.01 [b]	0.19 ± 0.04 [a]	0.10 ± 0.01 [b]
Esters								
ethyl acetate	880	MS, RI	fruity	5–5000 [d]	0.86 ± 0.11 [c]	1.12 ± 0.01 [a]	1.07 ± 0.11 [b]	0.92 ± 0.08 [bc]
butyl acetate	1078	MS, RI	pineapple	66 [c]	N.D.	0.04 ± 0.02 [a]	0.03 ± 0.03 [a]	N.D.
isoamyl acetate	1126	MS, RI	banana	2 [d]	N.D.	0.07 ± 0.01 [b]	0.13 ± 0.01 [a]	N.D.
methyl salicylate	1759	MS, RIL	minty	40 [d]	0.34 ± 0.07 [b]	1.28 ± 0.05 [a]	1.22 ± 0.04 [a]	1.45 ± 0.32 [a]
2-phenylethyl butanoate	1978	MS, RIL	rosy	N.A.	0.27 ± 0.04 [ab]	0.29 ± 0.01 [a]	0.22 ± 0.01 [c]	0.25 ± 0.02 [b]
Phenols								
2-methoxyphenol	1862	MS, RI	sweet	0.84 [b]	4.49 ± 0.81 [a]	0.35 ± 0.03 [c]	0.36 ± 0.01 [c]	0.53 ± 0.05 [b]
phenol	2008	MS, RI	phenolic	5900 [c]	1.63 ± 0.34 [a]	0.28 ± 0.01 [c]	0.29 ± 0.01 [c]	0.64 ± 0.05 [b]
2-methoxy-4-vinylphenol	2155	MS, RI	phenolic	5.1 [b]	1.33 ± 0.19 [a]	0.15 ± 0.01 [c]	0.13 ± 0.02 [c]	0.44 ± 0.09 [b]
eugenol	2176	MS, RIL	clove	6–30 [c]	2.22 ± 0.11 [d]	8.29 ± 1.22 [b]	6.32 ± 1.25 [c]	11.23 ± 1.96 [a]
S-containing compounds								
dimethyl sulfide	777	MS, RIL	garlic	0.84 [a]	5.71 ± 1.73 [c]	70.47 ± 6.99 [a]	45.39 ± 3.73 [b]	66.03 ± 11.74 [a]
dimethyl disulfide	1107	MS, RIL	fishy	1.1 [a]	N.D.	0.20 ± 0.07 [a]	N.D.	0.09 ± 0.04 [b]

Table 3. *Cont.*

Compounds	RI	Identification	Odor	Odor Threshold (μg/kg)	Relative Concentration (μg/kg)			
					SF	L.P.	P.A.	P.P.
N-containing compounds								
2-methylpyrazine	1258	MS, RIL	nutty, cocoa-like	60–100,000 [c]	0.89 ± 0.17 [a]	0.52 ± 0.04 [d]	0.68 ± 0.03 [b]	0.64 ± 0.01 [c]
2,3-dimethylpyrazine	1339	MS, RIL	nutty, cocoa-like	2500–35,000 [c]	0.37 ± 0.04 [a]	0.13 ± 0.01 [c]	0.24 ± 0.02 [b]	0.11 ± 0.02 [c]
2-ethyl-6-methylpyrazine	1377	MS, RIL	roasted	N.A.	0.46 ± 0.08 [a]	0.28 ± 0.01 [b]	0.48 ± 0.07 [a]	0.29 ± 0.01 [b]
trimethylpyrazine	1399	MS, RIL	baked potato	400–1800 [c]	0.13 ± 0.01 [a]	0.11 ± 0.01 [b]	0.14 ± 0.02 [a]	0.11 ± 0.01 [b]
tetramethylpyrazine	1472	MS, RIL	musty, fermented, coffee	1000–10,000 [c]	0.66 ± 0.05 [a]	0.49 ± 0.01 [b]	0.64 ± 0.04 [a]	0.48 ± 0.02 [b]
4-methyl-5-thiazoleethanol	2275	MS, RI	nutty	10,800 [c]	0.20 ± 0.02 [a]	0.10 ± 0.02 [b]	0.14 ± 0.05 [ab]	0.06 ± 0.00 [c]

Notes: Different letters denote significant differences in the same row ($p < 0.05$). MS, mass spectrum agreed with those of the authentic compound; RI, compound was confirmed by the retention index of the compound standard; RIL, compound was identified according to the retention index from the NIST Chemistry WebBook; N.D., not detected. [a] Threshold adapted from An et al. [43]; [b] threshold adapted from Czerny et al. [44]; [c] threshold adapted from Leffingwell and associates (web pages); [d] threshold adapted from Wang et al. [38].

For other compounds, the contents of ketones in LAB-inoculated Zhayu samples were significantly lower than those in the control, especially 2,3-butanedione with a buttery note and 3-hydroxy-2-butanone with a yogurt-like note. Not surprisingly, the LAB-inoculated fermentation dramatically increased the acids' contents in the Zhayu samples compared with the control, which was coincident with the sensory evaluation results (Figure 4d). Additionally, the Zhayu samples prepared with *L. plantarum* contained the highest acetic acid, butanoic acid, and 4-methylpentanoic acid. Meanwhile, esters, which were derived from the microbial esterification of acids with alcohols, also showed higher contents in Zhayu samples fermented with LAB than in the control. Esters usually had a fruity aroma, and they are important aroma compounds contributing to the unique odor of fermented products, such as dry-fermented sausages and fermented mandarin fish [8,46].

It was found that the inoculated fermentation with LAB promoted the generation of the compounds that derived from protein degradation, increasing the relative content of aromatic compounds (such as benzyl alcohol, phenylethyl alcohol, and methyl salicylate), phenols (except eugenol), and S-containing compounds. In particular, dimethyl sulfide with a garlic-like odor showed dramatically higher relative contents (45.39~70.47 μg/kg) in the Zhayu samples fermented with LAB than in the control (5.71 μg/kg). Moreover, the samples fermented with *L. plantarum* and *P. pentosaceus* contained more dimethyl sulfide and dimethyl disulfide. S-containing compounds were produced by the degradation of S-containing amino acids and the catabolism of aromatic [16], and they were also identified as major contributors to the typical fishy odor of fermented fish products [8]. The contents of Met in the LAB-inoculated Zhayu samples were also lower than those in the control (Table 2), suggesting that LAB could utilize Met to produce volatile S-containing compounds during fermentation [8,16]. However, some N-containing compounds, which were also generated by the protein degradation, decreased after fermentation with LAB (especially *L. plantarum* and *P. pentosaceus*), agreeing with the changes in TVB-N (Figure 3c). It was reported that N-containing compounds could be gradually consumed during the fermentation process of aquatic products [9]. Thus, it was presumed that the decrease in the N-containing compounds in the LAB-inoculated Zhayu samples was due to the further microbial decomposition and utilization of these compounds by microorganisms.

4. Conclusions

The Zhayu products fermented with LAB had lower pH values and TVB-N contents than the control, showing better safety qualities. Moreover, LAB-inoculated Zhayu samples contained more water-soluble proteins and crude fats as the inoculation amount exceeded 10^8 cfu/100 g, and fermentation with LAB decreased the hardness, springiness, and chewiness of Zhayu products, making the products more tender and soft. Additionally, the Zhayu prepared with LAB at the inoculation amount of 10^9 cfu/100 g presented better safety, nutritional, and textural qualities. LAB could also improve the flavor characteristics of Zhayu products. Zhayu samples fermented with *L. plantarum* and *P. acidilactici* presented the strongest sourness, while Zhayu samples prepared with *P. pentosaceus* showed the highest level of umami taste, consistent with the highest contents of water-soluble proteins and free amino acids with an umami taste (Asp and Glu). For odor characteristics, the addition of LAB decreased the relative contents of aliphatic aldehydes, (Z)-3-hexen-1-ol, and 1-octen-3-ol, subducting the earthy and fishy notes in the Zhayu samples, and the Zhayu samples fermented with *P. acidilactici* contained the lowest content of 1-octen-3-ol and the weakest earthy notes. In addition, LAB-inoculated fermentation enhanced the contents of terpenoids, acids, esters, and S-containing compounds, increasing the sour, pleasant, and unique odors of the Zhayu products. Especially for samples fermented with *L. plantarum*, the relative contents of acids and esters were the highest. Overall, the inoculation fermentation with LAB increased the acceptability of the comprehensive quality of Zhayu products. In further studies, the three kinds of LAB can be mixed and used together to exploit the advantages of each starter and improve the quality of fermented fish products.

Author Contributions: Conceptualization, Y.H. and X.H.; methodology, X.C. and L.C.; software, Y.A. and X.C.; validation, Y.A., X.C. and L.C.; formal analysis, Y.A. and X.C.; investigation, R.L.; resources, Y.H.; data curation, Y.A.; writing—original draft preparation, Y.A.; writing—review and editing, R.L.; visualization, Y.A.; supervision, X.H.; project administration, S.X.; funding acquisition, S.X. All authors have read and agreed to the published version of the manuscript.

Funding: This research was funded by the National Key R&D Program of China (grant number: 2018YFD0901003).

Institutional Review Board Statement: Not applicable.

Informed Consent Statement: Not applicable.

Data Availability Statement: Data are contained within the article.

Conflicts of Interest: The authors confirm that they have no conflict of interest with respect to the work described in this manuscript.

References

1. Mahanta, P.; Muzaddadi, A.U. Extension of shelflife of the fermented fish product, shidal by packaging in glass bottle and low temperature storage. *Indian J. Fish* **2013**, *60*, 135–143. [CrossRef]
2. Paludan-Müller, C.; Madsen, M.; Sophanodora, P.; Gram, L.; Møller, P.L. Fermentation and microflora of plaa-som, a Thai fermented fish product prepared with different salt concentrations. *Int. J. Food Microbiol.* **2022**, *73*, 61–70. [CrossRef]
3. Ijong, F.G.; Ohta, Y. Physicochemical and microbiological changes associated with Bakasang processing—A traditional Indonesian fermented fish sauce. *J. Sci. Food Agric.* **2015**, *71*, 69–74. [CrossRef]
4. Devi, K.R.; Deka, M.; Jeyaram, K. Bacterial dynamics during yearlong spontaneous fermentation for production of ngari, a dry fermented fish product of Northeast India. *Int. J. Food Microbiol.* **2015**, *199*, 62–71. [CrossRef] [PubMed]
5. Zeng, X.; Xia, W.; Yang, F.; Jiang, Q. Changes of biogenic amines in Chinese low-salt fermented fish pieces (Suan yu) inoculated with mixed starter cultures. *Int. J. Food Sci. Technol.* **2013**, *48*, 685–692. [CrossRef]
6. Gao, P.; Wang, W.; Jiang, Q.; Xu, Y.; Xia, W. Effect of autochthonous starter cultures on the volatile flavour compounds of Chinese traditional fermented fish (Suan yu). *Int. J. Food Sci. Technol.* **2016**, *51*, 1630–1637. [CrossRef]
7. Sun, Y.; Xu, Y.; Gao, P.; Xia, W.; Hua, Q.; Jiang, Q. Improvement of the quality stability of vacuum-packaged fermented fish (suanyu) stored at room temperature by irradiation and thermal treatments. *Int. J. Food Sci. Technol.* **2021**, *56*, 224–232. [CrossRef]
8. Wang, Y.; Shen, Y.; Wu, Y.; Li, C.; Huang, H. Comparison of the microbial community and flavor compounds in fermented mandarin fish (*Siniperca chuatsi*): Three typical types of chinese fermented mandarin fish products. *Food Res. Int.* **2021**, *144*, 110365. [CrossRef]
9. Shen, Y.; Wu, Y.; Wang, Y.; Li, L.; Yang, S. Contribution of autochthonous microbiota succession to flavor formation during chinese fermented mandarin fish (*Siniperca chuatsi*). *Food Chem.* **2021**, *348*, 129107. [CrossRef]
10. Yang, J.; Jiang, C.; Bao, R.; Liu, M.; Lin, X. Effects of flavourzyme addition on physicochemical properties, volatile compound components and microbial community succession of suanzhayu. *Int. J. Food Microbiol.* **2020**, *334*, 108839. [CrossRef]
11. Gao, P.; Jiang, Q.; Xu, Y.; Xia, W. Esterase activities of autochthonous starter cultures to increase volatile flavour compounds in Chinese traditional fermented fish (Suan yu). *Int. J. Food Prop.* **2017**, *20*, S663–S672. [CrossRef]
12. Zeng, X.; Xia, W.; Wang, J.; Jiang, Q.; Xu, Y.; Qiu, Y.; Wang, H. Technological properties of *Lactobacillus plantarum* strains isolated from Chinese traditional low salt fermented whole fish. *Food Control* **2014**, *40*, 351–358. [CrossRef]
13. Hua, Q.; Gao, P.; Xu, Y.; Xia, W.; Jiang, Q. Effect of commercial starter cultures on the quality characteristics of fermented fish-chili paste. *LWT* **2020**, *122*, 109016. [CrossRef]
14. Li, L.; Xu, Y. Influence of *Lactobacillus plantarum* on managing lipolysis and flavor generation of *staphylococcus xylosus* and *saccharomyces cerevisiae* in fish paste. *LWT* **2021**, *140*, 110709. [CrossRef]
15. Riebroy, S.; Benjakul, S.; Visessanguan, W.; Tanaka, M. Physical properties and microstructure of commercial Som-fug, a fermented fish sausage. *Eur. Food Res. Technol.* **2005**, *220*, 520–525. [CrossRef]
16. Feng, L.; Tang, N.; Liu, R.; Gong, M.; Wang, Z.; Guo, Y.; Wang, Y.; Zhang, Y.; Chang, M. The relationship between flavor formation, lipid metabolism, and microorganisms in fermented fish products. *Food Funct.* **2021**, *12*, 5685–5702. [CrossRef]
17. Tan, R.C.; Ouyang, J.M.; Lu, X.L.; Xiong, S.B. Fermentation conditions of Yuzha by inoculated *Lactobacillus plantarum* and *Pediococcus pentosaceus*. *Food Sci.* **2007**, *28*, 268–272, (In Chinese with English abstract).
18. Casquete, R.; Martin, A.; Benito, J.M.; Ruiz-Moyano, S.; Nevado, F.P. Impact of pre-selected autochthonous starter cultures on the flavor quality of iberian dry-fermented "salchichon" sausage with different ripening processes. *J. Food Sci.* **2011**, *76*, S535–S544. [CrossRef]
19. Fei, Y.T.; Liu, D.M.; Luo, T.H.; Chen, G.; Wu, H.; Li, L.; Yu, Y.G.; Li, W. Molecular characterization of *Lactobacillus plantarum* DMDL 9010, a strain with efficient nitrite degradation capacity. *PLoS ONE* **2014**, *9*, 113792. [CrossRef]
20. Ba, H.V.; Seo, H.W.; Seong, P.N.; Kang, S.M.; Kim, J.H. *Lactobacillus plantarum* (KACC 92189) as a potential probiotic starter culture for quality improvement of fermented sausages. *Korean J. Food Sci. An.* **2018**, *38*, 189–202. [CrossRef]

21. AOAC. *Association of Official Analytical Chemists International Official Methods of Analysis*, 16th ed.; AOAC: Arlington, VA, USA, 1997.
22. Cobb, B.F.; Alaniz, I.; Thompson, C.A. Biochemical and microbial studies on shrimp: Volatile nitrogen and amino nitrogen analysis. *J. Food Sci.* **1973**, *38*, 431–436. [CrossRef]
23. Shin, J.M.; Hwang, Y.O.; Tu, O.J.; Jo, H.B.; Kim, J.H.; Chae, Y.Z.; Park, S.K. Comparison of different methods to quantify fat classes in bakery products. *Food Chem.* **2013**, *136*, 703–709. [CrossRef] [PubMed]
24. Liao, E.; Xu, Y.; Jiang, Q.; Xia, W. Characterisation of dominant autochthonous strains for nitrite degradation of Chinese traditional fermented fish. *Int. J. Food Sci. Technol.* **2018**, *53*, 2633–2641. [CrossRef]
25. Lowry, O.H.; Rosebrough, N.J.; Farr, A.L.; Randall, R.J. Protein measurement with the Folin phenol reagent. *J. Biol. Chem.* **1951**, *193*, 265–275. [CrossRef]
26. *ISO 4120:2004(E)*; Sensory Analysis-Methodology-Triangle Test. International Organization for Standardization (ISO): Geneva, Switzerland, 2004.
27. *ISO 4121:2003*; Sensory Analysis-Guidelines for the Use of Quantitative Response Scales. International Organization for Standardization (ISO): Geneva, Switzerland, 2003.
28. Luo, X.; Xiao, S.; Ruan, Q.; Gao, Q.; An, Y.; Hu, Y.; Xiong, S. Differences in flavor characteristics of frozen surimi products reheated by microwave, water boiling, steaming, and frying. *Food Chem.* **2022**, *372*, 131260. [CrossRef] [PubMed]
29. Kováts, E. Gas-chromatographische Charakterisierung organischer Verbindungen. Teil 1: Retentionsindices aliphatischer Halogenide, Alkohole, Aldehyde und Ketone. *Helv. Chim. Acta* **1958**, *41*, 1915–1932. [CrossRef]
30. Ortiz-Rivera, Y.; Sanchez-Vega, R.; Gutierrez-Mendez, N.; Leon-Felix, J.; Acosta-Muniz, C.; Sepulveda, D.R. Production of reuterin in a fermented milk product by *Lactobacillus reuteri*: Inhibition of pathogens, spoilage microorganisms, and lactic acid bacteria. *J. Dairy Sci.* **2017**, *100*, 4258–4268. [CrossRef]
31. Paukatong, K.V.; Kunawasen, S. Hazard analysis and critical control points (HACCP) generic model for the production of Thai fermented pork sausage (Nham). *Berl. Munch. Tierarztl. Wochenschr.* **2011**, *114*, 327–330. [CrossRef]
32. Hu, Y.; Xia, W.; Ge, C. Characterization of fermented silver carp sausages inoculated with mixed starter culture. *LWT* **2008**, *41*, 730–738. [CrossRef]
33. Zhu, J.; Zhao, A.; Feng, L.; Gao, H. Quorum sensing signals affect spoilage of refrigerated large yellow croaker (*Pseudosciaena crocea*) by Shewanella baltica. *Int. J. Food Microbiol.* **2016**, *217*, 146–155. [CrossRef]
34. Xu, Y.; Xia, W.; Yang, F.; Kim, J.M.; Nie, X. Effect of fermentation temperature on the microbial and physicochemical properties of silver carp sausages inoculated with *Pediococcus pentosaceus*. *Food Chem.* **2010**, *118*, 512–518. [CrossRef]
35. Sun, Y.; Gao, P.; Xu, Y.; Xia, W.; Hua, Q.; Jiang, Q. Effect of storage conditions on microbiological characteristics, biogenic amines and physicochemical quality of low-salt fermented fish. *J. Food Prot.* **2020**, *83*, 1057–1065. [CrossRef]
36. Sanz, B.; Selgas, D.; Parejo, I.; Ordóñez, J.A. Characteristics of lactobacilli isolated from dry fermented sausages. *Int. J. Food Microbiol.* **1988**, *6*, 199–205. [CrossRef]
37. Montel, M.C.; Masson, F.; Talon, R. Bacterial role in flavour development. *Meat Sci.* **1998**, *49*, S111–S123. [CrossRef]
38. Wang, Z.; Xu, Z.; Sun, L.; Dong, L.; Du, M. Dynamics of microbial communities, texture and flavor in suan zuo yu during fermentation. *Food Chem.* **2020**, *332*, 127364. [CrossRef]
39. Maruji, Y.; Shimizu, M.; Murata, M.; Ando, M.; Sakaguchi, M.; Hirata, T. Multiple taste functions of the umami substances in muscle extracts of yellowtail and bastard halibut. *Fisheries Sci.* **2010**, *76*, 521–528. [CrossRef]
40. Nasri, N.; Beno, N.; Septier, C.; Salles, C.; Thomas-Danguin, T. Cross-modal interactions between taste and smell: Odour-induced saltiness enhancement depends on salt level. *Food Qual. Prefer.* **2011**, *22*, 678–682. [CrossRef]
41. Zhu, W.; Luan, H.; Bu, Y.; Li, J.; Zhang, Y. Changes in taste substances during fermentation of fish sauce and the correlation with protease activity. *Food Res. Int.* **2021**, *144*, 110349. [CrossRef]
42. Gao, P.; Li, L.; Xia, W.; Xu, Y.; Liu, S. Valorization of nile tilapia (*Oreochromis niloticus*) fish head for a novel fish sauce by fermentation with selected lactic acid bacteria. *LWT* **2020**, *129*, 109539. [CrossRef]
43. An, Y.; Qian, Y.L.; Magana, A.A.; Xiong, S.; Qian, M.C. Comparative characterization of aroma compounds in silver carp (*Hypophthalmichthys molitrix*), Pacific whiting (*Merluccius productus*), and Alaska pollock (*Theragra chalcogramma*) surimi by aroma extract dilution analysis, odor activity value, and aroma recombination studies. *J. Agric. Food Chem.* **2020**, *68*, 10403–10413. [CrossRef]
44. Czerny, M.; Christlbauer, M.; Christlbauer, M.; Fischer, A.; Granvogl, M.; Hammer, M.; Hartl, C.; Hernandez, N.M.; Schieberle, P. Re-investigation on odour thresholds of key food aroma compounds and development of an aroma language based on odour qualities of defined aqueous odorant solutions. *Eur. Food Res. Technol.* **2008**, *228*, 265–273. [CrossRef]
45. Li, J.; Tu, Z.; Zhang, L.; Lin, D.; Sha, X.; Zeng, K.; Wang, H.; Pang, J.; Tang, P. Characterization of volatile compounds in grass carp (*Ctenopharyngodon idellus*) soup cooked using a traditional Chinese method by GC-MS. *J. Food Process Pres.* **2017**, *41*, 12995. [CrossRef]
46. Sidira, M.; Kandylis, P.; Kanellaki, M.; Kourkoutas, Y. Effect of curing salts and probiotic cultures on the evolution of flavor compounds in dry-fermented sausages during ripening. *Food Chem.* **2016**, *201*, 334–338. [CrossRef]

foods

MDPI

Article

Physicochemical and Nutritional Characteristics of Cookies Prepared with Untapped Seaweed *Ulva intestinalis*: An Approach to Value Addition as a Functional Food

Md. Mohibbullah [1,*], Al Amin [1], Md. Abu Talha [1], Md. Abdul Baten [1], Md. Masud Rana [1], Ashfak Ahmed Sabuz [2], Asif Wares Newaz [1] and Jae-Suk Choi [3,*]

1 Department of Fishing and Post Harvest Technology, Sher-e-Bangla Agricultural University, Dhaka 1207, Bangladesh
2 Postharvest Technology Division, Bangladesh Agricultural Research Institute, Gazipur 1701, Bangladesh
3 Department of Seafood Science and Technology, The Institute of Marine Industry, Gyeongsang National University, 38 Cheondaegukchi-gil, Tongyeong-si 53064, Republic of Korea
* Correspondence: mmohib.fpht@sau.edu.bd (M.M.); jsc1008@gnu.ac.kr (J.-S.C.); Tel.: +880-2-44814069 (M.M.); +82-55-772-9142 (J.-S.C.)

Abstract: The present study was investigated to know the sensory, physicochemical, nutritional and fatty acid properties of seaweed-based cookies prepared with untapped seaweed *Ulva intestinalis* (UI) from Bangladesh coast. The cookies were formulated with different percentages of UI inclusions both in powdered (PUI) and fragmented (FUI) forms, in order to evaluate different quality attributes in prepared value-added cookies. In sensory analysis, seaweed inclusion levels of 1% PUI, 2.5% PUI, 1% FUI, 2.5% FUI and 5% FUI to cookies were acceptable by panelists. Considering the maximum percentage of seaweed inclusions, 2.5% PUI and 5% FUI were selected for further analysis. The results of physicochemical properties such as moisture content, spread factor, baking loss, pH, cookie density, color, texture properties, volatile basic nitrogen and thiobarbituric acid reactive species were within acceptable limits. In nutritional analysis, 2.5% PUI and 5% FUI cookies showed a remarkable and significant increase in lipid and ash contents, compared to untreated controls. Being the first report on fatty acids profile by UI from Bangladesh, among 24 fatty acids identified, the amount of total saturated, mono-unsaturated, omega-3 fatty acids and omega-6 fatty acids were reported to be 641.9 (36.2%), 563.7 (31.8%), 133.8 (7.6%) and 436.3 (24.6%) µg/g DW, respectively. The results suggest that cookies with 2.5% PUI and 5% FUI can be marketed as healthy foods for consumers.

Keywords: green algae; *Ulva intestinalis*; cookies; sensory evaluation; physicochemical and nutritional properties; functional food

Citation: Mohibbullah, M.; Amin, A.; Talha, M.A.; Baten, M.A.; Rana, M.M.; Sabuz, A.A.; Newaz, A.W.; Choi, J.-S. Physicochemical and Nutritional Characteristics of Cookies Prepared with Untapped Seaweed *Ulva intestinalis*: An Approach to Value Addition as a Functional Food. *Foods* **2023**, *12*, 205. https://doi.org/10.3390/foods12010205

Academic Editors: Tao Yin and Liu Shi

Received: 18 November 2022
Revised: 13 December 2022
Accepted: 23 December 2022
Published: 3 January 2023

1. Introduction

Seaweed refers to thousands of species of macroscopic, multicellular, marine algae. Seaweeds adhere to rocks in the intertidal zone, wash up on the beach, and float on the sea surface. Seaweeds are classified taxonomically into the following three groups: Chlorophyta, Rhodophyta, and Phaeophyta, corresponding to green, red and brown algae, respectively. More than 20,000 seaweeds are distributed around the world of which only 221 (1.1%) are commercially exploited, including 145 species for foods, including raw salad, curry, soup, pickles, cookies, etc., and 110 species for phycocolloids, including agar, agarose, carrageenan, algin, and mannitol [1,2]. Moreover, seaweeds have the potential to be used as medicine, cosmetics, animal feed, fish feed, fertilizers, soil conditioners, etc. [3]. A recent study also confirms that the use of sea grapes (*Caulerpa racemosa*) in cookies is a potential anti-aging novel-functional food [4].

In Bangladesh, on the south coast bordering the Bay of Bengal, from the Sundarban mangrove forest to the island of St. Martin, seaweeds are available from October to April [5].

More than 77 genera and approximately 250 species of seaweeds are found in the coastal and near-coastal waters of Bangladesh [6]. In Bangladesh, seaweed farming has been practiced in the tidal and shallow subtidal area characterized by a rocky coral substrate, sandy bottom with boulders, pebbles and broken shells as well as a sandy-muddy bottom that provides a suitable substrate and suitable habitats for the cultivation of various seaweeds.

Ulva intestinalis is a conspicuous bright grass-green seaweed, consisting of inflated irregularly constricted, tubular fronds that grow from a small discoid base. Fronds are usually unbranched and can be 10–30 cm or more in length and 6–18 mm in diameter, with the tips usually rounded [7]. *Ulva intestinalis* is remarkably euryhaline as it can grow in freshwater. However, there is evidence for the existence of genetic strains adapted to high and low salinity [8]. It can be found in a variety of habitats on all coastal plains. Given the right conditions, it can grow on rocks, mud, and even sand. This species is commonly found in areas where brackish water is most abundant and is also known as a common epiphyte found on other seaweeds. Availability of this species is reported from the coastal areas of Cox's Bazar and St. Martins Island, Bangladesh from October to April.

Ulva intestinalis is also of great commercial importance. Coastal people take this species as fodder and it is consumed only by the tribal community. A recent study reported that the proximate composition of *U. intestinalis* on a dry basis was crude protein (12.6%), carbohydrate (45.4%), crude fat (3.5%), crude fiber (17.1%) and Ash (12.7%) [9]. However, few studies have been found on the use of *Ulva intestinalis* as a feed ingredient and growth promoter [10,11]. Nowadays, seaweed has become a very versatile product and is widely used as a food for human consumption. Therefore, seaweed can be used as a functional food because it is characterized by low energy content, good sources of polysaccharides, high protein content, good supply of dietary fiber, high tocopherol, high fatty acids, and a rich source of distinguished minerals [12]. For the first time in Bangladesh, value-added seaweed products, such as seaweed-based cookies may be a new approach to market these untapped resources into valuable foods. Because of their unique taste and extended shelf life even at room temperature with appropriate packaging, cookies are considered popular foods regardless of countries in the world. However, the lack of valuable micronutrients and fiber in cookies due to the use of refined grains is of great concern to the health-conscious consumer. Therefore, the present study aimed to develop and evaluate the quality of seaweed inclusion in cookies with its two forms, including powdered *Ulva intestinalis* (PUI) and fragmented *Ulva intestinalis* (FUI), considering physicochemical and nutritional attributes in prepared value-added cookies.

2. Materials and Methods

2.1. Collection, Identification and Transportation of Seaweed

Ulva intestinalis was collected in March 2021 from the coast and tidal area of Nuniachara beach, Cox's Bazar (Latitude 21°28′27.79″ N and Longitude 91°58′23.00″ E). The seaweed was authenticated by a researcher at the Bangladesh Fisheries Research Institute (BFRI) who was implementing a project to develop and propagate the seaweed culture, and deposited as a voucher specimen of UI-NC:MM-2021-03 in the department of Fishing and Post Harvest Technology, Sher-e-Bangla Agricultural University, Bangladesh. Immediately after harvesting, the sample was placed in a large plastic container with sufficient seawater and transported to the laboratory for further processing.

2.2. Processing of Seaweed

Raw seaweed samples were properly washed with running water to remove sand and other debris from *Ulva intestinalis*. The seaweed was dried at room temperature (20–25 °C) for three to five days to reduce the moisture content. Then, dried seaweed was packed in plastic zipper bags and brought to a dark state. The dried seaweed were ground into powder in a mill (Mozart Mixer Grinder, Miyako, Dhaka, Bangladesh) with particle size of less than 50 μm in diameter and, on the other type, were made by hand-pressing into fragmented forms with uneven sizes from 500 μm to 3000 μm. Both samples were packed in airtight plastic zipper bags and stored at room temperature under dark condition for further use.

2.3. Preparation of Seaweed Cookie

The basic formulation for making seaweed-based cookies is outlined in Table 1. Seaweed powder has been substituted for wheat flour in varying amounts to fortify the nutritional content of cookies. In this study, 0 (control), 1, 2.5, 5 and 10% of the wheat flour were replaced with seaweeds and these compositions were selected based on previous reports of similar seaweed-based cookies [13,14]. All ingredients including wheat flour, seaweed powder, powdered sugar and butter were accurately weighed using an electric balance (FSH, A&D Company Ltd., Republic of Korea). Dry ingredients such as whole wheat flour (13% protein), sugar powder and seaweed powder were mixed together at room temperature. The butter was softened and then mixed with dry ingredients. The cookie dough was then rolled out with a rolling pin and cut to 5 mm thickness and 51 mm diameter with a cutter and baked in an oven at 180 °C for 12 min. The cookie was vacuum packed in high density polyethylene (HDPE) bags with a seal and placed in airtight containers at room temperature for further analysis.

Table 1. Ingredients of UI seaweed-based cookie.

Ingredients	Brand Name	Amount
Wheat flour	Teer	90~100%
Seaweed	Collected from Coastal Site	0~10%
Baking powder	Foster Clark's Baking Powder	0.5 g
Baking soda	Pure Baking Soda	0.05 g
Icing sugar	Fresh Refined Sugar	15 g
Salt	ACI PURE Salt	1 g
Milk powder	Diploma Instant Full Cream Milk Powder	8 g
Flavor	Foster Clark's Lemon Flavor	6 mL
Lemon juice	Extracted from locally available Lemon	2.5 mL
Butter	Aarong Dairy Butter	10 g
Oil	Rupchanda Fortified Soyabean Oil	25 mL

2.4. Sensory Evaluation of Cookies

The cookies containing 1, 2.5, 5 and 10% seaweed powder were subjected to a sensory evaluation along with the control biscuit (0%). This sensory test was conducted by 21 panelists, age between 25 and 40, belonging to the Faculty of Fisheries, Aquaculture and Marine Sciences, Sher-e-Bangla Agricultural University, Dhaka. The panelists were semi-trained and capable of discriminating the differences and communicating their sensory reactions as perceived by the sense of sight, taste and touch. The 10 g biscuit samples from each group were given to the panelists and asked to rate on a 9-point hedonic scale for the attributes of seaweed biscuit appearance, color, odor, texture, taste and overall acceptability. A hedonic scale of 1 to 9 points indicated differently as 1 = dislike extremely, 2 = dislike very much, 3 = dislike moderately, 4 = dislike slightly, 5 = neither like nor dislike (threshold point), 6 = like slightly, 7 = like moderately, 8 = like very much, and 9 = like extremely [15]. The samples were served on white dishes with mineral water, a stainless spoon, a paper cup, plain bread and pre-defined sensory evaluation paper with number rankings. Each sample group was evaluated at different times with the same panelists. All samples were encoded before serving and evaluated randomly. The sensorial properties were evaluated at 3 p.m. for a maximum of 2 h in a designated room. On the day of sensory evaluation, all cookies were freshly prepared at 10 a.m., cooled down at RT after baking, and vacuum-packed until sensory evaluation on the same day at 3 p.m.

2.5. Quality Assessment of Dough

The baking sample (5 g) was mixed in 45 mL of distilled water, homogenized for 2 min, and centrifuged (SHG-15D, SciLab, and Seoul, Republic of Korea) to collect the supernatant. The pH of the sample was measured using a pH meter (OHAUS STARTER 2100, Seoul, Republic of Korea). Dough or cookie density (g/cm^3) was measured by dividing the dough or cookie weight by a piece of dough or cookie used to form a biscuit as reported by Hadi Nezhad and Butler [16]. Moisture content was determined using the standard AOAC method [17]. The baking loss (%) was calculated by a comparison between the biscuit weight and the dough weight. The spread factor of the cookies was determined by dividing the diameter (mm) of a biscuit by its thickness (mm) according to the standard method of AACC [18].

2.6. Texture Analysis of Cookie

The texture of seaweed-based cookie was evaluated using the TA.XTPlus Texture Analyzer (Stable Micro Systems, Godalming, UK) equipped with Texture Exponent Lite software version 6.1.14.0 (Stable Micro System, Godalming, UK). During compression and extrusion with 12.5 mm diameter of a cylindrical probe at a rate of 1 mms^{-1}, number of attributes were measured including hardness (kg), adhesiveness (mJ), cohesiveness, springiness (mm) and gumminess (g), as described previously [19,20].

2.7. Color Analysis of Cookie

The color of the seaweed-based biscuit was evaluated with the Konica Minolta (Tokyo, Japan) CM-700d instrument. The surface color of each sample was calculated from previous reports [19,21,22]

2.8. Proximate Composition Analysis

Protein, lipid, carbohydrate, fiber, moisture and ash contents of seaweed cookies were determined using the standard AOAC method [17].

2.9. Total Volatile Basic Nitrogen (VBN) Analysis

The VBN value of the cookie was determined by a Conway microdiffusion method, as described by Mohibbullah et al. [19]. Briefly, 5 g of a ground sample was homogeneously mixed with 25 mL of distilled water. The filtrate was then mixed with potassium carbonate and incubated at 37 °C for 90 min. After titration with 0.01 N sodium hydroxide, the VBN value was calculated by the following equation:

$$\text{VBN (mg\%)} = 0.14 \times [(b-a) \times \text{f/W}] \times 100 \times \text{d} \tag{1}$$

where b = the amount of NaOH needed to titer Blank, a = the amount of NaOH needed to titer sample, f = 1.003 (NaOH), W = weight of the sample, d = dilution factor

2.10. Thiobarbituric Acid Reactive Substance (TBARS)

For the determination of TBARS, a total of 5 g of the ground sample was thoroughly mixed with 20% trichloroacetic acid solution, homogenized for 1 min and filtered using a Whatman No. 1 filter paper (55 mm), as described previously [23]. The filtrate was mixed with 0.005 M thiobarbituric acid solution, incubated at 95 °C for 30 min, and measured absorbance value at a wavelength of 530 mm. The TBARS was calculated by the following equation:

$$\text{TBARS (MDA mg/1000 g)} = [\text{Sample Absorbance} - \text{Blank (Water) Absorbance}] \times 5.2 \tag{2}$$

2.11. Soxhlet Extraction

Oil was obtained from finely ground seaweed. A 5 g sample was taken in a thimble and placed in a Soxhlet extractor. The extraction was carried out with diethyl ether (purity

99%) and heating at 45 °C for 5 h. Thereafter, the solvent was removed with a rotary evaporator at 40 °C and then with a vacuum nitrogen dryer until completely dry. The extracted yield of UI oil was 0.96% (*w*/*w*).

2.12. Fatty Ccid Analysis

Gas chromatography (Shimadzu Corp., Kyoto, Japan) equipped with a flame ionization detector (GC-FID) was used to identify and quantify the fatty acids from the seaweed sample. Fatty acid methylation of previously extracted oil (0.05 g) was performed according to the American Oil Chemists Society procedure, as followed by the addition of 3 mL 0.5 N NaOH, 3 mL boron trifluoride complex and 10% sodium chloride [24]. Fatty acid methyl esters (FAMEs) were filtered prior to GC injection. An Agilent HP-88 column (100 m, 0.25 mm, 0.2 m) with an oven temperature of 250 °C started from 120 °C, 1 min, 10 °C/min to 175 °C, 10 min, 5 °C/min to 210 °C, 5 min, 5 °C/min to 230 °C, 25 min, 5 °C/min. Helium with a split ratio of 1:20 was used as carrier gas. The FAMEs of the seaweed sample were identified and quantified by comparing standard fatty acids. FAMEs mixture of 37 components of the standard of C4–C24 saturated and unsaturated fatty acids (Sigma-Aldrich Inc., St. Louis, California, USA) were injected into GC/FID to identify the different fatty acids in methyl esterified oil sample extracted from seaweed, operated by windows 7-based software system (GC solution, version 2, Shimadzu Co., Kyoto, Japan). The average fatty acids content was taken from three different chromatograms of individual sample injections.

2.13. Statistical Analysis

The SPSS (statistical package for social sciences) 16 (SPSS 2010) statistical package was used for the analysis of the experimental results. The homogeneity of variance was examined using the Levene test. When Levene's test did not indicate significant deviations from the homogeneity of variance, the data obtained were analyzed by a one-way ANOVA test followed by Tukey and Duncan's multiple range tests to determine which pairs of the group comparison were significantly different ($p < 0.05$). Results were expressed as mean ± standard error (SE).

3. Results and Discussion

3.1. Comparison of Different Seaweed Percentage in Biscuit Preparation by Sensory Characteristics

Cookies of powdered and fragmented UI at different concentrations were shown in Figure 1A. One of the easiest and simplest ways to evaluate consumer products is sensory evaluation, which is associated with various sensorial parameters, such as color, odor, taste, and overall preference; scores for each attribute are based on a 9-point hedonic scale [24]. In the case of the radar chart (Figure 1Ba), the scores of PUI 1% and PUI 2.5% were found almost similar and close to PUI 0% (control), and the values were found on a hedonic scale greater than 6, respectively, which indicated 'like slightly' on the quality chart [14], whereas PUI 0% had an average mean value of sensory scores equal to or slightly greater than 7, which indicated 'like moderately'. In contrast, PUI 5% and PUI 10% showed significantly ($p < 0.05$) lower sensory scores compared to PUI 0%, PUI 1% and PUI 2.5%. Results indicated that there were a significant drop in taste, flavor and overall performance scores when more than 2.5% PUI was added, it might be due to the cause of the increasing fishy odor from seaweed materials. Due to the fact that, the scores remained equal or less than 5 on a hedonic scale, which indicated 'neither like nor dislike' or 'dislike slightly', respectively [15]. Moreover, in Figure 1Bb, FUI 1%, FUI 2.5%, and FUI 5% treated cookies showed similar sensory scores of greater than 6, which indicated 'like slightly' on the quality chart [15], whereas, PUI 0% showed slightly greater than 7 scores, which indicated 'like moderately' but a non-significant difference with FUI 1%, FUI 2.5% and FUI 5% groups. However, FUI 10% showed a significantly lower ($p < 0.05$) value compared to FUI 0%, FUI 1%, FUI 2.5%, and FUI 5%, respectively. The value was found an equal or less than 5 on a hedonic scale, which indicated 'neither like nor dislike' or 'dislike slightly', respectively [15].

The value we found for the sensory evaluation in our present study was higher than that of Oh et al. [13], where the author gave an average score of 5, which is due to the fishy odor of seaweed in prepared cookies. However, in our study, we were able to successfully suppress the dominance of the fishy smell in cookies with the inclusion of a high percentage of UI during preparation. Based on the sensory evaluation, we selected 2.5% PUI and 5% FUI seaweed for the preparation of value-added cookies. Hence, it is indicated that those value-added cookies can be accepted by the consumer and used as a functional food in the market.

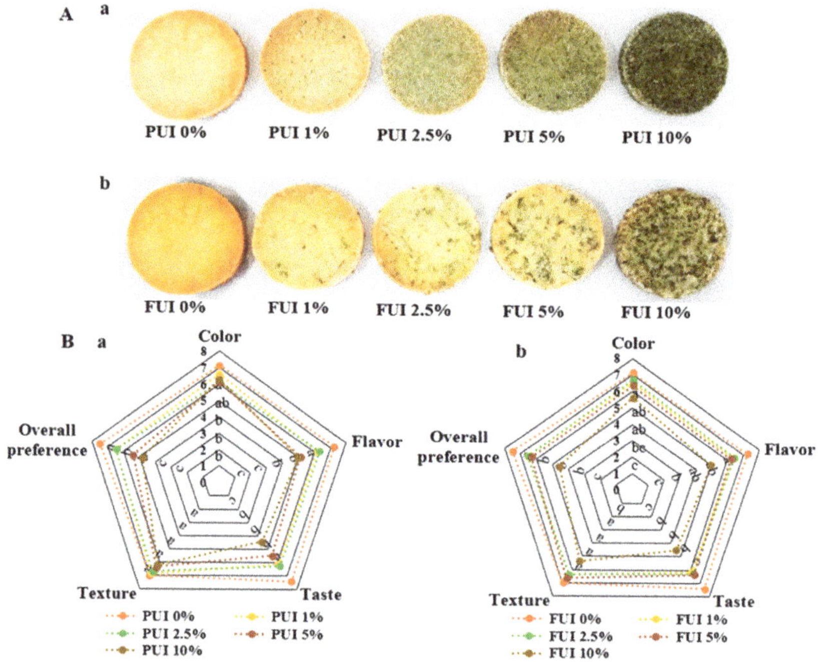

Figure 1. Comparison of different seaweed percentages in biscuit preparation for sensory characteristics. (**A**) Preparation of cookies using powdered UI (**a**) and fragmented UI (**b**). (**B**) Organoleptic analysis of cookies prepared using different concentrations of powdered (**a**) and fragmented (**b**) UI using a radar chart. Data represent the mean ± SE of 3 observations, where groups not sharing a letter are expressed as significantly different ($p < 0.05$).

3.2. *Quality Characteristics of Dough Prepared with PUI and FUI*

Table 2 examined the quality characteristics of dough prepared with UI seaweed. The pH value is used to indicate the food product quality during processing, as such product appearance, aroma, texture and taste might be affected with changes of pH values [19]. This study found significantly higher ($p < 0.05$) dough pH in control cookies than in PUI 2.5% and FUI 5% cookies. This could be due to the presence of polysaccharides and phenolic compounds containing carboxyl and sulfate groups in the cookies made from UI. Moreover, the ranges of pH among all groups were 6.51 to 6.59, indicating its acceptability for consumption. This study is evident in the findings of Oh et al. [13], where the author reported that the pH of the dough ranged from 6.96 to 7.18, when various seaweeds were added.

Table 2. Quality characteristics of value-added cookies with UI seaweed.

Physical Properties	UI 0%	PUI 2.5%	FUI 5%
Dough pH	6.59 ± 0.06 [a]	6.53 ± 0.05 [b]	6.51 ± 0.08 [b]
Dough density (g/cm^3)	1.33 ± 0.03 [a]	1.32 ± 0.02 [a]	1.33 ± 0.03 [a]
Cookie density (g/cm^3)	0.98 ± 0.03 [a]	0.94 ± 0.02 [a]	0.91 ± 0.05 [a]
Cookie moisture (%)	3.57 ± 0.03 [a]	3.35 ± 0.02 [a]	3.02 ± 0.10 [b]
Baking loss (%)	8.22 ± 0.52 [a]	7.84 ± 0.23 [a]	8.10 ± 0.28 [a]
Height (mm)	6.6 ± 0.24 [a]	6.6 ± 0.24 [a]	6.8 ± 0.20 [a]
Diameter (mm)	51.2 ± 0.83 [a]	51 ± 0.10 [a]	51.2 ± 0.83 [a]
Spread factor	7.81 ± 0.35 [a]	7.71 ± 0.33 [a]	7.56 ± 0.27 [a]
Adhesiveness (mJ)	0.42 ± 0.05 [a]	0.87 ± 0.15 [b]	0.72 ± 0.11 [bc]
Cohesiveness	0.45 ± 0.02 [a]	0.26 ± 0.02 [b]	0.16 ± 0.01 [c]
Springiness (mm)	3.05 ± 0.29 [a]	3 ± 0.49 [a]	1.75 ± 0.08 [b]
Gumminess (g)	360.89 ± 7.59 [a]	712.56 ± 75.52 [b]	457 ± 26.19 [a]

Data represent the mean ± SE of 3 observations, where groups not sharing a letter are expressed as significantly different ($p < 0.05$). UI: *Ulva intestinalis*, PUI: powdered *Ulva intestinalis*, FUI: fragmented *Ulva intestinalis*.

The density of the cookie dough is considered an indicator of the air trapped in the dough during mixing. Compared to the control cookies, a non-significant difference in dough and cookie (after baking) density was found in PUI 2.5% and FUI 5%, respectively, indicating no significant changes of cookie diameter among groups. Although dough density was not affected by the addition of *Hizikia fusiforme* up to 5% in cookies [13,25]. Moreover, the average dough density among all groups was 1.32 and, after baking, they were decreased to 0.94. It was due to the difference in weight loss between cookie dough and cookie after baking.

Moisture content is an important attribute for determining the quality of bakery products. Moisture content and water activity are recognized as key factors that directly influence the hardness of dry foods including cookies [26]. The moisture content of the cookies varied from 3.02 to 3.57% and the moisture content of FUI 5% was significantly lower ($p < 0.05$) than that of PUI 2.5% and control cookies, respectively. The current finding is in the range of Kabirullah et al. [27] analyzed the cookies and found a moisture content ranging from 4.06–4.97%.

Baking loss is the removal of moisture, which influences the texture and staling properties of baked products. No significant difference was observed in baking loss and the values were quite similar, ranging from 7.84 to 8.22 for all groups. This result agrees with the findings of Okpala et al. [28], where the author reported that the baking loss of different biscuit compositions varied between 7.14 and 10.39%. It has been reported that moisture content and baking loss are closely related to the moisture retention capacity of cookies [29]. According to Kotoki and Deka [30], baking loss was influenced by the dough ingredients present in cookies; although, there was not a significant difference in baking loss for either regular cookies or cookies with seaweed inclusions. Values for thickness and diameter of the cookies were found to range from 6.6 to 6.8 and 51 to 51.2, respectively, with no significant difference between groups. The spreading factor was found highest in the control, while the lowest value was found at FUI 5%, which is quite similar to the results of Oh et al. [13], where the author reported a higher spread factor in the control than in the treatments. This can be done by entrapping seaweed powder or fragments with other ingredients, and pulling them together into a coherent cookie mass; therefore, the spreading factor of the seaweed-supplemented groups was also found to be significantly lower than that of the control group.

The textural properties of cookie dough were further characterized by rheological tests of adhesiveness (the property of sticking together), cohesiveness (energy required for breaking down the internal structure), springiness (ability to spring back after first compression) and gumminess (energy required for breaking down semi-solid food for swallowing), as reported previously on semi-solid food items [31]. Comparing with control cookies, a significant increase of adhesiveness and gumminess were observed in PUI 2.5%

groups but not FUI 5%. Moreover, cohesiveness was significantly reduced in cookie dough prepared from PUI and FUI, as compared with the control. These changes in UI-added cookies may be due to the presence of various viscous materials present in seaweed [2]. However, the springiness of dough was significantly decreased in FUI 5% group, as because, it contained fragmented seaweed that disintegrates the dough materials in our study.

3.3. Textural Characteristics of Cookies Prepared with PUI and FUI

The texture properties of a food product are one of the most important physical quality attributes, by which, a consumer can easily evaluate [32]. Hardness is the textural property that is attracting more attention when evaluating baked goods as it is closely related to human perception of freshness [33]. Figure 2 shows the hardness of cookies made from UI 0%, PUI 2.5% and FUI 5% UI and the values were taken as 2.30, 2.82 and 2. 64 kg/cm^2 recorded. Although the values were not significantly different between treatments. This result agreed with the results of Mancebo et al. [34] where the author reported increased biscuit hardness from the addition of both insoluble and soluble fiber. This could be due to the changes in moisture content and network structure in these cookies caused by gums and polysaccharides contained in UI seaweed. In the case of PUI 2.5%, it was reported that higher hardness could be due to the higher solubility of powdered seaweed, whereas seaweed fragments have lower binding capacity and solubility.

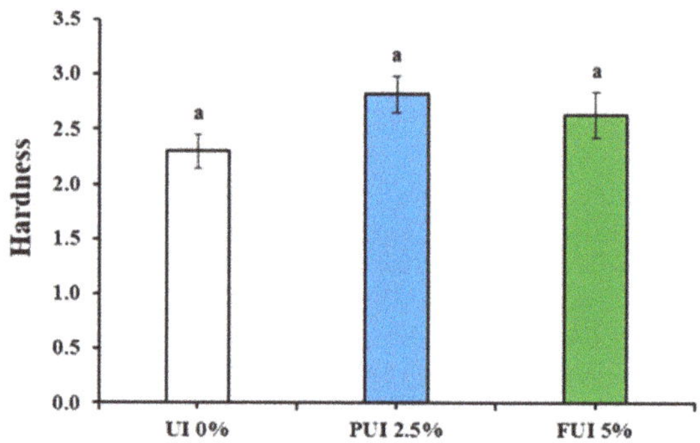

Figure 2. Hardness (kg/cm^2) comparison of cookies prepared with UI 0%, PUI 2.5%, and FUI 5% UI. Data represent the mean ± SE of 3 observations, where groups not sharing a letter are expressed as significantly different ($p < 0.05$).

3.4. Color Characteristics of Cookies Prepared with PUI and FUI

Seaweeds generally obtain chlorophyll or carotenoid pigments, providing them a unique green, brown, or red color, which are discolored during cooking and storage [35]. On the front side of the cookies, the L* (Lightness) value of PUI 2.5% was considerably higher compared to PUI 0% and FUI 5% but not statistically significant (Figure 3A). On the back side, L * (Lightness) was found higher in the control group and the significantly lowest value was recorded in FUI 5% (Figure 3B). This is indicative of the existence of a browner color on top of the cookie due to the cooking process and the contact with the hot air, since it has a higher temperature than the oven tray. The current finding was quite similar to those reported by Pereira et al. [36] and Adeola and Ohizua [37], where the authors reported lower L * (Lightness) in the top side of cookies when compared to the back sides. The a * (Redness) and b * (Yellowness) colors of cookies on both front and back sides were statistically significant in control (0% UI) cookies than that of seaweed cookies of PUI 2.5% and FUI 5%, respectively. This result is more or less similar to the observations made by Pereira et al. [36] but in contrast to those of Adeola and Ohizua [37]. This can be due to the difference in heat transfer, temperature, moisture content, and air velocity.

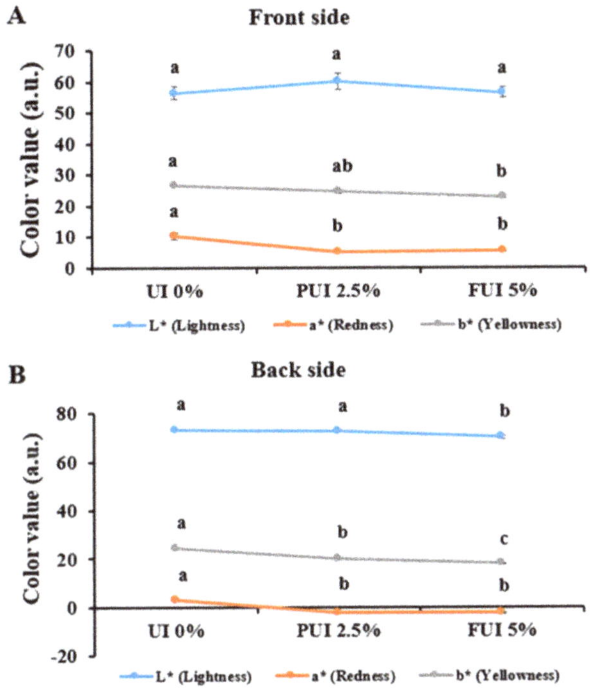

Figure 3. Changes in color values in both (**A**) front side and (**B**) back side of cookies prepared with UI 0%, PUI 2.5%, and FUI 5% UI. The dimension L* means lightness, with 100 for white and 0 for black appearance; a* indicates redness for positive value and greenness for negative value, b* indicates yellowness for positive value and blueness for a negative value. Data represent the mean ± SE of 3 observations, where groups not sharing a letter are expressed as significantly different ($p < 0.05$).

3.5. Nutritional Characteristics of Cookies Prepared with PUI and FUI

The nutritional value, particularly protein, lipid and ash, was significantly increased in cookies with PUI 2.5% and FUI 5% compared to control cookies (Figure 4). This study is similar with the study of Mamat et al. [38] as they found inclusion of seaweed improves the proximate composition of bakery product muffin. The protein content of cookies increased with the increase of UI concentrations. This result is similar to the findings of Kabirullah et al. [27] and Vijay [14], where the authors reported the protein percentage as 6.88–11.78% and 7.34–10.20%, respectively. Lipid content was found at 25.39% and 25.04% in PUI 2.5% and FUI 5% treated groups, respectively, and showed significantly higher ($p < 0.05$) compared to the control. This was similar to the findings of Kabirullah et al. [27] and Vijay [14], where the authors reported the lipid percentage as 23.02%, 5.66–26.67%, and 26.30–27.96%, respectively. An increase in lipid content might be an indication of the presence of PUFA [39]. Ash content varied from 1.45 to 4.81, where the highest value was noted in PUI 2.5% and found significantly higher ($p < 0.05$) compared to FUI 5% and control, respectively, indicating the incorporation of two different forms of seaweeds to cookies affected the ash content that might be the cause of differential particle sizes. Due to increased particle size of fragmented seaweed, water-holding and swelling capacities are high. In support, Jongaroontaprangsee et al. [40] reported similar findings where the outer leaves of cabbage were found to be lower in water-holding and swelling properties with decreasing particle size. In recent study, *U. intestinalis*, formerly known as *Enteromorpha intestinalis* from Bangladesh showed a considerable amount of mineral elements in dry matter basis (DM), among them, Ca (190.45 mg% DM), Fe (98.27 mg% DM), Cu (1.92 mg% DM), Zn (1.66 mg% DM), Na (12.33 mg% DM) and K (238.05 mg% DM) were the most dominant ones [9]. The study found higher minerals than the finding of Vijay [14], where the authors reported the ash percentage as 0.40–1.92%, respectively. These findings suggested that cookies from 2.5% PUI and 5% FUI have greater potential in the substantial level of essential minerals, thereby attracting consumer interest as UI cookies are more nutritious in comparison to conventional cookies. More precisely, it can be said that cookies made of 2.5% PUI was the best of the other inclusion level and resulted in more than double the ash content and almost similar lipid enhancement compared to 5% FUI cookies.

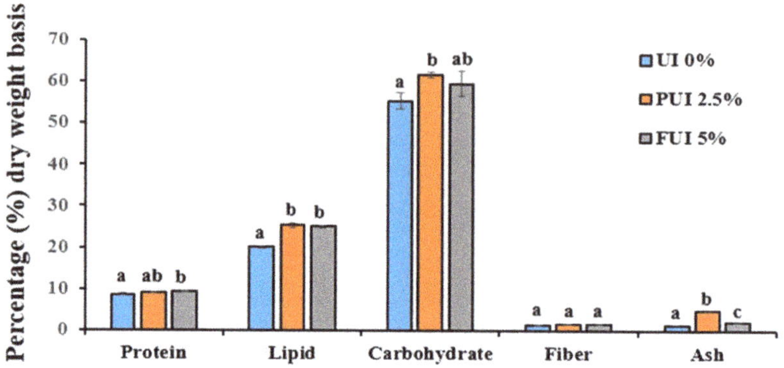

Figure 4. Nutritional composition of cookies prepared with UI 0%, PUI 2.5%, and FUI 5% UI. Data represent the mean ± SE of 3 observations, where groups not sharing a letter are expressed as significantly different ($p < 0.05$).

3.6. Freshness Characteristics of Cookies Prepared with PUI and FUI

Increased VBN value occurs when proteins are broken down by microorganisms or endogenous enzymes during processing and storage, which is an important indicator for freshness of food products. Moreover, cookies contain high levels of fat, leading to increased sensory attributes, but they are sources of unwanted compounds, such as lipid oxidation products, which can be monitored by TBARS values [23]. The VBN and TBARS values of fishery products for acceptance limit for human consumption have been reported as 30–35 mgN/100 g and 10–20 mg malonaldehyde/kg, respectively [19,21,41]. Both normal cookies and fortified with UI showed a very good and acceptable limit of chemical attributes in the contents of VBN (Figure 5A) and TBARS (Figure 5B). Moreover, only cookies in the 5% FUI group had a significant ($p < 0.05$) variation in reducing the VBN (0.40 ± 0.13) and TBARS (1.28 ± 0.03) values, compared with normal cookies. These results were in accordance with the previous study of Abraha et al. [41], who found in their cookies fortified with sturgeon fish fillet powder had acceptable VBN (0.47–1.09 mgN/100 g) and TBARS (0.13–1.27 mg malonaldehyde/kg) values. However, in a recent study conducted by Oh et al. [13] prepared cookies with various seaweeds from the Korean coast, VBN and TBARS values were not assessed during quality characterization. The results of the physicochemical characteristics of cookies indicated that UI addition may help to keep the freshness over longer periods of time by suppressing the various chemical reactions, possibly through the antioxidative and anti-microbial actions [42]. VBN contributes to the generation of ammonia, biogenic amines and other products of amino acid deamination and decarboxylation and it has a positive association with the activity of proteolytic enzymes present in food products [43]. Therefore, the addition of UI to cookies resulting in the decreased VBN values might be the cause of suppressing the proteolytic enzymes in cookies. It is well known that seaweeds possess a rich source of natural and bioactive antioxidants such as sulphated polysaccharides, phenolics and flavonoids, carotenoid pigments, dietary fiber etc. [9,12]. These antioxidant substances in seaweed would play a role in chelating free ions released from heat processing and, thus, inhibit degradation products of fats, collectively known as TBARS, when added to cookies.

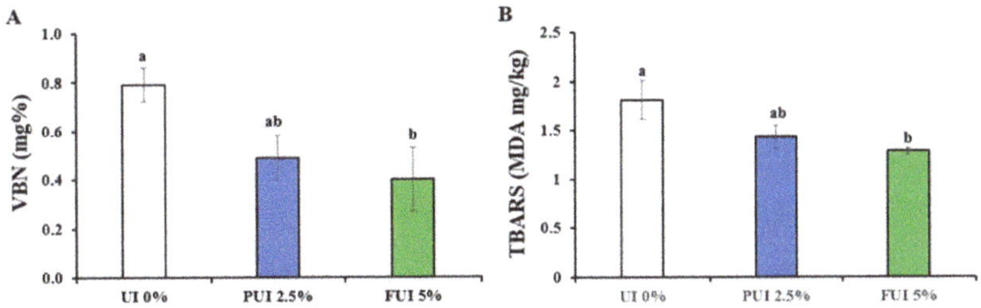

Figure 5. Changes in (**A**) VBN and (**B**) TBARS values in cookies prepared with UI 0%, PUI 2.5%, and FUI 5% UI. Data represent the mean ± SE of 3 observations, where groups not sharing a letter are expressed as significantly different ($p < 0.05$).

3.7. Fatty Acid Profiles of UI Seaweed

Since the earlier study of nutritional analysis showed high lipid contents of UI-treated cookies, here, we further evaluated the full fatty acid profile of UI to give potential insights into the health-beneficial effects after inclusion in cookies. Moreover, the present study reported for the first time, so far, on fatty acid profile of UI from Bangladesh coast. The fatty acid contents of UI were shown in Table 3 and the gas chromatographic profile of UI was also presented in Figure 6. A total of 24 fatty acids were identified in the oil-extracted sample of UI. The most abundant fatty acids were detected as pentadecenoic acid (286.86 ± 1.01 µg/g DW), followed by linoleic acid (224.52 ± 0.93 µg/g DW), myristic acid (165.72 ± 0.59 µg/g DW), pentadecylic acid (140.93 ± 0.30 µg/g DW), and so on. The amount of total saturated, mono-unsaturated, omega-3 fatty acids and omega-6 fatty acids were reported to be 641.84 (36.14%), 563.70 (31.74%), 133.81 (7.54%) and 436.30 (24.57%) µg/g DW, respectively. Seaweed is a significant source of polyunsaturated fatty acids [44]. The total polyunsaturated fatty acids were found to be 570 µg/g DW, which correspondent to 32.11%. Among individual polyunsaturated fatty acids identified in UI, linoleic acid (12.64%) was the most abundant one, followed by α-Linolenic acid (5.29%). The results suggested that the amount of total fatty acids in UI contributed the majority to the total lipid content analyzed previously. It is the first report on the fatty acids profiles of UI from Bangladesh. Among the green seaweed tested by Cardoso et al. [45], omega-3 and-6 fatty acids were found in *U. intestinalis* (14.6 and 10.7, respectively), *U. prolifera* (14.0 and 24.7 %, respectively) and *U. lactuca* (14.1 and 12.1, respectively); our findings were greatly supported by this study. Although, the amount of fatty acids in dry weight mass is absent. Moreover, another study conducted on green seaweed, *U. lactuca* was in agreement with our findings [46]. Since polyunsaturated fatty acids in UI possessed a substantial level offering a wide spectrum of health effects such as cardiovascular disease mitigation [47], neuroprotective [48] and neurotrophic [49] effects, and obesity [47] and diabetes [50] management, upon consumption, especially in the form of inclusion to dietary cookies in an approach of the present study.

Table 3. Fatty acid contents of UI dried powder.

Systematic Name	Common Name	Abbreviation	Amount (µg/g DW)
Decanoic acid	Caproic acid	C10:0	34.05 ± 0.23
Dodecanoic acid	Lauric acid	C12:0	45.27 ± 0.14
Tetradecanoic acid	Myristic acid	C14:0	165.72 ± 0.59
Pentadecylic acid		C15:0	140.93 ± 0.30
Hexadecanoic acid	Palmitic acid	C16:0	41.20 ± 0.46
Heptadecanoic acid	Margaric acid	C17:0	36.76 ± 0.12
Octadecanoic acid	Stearic acid	C18:0	76.04 ± 0.29
Docosanoic acid	Behenic acid	C22:0	91.49 ± 0.75
Tetracosanoic acid	Lignoceric acid	C24:0	10.38 ± 0.61

Table 3. *Cont.*

Systematic Name	Common Name	Abbreviation	Amount (μg/g DW)
∑SAFA			**641.84**
Pentadecenoic acid		C15:1	286.86 ± 1.01
Hexadecenoic acid	Palmitoleic acid	C16:1	83.96 ± 0.29
Cis-9-Octadecenoic acid	Oleic acid	C18:1 cis	140.14 ± 0.55
Trans-9-Octadecenoic acid	Elaidic acid	C18:1 trans	23.32 ± 0.10
Cis-11-Eicosenoic acid		C20:1	13.86 ± 0.25
Cis-13-Docosenoic acid	Erucic acid	C22:1	15.56 ± 0.63
∑MUFA			**563.70**
All cis-9,12,15-Octadecatrienoic acid	α -Linolenic acid	C18:3	94.06 ± 0.24
All cis-5,8,11,14,17-Eicosapentenoic acid	EPA	C20:5	24.42 ± 1.13
All cis-4,7,10,13,16,19-Docosahexenoic acid	DHA	C22:6	15.33 ± 0.23
∑PUFA (Omega-3)			**133.81**
All cis-9,12-Octadecadienoic acid	Linoleic acid	C18:2 cis	224.52 ± 0.93
All trans-9,12-Octadecadienoic acid	Linolelaidic acid	C18:2 trans	39.17 ± 0.21
All cis-6,9,12-Octadecatrienoic acid	γ-Linolenic acid	C18:3	37.08 ± 0.03
All cis-11,14-Eicosadienoic acid		C20:2	17.55 ± 0.03
All cis-8,11,14-Eicosatrienoic acid	Dihomogammalinolenic acid	C20:3	22.73 ± 1.10
All cis-5,8,11,14-Eicosatetraenic acid	Arachidonic acid	C20:4	95.26 ± 0.70
∑PUFA (Omega-6)			**436.30**

SAFA: saturated fatty acids, MUFA: mono-unsaturated fatty acids, PUFA: poly-unsaturated fatty acids.

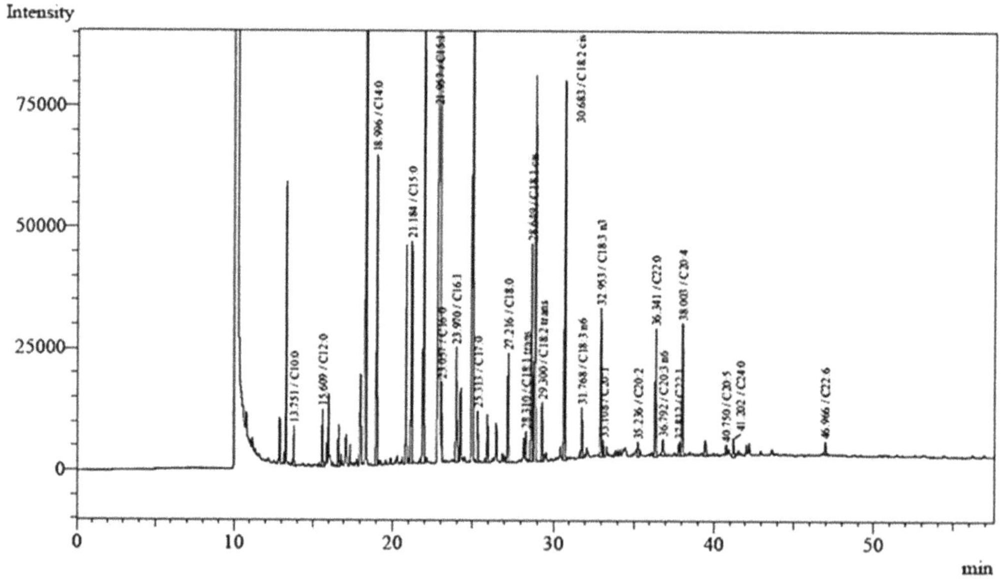

Figure 6. Gas chromatogram of injected FAMEs from UI with major fatty acids peaks.

4. Conclusions

The study was conducted to develop value-added health-functional seaweed-based cookies. The fishy smell was significantly reduced in cookies with 2.5% PUI and 5% FUI. It was found that the addition of 2.5% PUI and 5% FUI to the cookies was similar to the control cookies in terms of sensory properties. In addition, a significantly higher nutrient composition (protein, lipid and ash) was observed in seaweed-based cookies made from both PUI and FUI. At 2.5% PUI and 5% FUI, cookies had significant and acceptable physicochemical and structural properties. Furthermore, the addition of UI to cookies resulted in a high lipid content, the fatty acid profile of which was further analyzed, which was the first time for UI from Bangladesh coastline. UI has a significant content of polyunsaturated fatty acids, either in dry matter or as a percentage of total fatty acids, confirming its biofunctional properties when added to cookies. However, whether thermal processing could preserve the beneficial fatty acids in cookies and how long UI could be effective in preserving biscuit quality during storage should warrant for further study. These results suggest the possibility of developing baked goods with highly nutritious and healthy functional properties made from *Ulva intestinalis* available on the Bangladesh coast.

Author Contributions: Conceptualization, M.M.; methodology, M.M., M.A.T., A.A., M.A.B. and M.M.R.; validation, M.M. and J.-S.C.; formal analysis, M.M.; investigation, M.M., M.A.T., A.A. and A.A.S.; resources, M.M. and J.-S.C.; data curation, M.M., M.A.T., A.A., and M.A.B.; writing—original draft preparation, M.M., M.A.T., A.A. and M.A.B.; writing—review and editing, M.M., A.W.N. and J.-S.C. ; visualization M.M., M.A.T. and A.A.; supervision, M.M. and J.-S.C.; project administration, M.M. and J.-S.C.; funding acquisition, M.M. and J.-S.C. All authors have read and agreed to the published version of the manuscript.

Funding: This study was funded by Sher-e-Bangla Agricultural University Research System (SAURES) entitled "Development of Value Added Seaweed-based Health Functional Cookie with Improved Textural, Sensorial, and Nutritional Properties" [2021/2101 (82)].

Institutional Review Board Statement: Not applicable.

Informed Consent Statement: Not applicable.

Data Availability Statement: Data supporting the reported results are available upon request.

Acknowledgments: Our research was supported by the National Research Foundation of Korea (NRF) through Brain Pool Program (Md. Mohibbullah and Jae-Suk Choi: Grant No. 2021H1D3A2A01100053). Mohibbullah wishes to thank the Department of Fisheries, Cox's Bazar District Office, Bangladesh for their assistance in the collection of seaweed species.

Conflicts of Interest: The authors declare no competing financial interests.

References

1. Chennubhotla, V.; Rao, M.U.; Rao, K. Commercial importance of marine macro algae. *Seaweed Res. Utiln.* **2013**, *35*, 118–128.
2. Sahoo, D. *Farming the Ocean: Seaweeds Cultivation and Utilization*; Aravali: New Delhi, India, 2000.
3. Børresen, T. Seaweed as aquatic food. *J. Aquat. Food. Prod. Technol.* **2014**, *23*, 207. [CrossRef]
4. Ngadiarti, I.; Nurkolis, F.; Handoko, M.N.; Perdana, F.; Permatasari, H.K.; Taslim, N.A.; Mayulu, N.; Wewengkang, D.S.; Noor, S.L.; Batubara, S.C.; et al. Anti-aging potential of cookies from sea grapes in mice fed on cholesterol-and fat-enriched diet: In vitro with in vivo study. *Heliyon* **2022**, *8*, e09348. [CrossRef] [PubMed]
5. Sarkar, M.S.I.; Kamal, M.; Hasan, M.M.; Hossain, M.I. Present status of naturally occurring seaweed flora and their utilization in Bangladesh. *Res. Agric. Livest. Fish* **2016**, *3*, 203–216. [CrossRef]
6. FAO. *The State of World Fisheries and Aquaculture 2020*; Sustainability in Action: Rome, Italy, 2020.
7. Budd, G.C.; Pizzola, P. *Ulva Intestinalis. Gut Weed*; Marine Biological Association of the United Kingdom: Royal Citadel, UK, 2008. [CrossRef]
8. Reed, R.H.; Russell, G. Adaptation to salinity stress in populations of *Enteromorpha intestinalis* (L.) Link. *J. Estuarine Coast. Mar. Sci.* **1979**, *8*, 251–258. [CrossRef]
9. Hossain, M.T.; Sohag, A.A.M.; Haque, M.N.; Tahjib-Ul-Arif, M.; Dash, R.; Chowdhury, M.T.H.; Hossain, M.A.; Moon, I.S.; Hannan, M.A. Nutritional Value, Phytochemical Profile, Antioxidant Property and Agar Yielding Potential of Macroalgae from Coasts of Cox's Bazar and St. Martin's Island of Bangladesh. *J. Aquat. Food. Prod. Technol.* **2021**, *30*, 217–227. [CrossRef]
10. Siddik, M.A.B.; Anh, N.T.N. Preliminary assessment of the Gut weed *Ulva intestinalis* as food for herbivorous fish. *Int. Aquat. Res.* **2015**, *7*, 41–46. [CrossRef]
11. Mondal, K.; Bhattacharyya, S.B.; Mitra, A. Marine algae *Enteromorpha intestinalis* acts as a potential growth promoter in prawn feed. *World J. Pharmaceutical Res.* **2014**, *3*, 764–775.
12. Peñalver, R.; Lorenzo, J.M.; Ros, G.; Amarowicz, R.; Pateiro, M.; Nieto, G. Seaweeds as a functional ingredient for a healthy diet. *Mar. Drugs* **2020**, *18*, 301. [CrossRef]
13. Oh, H.; Lee, P.; Kim, S.Y.; Kim, Y.-S. Preparation of cookies with various native seaweeds found on the Korean coast. *J. Aquat. Food Prod. Technol.* **2020**, *29*, 167–174. [CrossRef]

14. Vijay, K.; Balasundari, S. Harnessing Tropical Brown Seaweed to Derive Nutritional and Functional Seaweed Based Products. Ph.D. Thesis, Tamil Nadu Fisheries University, Tamil Nadu, India, 2017.
15. Peryam, D.R.; Pilgrim, F.J. Hedonic scale method of measuring food preferences. *J. Food Technol.* **1957**, *11*, 9–14.
16. HadiNezhad, M.; Butler, F. Effect of flour type and dough rheological properties on cookie spread measured dynamically during baking. *J. Cereal Sci.* **2009**, *49*, 178–183. [CrossRef]
17. AOAC. *Official methods of analysis*, 16th ed.; Association of official analytical chemists: Washington, DC, USA, 1995.
18. AACC. *Approved Methods of the American Association of Cereal Chemists*, 10th ed.; American Association of Cereal Chemists: Saint Paul, MN, USA, 2000.
19. Mohibbullah, M.; Won, N.E.; Jeon, J.H.; An, J.H.; Park, Y.; Kim, H.; Bashir, K.M.I.; Park, S.M.; Kim, Y.S.; Yoon, S.J.; et al. Effect of superheated steam roasting with hot smoking treatment on improving physicochemical properties of the adductor muscle of pen shell (*Atrina pectinate*). *Food Sci. Nutr.* **2018**, *6*, 1317–1327. [CrossRef] [PubMed]
20. Khan, M.H.H.; Molla, M.M.; Sabuz, A.A.; Chowdhury, M.G.F.; Alam, M.; Choudhury, A.K.; Sarker, P.C. Effect of Orange Peel Concentration on the Development of Sapota Marmalade in Terms of Proximate and Nutritional Composition and Consumer Preferences. *J. Agric. Sci. Food Sci. Technol.* **2021**, *7*, 57–65. [CrossRef]
21. Baten, M.A.; Won, N.E.; Sohn, J.H.; Kim, J.S.; Mohibbullah, M.; Choi, J.S. Improvement of Sensorial, Physicochemical, Microbiological, Nutritional and Fatty Acid Attributes and Shelf Life Extension of Hot Smoked Half-Dried Pacific Saury (*Cololabis saira*). *Foods* **2020**, *9*, 1009. [CrossRef]
22. Chen, Y.; Wu, S.; Pan, S. Effect of water-soluble chitosan in combination with glutathione on the quality of pen shell adductor muscles. *Int. J. Biol. Macromol.* **2015**, *72*, 1250–1253. [CrossRef]
23. Rana, M.M.; Mohibbullah, M.; Won, N.E.; Baten, M.A.; Sohn, J.H.; Kim, J.-S.; Choi, J.-S. Improved hot smoke processing of chub mackerel (*Scomber japonicus*) promotes sensorial, physicochemical and microbiological characteristics. *Applied Sciences* **2021**, *11*, 2629. [CrossRef]
24. Firestone, D. *AOCs Official Method Ce 2–66, Preparation of Methyl Esters of Fatty Acids. Official Methods and Recommended Practices*, 5th ed.; Oil Chemists' Society: Champaign, IL, USA, 1998.
25. Oh, Y.-J.; Choi, K.-S. Effects of steam-dried *Hizikia fusiformis* powder on the quality characteristics in wet noodles. *Culin. Sci. Hosp. Res.* **2006**, *12*, 206–221.
26. Romani, S.; Balestra, F.; Angioloni, A.; Rocculi, P.; Dalla Rosa, M. Physico-chemical and electronic nose measurements on the study of biscuit baking kinetics. *Italian J. Food Sci.* **2012**, *24*, 32–40.
27. Kabirullah, M.; Ahmed, R.; Khan, S.; Hossain, M.; Mojibur Rahman, A.; Azizul Islam Kazi, M.; Moazzam Hossain, M. Analysis of Nutrients of Bangladesh Processed Foods Part-I: Different types of biscuits. *Bangladesh J. Sci. Ind. Res.* **1995**, *30*, 121–130.
28. Okpala, L.; Okoli, E.; Udensi, E. Physico-chemical and sensory properties of cookies made from blends of germinated pigeon pea, fermented sorghum, and cocoyam flours. *Food Sci. Nutr.* **2013**, *1*, 8–14. [CrossRef] [PubMed]
29. Thongram, S.; Tanwar, B.; Chauhan, A.; Kumar, V. Physicochemical and organoleptic properties of cookies incorporated with legume flours. *Cogent. Food Agric.* **2016**, *2*, 1172389. [CrossRef]
30. Kotoki, D.; Deka, S.C. Baking loss of bread with special emphasis on increasing water holding capacity. *J. Food Sci. Tech.* **2010**, *47*, 128–131. [CrossRef] [PubMed]
31. Yarnpakdee, S.; Benjakul, S.; Kingwascharapong, P. Physico-chemical and gel properties of agar from *Gracilaria tenuistipitata* from the lake of Songkhla, Thailand. *Food Hydrocoll.* **2015**, *51*, 217–226. [CrossRef]
32. Sutikno, L.A.; Bashir, K.M.I.; Kim, H.; Park, Y.; Won, N.E.; An, J.H.; Jeon, J.-H.; Yoon, S.-J.; Park, S.-M.; Sohn, J.H. Improvement in Physicochemical, Microbial, and Sensory Properties of Common Squid (*Todarodes pacificus* Steenstrup) by Superheated Steam Roasting in Combination with Smoking Treatment. *J. Food Qual.* **2019**, *2019*, 1–15. [CrossRef]
33. Lee, J.-H.; Kim, I.-Y. Consumer perception and sensory characteristics of cookies incorporated with strawberry powder. *Prev. Food Sci. Nutr.* **2009**, *14*, 66–70. [CrossRef]
34. Mancebo, C.M.; Rodríguez, P.; Martínez, M.M.; Gómez, M. Effect of the addition of soluble (nutriose, inulin and polydextrose) and insoluble (bamboo, potato and pea) fibres on the quality of sugar-snap cookies. *Int. J. Food Sci. Technol.* **2018**, *53*, 129–136. [CrossRef]
35. Pina, A.; Costa, A.; Lage-Yusty, M.; López-Hernández, J. An evaluation of edible red seaweed (*Chondrus crispus*) components and their modification during the cooking process. *LWT-Food Sci Technol* **2014**, *56*, 175–180. [CrossRef]
36. Pereira, D.; Correia, P.; Guiné, R. Analysis of the physical-chemical and sensorial properties of Maria type cookies. *Acta. Chim. Slov.* **2013**, *6*, 269–280. [CrossRef]
37. Adeola, A.A.; Ohizua, E.R. Physical, chemical, and sensory properties of biscuits prepared from flour blends of unripe cooking banana, pigeon pea, and sweet potato. *Food Sci. Nutr.* **2018**, *6*, 532–540. [CrossRef]
38. Mamat, H.; Akanda, J.M.H.; Zainol, M.K.; Ling, Y.A. The influence of seaweed composite flour on the physicochemical properties of muffin *J. Aquat. Food. Prod. Technol.* **2018**, *27*, 635–642. [CrossRef]
39. Nunes, N.; Rosa, G.P.; Ferraz, S.; Barreto, M.C.; de Carvalho, M. Fatty acid composition, TLC screening, ATR-FTIR analysis, anti-cholinesterase activity, and in vitro cytotoxicity to A549 tumor cell line of extracts of 3 macroalgae collected in Madeira. *J. Appl. Phycol.* **2020**, *32*, 759–771. [CrossRef]

40. Jongaroontaprangsee, S.; Tritrong, W.; Chokanaporn, W.; Methacanon, P.; Devahastin, S.; Chiewchan, N. Effects of drying temperature and particle size on hydration properties of dietary fiber powder from lime and cabbage by-products. *Int. J. Food Prop.* **2007**, *10*, 887–897. [CrossRef]
41. Abraha, B.; Mahmud, A.; Admassu, H.; Habte-Tsion, H.-M.; Xia, W.; Yang, F. Production of biscuit from Chinese sturgeon fish fillet powder (*Acipeneser sinensis*): A snack food for children. *J. Aquat. Food. Prod. Technol.* **2018**, *27*, 1048–1062. [CrossRef]
42. Corsetto, P.A.; Montorfano, G.; Zava, S.; Colombo, I.; Ingadottir, B.; Jonsdottir, R.; Sveinsdottir, K.; Rizzo, A.M. Characterization of antioxidant potential of sea-weed extracts for enrichment of convenience food. *Antioxidants* **2020**, *9*, 249. [CrossRef]
43. Bekhit, A.E.D.A.; Holman, B.W.; Giteru, S.G.; Hopkins, D.L. Total volatile basic nitrogen (TVB-N) and its role in meat spoilage: A review. *Trends Food Sci. Technol.* **2021**, *109*, 280–302. [CrossRef]
44. Mohibbullah, M.; Abdul Hannan, M.; Park, I.S.; Moon, I.S.; Hong, Y.K. The Edible Red Seaweed *Gracilariopsis chorda* Promotes Axodendritic Architectural Complexity in Hippocampal Neurons. *J. Med. Food* **2016**, *19*, 638–644. [CrossRef]
45. Cardoso, C.; Ripol, A.; Afonso, C.; Freire, M.; Varela, J.; Quental-Ferreira, H.; Pousao-Ferreira, P.; Bandarra, N. Fatty acid profiles of the main lipid classes of green seaweeds from fish pond aquaculture. *Food Sci. Nutr.* **2017**, *5*, 1186–1194. [CrossRef]
46. Oucif, H.; Benaissa, M.; Ali Mehidi, S.; Prego, R.; Aubourg, S.P.; Abi-Ayad, S.-M.E.-A. Chemical composition and nutritional value of different seaweeds from the west Algerian coast. *J Aquat. Food. Prod. Technol.* **2020**, *29*, 90–104. [CrossRef]
47. Sokoła-Wysoczańska, E.; Wysoczański, T.; Wagner, J.; Czyż, K.; Bodkowski, R.; Lochyński, S.; Patkowska-Sokoła, B. Polyunsaturated fatty acids and their potential therapeutic role in cardiovascular system disorders—A review. *Nutrients* **2018**, *10*, 1561. [CrossRef]
48. Mohibbullah, M.; Hannan, M.A.; Choi, J.Y.; Bhuiyan, M.M.; Hong, Y.K.; Choi, J.S.; Choi, I.S.; Moon, I.S. The Edible Marine Alga *Gracilariopsis chorda* Alleviates Hypoxia/Reoxygenation-Induced Oxidative Stress in Cultured Hippocampal Neurons. *J. Med. Food* **2015**, *18*, 960–971. [CrossRef] [PubMed]
49. Mohibbullah, M.; Choi, J.S.; Bhuiyan, M.M.H.; Haque, M.N.; Rahman, M.K.; Moon, I.S.; Hong, Y.K. The Red Alga *Gracilariopsis chorda* and Its Active Constituent Arachidonic Acid Promote Spine Dynamics via Dendritic Filopodia and Potentiate Functional Synaptic Plasticity in Hippocampal Neurons. *J. Med. Food* **2018**, *21*, 481–488. [CrossRef] [PubMed]
50. Elshani, B.; Kotori, V.; Daci, A. Role of omega-3 polyunsaturated fatty acids in gestational diabetes, maternal and fetal insights: Current use and future directions. *J. Matern-Fetal Neonatal Med.* **2021**, *34*, 124–136. [CrossRef] [PubMed]

foods

Article

Downregulation of Peroxidase Activity of Platinum Cube Enables Minute–Time Scale Colorimetric Signaling of Hypoxanthine for Fish Freshness Monitoring

Xiaoming Ma [1,2], Tingting Feng [1], Peng Zhang [1], Hui Zhang [1], Xuan Hu [1], Yuying Yang [1], Zhen Wang [1], Huifang Zhang [1], Dong Peng [1], Xun Li [1,*] and Jianguo Xu [3,*]

1 Key Laboratory of Organo-Pharmaceutical Chemistry of Jiangxi Province, School of Chemistry and Chemical Engineering, Gannan Normal University, Ganzhou 341000, China
2 Fujian Province-Indonesia Marine Food Joint Research and Development Center, Fujian Polytechnic Normal Univeristy, Fuzhou 350300, China
3 School of Food and Biological Engineering, Hefei University of Technology, Hefei 230009, China
* Correspondence: 0900001@gnnu.edu.cn (X.L.); jgxu-sfse@hfut.edu.cn (J.X.)

Abstract: Due to its unique biological composition, aquatic products, especially fish, are extremely perishable compared to other muscle products. Herein, we proposed an artificial nanozyme-based colorimetric detection of hypoxanthine (Hx), the indicator of fish freshness, in a minute–time scale without the assistance of a natural enzyme (hypoxanthine oxidase). The principle is based on the interaction between Hx and polyvinylpyrrolidone-modified platinum cubic nanomaterials (PVP-PtNC), in which the catalytic active sites of PVP-PtNC's surface were blocked by Hx. This causes the downregulation of PVP-PtNC's catalytic ability and weakened its ability to catalyze the oxidization of 3,3′,5,5′-Tetramethylbenzidine (TMB) by H_2O_2. Accordingly, the decrease in the UV–vis absorption and the weakening of the colorimetric reaction color is proportional to the Hx concentration. On this basis, a target-triggered colorimetric method for detecting Hx is developed for fish freshness monitoring with a fast detection speed, low cost, high accuracy, and simplified operation. Experiments reveal that the correlation response of Hx is from 0.5 μM to 10 mM with a limit of detection of 0.16 μM. In particular, the Hx detected from real fish indicates that the method possesses a promising potential for practical application. All of these features are expected to promote the development of online detection tools for food safety monitoring.

Keywords: food safety; nanozyme; fish freshness; platinum nanocube; hypoxanthine

Citation: Ma, X.; Feng, T.; Zhang, P.; Zhang, H.; Hu, X.; Yang, Y.; Wang, Z.; Zhang, H.; Peng, D.; Li, X.; et al. Downregulation of Peroxidase Activity of Platinum Cube Enables Minute–Time Scale Colorimetric Signaling of Hypoxanthine for Fish Freshness Monitoring. *Foods* **2023**, *12*, 291. https://doi.org/10.3390/foods12020291

Academic Editors: Tao Yin and Liu Shi

Received: 5 December 2022
Revised: 31 December 2022
Accepted: 5 January 2023
Published: 8 January 2023

1. Introduction

As one of the most important components of the modern food industry, aquatic products are especially favored by humans because they not only offer essential calories to the human body but also provide multiple nutrients such as protein and docosahexaenoic acid [1]. According to the official report by fishery and aquaculture products fishery and aquaculture products (FAPs), the consumption of aquatic products is increasing steadily in people's daily diets [2]. China has ranked first in fisheries and aquatic products in the world [3], most of which is consumed by the domestic market and China is responsible for most of the increase in world fish consumption. However, aquatic products, especially fish, are highly perishable with a limited shelf-life after slaughter due to the ready occurrence of protein degradation, lipid oxidation, and changes in odor, flavor, and texture [4,5]. The aquatic product spoilage thus stimulates the accurate freshness analysis of fish. Traditionally, consumers mainly rely on a sensory approach to judge the organoleptic features of fish according to their perceptions and experiences, which is not reliable and is subjective for freshness evaluation and might even cause severe food safety incidents if the food had been spoiled for a long time. Therefore, it is imperative to develop accurate analytical methods for accurately monitoring aquatic products' freshness.

With the rapid development of aquatic product science and the advancing of modern analytical chemistry, the methods adopted for aquatic product freshness, exemplified by fish, identification has been at the biochemical and microbiological level. For instance, the growth of microorganisms in fish tissues leads to the degradation of adenosine triphosphate (ATP) but the right accumulation of a reaction intermedium of hypoxanthine (Hx) [6,7]. This transform means that Hx can be used as an important indicator to reveal the freshness of fish since its accumulation is proportional to the storage time. Currently, to meet the freshness standard, conventional methods including mass spectrometry technology [8,9] near-infrared spectra [10], electronic nose [11], electronic tongue technology [12], fluorescence [13–15], computer vision technology [13,16], and gas chromatography–mass spectrometry (GC–MS) technology [15] have been already been applied to realize the measuring of Hx. Although these methods are highly repetitive, accurate, and reliable, the request for expensive equipment, professional personnel, and specialized laboratory limits their wide application. The ask for a user-friendly detection method with fast detection speed, low cost, high accuracy, and simplified operation is continually demanded.

In the past decade, nanozymes, as a new generation of artificial enzymes with highly effective enzyme-like properties, come into focus and achieve great progress. The landmark work is reported by Yan's group in 2007. In their work, they first found the intrinsic peroxidase-like activity of ferromagnetic nanoparticles [17]. From that time, several nanozymes (e.g., Au nanoclusters [18,19], WS_2 nanosheets [20], graphene oxide [21], and carbon dots [22]) have been explored and applied for building biosensors in responses to a variety of target analytes including proteins, nucleic acids, metal ions, inorganic and organic compounds, and so on [23–27]. In particular, noble metal nanomaterials have become one of the hot topics due to their unique chemical properties and excellent catalytic activity. Moreover, compared with natural enzymes that are difficult to prepare, easy to deactivate, and expensive to use, artificial nanozymes can be easily prepared with low cost, high stability, and especially, high catalytic ability [26]. These advantages allow the boom of nanozyme-based sensors for biochemical analysis, which might also be useful for us to build a Hx-responsive online detection method with high assay performance.

The conventional Hx sensing platform typically requires xanthine oxidase (XOD) to convert Hx to H_2O_2 and uric acid in the presence of oxygen. The generated H_2O_2 can be further catalyzed and produce hydroxyl radicals (•OH) via the catalysis of peroxidase-mimicking nanozyme and oxidize the chromogenic substrate from colorless to colored [28,29]. For instance, Zhang et al. demonstrated an efficient Hx-sensing platform based on the peroxidase-mimicking activity of Fe-doped polydopamine (Fe-PDA) [30]. In the presence of Hx, XOD can quantitatively convert Hx to H_2O_2, and the generated H_2O_2 can further oxidize the colorless 3,3′,5,5′-Tetramethylbenzidine (TMB) to blue oxTMB with the catalysis of Fe-PDA. However, these biosensors still need the assistance of the natural enzyme (XOD), which meant it was difficult to avoid the drawbacks of conventional enzymes such as their instabilities under harsh conditions.

Taking the above into consideration, the research objective is to construct a conventional colorimetric sensing platform for fish freshness determination. Herein, we hypothesized a colorimetric method for the determination of Hx based on the downregulation of peroxidase activity of the polyvinylpyrrolidone-modified platinum cubic nanomaterials (PVP-PtNC). This strategy lies in our finding that the Hx can block the active catalytic sites on the PVP-PtNC surface so that the ultra-high catalytic ability of PVP-PtNC can be downregulated directly by Hx. Once encountered with the H_2O_2 to catalyze the oxidation of TMB, a typical peroxidase substrate, the significantly weakened peroxidase activity would powerfully inhibit the generation of oxidized blue products. Therefore, this sensing platform can realize the quantitative determination of Hx without the assistance of a natural enzyme (XOD). Our results showed that the morphology of PVP-PtNC was well demonstrated by transmission electron microscopy (TEM). The decrease in the UV–vis absorption and the weakening of the colorimetric reaction color is proportional to the Hx concentration. Moreover, benefiting from the wide linear response range, low detection limit, high specificity, and repeatability, this colorimetric nanosensing method has been successfully applied to analyze real fish samples in only 10 min. Such a nanozyme-based sensing platform with its minute-time scale detection merit would thus open a new avenue for the online monitoring of aquatic products' freshness.

2. Materials and Methods

2.1. Materials and Instruments

Hypoxanthine (Hx), uric acid, dopamine, and 3,3′,5,5′-tetramethylbenzidine hydrochloride (TMB) were purchased from Shanghai bioengineering technology co. LTD. (Shanghai, China). Hydrochloric acid (HCl), acetic acid (CH_3COOH), sodium acetate (CH_3COONa), sodium hydroxide (NaOH), Glucose (Glucose), hydrogen peroxide (H_2O_2, 30 wt%), ascorbic acid ($C_6H_8O_6$), zinc chloride ($ZnCl_2$), magnesium chloride ($MgCl_2$), Calcium chloride ($CaCl_2$), silver nitrate ($AgNO_3$), potassium chloroplatinate hexahydrate (IV) ($K_2PtCl_6{\cdot}6H_2O$), potassium bromide (KBr), polyethylene pyrrolidone (PVP, MW $\approx$ 50,000), and ethylene glycol (EG) were purchased from Sigma Aldrich (St. Louis, MO, USA). Live fish were purchased from the supermarket located in the Gannan Normal University (Jiangxi, China). All reagents were used directly without further purification.

The UV–vis spectrophotometer (UV-1780, Shimadzu, Kyoto, Japan) was used to record UV–vis absorption. The Precision Electronic Balance (BSA2245, Sartorius, Goettingen, Germany) was used to weigh reagents. Transmission electron microscope (Titan G260-300, FEI Company, Hillsboro, OR, USA) was used to characterize the morphology of nanoparticles. The concentration of Platinum element was determined with the inductively coupled plasma emission spectrometry (ICP-OES) (Ultima2, HORIBA Jobin Yvon, Longjumeau, France). The High-Speed refrigerated centrifuge (Sorvall ST 16R, Thermo Scientific, Waltham, MA, USA) was used to separate and enrich samples. The water preparation apparatus (Milli-Q, Millipore Corporation, Bedford, MA, USA) was used to prepare ultrapure water.

2.2. Synthesis of PVP-PtNC

The PVP-PtNC nanomaterials were synthesized according to literature reports [31]. First, 20 mg KBr and 40 mg PVP were added into a 3.5 mL glycol (EG) solution. After dissolving, the mixture was heated to reflux (about 180 °C) for 15 min under an oil bath. This is followed by adding 0.5 mL $K_2PtCl_6{\cdot}6H_2O$ solution (40 mg/mL, EG dissolved) to react with the mixture for another 20 min. Subsequently, the resultant solution was immediately cooled with an ice bath. The excess PVP was removed by centrifugation and the final reaction products were washed with acetone and deionized water 3 times to obtain the purified PVP-PtNC, which was stored in 4 mL ultrapure water at 4 °C before usage. The concentration of platinum was measured to be 0.473 g/L in PVP-PtNC by ICP-OES.

2.3. PVP-PtNC-Based Colorimetric Sensing of Hx

Prior to detecting Hx, the standard Hx solutions were prepared by weighing 0.0136 g Hx and dissolving it in 0.4 mL of 1 M HCl since Hx is poorly soluble in neutral conditions. The resultant mixture was then added with 9.6 mL of 0.05 M Sodium acetate-Acetic acid (NaAC-HAC) buffer (pH = 4) to obtain the storing solution of Hx (10 mM). Other concentrations of Hx were obtained by diluting the storing solution of Hx with 0.05 M NaAC-HAC buffer (pH = 4). To conduct the colorimetric detection of Hx, the PVP-PtNC (100 µL, 5 µM) and Hx solutions at different concentrations were mixed in 0.05 M NaAC-HAC buffer (pH = 4) and incubated at room temperature for 30 min. The resultant mixture (100 µL) was added with H_2O_2 (25 µL, 1 M) and TMB (25 µL, 15 mM) to carry out the colorimetric catalysis reaction at room temperature for 10 min. The visible absorption spectra from 400 to 800 nm were measured by a UV–vis spectrophotometer and the peak absorption value at 651 nm was recorded for evaluating its sensing performance.

2.4. *Processing of Real Fish Samples*

Firstly, we took fresh live fish and removed their scales, skin, and bones to obtain the fresh fish meat, which is then processed into minced meat by grinding treatment and spoilage-treated by placing them at room temperature. After 12 h, 5.0 g minced fish meat was added with H_2SO_4 (10 mL, 0.06 M) solution to make them digested, and then added with 39 mL of ultrapure water to break up the fine tissue by homogenate processing. The resultant solution was placed in a shaker for 4 h at room temperature. After the shaking was completed, the residue was filtered with filter paper, and the maintained solution phase was filtered twice with an organic filter followed by adding NaOH (1 mL, 6 M) to adjust its pH to 4.0. The final sample was placed in the refrigerator and stored at 4 °C until use.

3. Results and Discussion

3.1. *Principle of the Proposed Colorimetric Sensor*

Figure 1 presents the working diagram of the PVP-PtNC-based colorimetric sensor to detect Hx. One can find that, in the absence of Hx, the PVP-PtNC can maximally maintain its peroxidase activity. When exposed to the mixture of H_2O_2 and TMB, the oxidization of TMB by H_2O_2 can be immediately performed by the catalytic peroxidase activity of PVP-PtNC, in turn resulting in an obvious solution color change and a significant enhancement of the UV–vis absorption for the Hx-absent negative sample. However, in the presence of Hx, the surface with catalytic active sites on PVP-PtNC can be coated with Hx because of the physical absorption of Hx on PVP-PtNC. The catalytic activity of PVP-PtNC is thus downregulated with the net decrease in its active sites. Upon being introduced with H_2O_2 and TMB, the oxidization of TMB by H_2O_2 cannot be strongly executed anymore. We can only observe a weakened color change or even a colorless solution of the Hx-presented positive sample. In this case, the signal absorption is also decreased. From the working principle, it is seen that the color change degree and the decrease in the UV–vis absorption are based on the Hx concentration since the more the Hx presented, the more active sites of the PVP-PtNC will be occupied. The color change can be easily observed by the naked eye, while the intensity change of UV–vis can be monitored by a UV–vis spectrometer.

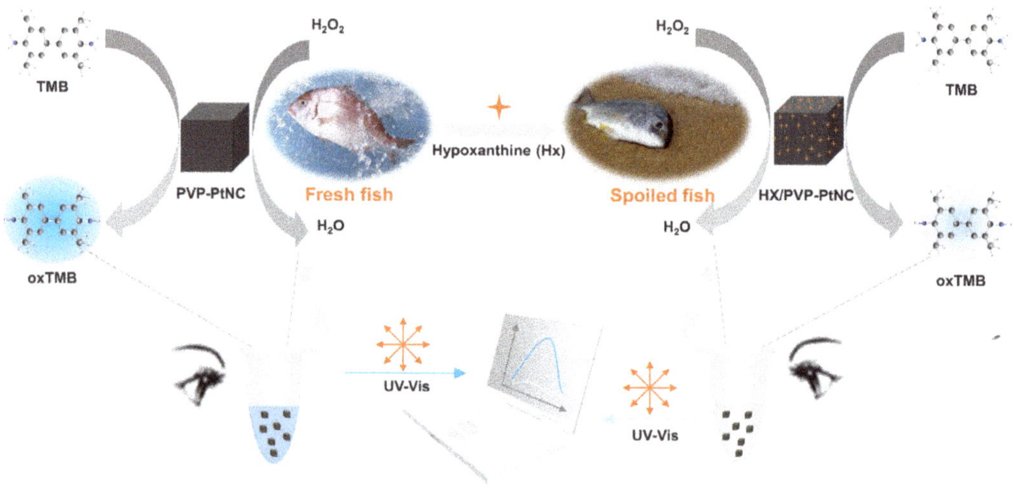

Figure 1. Schematic diagram of the working principle for the PVP-PtNC-based colorimetric detection of Hx.

3.2. Feasibility Demonstration for Hx Analysis

To validate the feasibility of Hx analysis, the strong peroxidase activity of the platinum cube is the prerequisite for subsequent studies, which should be confirmed first. As shown in sample a of Figure 2A, the mixture of TMB and H_2O_2 showed a very weak color, suggesting that the autocatalysis of TMB by H_2O_2 is very difficult. After added with PVP-PtNC, as shown in sample b, we can excitingly find that the solution was changed to dark blue. This phenomenon confirms the strong peroxidase activity of PVP-PtNC. On this basis, we tested the catalysis system for Hx sensing in sample c. As expected, the pre-incubation of Hx with PVP-PtNC, and then, using the Hx-coated PVP-PtNC to react with the mixture of TMB and H_2O_2 was unable to strongly catalyze the oxidization of TMB by H_2O_2 since sample c only showed a light blue. The big difference in samples b and c demonstrate the downregulation of the peroxidase activity of PVP-PtNC by Hx, which also evidences the availability of this sensing method for Hx analysis. We also further measured the UV–vis absorption spectra of samples a, b, and c in Figure 2B. It can be seen that the oxidization of TMB by H_2O_2 displayed the weakest UV–vis absorption in line a. The further introduction of PVP-PtNC caused a significantly improved peak absorption intensity in line b. However, when used to analyze Hx, the peak absorption was obviously decreased in line c. The variation tendency of the UV–vis absorption spectra in Figure 2B is consistent with the color changes in Figure 2A, confirming further the feasibility of Hx detection.

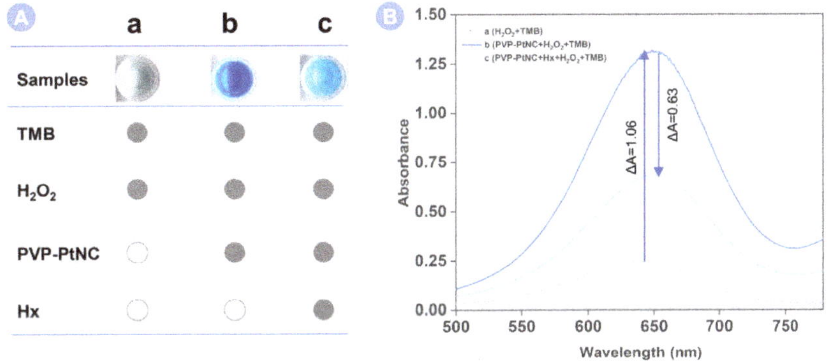

Figure 2. (**A**) Photographed images and (**B**) UV–vis absorption spectra of (a) TMB + H_2O_2, (b) PVP-PtNC + TMB + H_2O_2, and (c) PVP-PtNC + Hx + TMB + H_2O_2.

3.3. Characterization of as-Synthesized PVP-PtNC

In this study, the successful colorimetric sensing of Hx is based on the application of PVP-PtNC. We, therefore, need to first characterize the as-synthesized PVP-PtNC. As shown in Figure S1A, the synthesized PVP-PtNC were all cube-shaped and uniformly dispersed. The average particle size is about 7.3 nm. Figure S1B is a high-resolution TEM image of a single PVP-PtNC nanomaterial, which clearly showed the corresponding lattice fringe and the lattice fringe spacing. These results indicate that PVP-PtNC has been successfully prepared.

The Michaelis constant (*K*m) value is a key indicator of enzymatic efficiency that indicates the affinity between enzymes and substrates, in which the lower *K*m value signifies a higher affinity [26]. To determine the peroxidase-like activity of the PVP-PtNC, a steady-state kinetic assay was carried out at room temperature. The Michaelis–Menten curves were obtained and fitted to the double-reciprocal plot (Figure S2A,B) and the *K*m of the PVP-PtNC toward TMB was calculated to be 2.22×10^{-5} M. The *K*m of the PVP-PtNC toward TMB was lower than that of HRP, suggesting that PtNC has a higher binding affinity to TMB compared with HRP (Table S1). The *K*m of the PVP-PtNC toward H_2O_2 was calculated by plotting the initial reaction velocities against the H_2O_2 concentration curve (Figure S2C) and then fitted to the double-reciprocal plot (Figure S2D). The Km was 3.92×10^{-3} M, indicating that PtNC had a higher binding affinity to H_2O_2 compared with HRP. Therefore, PtNC was a promising peroxidase candidate for the peroxidase-like enzymatic reaction.

3.4. Exploring the Interaction of PVP-PtNC with Hx

To deeply understand the interaction between PVP-PtNC and Hx, the charge states of Pt species are investigated by XPS. As seen in Figure 3A, the PVP-PtNC showed four peaks at 71.03, 72.33, 74.26, and 75.74 eV. According to the literature, the binding energies for Pt (0) $4f_{7/2}$ and $4f_{5/2}$ are 71.1 and 74.4 eV, respectively [32,33]. The peaks of PVP-PtNC at 71.03 and 75.74 eV can be assigned to Pt (0) $4f_{7/2}$ and $4f_{5/2}$, respectively. In the absence of target Hx, the fraction of Pt (II) and Pt (0) is determined as 34.6% and 65.4%, respectively. Interestingly, the addition of Hx to react with PVP-PtNC decreased the content of Pt (II) because the fraction of Pt (II) and Pt (0) is determined as 83.9% and 16.1% in Figure 3B. The comparative results indicate the reduction of Pt (II) by Hx. In addition, the zeta potential of the synthesized PVP-PtNC and Hx solution was −13.7 mV and −9.4 mV, respectively, and their mixture changed the zeta potential to –15.4 mV in Figure 3C. Therefore, Hx potentially regulates the peroxidase-like activity of PVP-PtNC through electrostatic adsorption and the in situ reduction interaction between Hx and PVP-PtNC.

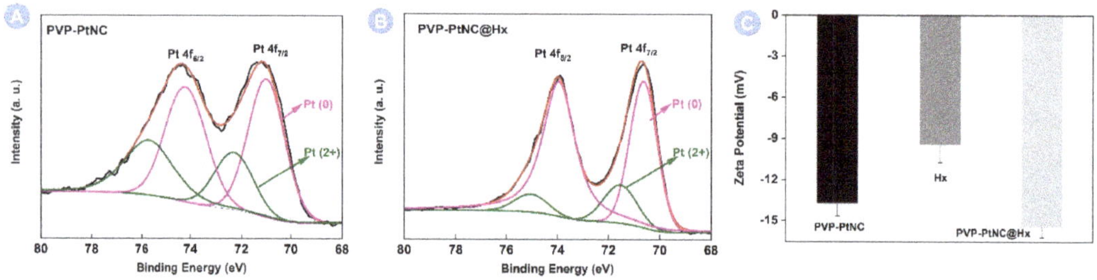

Figure 3. Pt 4f XPS spectra of PVP-PtNC before (**A**) and after (**B**) incubated with Hx. Black, red, magenta, and green lines represent the raw curve, the fitted curve, the Pt (0), and the Pt (II) components' curves, respectively. (**C**) Zeta potential measurements of PVP-PtNC, Hx, and their mixture.

3.5. Optimization of Experimental Conditions

To achieve the best assay performance, the experimental conditions that are closely related to the catalytical reaction such as the reaction time, the TMB concentration, the H_2O_2 concentration, and the pH of the reaction solution, were investigated in Figure 4. The dynamic monitoring of the peak absorption as a function of the reaction time is shown in Figure 4A. The peak absorption was initially increased and then leveled off indicating the optimal reaction time is 10 min. Likewise, when the TMB concentration in Figure 4B was greater than 15 mM and the H_2O_2 concentration was greater than 1 M in Figure 4C, the peak absorption also reached a plateau, suggesting the concentration of TMB at 15 mM and the H_2O_2 at 1 M are optimal for building the sensing system. Last but not least, we optimized the pH of the reaction solution. As explored in Figure 4D, the employing of 0.1 M acetic acid/sodium acetate buffer to control the pH value of the reaction solution as 4 can achieve the highest signal output. Below or above this value, the peak absorption has deteriorated. We thus controlled the pH value of the reaction solution as 4 to investigate the Hx detection.

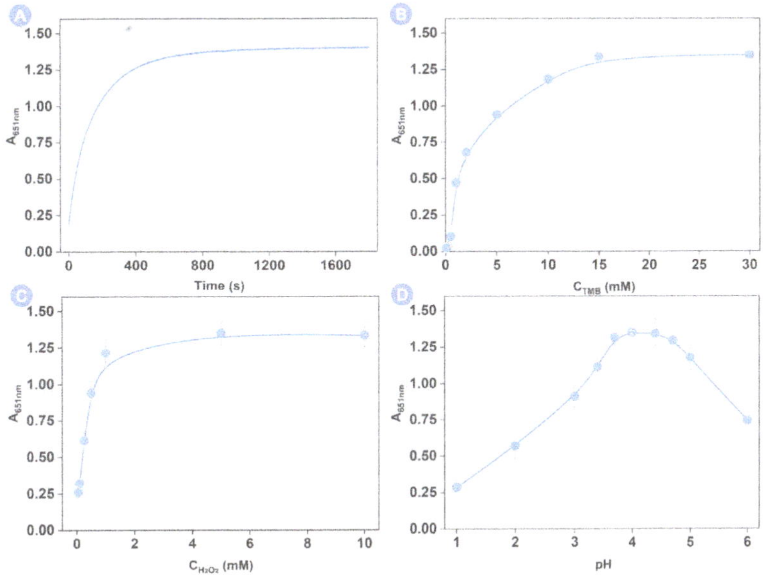

Figure 4. Influences of (**A**) the reaction time, (**B**) the concentration of TMB, (**C**) the concentration of H_2O_2, and (**D**) the pH of the reaction solution on the catalytic reaction between TMB and H_2O_2. Error bars were obtained from three repetitive experiments.

3.6. Investigation for Quantitative Detection of Hx

Under optimized conditions (Figure 4), the developed colorimetric sensor was used to detect Hx based on the reverse regulation of peroxidase activity of PVP-PtNC. As shown in Figure 5A, with the increase in the Hx concentrations, the UV–vis absorption spectra gradually decreased. This is reasoned by the inhibiting effect of Hx on the PVP-PtNC catalytic activity. Figure 5B plotted the peak absorption at 651 nm dependent on the Hx concentration. In the Hx concentration range from 0 to 10 mM, the four-parameters logistic relationship was obtained between the peak absorption value (A_{651}) and the concentration of the Hx. The equation was $y = 1.3005/[1 + (x/0.1822)^{0.5254}] + 0.077$ ($R^2 = 0.9958$), where y was the absorbance intensity at 651 nm, and x was the Hx concentration. The limit of detection (LOD) calculated by 3σ/slope is 0.16 μM. Compared with reported fluorescent and electrochemical detection methods or paper-based device in Table S2, the current colorimetric Hx detection method not only shows a lower LOD but also is easy to operate and able to output detectable signals quickly. A previous study has related the concentration of HX to freshness, suggesting that the fish is fresh if the HX concentration is below 250 μM and spoiled at >630 μM [34]. As shown in Figure S3, the colorimetric images show that fresh fish and rotten fish have distinct color differences. Therefore, the colorimetric signal output pattern does not require the use of expensive instruments and professional operators, which can realize semi-quantitative detection directly based on the naked eye.

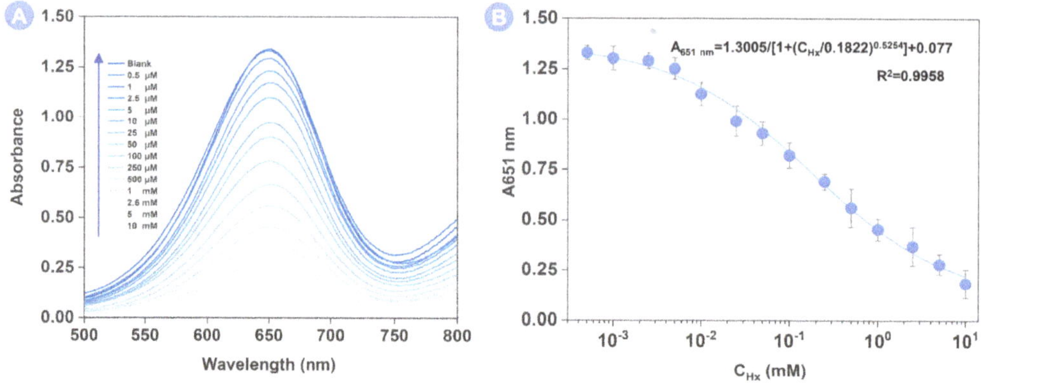

Figure 5. (**A**) Typical UV–vis spectra collected from the PVP-PtNC-based sensing system introduced with Hx at a concentration range from top to bottom: 0, 0.5 μM, 1.0 μM, 2.5 μM, 5.0 μM, 10 μM, 25 μM, 50 μM, 100 μM, 250 μM, 500 μM, 1 mM, 2.5 mM, 5 mM, 1 mM, and 10 mM. (**B**) The dynamic relationship of the peak absorption intensity against the Hx concentration. Error bars were obtained from three repetitive experiments.

3.7. *Investigation for the Qualitative Detection of Hx*

As well as the quantitative detection of Hx, the qualitative detection of Hx is another vital point for evaluating the assay ability of this method [35]. To examine this specificity, non-target analytes including glucose, uric acid, dopamine, Na^+, K^+, Zn^{2+}, Mg^{2+}, Cl^-, and NO_3^- were selected and used to challenge this system. It is worth considering that the metal ions were selected as analytes because some reports have found that the catalysis activity of PVP-PtNC can also be influenced by some metal ions. These metal ions may also exist in fish. These results gathered in Figure 6 show that compared with the negative sample having a high signal absorption, the PVP-PtNC-based sensing system is only Hx responsive considering the significant decrease in the signal. However, the presence of any other non-target analytes, even with a concentration five times higher than Hx, cannot induce any downregulation of the UV–vis signal. The comparative responses demonstrate the excellent specificity of our nanosensing method for the qualitative detection of Hx.

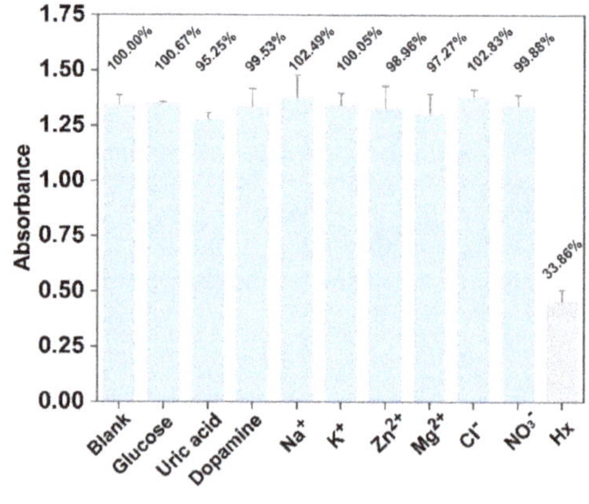

Figure 6. Peak absorption values of the sensing system in the absence and presence of glucose, uric acid, dopamine, Na^+, K^+, Zn^{2+}, Mg^{2+}, Cl^-, NO_3^-, and Hx, respectively. The concentration of Hx was 1 mM, while other non-target species is 5 mM. Error bars were obtained from three repetitive experiments.

3.8. Practicability Investigation

After demonstrating the ability for the quantitative and qualitative detection of Hx, we finally applied the PVP-PtNC-based sensing method to detect Hx from real fish samples, which were prepared by spiking standard Hx at different concentrations into two batches of spoiled fish meat. The pretreat of the fish meat is mentioned above. As shown in Table 1, no matter if the fish samples spoiled for 12 h or 24 h, the spiked Hx at indicated concentrations of 10, 100, and 1000 μM all exhibited favorable recovery rates, which were in the range of 102.6–106.0% and 100.9–101.9%, respectively. Meanwhile, the relative standard derivation (RSD) values were also acceptable. These results affirmed the application potential of this method to detect the amount of Hx in fish meat for monitoring fish freshness.

Table 1. Recoveries of the Hx in real fish samples.

Sample	12 h			24 h		
	Spiking (μM)	Found (μM)	Recovery (%)	Spiking (μM)	Found (μM)	Recovery (%)
1	10	10.6	106.0	10	10.15	101.5
2	100	103.5	103.5	100	101.9	101.9
3	1000	1025.6	102.6	1000	1009.7	100.9

4. Conclusions

In summary, we developed a colorimetric method for the ultrafast signaling of Hx based on the target-triggered reverse regulation of peroxidase activity of PVP-PtNC without the assistance of a natural enzyme. This work shows that the colorimetric method for Hx detection has the following advantages: first, Hx was used to directly regulate the catalytic activity of the nanozyme, which reduced the detection cost and realized the one-step Hx detection with only 10 min, and has been successfully applied to monitor the freshness of live fish; second, the colorimetric method is easy to operate without the assistance of advanced expensive equipment; third, this colorimetric method can detect Hx with high sensitivity and specificity in a wide detection range. We expect this method to have a good application prospect in the freshness detection of fish samples for ensuring food safety.

Supplementary Materials: The following are available online at https://www.mdpi.com/article/10.3390/foods12020291/s1, Figure S1: (A) TEM image of PVP-PtNCs to show its uniformity and cube shape, (B) Image of single PVP-PtNC taken by a high-resolution TEM, Figure S2: The kinetic assays of PVP-PtNC as artificial peroxidase-like catalysts for oxidation of TMB by H_2O_2. (A) Graphic representation of the plots of initial rate (ν) vs. TMB concentration; (B) double-reciprocal plot generated from (A); (C) Graphic representation of ν vs. H_2O_2 concentration; (D) double-reciprocal plot generated from (C), Figure S3: The colorimetric images of as-proposed method under different concentrations of Hx, Table S1: Comparison of PVP-PtNC kinetic parameters and HRP, Table S2: Comparison of reported methods with the current one for Hx detection. References [17,29,36–40] are cited in the supplementary materials.

Author Contributions: Conceptualization, X.M. and T.F.; validation, H.Z. (Hui Zhang) and X.H.; formal analysis, D.P.; investigation, H.Z. (Hui Zhang) and Z.W.; resources, H.Z. (Huifang Zhang) and X.L.; writing—original draft preparation, X.M. and Z.W.; writing—review and editing, J.X.; visualization, P.Z. and Y.Y.; supervision, H.Z. (Huifang Zhang); project administration, X.M.; funding acquisition, X.M. and X.L. All authors have read and agreed to the published version of the manuscript.

Funding: This work was financially supported by the National Natural Science Foundation of China (Grant No. 22264002 and 21964003), the project (Grant No. YC2021-S713) supported by Jiangxi Province Training Program of Innovation and Entrepreneurship for master, the National College Students' innovation and entrepreneurship training program (Grant No. 202110418002), the Science and Technology Project of the Education Department of Jiangxi Province of China (Grant No. GJJ190775), the Key Research and Development Program of Anhui Province (Grant No. 202004h07020029), the Fujian Key Laboratory of Functional Marine Sensing Materials, Minjiang University (Grant No. MJUKF-FMSM202105).

Institutional Review Board Statement: Not applicable.

Informed Consent Statement: Not applicable.

Data Availability Statement: The data are available from the corresponding author.

Conflicts of Interest: The authors declare that they have no competing interests.

References

1. Nawaz, A.; Li, E.; Irshad, S.; Xiong, Z.; Xiong, H.; Shahbaz, H.M.; Siddique, F. Valorization of fisheries by-products: Challenges and technical concerns to food industry. *Trends Food Sci. Technol.* **2020**, *99*, 34–43. [CrossRef]
2. Lucas, S.; Soler, L.-G.; Irz, X.; Gascuel, D.; Aubin, J.; Cloâtre, T. The environmental impact of the consumption of fishery and aquaculture products in France. *J. Clean. Prod.* **2021**, *299*, 126718. [CrossRef]
3. Naylor, R.L.; Hardy, R.W.; Buschmann, A.H.; Bush, S.R.; Cao, L.; Klinger, D.H.; Little, D.C.; Lubchenco, J.; Shumway, S.E.; Troell, M. A 20-year retrospective review of global aquaculture. *Nature* **2021**, *591*, 551–563. [CrossRef] [PubMed]
4. Duarte, A.M.; Silva, F.; Pinto, F.R.; Barroso, S.; Gil, M.M. Quality Assessment of Chilled and Frozen Fish-Mini Review. *Foods* **2020**, *9*, 1739. [CrossRef]
5. Gokoglu, N. Novel natural food preservatives and applications in seafood preservation: A review. *J. Sci. Food Agric.* **2019**, *99*, 2068–2077. [CrossRef] [PubMed]
6. Li, D.; Qin, N.; Zhang, L.; Li, Q.; Prinyawiwatkul, W.; Luo, Y. Degradation of adenosine triphosphate, water loss and textural changes in frozen common carp (Cyprinus carpio) fillets during storage at different temperatures. *Int. J. Refrig.* **2019**, *98*, 294–301. [CrossRef]
7. Pan, C.; Chen, S.; Hao, S.; Yang, X. Effect of low-temperature preservation on quality changes in Pacific white shrimp, Litopenaeus vannamei: A review. *J. Sci. Food Agric.* **2019**, *99*, 6121–6128. [CrossRef]
8. Shi, C.; Guo, H.; Wu, T.; Tao, N.; Wang, X.; Zhong, J. Effect of three types of thermal processing methods on the lipidomics profile of tilapia fillets by UPLC-Q-Extractive Orbitrap mass spectrometry. *Food Chem.* **2019**, *298*, 125029. [CrossRef]
9. Chang, W.C.; Wu, H.Y.; Yeh, Y.; Liao, P.C. Untargeted foodomics strategy using high-resolution mass spectrometry reveals potential indicators for fish freshness. *Anal. Chim. Acta* **2020**, *1127*, 98–105. [CrossRef]
10. Zhou, J.; Wu, X.; Chen, Z.; You, J.; Xiong, S. Evaluation of freshness in freshwater fish based on near infrared reflectance spectroscopy and chemometrics. *LWT* **2019**, *106*, 145–150. [CrossRef]
11. Wu, L.; Pu, H.; Sun, D.-W. Novel techniques for evaluating freshness quality attributes of fish: A review of recent developments. *Trends Food Sci. Technol.* **2019**, *83*, 259–273. [CrossRef]
12. Chang, L.Y.; Chuang, M.Y.; Zan, H.W.; Meng, H.F.; Lu, C.J.; Yeh, P.H.; Chen, J.N. One-Minute Fish Freshness Evaluation by Testing the Volatile Amine Gas with an Ultrasensitive Porous-Electrode-Capped Organic Gas Sensor System. *ACS Sens.* **2017**, *2*, 531–539. [CrossRef]
13. Weng, X.; Luan, X.; Kong, C.; Chang, Z.; Li, Y.; Zhang, S.; Al-Majeed, S.; Xiao, Y. A Comprehensive Method for Assessing Meat Freshness Using Fusing Electronic Nose, Computer Vision, and Artificial Tactile Technologies. *J. Sens.* **2020**, *2020*, 8838535. [CrossRef]
14. Gao, X.; Liu, J.; Zhuang, X.; Tian, C.; Luan, F.; Liu, H.; Xiong, Y. Incorporating copper nanoclusters into a zeolitic imidazole framework-90 for use as a highly sensitive adenosine triphosphate sensing system to evaluate the freshness of aquatic products. *Sens. Actuators B Chem.* **2020**, *308*, 127720. [CrossRef]
15. Li, P.; Geng, J.; Li, H.; Niu, Z. Fish meal freshness detection by GBDT based on a portable electronic nose system and HS-SPME–GC–MS. *Eur. Food Res. Technol.* **2020**, *246*, 1129–1140. [CrossRef]
16. Mohammadi Lalabadi, H.; Sadeghi, M.; Mireei, S.A. Fish freshness categorization from eyes and gills color features using multi-class artificial neural network and support vector machines. *Aquac. Eng.* **2020**, *90*, 102076. [CrossRef]
17. Gao, L.; Zhuang, J.; Nie, L.; Zhang, J.; Zhang, Y.; Gu, N.; Wang, T.; Feng, J.; Yang, D.; Perrett, S.; et al. Intrinsic peroxidase-like activity of ferromagnetic nanoparticles. *Nat. Nanotechnol.* **2007**, *2*, 577–583. [CrossRef] [PubMed]
18. Song, Y.; Qiao, J.; Liu, W.; Qi, L. Norfloxacin detection based on the peroxidase-like activity enhancement of gold nanoclusters. *Anal. Bioanal. Chem.* **2021**, *413*, 979–985. [CrossRef]
19. Jin, L.; Meng, Z.; Zhang, Y.; Cai, S.; Zhang, Z.; Li, C.; Shang, L.; Shen, Y. Ultrasmall Pt Nanoclusters as Robust Peroxidase Mimics for Colorimetric Detection of Glucose in Human Serum. *ACS Appl. Mater. Interfaces* **2017**, *9*, 10027–10033. [CrossRef]
20. Tang, Y.; Hu, Y.; Yang, Y.; Liu, B.; Wu, Y. A facile colorimetric sensor for ultrasensitive and selective detection of Lead(II) in environmental and biological samples based on intrinsic peroxidase-mimic activity of WS_2 nanosheets. *Anal. Chim. Acta* **2020**, *1106*, 115–125. [CrossRef]
21. Zhang, J.; Wu, S.; Ma, L.; Wu, P.; Liu, J. Graphene oxide as a photocatalytic nuclease mimicking nanozyme for DNA cleavage. *Nano Res.* **2020**, *13*, 455–460. [CrossRef]
22. Lopez-Cantu, D.O.; Gonzalez-Gonzalez, R.B.; Melchor-Martinez, E.M.; Martinez, S.A.H.; Araujo, R.G.; Parra-Arroyo, L.; Sosa-Hernandez, J.E.; Parra-Saldivar, R.; Iqbal, H.M.N. Enzyme-mimicking capacities of carbon-dots nanozymes: Properties, catalytic mechanism, and applications—A review. *Int. J. Biol. Macromol.* **2022**, *194*, 676–687. [CrossRef] [PubMed]

23. Unnikrishnan, B.; Lien, C.W.; Chu, H.W.; Huang, C.C. A review on metal nanozyme-based sensing of heavy metal ions: Challenges and future perspectives. *J. Hazard. Mater.* **2021**, *401*, 123397. [CrossRef] [PubMed]
24. Sharifi, M.; Hosseinali, S.H.; Yousefvand, P.; Salihi, A.; Shekha, M.S.; Aziz, F.M.; JouyaTalaei, A.; Hasan, A.; Falahati, M. Gold nanozyme: Biosensing and therapeutic activities. *Mater. Sci. Eng. C* **2020**, *108*, 110422. [CrossRef] [PubMed]
25. Jiang, D.; Ni, D.; Rosenkrans, Z.T.; Huang, P.; Yan, X.; Cai, W. Nanozyme: New horizons for responsive biomedical applications. *Chem. Soc. Rev.* **2019**, *48*, 3683–3704. [CrossRef] [PubMed]
26. Wang, J.; Huang, R.; Qi, W.; Su, R.; Binks, B.P.; He, Z. Construction of a bioinspired laccase-mimicking nanozyme for the degradation and detection of phenolic pollutants. *Appl. Catal. B Environ.* **2019**, *254*, 452–462. [CrossRef]
27. Yuan, L.; Gan, Z.; Fan, Y.; Ding, F.; Xu, X.; Chen, X.; Zou, X.; Zhang, W. Thermal-controlled active sensor module using enzyme-regulated UiO-66-NH_2/MnO_2 fluorescence probe for total organophosphorus pesticide determination. *J. Hazard. Mater.* **2022**, *436*, 129111. [CrossRef]
28. Yang, Q.; Cui, X.; Qin, Y.; Lei, T.; He, Y.; Song, G. Cu nanoclusters decorated Ti_3C_2 nanosheets composite with tetraenzyme mimic activities and the application for smartphone-assisted detection of hypoxanthine. *Anal. Chim. Acta* **2022**, *1232*, 340494. [CrossRef]
29. Chen, J.; Lu, Y.; Yan, F.; Wu, Y.; Huang, D.; Weng, Z. A fluorescent biosensor based on catalytic activity of platinum nanoparticles for freshness evaluation of aquatic products. *Food Chem.* **2020**, *310*, 125922. [CrossRef]
30. Zhang, Y.; Gao, X.; Ye, Y.; Shen, Y. Fe-Doped polydopamine nanoparticles with peroxidase-mimicking activity for the detection of hypoxanthine related to meat freshness. *Analyst* **2022**, *147*, 956–964. [CrossRef]
31. Gao, Z.; Liu, G.G.; Ye, H.; Rauschendorfer, R.; Tang, D.; Xia, X. Facile Colorimetric Detection of Silver Ions with Picomolar Sensitivity. *Anal. Chem.* **2017**, *89*, 3622–3629. [CrossRef] [PubMed]
32. Zou, J.-J.; Zhang, Y.-p.; Liu, C.-J. Reduction of supported noble-metal ions using glow discharge plasma. *Langmuir* **2006**, *22*, 11388–11394. [CrossRef]
33. Wang, Z.; Xiao, B.; Lin, Z.; Xu, Y.; Lin, Y.; Meng, F.; Zhang, Q.; Gu, L.; Fang, B.; Guo, S. $PtSe_2$/Pt heterointerface with reduced coordination for boosted hydrogen evolution reaction. *Angew. Chem. Int. Ed.* **2021**, *133*, 23576–23581. [CrossRef]
34. Chen, Z.; Lin, Y.; Ma, X.; Guo, L.; Qiu, B.; Chen, G.; Lin, Z. Multicolor biosensor for fish freshness assessment with the naked eye. *Sens. Actuators B Chem.* **2017**, *252*, 201–208.
35. Avan, A.N.; Karakaş, Ö.; Demirci-Çekiç, S.; Apak, R. Enzymatic determination of hypoxanthine in fish samples as a freshness indicator using the CUPRAC colorimetric sensor. *Enzym. Microb. Technol.* **2023**, *162*, 110137.
36. Zhang, Z.; Kwok, R.T.K.; Yu, Y.; Tang, B.Z.; Ng, K.M. Aggregation-induced emission luminogen-based fluorescence detection of hypoxanthine: A probe for biomedical diagnosis of energy metabolism-related conditions. *J. Mater. Chem. B* **2018**, *6*, 4575–4578. [CrossRef]
37. Mustafa, F.; Andreescu, S. Paper-Based Enzyme Biosensor for One-Step Detection of Hypoxanthine in Fresh and Degraded Fish. *ACS Sens.* **2020**, *5*, 4092–4100. [CrossRef]
38. Chen, P.-C.; Li, Y.-C.; Ma, J.-Y.; Huang, J.-Y.; Chen, C.-F.; Chang, H.-T. Size-tunable copper nanocluster aggregates and their application in hydrogen sulfide sensing on paper-based devices. *Sci. Rep.* **2016**, *6*, 1–9. [CrossRef]
39. Albelda, J.A.V.; Uzunoglu, A.; Santos, G.N.C.; Stanciu, L.A. Graphene-titanium dioxide nanocomposite based hypoxanthine sensor for assessment of meat freshness. *Biosens. Bioelectron.* **2017**, *89 Pt 1*, 518–524. [CrossRef]
40. Liao, L.; Xing, Y.; Xiong, X.; Gan, L.; Hu, L.; Zhao, F.; Tong, Y.; Deng, S. An electrochemical biosensor for hypoxanthine detection in vitreous humor: A potential tool for estimating the post-mortem interval in forensic cases. *Microchem. J.* **2020**, *155*, 104760. [CrossRef]

foods

MDPI

Article

Effects of Sous Vide Cooking on the Physicochemical and Volatile Flavor Properties of Half-Shell Scallop (*Chlamys farreri*) during Chilled Storage

Yuexiang Zhan [1,2], Chuanhai Tu [1,2,*], Huili Jiang [1,2], Soottawat Benjakul [3], Jilong Ni [2], Kaixuan Dong [2] and Bin Zhang [1,2,*]

1 Pisa Marine Graduate School, Zhejiang Ocean University, Zhoushan 316022, China
2 Key Laboratory of Health Risk Factors for Seafood of Zhejiang Province, College of Food Science and Pharmacy, Zhejiang Ocean University, Zhoushan 316022, China
3 International Center of Excellence in Seafood Science and Innovation, Faculty of Agro-Industry, Prince of Songkla University, Hat Yai 90110, Thailand
* Correspondence: tu2021@zjou.edu.cn (C.T.); zhangbin@zjou.edu.cn or zhangbin_ouc@163.com (B.Z.); Tel.: +86-0580-255-4781 (B.Z.)

Abstract: This study explored the effects of sous vide (SV) cooking treatments on the physicochemical quality and volatile flavor of half-shell scallop (*Chlamys farreri*) during 30 d of chilled storage. The vacuum-packed scallop samples were cooked at 70 °C (SV-70) and 75 °C (SV-75) and maintained for 30 min. The samples were compared with the positive control (cooked at 100 °C for 10 min, CK). The results indicate that the total volatile basic nitrogen (TVBN), pH, texture, and malondialdehyde (MDA) content gradually increased, while the myofibrillar protein (MP) extraction rate of the CK, SV-70, and SV-75 samples significantly decreased with increasing chilled storage time. Significantly, the SV cooking treatments maintained a much higher water-holding capacity of scallop muscle, compared with the conventional cooking process at 100 °C. Additionally, the SV-75 cooking treatment maintained relatively stable TVBN, pH, and MDA content, springiness, and shearing force properties of scallop samples, especially during 0–20 d of storage. Volatile flavor analysis showed that a total of 42 volatile organic compounds (VOCs) were detected in the scallop samples, and there were no considerable differences in these VOCs between the CK and SV-75 cooked samples (0 d). Overall, the SV cooking treatments effectively maintained acceptable and stable physicochemical and volatile flavor properties of half-shell scallop samples during chilled storage.

Keywords: sous vide; scallop; muscle quality; volatile compounds; chilled storage

Citation: Zhan, Y.; Tu, C.; Jiang, H.; Benjakul, S.; Ni, J.; Dong, K.; Zhang, B. Effects of Sous Vide Cooking on the Physicochemical and Volatile Flavor Properties of Half-Shell Scallop (*Chlamys farreri*) during Chilled Storage. *Foods* **2022**, *11*, 3928. https://doi.org/10.3390/foods11233928

Academic Editors: Edel Oddny Elvevoll, Tao Yin and Liu Shi

Received: 3 October 2022
Accepted: 1 December 2022
Published: 5 December 2022

1. Introduction

Scallop (*Chlamys farreri*) is very popular among consumers because of its relatively large size, fast growth, rich nutrition, and delicious taste. According to a report by the China Fishery Statistics Yearbook [1], the mariculture production of scallops in 2020 reached 1.828 million tons, and scallops were mainly produced in Shandong, Liaoning, and Guangdong Provinces of China. Due to their high moisture and protein content, fresh bivalve molluscs such as mussels and scallops are highly perishable after death, and their rapid deterioration is mainly caused by the work of endogenous enzymes and microorganisms during storage [2,3]. Therefore, scallops are commonly processed in the form of dried, cooked, and/or frozen products [4,5]. Dried scallop products have higher requirements for their storage conditions, and the meat is very rough and hard. Refrigerated scallop products are also susceptible to deterioration via protein denaturation, oxidation, bacterial invasion, and protease degradation during long-term cold storage. The traditional 100 °C heating method is the most common processing method, but it induces significant drip loss of nutrients and water, decreases tenderness, and produces an overripe taste in the scallop products [6]. The shortcomings of this traditional cooking method are well known, and further research is required to improve the cooking process for household consumption and industrial production.

Sous vide (SV), a new type of processing method, is a controlled cooking method of vacuum-packaged food, using water or steam as the heating medium under suitable temperatures and time conditions [7]. It is widely used for cooking foie gras and beef in European countries. In recent years, it has been applied to aquatic products to improve their taste and nutritional value, such as crab lump meat [vii], large yellow croaker [8], and European sea bass [9]. Significantly, this low-temperature heating treatment reduces the loss of nutritional compounds and inactivates enzymes and microorganisms in the products. Moreover, the SV-cooked products are vacuum-packed, which helps to prevent exogenous bacteria contamination, oxidation reactions, and the presence of other contaminants during storage. Zavadlav et al. [10] showed that SV-cooked salmon could be stored at 4 °C for up to 24 d, and the shelf life of SV-cooked cod was extended to 32 d under chilled conditions.

In the case of scallop products, consumers usually prefer to use their shells as containers when the scallop products are processed, cooked, and/or consumed. However, there are few studies on SV-processed bivalve molluscs, especially for half-shell scallop products. There is no large-scale industrial preparation and production of ready to eat half-shell scallops. In this study, half-shell scallops treated with different cooking conditions were evaluated during storage. We aimed to explore the changes in the physicochemical and volatile flavor properties of scallop products. This work provides theoretical support for the implementation of industrial-scale SV preparation of marine bivalve molluscs, such as scallops, mussels, and clams.

2. Materials and Methods

2.1. Chemical Reagents

Boric acid, sodium chloride, absolute ethanol, acetic acid, disodium hydrogen phosphate, trichloroacetic acid (TCA), and sodium dihydrogen phosphate were obtained from Sinopharm Chemical Reagent Co., Ltd. (Suzhou, China). Magnesium oxide was purchased from Aladdin Biochemical Technology Co., Ltd. (Shanghai, China). A malondialdehyde (MDA) assay kit was obtained from the Nanjing Jiancheng Institute of Biological Engineering (Nanjing, China). All chemicals used in this study were of analytical grade.

2.2. Scallop Samples and Treatments

Live scallop (*Chlamys farreri*) samples (an average weight of 65 ± 5 g, an average length of 6.3 ± 0.5 cm, and an average height of 3.0 ± 0.4 cm; 150 individuals) were obtained from a local aquatic product market in Zhoushan (Zhejiang Province, China; harvested in May 2021), and were packed in an insulated box filled with crushed ice and transported to the lab within one hour. All scallop samples were cleaned with tap water to remove the sand and dirt on the surface of the shell. Next, the adductor muscle was removed from one side of the shell using a sterile scalpel, and the entire scallop muscle was contained within a half-shell. The mantle and viscera were manually removed. Subsequently, the obtained half-shell scallops were packaged independently in a heat-resistant polyethylene (PE) bag (12 × 17 cm; obtained from Hebei Wangshi Packaging Co., Ltd., Shijiazhuang, China). The samples were vacuum-packed in a vacuum sealer (DZ-400-2S, Xinqi Machinery Group Co., Ltd. Shanghai, China). The sealed half-shell scallops were randomly divided into three groups, namely an SV-70 group (immersed in a water bath at 70 °C for 30 min; 50 individuals), an SV-75 group (immersed in a water bath at 75 °C for 30 min; 50 individuals), and a positive control group (CK; immersed in a water bath at 100 °C for 10 min; 50 individuals). Next, the cooked samples were rapidly cooled in an ice water bath for 10 min with continuous stirring. Finally, the obtained samples were stored in a refrigerator at 4 °C for 30 d, and the samples were removed and measured every 5 d.

2.3. Total Volatile Basic Nitrogen (TVBN) Content and pH Analysis

TVBN content in the scallop adductor muscle was determined via a method reported by Han et al. [11]. The results were calculated as milligrams N/100 g muscle. The pH value was determined by dipping a pH electrode (E-201F, INESA Scientific Instrument Co., Ltd., Shanghai, China) into the homogenate of the adductor muscle in 0.85% NaCl solution.

2.4. Weight Loss Analysis

After removing the water on the surface of the shell and muscle, the half-shell scallop samples were weighed via a method reported by Ortuño et al. [12]. Weight loss was calculated according to the difference in weight before (M_1) and after the storage (M_2), according to the following formula:

$$\text{Weight loss (\%)} = (M_2 - M_1)/M_1 \times 100. \quad (1)$$

2.5. Texture Analysis

The springiness of the muscle samples was measured using a TMS-Pilot texture analyzer (TMS-Pilot, FTC, VA, USA) coupled with a 50 mm diameter cylindrical probe, according to the method reported by Zhu et al. [13]. The testing was performed in two consecutive cycles at 30% compression, a constant speed of 1 mm/s, and a trigger point of 0.06 N. Each assay of each sample was repeated six times. The shearing force of the muscle samples was determined by using a Warner–Bratzler blade at a constant speed of 1 mm/s. The muscle samples were sheared perpendicularly along the fiber direction [14].

2.6. Malondialdehyde Content Analysis

MDA content in the scallop muscle was measured using an MDA assay kit according to a method described by Zhu et al. [15]. The homogenate (10,000 r/min for 60 s; by using a GM200 tissue grinder (Shanghai Instruments and Equipment Co., Ltd., Shanghai, China)) of muscle samples was combined with trichloroacetic acid (TCA) solution and mixed with thiobarbituric acid (TBA). Next, the obtained mixture was heated in boiling water for 40 min and then cooled with cold water to room temperature. After centrifugation at 4000× g for 10 min, the absorbance of the supernatant was measured at 532 nm using a U-2600 UV-Vis spectrophotometer (Shimadzu (China) Co., Ltd., Shanghai, China). The results were expressed as mg MDA/kg of muscle.

2.7. Myofibrillar Protein (MP) Extraction Rate Analysis

The scallop muscle was mixed with 5 volumes of phosphate buffer solutions (50 mmol/L, pH 7.2) and homogenized for 60 s in a T18 Ultra-Turrax homogenizer (IKA, Baden-Württemberg, Germany). The homogenate (10,000 r/min for 60 s) was centrifuged at 8000× g for 10 min at 4 °C (5424R, Eppendorf, Saxony, Germany). The obtained sediment was collected and extracted again using the same buffer solutions. Next, the pooled precipitate was further homogenized in 10 volumes of ice-cold phosphate buffer solutions (50 mmol/L, containing 0.6 mol/L NaCl, pH 7.2). Then, the mixture was centrifuged at 8000× g for another 10 min at 4 °C. The collected supernatant was recognized as the MP extract [16]. The content of MPs in scallop muscle was measured using the Bradford method. The MP extraction rate was calculated according to the difference in the content of MPs from the fresh sample (C_1) and cooked samples (C_2) with the same mass, with the formula [17]:

$$\text{MP extraction rate (\%)} = C_2/C_1 \times 100. \quad (2)$$

2.8. Volatile Compound Analysis

In this experiment, the fresh scallop (raw), 100 °C-cooked samples stored for 0 d (CK-0), SV-75 samples stored for 0 d (SV-75-0), 15 d (SV-75-15), and 30 d (SV-75-30) (as representatives) were selected to determine the composition of volatile organic compounds in the scallop muscle via gas chromatography (GC) coupled with a high-resolution ion mobility spectrometry (IMS) (FlavourSpec© 1H1-00128, G.A.S, Dortmund, Germany). The scallop muscle (2.00 g) was transferred into a 20 mL headspace bottle and incubated at 75 °C for 15 min. Then, 500 μL headspace gas was injected into an MXT-5 column (15 m × 0.53 mm × 1.0 μm) using nitrogen as the carrier gas [18]. The flow rate of the carrier gas was as follows: 2 mL/min maintained for 2 min, 10 mL/min maintained for 10 min, 100 mL/min maintained for 20 min, and 150 mL/min maintained for 30 min. Next, the separated compounds were ionized in the IMS ionization chamber at 45 °C [19]. The volatile compounds were identified by comparing their mass spectra with those stored in the National Institute of Standards and Technology (NIST) spectral database and the Ion Mobility Spectrometry spectral database. The changes in the peak intensity reflected the changes in the volatile compound content between different scallop samples.

2.9. Data Analysis

Three parallels were performed for each sample, except for the springiness and shearing force determinations (six parallel measurements). Significance analysis was executed using SPSS v26.0 (SPSS Inc., Chicago, IL, USA) at the significance level of $p < 0.05$. Data are displayed as the mean ± standard deviation (SD). Fingerprint and principal component analyses (PCAs) were performed using VOCal software (version 0.1.1, G.A.S., Dortmund, Germany).

3. Results and Discussion

3.1. TVBN and pH Analysis

TVBN content is an important indicator to characterize the degree of protein degradation, and it is also used to evaluate the freshness of muscle products [20]. As shown in Figure 1A, the TVBN content of three groups of scallop samples increased with extended storage time. The initial TVBN contents of the SV-70 and SV-75 samples were comparatively lower than that of the positive control (CK) samples. Generally, the rapid increases in the TVBN content of scallop muscle were attributed to the growth of spoilage bacteria and associated metabolites, as well as the actions of endogenous enzymes during processing and storage. During initial storage, the retarded TVBN values of SV-70 and SV-75 samples were mainly due to oxygen deficiency and heating treatments, while the TVBN values increased remarkably during the following 5–15 d of storage, suggesting that the ammonia and amine nitrogenous substances significantly accumulated in muscle tissues, resulting from the decarboxylation and deamination of amino acids, induced by microorganisms, metabolic enzymes, and protein/lipid oxidation. Although the TVBN values of the SV-70 and SV-75 samples on day 30 increased to 7.60 and 8.92 mg/100 g of muscle, respectively, they were much lower than the acceptable limit (≤30 mg/100 g), as defined by the European Commission. These results are also consistent with the previous report for European seabass (*Dicentrarchus labrax*) fillets and Atlantic salmon (*Salmo salar*) slices by Kritikos et al. [21]. Herein, the SV cooking method inhibited the degradation of muscle proteins induced by microorganisms and endogenous/exogenous enzymes to a certain extent. Moreover, the method prolonged the shelf life of scallop products, although the effect was significantly lower than that observed in the CK samples, especially during prolonged storage periods [vii]. SV will reduce the number of bacteria as it is a heat treatment. Pino-Hernández et al. [22] indicated that SV cooking (60 °C) significantly reduced the number of bacteria in pirarucu (*Arapaima gigas*) fillets. Humaid et al. [23] also illustrated that SV cooking (65 °C) extended the shelf life of lobster (*Homarus americanus*) tails mainly by inhibiting microbial activity. Additionally, these results of TVBN content and their variations are similar to those of the current study. The microbiological test of the half-shell scallops during storage is still in progress using high-throughput sequencing procedures, including a 16S rRNA analysis and an 18S rDNA analysis.

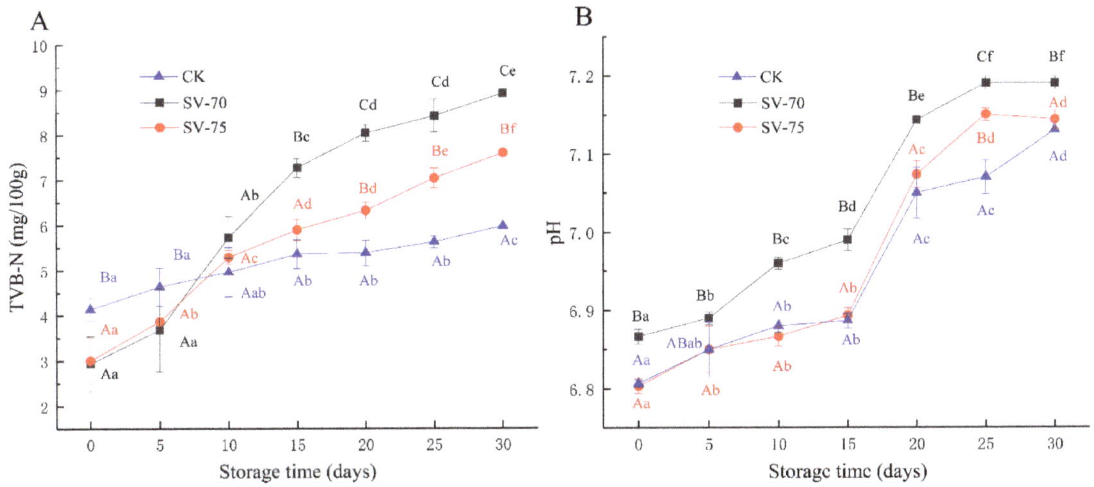

Figure 1. Changes in TVBN content (**A**) and pH value (**B**) of scallop samples with different treatments during 30 d of chilled storage. Different lowercase letters in the same color for the same point plot indicate significant difference ($p < 0.05$), and different uppercase letters in the same storage time indicate significant difference ($p < 0.05$).

As illustrated in Figure 1B, the changes in pH values were similar to those of the TVBN content of scallop samples during 30 d of chilled storage. The results indicate that the pH values of SV-cooked scallops were relatively stable during the initial 0–15 d of storage, while significant increases were observed during the following 15–25 d ($p < 0.05$). These changes were attributed to the increased abundance of spoilage bacteria and the accumulation of alkaline substances in muscle tissues [vii]. The pH values of the SV-75 samples were similar to that of CK, but significantly ($p < 0.05$) lower than those of the SV-70 samples on day 30; it was suggested that SV cooking at 75 °C likely hindered the growth and reproduction of bacteria, as well as limited the activity of endogenous/exogenous enzymes during chilled storage [24].

3.2. MP Extraction Rate Analysis

The extraction rate of MPs can be used to evaluate the denaturation degree of muscle samples after thermal processing. Muscle tissues were completely denatured when the extraction rate of MPs was less than 10% [17]. As depicted in Figure 2, the MP extraction rate of CK, SV-70, and SV-75 samples were all less than 10% after the cooking treatments, indicating that the mild SV conditions were acceptable to denature the MPs in the scallop samples. Moreover, the MP extraction rate (MP denaturation) of the SV-70 and SV-75 samples was significantly ($p < 0.05$) higher than that of the CK samples during 0–30 d of storage, and this was mainly due to the application of low temperatures. It was suggested that mild SV cooking treatment comparatively maintained the MP structure and function to a certain extent and reduced the degree of MP denaturation, which was beneficial for the preservation of the texture and water-holding capacity of cooked muscle tissues during storage. The lateral shrinkage of muscle fibers commonly occurs between 45 °C and 65 °C in response to myosin denaturation, while longitudinal shrinkage is mainly associated with actin denaturation, which occurs between 70 °C and 75 °C [25]. In the current study, the SV-cooked samples at 70 °C and 75 °C showed different MP extraction rates, which were also in agreement with the different denaturation temperatures of the MPs in the muscle. The MP extraction rate of the three sample groups decreased continuously with prolonged storage, and this could be attributed to the inhibition of aerobic bacteria and endogenous enzymes during chilled storage [26]. In addition, the SV-70 samples showed a higher decrease in the MP extraction rate than that of SV-75 samples, indicating that SV-70 cooking might be less effective in limiting the activation of spoilage bacteria and endogenous enzymes, as compared to SV-75 cooking treatments.

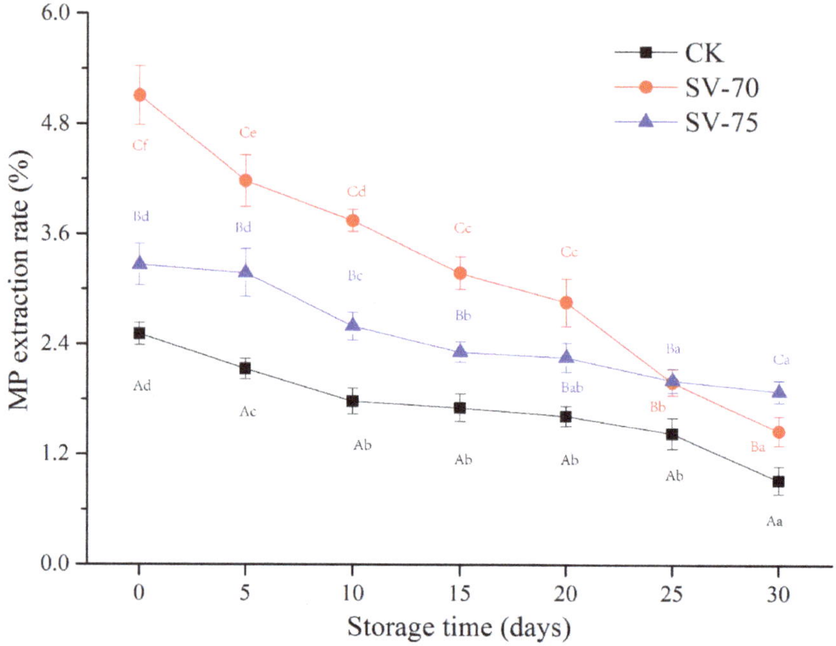

Figure 2. Changes in the MP extraction rate of scallop muscle with different treatments during 30 d of chilled storage. Different lowercase letters in the same color for the same point plot indicate significant difference ($p < 0.05$), and different uppercase letters in the same storage time indicate significant difference ($p < 0.05$).

3.3. Weight Loss Analysis

As shown in Figure 3, the CK group had the highest weight loss (cooking loss) during the storage period, which was mainly induced by high temperature during the boiling process. High-temperature stress caused serious denaturation and destruction of myofibrillar proteins in muscle tissues, resulting in shrinkage of muscle fibers and collagen and subsequent extrusion of water molecules from intracellular and intercellular space [27]. For the 100 °C-cooked samples, a large amount of drip loss containing some water-soluble nutrients was a clear shortcoming, which greatly affected the sensory and nutritional values of scallop muscle products. By comparison, the weight loss of SV-70 and SV-75 samples was always lower than that of the CK samples over 30 d of storage. This suggests that the SV cooking process effectively reduced the drip loss of scallop muscle and maintained a higher water-holding capacity of scallop muscle tissues during chilled storage [28]. During 10–30 d of storage, the weight loss of SV-70 samples was comparatively higher than that of the SV-75 samples, indicating a rapid decrease in quality occurred in the SV-70 samples, and this was most likely due to the growth and reproduction of spoilage bacteria over time. It was also suggested that the heating treatment at 70 °C might not be as sufficient as the 75 °C treatment to maintain the scallop muscle quality and limit the activity of bacteria during 30 days of storage. These findings are consistent with the results of previous TVBN and pH measurements, which also indicate the 75 °C treatment is better for maintaining the scallop muscle quality than the 70 °C treatment. Importantly, the SV-75 treatments showed positive effects on the water-holding capacity (juiciness characteristic) of scallop muscle tissues during chilled storage.

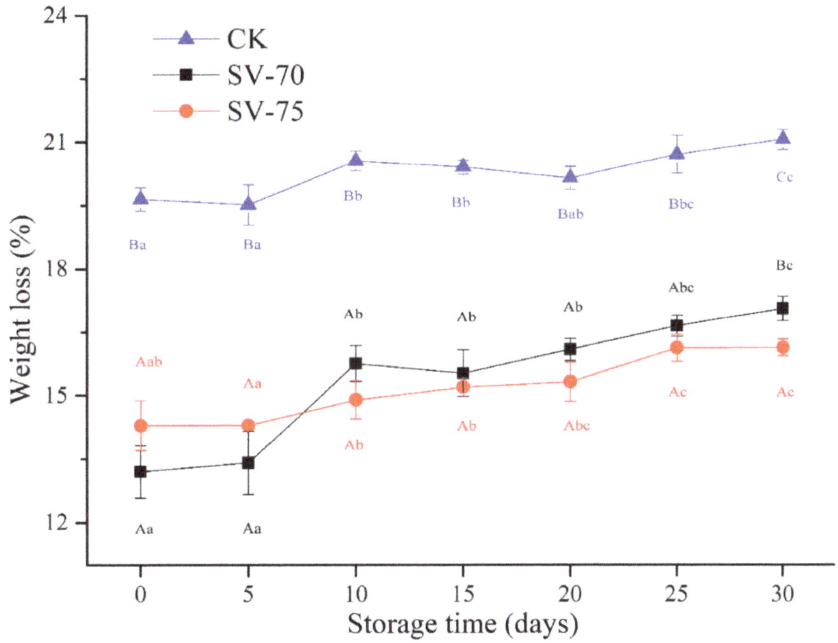

Figure 3. Changes in weight loss of scallop muscle with different treatments during 30 d of chilled storage. Different lowercase letters in the same color for the same point plot indicate significant difference ($p < 0.05$), and different uppercase letters in the same storage time indicate significant difference ($p < 0.05$).

3.4. Springiness and Shearing Force Analysis

Springiness indicates the ability of muscle tissue to recover its deformation within a certain period. As shown in Table 1, the springiness of each of the three groups of scallop samples showed a decreasing trend during the storage time. This may be due to the structural alterations of the muscle fibers, as well as its connective tissues, leading to a decline in springiness. Previously, Zhao et al. [29] illustrated that the degradation of myofibrillar proteins in large yellow croaker (*Pseudosciaena crocea*) negatively affected the springiness properties of muscle tissues during storage. Significantly, the springiness values of the SV-70 and SV-75 samples were comparatively lower than those of the CK samples during storage, which were associated with the denaturation levels in different cooked scallop samples. The CK and SV-75 treatments caused thermal denaturation (e.g., cross-linking, uncoiling, and aggregation) of muscle proteins (including collagen, myofibrillar, and sarcoplasmic proteins) and shrinkage of connective tissues, which resulted in a comparative delicate texture [6]. There was no significant difference ($p > 0.05$) between the springiness of the CK and SV-75 samples, both of which were significantly better than the SV-70 samples ($p < 0.05$) during the entire storage period. Continuous deterioration of the springiness metric occurred in the SV-70 samples over 30 d of storage, which might be due to the resuscitation of anaerobic bacteria and the intrinsic biological factors involved in endogenous enzymes (e.g., calpain and cathepsin) [28].

Table 1. Changes in springiness and shearing force of scallop muscle with different treatments during 30 d of chilling storage.

Texture Property	Group	Chilled Storage Period 0 d	5 d	10 d	15 d	20 d	25 d	30 d
Springiness (mm)	CK	^{A}1.36 ± 0.02 ^{a}	^{A}1.31 ± 0.03 ^{ab}	^{A}1.28 ± 0.02 ^{b}	^{A}1.26 ± 0.02 ^{bc}	^{A}1.21 ± 0.03 ^{cd}	^{A}1.18 ± 0.01 ^{de}	^{A}1.13 ± 0.01 ^{e}
	SV-70	^{B}1.19 ±0.04 ^{a}	^{B}1.14 ± 0.04 ^{ab}	^{B}1.11± 0.02 ^{bc}	^{B}1.09 ± 0.02 ^{bc}	^{B}1.04 ± 0.02 ^{cd}	^{B}1.01 ± 0.02 ^{de}	^{B}0.96 ± 0.02 ^{e}
	SV-75	^{A}1.29 ± 0.05 ^{Ba}	^{A}1.25 ± 0.01 ^{ab}	^{A}1.23 ± 0.03 ^{abc}	^{A}1.21 ± 0.03 ^{bc}	^{A}1.17 ± 0.01 ^{cd}	^{A}1.15 ± 0.01 ^{de}	^{A}1.10 ± 0.03 ^{e}
Shearing force (N)	CK	^{A}7.79 ± 0.25 ^{e}	^{A}9.23 ± 0.26 ^{d}	^{A}9.80 ± 0.23 ^{c}	^{A}10.04 ± 0.17 ^{bc}	^{A}10.39 ± 0.10 ^{b}	^{A}11.99 ± 0.07 ^{a}	^{A}9.31 ± 0.05 ^{d}
	SV-70	^{B}3.06 ± 0.17 ^{e}	^{B}3.56 ± 0.14 ^{cd}	^{B}3.77 ± 0.14 ^{b}	^{B}4.33 ± 0.28 ^{a}	^{C}3.57 ± 0.16 ^{c}	^{C}3.40 ± 0.23 ^{cde}	^{C}3.11 ±0.14 ^{de}
	SV-75	^{B}3.29 ± 0.33 ^{d}	^{B}3.64 ± 0.24 ^{cd}	^{B}3.89 ± 0.12 ^{bc}	^{B}4.30 ± 0.05 ^{ab}	^{B}4.49 ± 0.10 ^{a}	^{B}4.76 ± 0.29 ^{a}	^{B}3.62 ±0.12 ^{cd}

Different uppercase letters in the same column for the same parameter indicate significant difference ($p < 0.05$), and different lowercase letters in the same row indicate significant difference ($p < 0.05$).

Shearing force is related to the tenderness and juiciness of muscle tissues, and it directly affects the consumer acceptance of meat products [30]. During chilled storage, the shearing force of three groups of scallop samples increased first and then decreased, and this trend was closely related to the denaturation of myofibrillar proteins, the actions of endogenous enzymes, and/or the microbial activity and its metabolites [31]. For CK samples, treatment at 100 °C induced a large amount of weight loss (drip loss) and a decrease in the water-holding capacity of scallop muscle tissues. This led to an increase in shearing force (tougher texture) in the cooked muscle samples. By comparison, the SV-70 and SV-75 samples exhibited significantly improved shearing force (tenderness). The SV treatments with considerably milder temperatures limited the heat damage to the muscle proteins and connective tissues, as well as reduced the liquid loss and water-soluble nutrients. Thus, the textural properties were improved, compared with the conventional 100 °C cooking process [32]. Additionally, SV heating provided a uniform and efficient transfer of heat from outside to inside the muscle in a water bath and ensured the preservation of volatile flavor compounds under vacuum packaging conditions.

3.5. MDA Content Analysis

As a secondary product of lipid oxidation, the content of MDA is commonly used to reflect the development and degree of lipid oxidation in muscle products. The variations in MDA content in the scallop muscle samples during chilled storage are shown in Figure 4. During the initial 0–15 d, the MDA content of all samples was relatively stable. The MDA content significantly accelerated during the following 20–30 d of storage, which suggests considerable lipid oxidation in muscle tissues during storage. The heating treatments induced the dissociation of heme proteins and promoted the disruption of cell membranes and the liberation of free iron. These changes advanced lipid oxidation in muscle tissues during chilled storage [33]. As heating temperature and time increased, the development of lipid oxidation in the muscle samples also increased, which in turn caused undesirable changes in the texture, water-holding capacity, and nutritive values [34]. In the current study, the mild SV treatments effectively hindered the lipid oxidation reaction under the limited oxygen conditions during storage, compared with the CK samples. Similar observations were also found in largemouth bass (*Micropterus salmoides*) [35] and Segureño lamb meat [12], in which samples were pretreated with the SV cooking treatments. Díaz et al. [36] determined that the MDA value of SV-processed salmon was maintained at 2.30 mg MDA/kg after 10 weeks of frozen storage, which was not sufficient to detect rancidity in the salmon products. Bongiorno et al. [37] indicated that SV cooking at 85 °C for 10 min with salt brine resulted in being able to maintain the quality of mussels and extend their shelf life to 21 days under chilled conditions, while the mussels cooked traditionally (at 90 °C for 10 min) showed a shelf life of about 14 days. These findings suggest that the SV cooking method significantly improved the oxidative stability of lipid species, compared to the traditional cooking method.

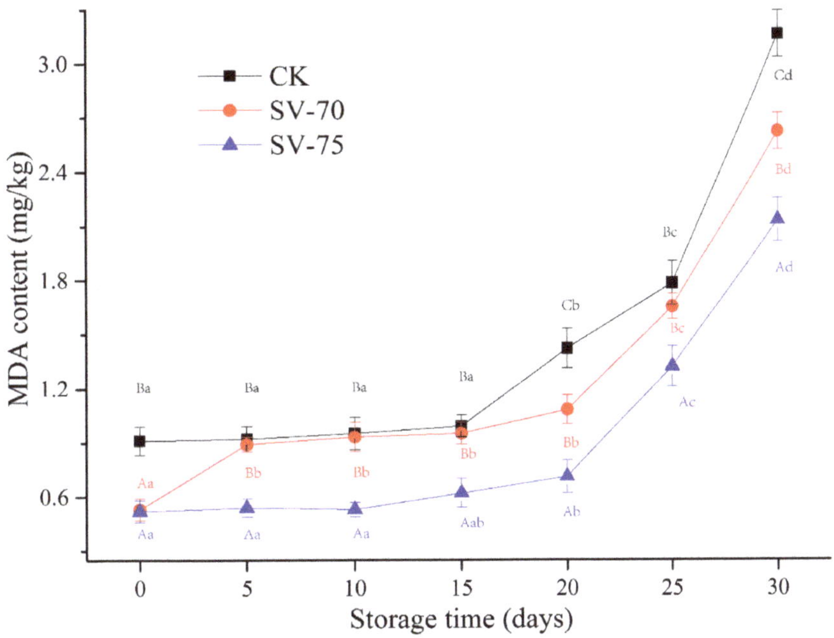

Figure 4. Changes in MDA content of scallop muscle with different treatments during 30 d of chilled storage. Different lowercase letters in the same color for the same point plot indicate significant difference ($p < 0.05$), and different uppercase letters in the same storage time indicate significant difference ($p < 0.05$).

3.6. Volatile Organic Compound (VOC) Analysis

The changes in volatile organic compounds (VOCs) were investigated in the SV-75 samples (as representatives; SV-75-treated samples presented better stability than the other samples) during chilled storage (Figure 5). The GC-IMS spectrum of the fresh scallop (raw) was used as a reference, and it was deducted from the other GC-IMS spectra of the scallop samples. In the plot (Figure 5A), the white color represents the same signal strength of the VOCs; the blue color represents the lower signal strength of the VOCs; and the red color represents the higher signal strength of the VOCs in the samples, compared with the fresh scallop (raw). The results show considerable differences in the VOCs between the fresh and cooked scallop samples, with no apparent differences between the CK-0 and SV-75-0 samples. This indicates that the SV cooking method effectively promoted the formation of the VOCs (flavor compounds) in muscle tissues, which was similar to the 100 °C thermal processing treatments. In comparison with the SV-75-0 samples, the composition and content of the VOCs were greatly altered in the scallop samples (SV-75-30) after 30 d of chilled storage, and this was likely due to the growth and reproduction of residual bacteria and the hydrolysis of endogenous enzymes in muscle tissues during long-term storage [38].

The fingerprinting plot (Figure 5B) of the VOCs was generated to visualize the differences between the different scallop samples. In this study, a total of 42 VOCs were determined in the scallop samples, including 13 aldehydes, 6 alcohols, 3 ketones, 2 acetates, 1 amine, and 17 unidentified compounds (Table 2). Herein, aldehydes mainly resulted from the oxidation of unsaturated fatty acids (such as valeraldehyde, heptyl aldehyde, and caprylic aldehyde), triglycerides, and the degradation of several amino acids [39]. Since the sensory threshold of aldehyde compounds is relatively low, they readily react with other substances to advance the characteristic meat flavors [40]. Among the alcohol components, saturated alcohols have high sensory thresholds and provide small contributions to the overall flavor, while unsaturated alcohols have lower sensory thresholds and present mushroom and metal-like aromas. Their contribution to the overall flavor of meat products is significant [41].

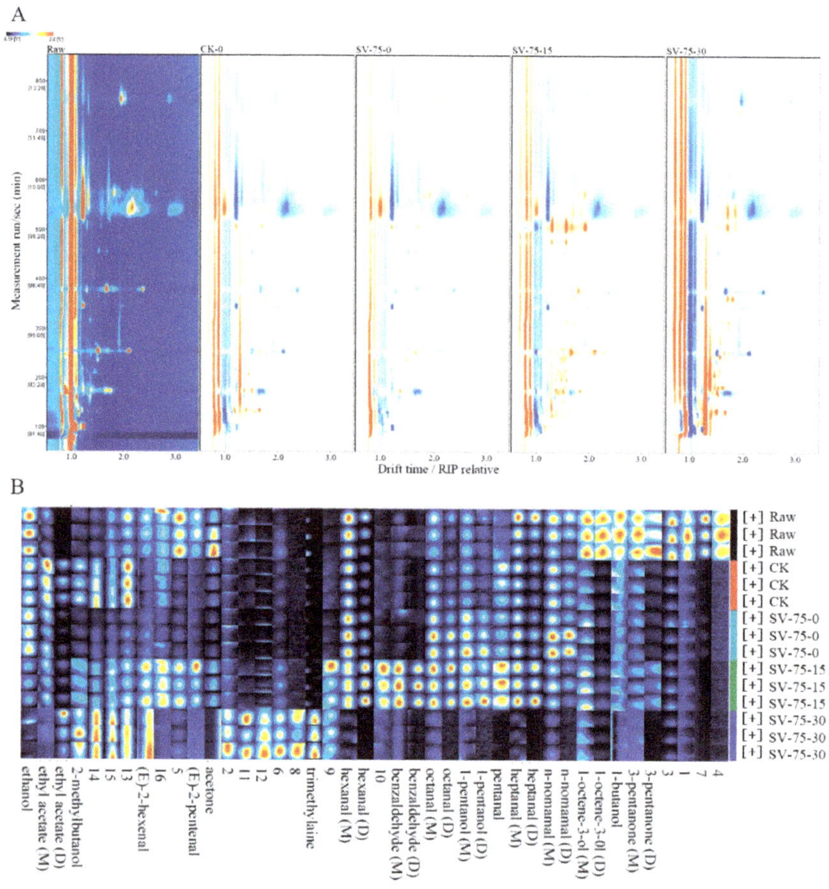

Figure 5. Two-dimensional spectrum (**A**) and fingerprinting plot (**B**) of volatile organic compounds in scallop muscle. Raw, fresh scallops; CK, scallop samples cooked at 100 °C for 10 min and stored for 0 d; SV-75-0, SV-75-15, and SV-75-30, scallop samples cooked at 75 °C for 30 min and stored for 0, 15, and 30 d, respectively.

Compared with fresh samples, the contents of 3-pentanone, 1-butanol, and 1-octene-3-ol significantly decreased in the cooked scallop samples (CK-0 and SV-75-0), due to the thermal heating treatments. Compared with the CK-0 samples, the contents of ethyl acetate, 2-methylbutanol, 2-pentenal, and acetone were comparatively lower than in the SV-75-0 samples, which suggests the fruity aromas, fishy smell, and fat aroma were reduced in the SV-cooked meat samples. In addition, the intensity of nonanal and ethyl acetate in the SV-cooked scallop samples decreased with a prolonged storage period, and the contents of trimethylamine, 2-hexenal, and 2-methyl butanol increased during 30 d of storage. These results indicate that the fruity, nutty, honey, and citrus aromas were weakened, while the fishy and rancid odors developed during storage, and this was due to the deterioration in muscle tissues caused by the specific spoilage bacteria and several endogenous cathepsins [42]. Overall, the changes in VOCs in the SV-75 scallop samples were relatively small during

0 to 15 d of storage, which indicated that the quality of these samples was comparatively acceptable and stable during this time frame. These results were in agreement with the above TVBN, pH, and MDA results.

Table 2. The volatile organic compounds (except for the unidentified compounds) in scallop muscle with different treatments.

Compounds	Molecular Formula	Peak Intensity				
		Raw	CK	SV-75-0	SV-75-15	SV-75-30
n-Nonanal (M)	$C_9H_{18}O$	1356 ± 143 [b]	1344 ± 70 [b]	1661 ± 280 [a]	1146 ± 91 [b]	581 ± 65 [c]
n-Nonanal (D)	$C_9H_{18}O$	232 ± 68 [ab]	205 ± 18 [bc]	341 ± 54 [a]	173 ± 16 [bc]	69 ± 3 [d]
Octanal (M)	$C_8H_{16}O$	868 ± 74 [b]	985 ± 41 [b]	1341 ± 140 [a]	1557 ± 76 [a]	543 ± 14 [c]
Octanal (D)	$C_8H_{16}O$	171 ± 23 [c]	166 ± 20 [c]	302 ± 51 [b]	409 ± 43 [a]	70 ± 6 [d]
Benzaldehyde (M)	C_7H_6O	449 ± 67 [bc]	468 ± 32 [bc]	367 ± 31 [c]	2008 ± 53 [a]	570 ± 41 [b]
Benzaldehyde (D)	C_7H_6O	341 ± 40 [b]	170 ± 15 [c]	210 ± 41 [bc]	1399 ± 107 [a]	195 ± 13 [bc]
Heptanal (M)	$C_7H_{14}O$	1176 ± 129 [b]	902 ± 55 [b]	991 ± 91 [b]	1284 ± 91 [a]	372 ± 19 [c]
Heptanal (D)	$C_7H_{14}O$	395 ± 43 [b]	210 ± 26 [c]	252 ± 50 [c]	552 ± 84 [a]	38 ± 2 [d]
Hexanal (M)	$C_6H_{12}O$	1457 ± 64 [a]	1082 ± 57 [b]	1431 ± 59 [a]	1632 ± 25 [a]	364 ± 28 [c]
Hexanal (D)	$C_6H_{12}O$	897 ± 40 [b]	492 ± 58 [c]	866 ± 70 [b]	2268 ± 139 [a]	92 ± 21 [d]
2-Pentenal (E)	C_5H_8O	197 ± 19 [a]	48 ± 6 [b]	77 ± 1 [b]	217 ± 26 [a]	37 ± 3 [b]
Pentanal	$C_5H_{10}O$	96 ± 13 [c]	187 ± 26 [b]	150 ± 10 [b]	270 ± 10 [a]	22 ± 2 [d]
2-Hexenal (E)	$C_6H_{10}O$	90 ± 11 [c]	60 ± 5 [cd]	58 ± 1 [d]	139 ± 14 [b]	192 ± 22 [a]
1-Pentanol (M)	$C_5H_{12}O$	380 ± 23 [b]	335 ± 20 [b]	575 ± 49 [a]	603 ± 51 [a]	188 ± 9 [c]
1-Pentanol (D)	$C_5H_{12}O$	48 ± 6 [b]	37 ± 5 [bc]	86 ± 6 [a]	103 ± 11 [a]	23 ± 1 [c]
Ethanol	C_2H_6O	885 ± 94 [a]	682 ± 48 [b]	797 ± 22 [ab]	421 ± 15 [c]	223 ± 10 [d]
1-Butanol	$C_4H_{10}O$	266 ± 21 [a]	141 ± 10 [b]	141 ± 9 [b]	131 ± 5 [bc]	103 ± 18 [c]
1-Octene-3-ol (M)	$C_8H_{16}O$	1877 ± 46 [a]	1002 ± 45 [c]	1013 ± 98 [c]	1466 ± 50 [b]	1092 ± 57 [c]
1-Octene-3-ol (D)	$C_8H_{16}O$	2233 ± 118 [a]	297 ± 36 [c]	310 ± 63 [bc]	467 ± 56 [b]	287 ± 24 [c]
2-Methylbutanol	$C_5H_{12}O$	124 ± 23 [c]	338 ± 16 [b]	122 ± 14 [c]	359 ± 24 [b]	479 ± 32 [a]
3-Pentanone (M)	$C_5H_{10}O$	336 ± 24 [a]	150 ± 7 [c]	180 ± 5 [b]	174 ± 2 [bc]	60 ± 7 [d]
3-Pentanone (D)	$C_5H_{10}O$	418 ± 82 [a]	69 ± 7 [c]	61 ± 5 [c]	174 ± 18 [b]	25 ± 5 [d]
Acetone	C_3H_6O	3713 ± 68 [a]	1383 ± 90 [c]	720 ± 44 [d]	2271 ± 92 [b]	2182 ±44 [b]
Ethyl acetate (M)	$C_4H_8O_2$	246 ± 52 [c]	692 ± 72 [a]	396 ± 54 [b]	420 ± 30 [b]	118 ± 17 [d]
Ethyl acetate (D)	$C_4H_8O_2$	41 ± 7 [e]	354 ± 27 [b]	87 ± 19 [d]	157 ± 25 [c]	577 ± 60 [a]

Different letters in the same row indicate significant differences at $p < 0.05$.

4. Conclusions

The physicochemical quality and volatile flavor profile of scallop muscle cooked at 75 °C for 30 min were investigated during 30 d of chilled storage. The results show that the physicochemical (TVBN content, pH, MDA content, and texture) and VOC contents of scallop samples treated with traditional heating at 100 °C for 10 min (CK), SV cooking at 70 °C for 30 min (SV-70), and SV cooking at 75 °C for 30 min (SV-75) gradually deteriorated with a prolonged storage period. Importantly, the SV cooking treatments significantly improved the water-holding capacity of scallop samples, as indicated by less weight loss, and maintained a higher MP extraction rate, compared with the CK samples. In addition, there were no significant differences between the VOC contents of SV-75 and CK samples after treatment. Here, the SV-75 samples effectively preserved the quality of half-shell scallops and extended their shelf life to 15–20 d of chilled storage, compared with the SV-70 samples. Scallop products processed by SV cooking and marketed under chilled conditions may be a promising choice for "ready to cook" or "ready to eat" seafood, with high nutritional value and flavor.

Author Contributions: Conceptualization, B.Z., C.T., and S.B.; writing—original draft preparation, Y.Z. and H.J.; writing—review and editing, B.Z., C.T., J.N., K.D., and S.B.; project administration, C.T.; funding acquisition, B.Z. All authors have read and agreed to the published version of the manuscript.

Funding: This research was funded by the National key research and development program of China, grant number 2021YFD2100504; the Zhejiang Leading Training Program, grant number 2020R52027; the Fundamental Research Funds for Zhejiang Province, grant number 2021JD005; and the College Students' Science and Technology Innovation Program (new seedling talent Program) Project of Zhejiang Province, grant number 2021R411032.

Data Availability Statement: The data presented in this study are available on request from the corresponding author.

Conflicts of Interest: The authors declare that they have no known competing financial interests or personal relationships that could have appeared to influence the work reported in this paper.

References

1. Liu, X.Z.; Cui, L.F.; Li, S.M.; Han, X.; Jiang, K.Y.; Yuan, X.C.; Yu, X.J.; Wang, D.; Wu, F.X.; Song, D.D.; et al. *2021 China Fishery Statistical Yearbook*; China Agriculture Press: Beijing, China, 2022.
2. Parlapani, F.; Syropoulou, F.; Tsiartsafis, A.; Ekonomou, S.; Madesis, P.; Exadactylos, A.; Boziaris, I. HRM analysis as a tool to facilitate identification of bacteria from mussels during storage at 4 °C. *Food Microbiol.* **2020**, *85*, 103304. [CrossRef]
3. Wei, H.; Tian, Y.; Lin, Y.; Maeda, H.; Yamashita, T.; Yu, K.; Takaki, K.; Yuan, C. Condition-dependent adenosine monophosphate decomposition pathways in striated adductor muscle from Japanese scallop (*Patinopecten yessoensis*). *J. Food Sci.* **2020**, *85*, 1462–1469. [CrossRef]
4. Zhu, Z.; Zhao, Y.; Zhang, Y.; Wu, X.; Liu, J.; Shi, Q.; Fang, Z. Effects of ultrasound pretreatment on the drying kinetics, water status and distribution in scallop adductors during heat pump drying. *J. Sci. Food Agric.* **2021**, *101*, 6239–6247. [CrossRef]
5. Liu, B.; Liu, Z.-Q.; Li, D.Y.; Yu, M.-M.; Liu, Y.-X.; Qin, L.; Zhou, D.-Y.; Shahidi, F.; Zhu, B.-W. Action of endogenous proteases on texture deterioration of the bay scallop (*Argopecten irradians*) adductor muscle during cold storage and its mechanism. *Food Chem.* **2020**, *323*, 126790. [CrossRef] [PubMed]
6. Kathuria, D.; Dhiman, A.K.; Attri, S. Sous vide, a culinary technique for improving quality of food products: A review. *Trends Food Sci. Technol.* **2022**, *119*, 57–68. [CrossRef]
7. Olatunde, O.O.; Benjakul, S. Sous-vide cooking as a systematic approach for quality maintenance and shelf-life extension of crab lump meat. *LWT-Food Sci. Technol.* **2021**, *142*, 111004. [CrossRef]
8. Qiu, X.; Wu, Y. Application of Taguchi method to improve the sous vide processed large yellow croaker (*Larimichthys crocea*) fillet product quality during cold storage. *J. Food Process. Preserv.* **2021**, *45*, e15565. [CrossRef]
9. Nieva-Echevarría, B.; Manzanos, M.J.; Goicoechea, E.; Guillén, M.D. Changes provoked by boiling, steaming and sous-vide cooking in the lipid and volatile profile of European sea bass. *Food Res. Int.* **2017**, *99*, 630–640. [CrossRef]
10. Zavadlav, S.; Blažić, M.; Van de Velde, F.; Vignatti, C.; Fenoglio, C.; Piagentini, A.M.; Pirovani, M.E.; Perotti, C.M.; Bursać Kovačević, D.; Putnik, P. Sous-vide as a technique for preparing healthy and high-quality vegetable and seafood products. *Foods* **2021**, *9*, 1537. [CrossRef] [PubMed]
11. Han, J.; Liu, Y.; Zhu, L.; Liang, R.; Dong, P.; Niu, L.; Hopkins, D.L.; Luo, X.; Zhang, Y. Effects of spraying lactic acid and peroxyacetic acid on the quality and microbial community dynamics of vacuum skin-packaged chilled beef during storage. *Food Res. Int.* **2021**, *142*, 110205. [CrossRef] [PubMed]
12. Ortuño, J.; Mateo, L.; Rodríguez-Estrada, M.T.; Bañón, S. Effects of sous vide vs grilling methods on lamb meat colour and lipid stability during cooking and heated display. *Meat Sci.* **2021**, *171*, 108287. [CrossRef] [PubMed]
13. Zhu, Y.; Yan, Y.; Yu, Z.; Wu, T.; Bennett, L.E. Effects of high pressure processing on microbial, textural and sensory properties of low-salt emulsified beef sausage. *Food Control* **2022**, *133*, 108596. [CrossRef]
14. Chen, L.; Jiao, D.; Liu, H.; Zhu, C.; Sun, Y.; Wu, J.; Zheng, M.; Zhang, D. Effects of water distribution and protein degradation on the texture of high pressure-treated shrimp (*Penaeus monodon*) during chilled storage. *Food Control* **2022**, *132*, 108555. [CrossRef]
15. Zhu, W.; Zhang, C.; Tan, K.; Wang, B.; Huang, R.; Wen, J.; Xu, B.; Liu, X.; Lichu, L.; Zheng, H. Variation of lipids and fatty acids in noble scallop Chlamys nobilis under low temperature stress. *Aquaculture* **2022**, *554*, 738121. [CrossRef]
16. Li, D.Y.; Tan, Z.F.; Liu, Z.Q.; Wu, C.; Liu, H.L.; Guo, C.; Zhou, D.Y. Effect of hydroxyl radical induced oxidation on the physicochemical and gelling properties of shrimp myofibrillar protein and its mechanism. *Food Chem.* **2021**, *351*, 129344. [CrossRef] [PubMed]
17. Wu, Z.X.; Li, D.Y.; Shen, M.; Wang, Z.Y.; Wang, Z.W.; Liu, Y.X.; Bai, Y.H.; Zhou, D.Y. Effect of different sous-vide cooking conditions on textural properties, protein physiochemical properties and microstructure of scallop (*Argopecten irradians*) adductor muscle. *Food Chem.* **2022**, *394*, 133470. [CrossRef]
18. Li, Y.; Jia, S.; Hong, H.; Zhang, L.; Zhuang, S.; Sun, X.; Liu, X.; Luo, Y. Assessment of bacterial contributions to the biochemical changes of chill-stored blunt snout bream (*Megalobrama amblycephala*) fillets: Protein degradation and volatile organic compounds accumulation. *Food Microbiol.* **2020**, *91*, 103495. [CrossRef]
19. Xie, Q.; Xu, B.; Xu, Y.; Yao, Z.; Zhu, B.; Li, X.; Sun, Y. Effects of different thermal treatment temperatures on volatile flavour compounds of water-boiled salted duck after packaging. *LWT-Food Sci. Technol.* **2022**, *154*, 112625. [CrossRef]
20. Qiao, L.; Tang, X.; Dong, J. A feasibility quantification study of total volatile basic nitrogen (TVB-N) content in duck meat for freshness evaluation. *Food Chem.* **2017**, *237*, 1179–1185. [CrossRef]

21. Kritikos, A.; Aska, I.; Ekonomou, S.; Mallouchos, A.; Parlapani, F.F.; Haroutounian, S.A.; Boziaris, I.S. Volatilome of chill-stored European seabass (*Dicentrarchus labrax*) fillets and Atlantic salmon (*Salmo salar*) slices under modified atmosphere packaging. *Molecules* **2020**, *25*, 1981. [CrossRef]
22. Pino−Hernández, E.; da Costa, W.A.; Araujo, E.A.F.; Villa, P.M.; Lourenço, L.D.F.H.; Junior, R.D.C. Influence of grilling pretreatment and optimization of sous vide processing parameters on the physicochemical and microbiological quality of pirarucu fillet. *Food Sci. Technol. Int.* **2021**, *27*, 84–96. [CrossRef]
23. Humaid, S.; Nayyar, D.; Bolton, J.; Perkins, B.; Skonberg, D.I. Refrigerated shelf-life evaluation of high pressure processed, raw and sous vide cooked lobster. *High Press. Res.* **2020**, *40*, 444–463. [CrossRef]
24. Yang, X.; Wang, H.; Badoni, M.; Zawadski, S.; McLeod, B.; Holman, D.; Uttaro, B. Effects of a novel three-step sous-vide cooking and subsequent chilled storage on the microbiota of beef steaks. *Meat Sci.* **2020**, *159*, 107938. [CrossRef] [PubMed]
25. Supaphon, P.; Kerdpiboon, S.; Vénien, A.; Loison, O.; Sicard, J.; Rouel, J.; Astruc, T. Structural changes in local Thai beef during sous-vide cooking. *Meat Sci.* **2021**, *175*, 108442. [CrossRef]
26. González-Fandos, E.; Villarino-Rodríguez, A.; García-Linares, M.C.; García-Arias, M.T.; García-Fernández, M.C. Microbiological safety and sensory characteristics of salmon slices processed by the sous vide method. *Food Control* **2005**, *16*, 77–85. [CrossRef]
27. Roldan, M.; Antequera, T.; Hernandez, A.; Ruiz, J. Physicochemical and microbiological changes during the refrigerated storage of lamb loins sous-vide cooked at different combinations of time and temperature. *Food Sci. Technol. Int.* **2015**, *21*, 512–522. [CrossRef]
28. Ismail, I.; Hwang, Y.-H.; Bakhsh, A.; Lee, S.-J.; Lee, E.-Y.; Kim, C.-J.; Joo, S.-T. Control of sous-vide physicochemical, sensory, and microbial properties through the manipulation of cooking temperatures and times. *Meat Sci.* **2020**, *188*, 108787. [CrossRef]
29. Zhao, J.; Li, J.; Wang, J.; Lv, W. Applying different methods to evaluate the freshness of large yellow croacker (*Pseudosciaena crocea*) fillets during chilled storage. *J. Agric. Food Chem.* **2012**, *60*, 11387–11394. [CrossRef]
30. Ji, X.; Luo, X.; Zhu, L.; Mao, Y.; Lu, X.; Chen, X.; Hopkins, D.L.; Zhang, Y. Effect of medium voltage electrical stimulation and prior ageing on beef shear force during superchilled storage. *Meat Sci.* **2021**, *172*, 108320. [CrossRef]
31. Jiang, Q.; Gao, P.; Liu, J.; Yu, D.; Xu, Y.; Yang, F.; Wang, B.; Yu, P.; Xia, W. Endogenous proteases in giant freshwater prawn (*Macrobrachium rosenbergii*): Changes and its impacts on texture deterioration during frozen storage. *Int. J. Food Sci. Technol.* **2021**, *56*, 5824–5832. [CrossRef]
32. Bhat, Z.F.; Morton, J.D.; Zhang, X.; Mason, S.L.; Bekhit, A.E.A. Sous-vide cooking improves the quality and *in-vitro* digestibility of Semitendinosus from culled dairy cows. *Food Res. Int.* **2020**, *127*, 108708. [CrossRef] [PubMed]
33. Cropotova, J.; Mozuraityte, R.; Standal, I.B.; Rustad, T. Assessment of lipid oxidation in Atlantic mackerel (*Scomber scombrus*) subjected to different antioxidant and sous-vide cooking treatments by conventional and fluorescence microscopy methods. *Food Control* **2019**, *104*, 1–8. [CrossRef]
34. Roldan, M.; Antequera, T.; Armenteros, M.; Ruiz, J. Effect of different temperature-time combinations on lipid and protein oxidation of sous-vide cooked lamb loins. *Food Chem.* **2014**, *149*, 129–136. [CrossRef] [PubMed]
35. Wan, J.; Cao, A.; Cai, L. Effects of vacuum or sous-vide cooking methods on the quality of largemouth bass (*Micropterus salmoides*). *Int. J. Gastron. Food Sci.* **2019**, *18*, 100181. [CrossRef]
36. Díaz, P.; Garrido, M.D.; Banon, S. Spoilage of sous-vide cooked salmon (*Salmo salar*) stored under refrigeration. *Food Sci. Technol. Int.* **2011**, *17*, 31–37. [CrossRef]
37. Bongiorno, T.; Tulli, F.; Comi, G.; Sensidoni, A.; Andyanto, D.; Iacumin, L. Sous vide cook-chill mussel (*Mytilus galloprovincialis*): Evaluation of chemical, microbiological and sensory quality during chilled storage (3 °C). *LWT-Food Sci. Technol.* **2018**, *91*, 117–124. [CrossRef]
38. Huang, J.; Zhou, Y.; Chen, M.; Huang, J.; Li, Y.; Hu, Y. Evaluation of negative behaviors for single specific spoilage microorganism on little yellow croaker under modified atmosphere packaging: Biochemical properties characterization and spoilage-related volatiles identification. *LWT-Food Sci. Technol.* **2021**, *140*, 110741. [CrossRef]
39. Zou, Y.; Kang, D.; Liu, R.; Qi, J.; Zhou, G.; Zhang, W. Effects of ultrasonic assisted cooking on the chemical profiles of taste and flavor of spiced beef. *Ultrason. Sonochem.* **2018**, *46*, 36–45. [CrossRef]
40. Xu, Y.; Li, L.; Mac Regenstein, J.; Gao, P.; Zang, J.; Xia, W.; Jiang, Q. The contribution of autochthonous microflora on free fatty acids release and flavor development in low-salt fermented fish. *Food Chem.* **2018**, *256*, 259–267. [CrossRef]
41. Zhang, J.; Cao, J.; Pei, Z.; Wei, P.; Xiang, D.; Cao, X.; Shen, X.; Li, C. Volatile flavour components and the mechanisms underlying their production in golden pompano (*Trachinotus blochii*) fillets subjected to different drying methods: A comparative study using an electronic nose, an electronic tongue and SDE-GC-MS. *Food Res. Int.* **2019**, *123*, 217–225. [CrossRef]
42. Jia, S.; Li, Y.; Zhuang, S.; Sun, X.; Zhang, L.; Shi, J.; Hong, H.; Luo, Y. Biochemical changes induced by dominant bacteria in chill-stored silver carp (*Hypophthalmichthys molitrix*) and GC-IMS identification of volatile organic compounds. *Food Microbiol.* **2019**, *84*, 103248. [CrossRef] [PubMed]

foods

MDPI

Article

Effects of Weak Acids on the Microbiological, Nutritional and Sensory Quality of Baltic Herring (*Clupea harengus membras*)

Nora Logrén [1,*], Jaakko Hiidenhovi [2], Tanja Kakko [3], Anna-Liisa Välimaa [4], Sari Mäkinen [2], Nanna Rintala [1], Pirjo Mattila [5], Baoru Yang [3] and Anu Hopia [1]

1 Functional Foods Forum, Faculty of Medicine, University of Turku, 20014 Turku, Finland; nanna.rintala@utu.fi (N.R.); anuhop@utu.fi (A.H.)
2 Food Processing and Quality, Production Systems, Natural Resources Institute Finland (Luke), Myllytie 1, 31600 Jokioinen, Finland; jaakko.hiidenhovi@luke.fi (J.H.); sari.makinen@luke.fi (S.M.)
3 Food Chemistry and Food Development, Department of Life Technologies, University of Turku, 20014 Turku, Finland; tatese@utu.fi (T.K.); bayang@utu.fi (B.Y.)
4 Food Processing and Quality, Production Systems, Natural Resources Institute Finland (Luke), Paavo Havaksen tie 3, 90570 Oulu, Finland; anna-liisa.valimaa@luke.fi
5 Food Processing and Quality, Production Systems, Natural Resources Institute Finland (Luke), Itäinen Pitkäkatu 4, 20520 Turku, Finland; mattilapirjoh@gmail.com
* Correspondence: nodalo@utu.fi; Tel.: +358-50-4739649

Abstract: Baltic herring (*Clupea harengus membras*) pickled in vinegar is a common product in the Nordic countries. Other weak acids are used to cook and preserve fish in other food cultures. The aim of this study was to evaluate the potential of weak acids to produce safe and nutritious pickled fish products with varying sensory properties. The influence of acetic, citric, lactic, malic, and tartaric acids on the preservability and quality of pickled and marinated Baltic herring was studied by measuring microbiological quality, pH, chemical composition, and lipid oxidation and by sensory profiling. Pickling with these acids with pH levels of 3.7–4.2 resulted in pickled Baltic herring products with high microbiological quality. The results of the chemical analysis of the samples indicated that pickling and storage on marinade influenced the chemical composition of fish. The most significant changes in chemical composition were the increase in moisture and decrease in protein content of the samples during storage. Fat content decreased during the storage period in acetic acid and malic acid samples. All tested acids inhibited lipid oxidation for one month, but at three and four month time points, the content of oxidation products increased except in the samples pickled with tartaric acid. The highest oxidation level was observed in the case of citric acid and the lowest with tartaric acid. The results indicate that replacing acetic acid with other weak acids frequently used in the food industry results in pickled and marinated fish products with novel and milder sensory profiles.

Keywords: pickling; fish; sensory profile; organic acids; chemical composition; lipid oxidation; microbiological safety

Citation: Logrén, N.; Hiidenhovi, J.; Kakko, T.; Välimaa, A.-L.; Mäkinen, S.; Rintala, N.; Mattila, P.; Yang, B.; Hopia, A. Effects of Weak Acids on the Microbiological, Nutritional and Sensory Quality of Baltic Herring (*Clupea harengus membras*). *Foods* **2022**, *11*, 1717. https://doi.org/10.3390/foods11121717

Academic Editors: Tao Yin and Liu Shi

Received: 29 April 2022
Accepted: 10 June 2022
Published: 12 June 2022

1. Introduction

Globally, acid treatment is a widely applied method to both cook and preserve different types of fish products. Pickling with the help of acid and salt is among the oldest preservation techniques for fish. The acid and salt treatment tenderizes the fish muscles in a similar manner to heat and inhibits the growth of spoiling bacteria [1]. In Europe, the most common acid used for producing pickled fish is acetic acid. Vinegar-based fish pickled with various herbs, spices and other seasonings represent a wide range of products. The preparation of canned Baltic herring (*Clupea harengus membras)* in Finland is based on pickling the herring in acetic acid. Although vinegar is the most common added acidic pickling agent, acidity can also be achieved through the production of acid by fermenting acetic and lactic acid bacteria. Lemon and lime juice rich in citric acid are used in many traditional fish recipes such as in the originally Peruvian dish, ceviche. In fermented fish products, such as Swedish surströmming, lactic acid produced by lactic acid bacteria is the source of acidity.

Under acidic conditions, the growth of food-borne pathogens and food spoilage microbes is inhibited. *Listeria monocytogenes* is a food-borne pathogen that might cause listeriosis, a disease that can even lead to death. [2]. The bacterium is ubiquitous in the environment, with ready-to-eat (RTE) fish and fishery products being a particularly potential vehicle for the microbes [3]. The growth of *L. monocytogenes* may vary depending on the strain and matrix. In a broth media, the lower growth limit is at pH 4.0–4.3 [4]. Organic acids have been widely used in the food industry due to their antimicrobial nature [5].

In addition to inhibition of bacterial growth, a low pH also changes the muscle structure as protein denaturation is dependent on the pH. Protein is most prone to denaturation at a pH below or above the isoelectric point due to the increase in the net charge of the protein, inducing protein-protein repulsion, and unfolding and swelling. The swelling of the muscle tissue caused by a low pH is likely to increase tenderness [6,7]. At the isoelectric point, protein is most resistant to denaturation [8]. Most previous studies on the acid treatment of muscle foods using weak organic acids have only included comparisons of different concentrations of the same acid [9–11], and/or the same concentrations of different acids [9]; however, few studies have addressed the effect of different acids at a constant pH, despite the significance of pH in the process.

The pickling and marinating process affects the texture of meat via the protein degrading enzyme activity. It has been shown that during the marinating process the amount of nitrogen containing compounds decreases in Baltic herring fillets and carcasses, and primarily increases in the marinating solution [12]. This indicates that there is some proteolytic activity in the solutions during the pickling and marinating processes. The proteolysis catalyzed by, e.g., cathepsin D may increase the tenderness [7]. Cathepsins, found in lysosomes [8], are protein hydrolyzing proteases that break proteins down into smaller molecules of nitrogen containing compounds such as amino acids. Cathepsin D, an aspartic peptidase, is one of the most abundant cathepsin in lysosomes. It is stable at pH 3–5 [13,14] and most active at pH 2–4 [15] and it has been found to be in an active form in brine after marinating frozen-thawed Atlantic herring (*Clupea harengus*) for 7 days with a 5% acetic acid [16]. Burke and Monahan [9] suggest that collagen solubilization also contributes to tenderization after the marinating process and that there are two mechanisms by which collagen molecules may solubilize. The first mechanism suggested is peptide bond hydrolysis which increases as the pH decreases, and, thus, the collagen solubilizes. The second mechanism suggests that collagen solubilizes through a breakdown of the cross-links that are sensitive to pH, heat, or denaturing agents.

Dark-muscled fish such as Baltic herring are prone to oxidation due to the high content of heme proteins, known to be significant endogenous pro-oxidants in fish [17]. Baltic herring is high in polyunsaturated fatty acids (PUFA), with as high as approximately one third of all the fatty acids being polyunsaturated [18,19]. Decreasing the pH of fish can increase lipid oxidation due to the deoxygenation of hemoglobin via the Root and Bohr effects, since deoxyhemoglobin is considered a better catalyst of lipid oxidation than oxyhemoglobin [17]. Moreover, the treatment of fish with organic acids has even been seen to produce a protective effect regarding oxidation [20,21].

Fish products pickled with vinegar have a characteristic flavor profile in which acidity and vinegary flavor dominate. The pickling process has an effect on the formation of taste and flavor compounds in the fish. Free amino acids produced by proteolysis may have sweet, sour, bitter, and umami taste properties [22] that can affect the flavor properties of the pickled and marinated herring products. Furthermore, the lipid oxidation products of unsaturated fatty acids in fish contribute to the flavor. However, the research regarding the sensorial quality of pickled Baltic herring products is limited. The topic has been studied with, e.g., acceptance assessment [11,23], hedonic evaluation [24,25] and by using the "Just about right–scale" [25]. The sensory properties of traditionally pickled and marinated Atlantic and Baltic herring have been studied by sensory profiling [26,27]. The traditional flavor profile is preferred by certain consumer groups. However, the consumption of Baltic herring has decreased by one third in the last 40 years. In order to obtain new consumers for pickled Baltic herring, products with milder sensory properties might be attractive.

To summarize, the chemical, microbiological and sensory studies on the pickling of fish have mostly been focused on pickling with acetic acid and the potential of other weak acids commonly used in food production has not been systematically documented for fish. The aim of this study was to evaluate the potential of food grade weak acids other than the commonly used acetic acid in order to develop pickled Baltic herring products with novel sensory profiles. The acids studied were citric, lactic, malic, and tartaric acid, acetic acid being the reference. The quality and preservability of the samples were studied by measuring the pH, composition, lipid oxidation, and microbiological quality and by sensory profiling.

2. Materials and Methods

2.1. Raw Material and Sample Preparation

The skinned Baltic herring fillets were provided by Martin Kala Ltd. (Turku, Finland). The fish were caught in the southern part of the Gulf of Bothnia on 27 October 2020 and the skinless fillets were delivered to Camilla's Fiskdelikatess Ltd. (Molpe, Finland) within 2 days of being caught and kept fresh on ice. Food grade acids, citric acid monohydrate, malic acid, lactic acid, and tartaric acid (Vinoferm®, Brouwlan, Beverlo, Belgium), were purchased from Juomatarviketukku Lappo Ltd. (Piikkiö, Finland). Acetic acid (Berner Ltd., Helsinki, Finland) was provided by Camilla's Fiskdelikatess Ltd.

The concentration and pH values of the acid solutions are presented in Table 1. The suitable concentration for acetic acid, citric acid, lactic acid, malic acid, and tartaric acid for the pickling and marinating solutions was determined by pre-tests, since the addition of fish increased the pH. This is due to the different strengths and buffering capacities (ability to maintain acidic pH after addition of the fish) of the organic acids. The aim was to achieve a pH of <4 at the end of pickling. The target of pH < 4 was chosen since it is a typical pH for fish preservatives treated with acetic acid, and is considered a safe pH with regard to limiting the growth of food pathogens, including *L. monocytogenes* [5]. Sodium benzoate was also added to the marinating solutions as a preservative.

Table 1. Concentration and pH values of the pickling and marinade solutions.

Acid	Concentration in Pickling Solution (g/L)	Concentration in Marinade Solution (g/L)	pH of Pickling Solution	pH at the End of Pickling
Acetic acid	50.0	0.67	2.40 ± 0.00	3.85 ± 0.01
Citric acid	20.0	0.17	1.90 ± 0.01	3.63 ± 0.01
Lactic acid	20.0	0.16	2.10 ± 0.01	3.56 ± 0.01
Malic acid	20.0	0.14	2.03 ± 0.01	3.53 ± 0.01
Tartaric acid	21.5	0.11	1.78 ± 0.01	3.24 ± 0.01

The samples were produced at Camilla's Fiskdelikatess Ltd., whereas the training samples for the sensory evaluations were prepared in the test kitchen of the Functional Foods Forum, University of Turku (Turku, Finland) (ISO-8589:1988). The concentration of salt in the pickling solutions was 5.0% (*w*/*v*). The concentrations of sugar and sodium benzoate in the marinating solutions were from the commercial recipes of Camilla's Fiskdelikatess Ltd. The fish-to-solution ratio in the pickling solutions of the training and actual samples was 1:1.4, and 1:1.5 (*w*/*v*), respectively, and in the marinating solutions of the training and actual samples it was 1:1.2, and 1:1.3 (*w*/*v*), respectively. The pickling process was conducted in 150 l plastic barrels commonly used in the manufacturing of pickled herring products. The pickling was conducted at 2–3 °C for 24 ± 3 h. The fillets were removed from the pickling solution and placed on a grating to drain the excess solution. The fillets were weighted down in glass jars with a grating to keep the herring under the marinade. The marinade was added, and the jars were vacuum packed. The samples were kept sealed at 3–6 °C until analysis. The samples were analyzed at 0, 1, 3, and 4 month time points after preparation except for the sensory profile, which was created and reviewed twice after 0, 1, and 4 months, respectively. The 0 month's analysis was started after the fish had been pickled and placed in the marinating solutions.

2.2. Microbiological Quality

The number of L. monocytogenes from the fresh samples (at the time point 0 month) was determined according to the qualitative method SFS-EN ISO 11290-1:2017 in an accredited laboratory SeiLab Ltd. (Seinäjoki, Finland). The horizontal method is used for the detection and enumeration of *L. monocytogenes* and of Listeria spp. in food and feed [28]. The result was reported as found/not found per 25 g of the sample. At the time points 1 month and 4 months the number of bacteria was analyzed by the qualitative method Vidas LMX (*) Vidas Express (LMX), ISO 11290-2:2017 in the accredited laboratory ScanLab Ltd. (Oulu, Finland). The result was reported as found/not found per 25 g of the sample.

The number of sulfite-reducing bacteria was determined according to the method of the Nordic Committee on Food Analysis NMKL 56:2015 NMKL Method No. 56, 5. Ed., 2015 in ScanLab Ltd. (Oulu, Finland). The result was reported as the number of viable anaerobic sulfite-reducing bacteria per gram of the samples, i.e., the number of colonies forming units per gram of sample (cfu/g). The results were expressed on a logarithmic scale.

The number of aerobic microorganisms at 30 °C were determined by the colony count method according to the NMKL Method No.86, 5th Ed., 2013 used by ScanLab Ltd. The result was reported as cfu/g expressed on a logarithmic scale.

The number of aerobic psychro-trophic microorganisms, the number of yeasts, and the number of hydrogen sulfide-producing bacteria were determined in the Natural Resources Institute Finland (Luke) (Oulu, Finland). For all these analyses, 10 g of sample was mixed with 90 mL of sterile Maximum Recovery Diluent (Lab M Ltd., Heywood, UK), and homogenized by a Bagmixer 400 (Interscience, Saint Nom la Bretêche, France) for 1 min. A ten-fold serial dilution was also prepared with the same media. After incubation on the appropriate agar under the microbial specific growth conditions the number of colonies on the agar plates was counted. The result was reported as cfu/g expressed on a logarithmic scale.

The number of aerobic psychro-trophic microorganisms were determined by the colony counting method according to the NMKL Method No.86, 5th Ed., 2013 in the Natural Resources Institute Finland. Using this method, the dilutions of the samples are pipetted into Petri dishes, and Plate Count Agar (PCA) (Lab M Ltd., Heywood, UK) was poured into Petri dishes and the suspension was mixed. The plates were incubated under aerobic conditions at 6.5 °C ± 1.0 for 10 days.

The number of yeasts were measured by a dilution plating technique according to the Nordic Committee on Food Analysis NMKL Method No.98, 4th Ed.,2005 using oxytetracycline-glucose yeast extract agar (O.G.Y.E.) (Lab M Ltd., Heywood, UK). The dilutions of the samples were spread on the O.G.Y.E. agar surface in Petri dishes. The agar plates were incubated under aerobic conditions at 25.0 °C ± 1.0 for 5 days.

The number of hydrogen sulfide-producing bacteria were measured by Iron Agar (Lyngby) (Oxoid Ltd., Hampshire, UK) with a 4% w/v solution of L-cysteine hydrochloride monohydrate (Sigma-Aldrich, St. Louis, MO, USA). The media was prepared according to the manufacturer's instructions. The dilutions of the samples were spread on the Lyngby agar surface in Petri dishes. The plates with *Shewanella putrefaciens* ATCC 8071 produce black colonies and *Pseudomonas fluorescens* and ATCC 13525 which produces white colonies on the Lyngby agar was used as the reference microorganisms. The agar plates were incubated under anaerobic conditions at 25.0 °C ± 1.0 for 48 h. The black colonies, with characteristics that were comparable to the colonies of *S. putrefaciens* ATCC, 8071 were counted as hydrogen sulfide-producing bacteria.

2.3. pH

The pH values in the solution of the canned samples were measured at 0, 1, 3, and 4 months. Measurements were conducted in duplicate from three different cans using a FiveEasy pH meter equipped with an LE409-DIN electrode (Mettler Toledo, Columbus, OH, USA).

2.4. Composition

The moisture and ash contents in the samples were determined by using a TGA701 Thermo-gravimetric Analyzer (Leco Corporation, St. Joseph, MI, USA). The TGA measured weight loss as a function of temperature (105 °C and 650 °C for moisture and ash, respectively) under controlled conditions, and automatically determined the moisture and ash contents in the samples with the selected program. The nitrogen content was determined with an in-house Kjeldahl method based on ISO 20483, ISO 5983-2 and AOAC 2011.11 methods by using a Kjeltec TM8400 analyzer (Foss Analytical Ltd., Höganäs, Sweden). A conversion factor of 6.25 was used to calculate total protein content. The total fat content was determined using the SoxCap TM 2047 in combination with the Soxtec TM 2050 extraction system (Foss Ltd., Hillerød, Denmark) with a preparatory acid hydrolysis step and diethyl ether extraction according to ISO 6492. The total carbohydrate content was calculated with the following formula:

$$\text{carbohydrates (\% fresh weight, FW)} = 100 - \text{moisture (\%)} - \text{protein content (\%FW)} - \text{crude fat (\%FW)} - \text{ash (\%FW)}. \quad (1)$$

2.5. Lipid Oxidation

The effect of the different acids on lipid oxidation in the Baltic herring fillets was assessed by measuring the thio-barbituric acid reactive substances (TBARS) in the fish samples. The TBARS in the fish samples were measured after 0, 1, 2, and 4 months of storage. Prior to the analysis, the Baltic herring samples were subjected to alkaline hydrolysis which releases malondialdehyde (MDA) from proteins according to Mäkinen et al. [29]. Briefly, the samples were homogenized, and three 100 mg subsamples of each homogenized sample were taken for alkaline hydrolysis. The hydrolysis was conducted by mixing the subsamples with NaOH and incubating the suspensions in a 60 °C water bath for 30 min. After the hydrolysis, the proteins were precipitated with sulfuric acid and trichloroacetic acid (TCA), and supernatants were collected after centrifugation. The supernatants were then reacted with thio-barbituric acid (TBA) to form MDA-TBA adducts with a pink pigment. Samples were then analyzed with UHPLC at 532 nm. The concentration of MDA was calculated using the MDA-TBA standard curve (12.5–800 μM). Samples were analyzed in triplicates. Chromatographic analyses were performed in duplicate from each of the subsamples ($n \geq 6$). Results are expressed as mean ± SD.

2.6. Sensory Profile

The sensory properties of the pickled and marinated Baltic herring samples were analyzed with a generic descriptive method. A panel (n = 10, 10 women) was recruited from a group of trained sensory panelists. The panelists had previous experience in sensory evaluation of various food samples, and the acuity of their senses had been evaluated. The panel was then specifically trained in analyzing marinated Baltic herring samples during four training sessions. The training of the panel at 0 month time point was done with samples prepared earlier but with the same procedure as the actual samples (Table S1). Due to the prevailing COVID-19 pandemic situation the training sessions were held with only 3–4 people present per session except at the 4-month evaluation where only 2 panelists took part in the training simultaneously. A list of sensory attributes was created by the sensory panel to describe the characteristics of the pickled herring samples. Suitable reference samples were defined to reflect the corresponding sensory attributes and their intensities. The attributes were evaluated on a scale from 0–10 verbally anchored at the ends (0 = not at all, 10 = very strong). The attributes and reference samples are presented in Table 2.

Table 2. The attributes and reference samples used in the sensory profiling. The intensities on a scale from 0–10 (0 = not at all, 10 = very strong) of the reference samples are in parenthesis.

Sensory Properties	Attribute	Reference Sample (Intensity)
Odor	Total intensity	Verbal instructions
	Sweet	Felix American pickled cucumber (5)
	Vinegar	1.25% (*v*/*v*) acetic acid (5)
	Fish	Verbal instructions
	Stuffy	Verbal instructions
	Rancid	Verbal instructions
	Metallic [1]	Verbal instructions
Texture	Crumbly	Pirkka salad cheese, lactose free (9)
	Moistness	Pirkka salad cheese, lactose free (3)
Taste	Total intensity	Verbal instructions
	Sweet	2.0% (*w*/*v*) sucrose (5)
	Salty	0.2% (*w*/*v*) NaCl (5)
	Umami	0.1% and 0.2% (*w*/*v*) L-Glutamic acid monosodium salt monohydrate (5 and 8, respectively) [2]
	Sour	0.04% (*w*/*v*) citric acid (5)
	Bitter	0.04% (*w*/*v*) caffeine (6)
Flavor	Vinegar	0.63% (*v*/*v*) acetic acid (6)
	Fish	Verbal instructions
	Stuffy	Verbal instructions
	Rancid	Verbal instructions
	Metallic	Verbal instructions

[1] Metallic odor was evaluated at 1 and 4 months. [2] 0.1% (*w*/*v*) L-Glutamic acid monosodium salt monohydrate was included as a reference sample of umami taste 1- and 4-month's evaluations.

Sensory evaluations were carried out during four sessions in the sensory laboratory of the Functional Foods Forum, University of Turku (ISO-8589:1988). The samples were coded with random three-digit numbers and offered at room temperature in randomized order. Compusense® Cloud software (Version 21.0, Compusense Inc., Guelph, ON, Canada) was used for data collection. To confirm the safety of the products evaluated, the number of *L. monocytogenes*, the number of aerobic microorganisms and the number of sulfite-reducing bacteria (see 2.4. Microbiological quality) were determined before the time point of each sensory evaluation to ensure the microbiological safety of the samples.

The first sensory evaluation at 0 months was held on day 6 after the pickling of the herring. The 1- and 4-month sensory evaluations were carried out in a similar manner after five and 17–18 weeks of the sample preparation, respectively. For the 1- and 4-month evaluations, the same sensory panel was recruited, with nine of the ten panelists able to participate at 1 month and six at 4 months. An additional sensory evaluation was held to collect sufficient data at the 4-month evaluation. A training session was held before the subsequent sessions in order to recall and discuss the attributes before the evaluations.

2.7. Statistical Analysis

Statistically significant differences between pH values were determined using IBM SPSS Statistics 25 software and the statistical analyses for the chemical composition, lipid oxidation, and sensory analyses were executed with IBM SPSS Statistics 27 (Armonk, NY, The United States). The limit for the statistical significance level was set at $p < 0.05$. Statistically significant differences between treatments were calculated with a one-way analysis of variance (ANOVA) test and Tukey's honestly significant difference post-hoc test. Tamhane's T2 post-hoc test was also used in the statistical analysis of the sensory analysis depending on the equivalence of variance.

Differences in pH values and sensory properties of each acid treatment between the time points were tested with paired-samples t-test and repeated ANOVA measures depending on the number of variables. Bonferroni and Šidák corrections were used to correct the *p*-values. A paired samples *t*-test was conducted to compare the results of the chemical analysis at different time points within the same acid treatment.

3. Results

3.1. Microbiological Quality

L. monocytogenes bacteria was not found in any samples tested. The detection limits were 1 cfu/25 g of the sample. The number of sulfite-reducing Clostridia in all the tested samples was below the detection limit of <10 cfu/g.

The number of aerobic microorganisms was generally very low in all the tested samples (Table 3). The highest count was observed in the fresh (0 months) citric acid (2.6 log cfu/g) followed by lactic acid (2.0 log cfu/g), malic acid (1.9 log cfu/g), and tartaric acid (1.5 log cfu/g) samples. The acetic acid sample had the lowest bacterial count (1.0 log cfu/g) of the samples tested. The number of aerobic microorganisms had already decreased after one month of storage and remained low during the 4-month storage period. The bacterial count was below the detection limit of <1.0 log cfu/g in the acetic acid and tartaric acid samples. In the citric acid sample, the bacterial count decreased by 1 log, and in the lactic acid and citric acid samples by 0.6 log.

Table 3. Aerobic microorganisms (cfu/g log_{10}) in the pickled and marinated fish fillet samples determined at 30 °C after 0, 1, 3, and 4 months of storage.

Acid	0 Months	1 Month	3 Months	4 Months
Acetic acid	1.00	<1.00	n.d. [1]	<1.00
Citric acid	2.58	1.00	n.d. [1]	1.00
Lactic acid	2.01	1.40	n.d. [1]	1.40
Malic acid	1.85	1.30	n.d. [1]	1.30
Tartaric acid	1.48	<1.00	n.d. [1]	<1.00

[1] not determined.

The number of aerobic psychro-trophic microorganisms and the number of yeasts in all the tested samples were below the detection limit of <10 cfu/g. The number of hydrogen sulfide-producing bacteria in the fresh citric acid sample was 10 cfu/g. The number of bacteria in all the other samples tested was below the detection limit of <10 cfu/g.

3.2. pH

The pH values of the canned samples at the different timepoints are presented in Table 4. Despite the high pH of some of the marinade solutions (Table 1), a pH of <4 was mostly achieved in the canned samples. With most of the acids, the pH increased slightly over the first three months of storage.

Table 4. Measured pH values after 0, 1, 3, and 4 months of storage.

Acid	0 Months	1 Month	3 Months	4 Months
Acetic acid	3.78 ± 0.06^{abA}	3.83 ± 0.03^{abA}	3.91 ± 0.03^{abA}	3.80 ± 0.08^{Aa}
Citric acid	4.02 ± 0.04^{Ca}	4.13 ± 0.05^{Cb}	4.14 ± 0.08^{Cb}	4.19 ± 0.12^{Cab}
Lactic acid	3.96 ± 0.02^{Ca}	3.98 ± 0.06^{bcA}	4.12 ± 0.04^{Cb}	4.09 ± 0.11^{bcAB}
Malic acid	3.87 ± 0.01^{Ba}	3.86 ± 0.09^{abA}	4.02 ± 0.05^{bcB}	3.93 ± 0.04^{abAB}
Tartaric acid	3.68 ± 0.09^{Aa}	3.79 ± 0.17^{Aab}	3.92 ± 0.18^{Ab}	3.87 ± 0.12^{Ab}

a–c Different letter indicates statistically significant difference between samples at each time point's evaluation. A–C Different letter within an attribute indicates statistically significant difference between each time point's evaluation of each acid sample.

3.3. Composition

Initially, the moisture content was a little over 50% in all samples and gradually increased during the 4-month storage to reach about 60% in the acetic acid and citric acid samples and 56–58% in the other samples (Figure 1).

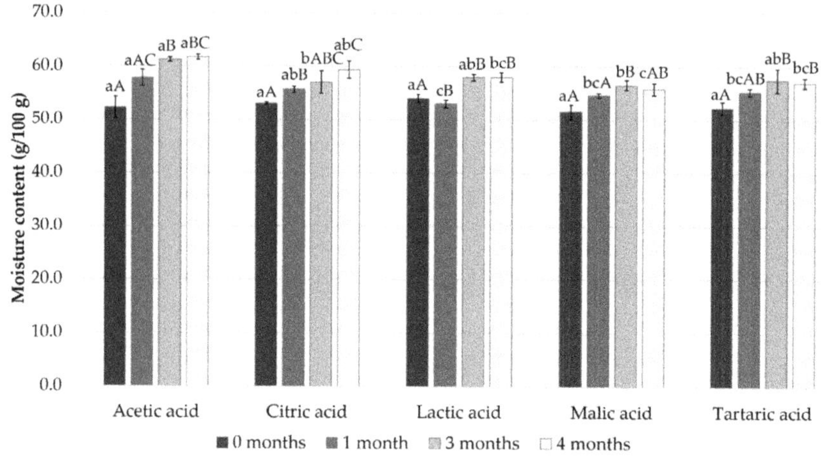

Figure 1. Measured moisture content after 0, 1, 3, and 4 months of storage. Different small letters within the same month and capital letters within the same acid indicate a statistically significant difference on a 95% confidence level.

The changes in protein content in Baltic herring fillets during the marinating process are shown in Figure 2. The initial protein content of the samples varied from 13.5% in the acetic acid sample to 14.7% in the tartaric acid sample. All the acids had a significant effect on the protein content from day 0 and this further increased during the storage in marinade. After the 4 months storage period the protein content levels decreased most significantly in the acetic acid sample, which went down to 7.8%; in comparison the protein content of the malic acid and tartaric acid samples remained highest, 12.9 and 12.6%, respectively.

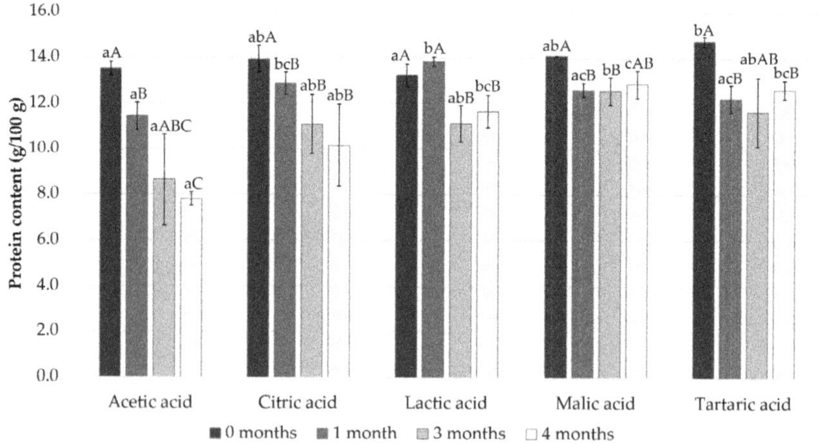

Figure 2. Measured protein content after 0, 1, 3, and 4 months of storage. Different small letters within the same month and capital letters within the same acid indicate a statistically significant difference on a 95% confidence level.

The carbohydrate content of the samples was approximately 30% in the samples immediately after the sample preparation and decreased slightly during the storage period to 26–28% in all the samples (Figure 3).

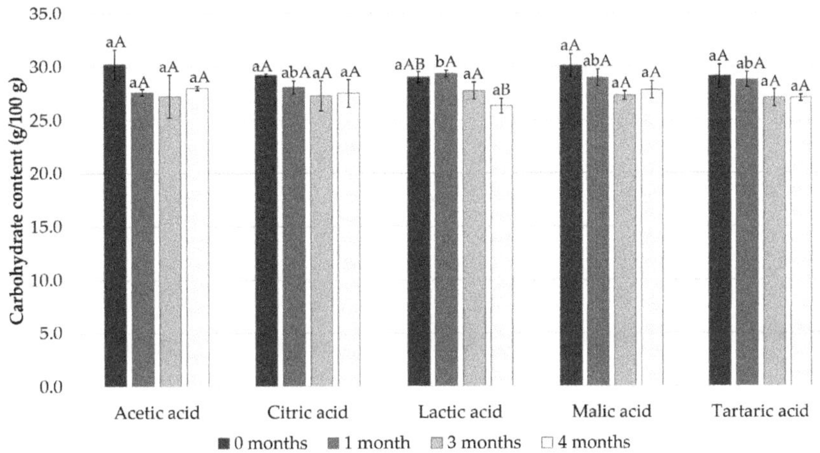

Figure 3. Calculated carbohydrate content after 0, 1, 3, and 4 months of storage. Different small letters within the same month and capital letters within the same acid indicate a statistically significant difference on a 95% confidence level.

The fat content of the Baltic herring fillets samples on the different timepoints are presented in Figure 4. At the beginning of the storage period, the fat content of the fillets was approximately 3% and decreased during the storage in all the samples except the lactic acid sample. The natural variation in the fat analyses was high. In the acetic acid samples, the fat content decreased the most, from 3.2% to 1.4%.

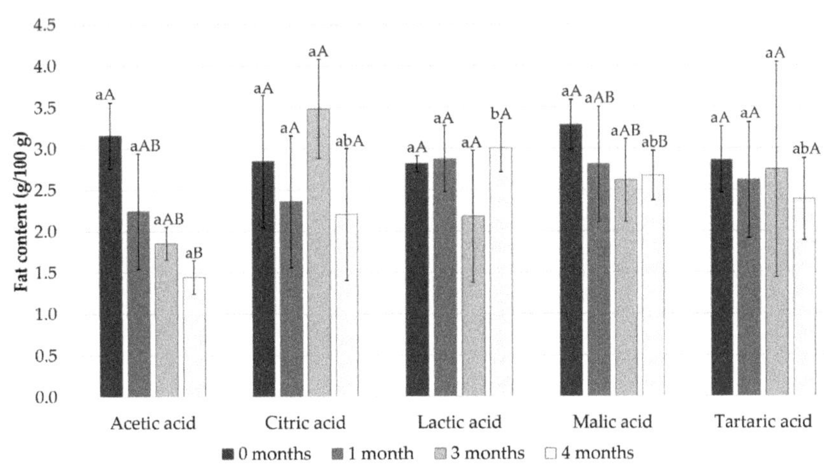

Figure 4. Measured fat content after 0, 1, 3, and 4 months of storage. Different small letters within the same month and capital letters within the same acid indicate a statistically significant difference on a 95% confidence level.

3.4. Lipid Oxidation

Fish fat is a highly specific nutritious element, due specifically to its high content of PUFAs. MDA has been widely used to represent the degree of lipid oxidation in various meat products. The production of MDA occurs in the second phase of lipid oxidation, during which peroxide is oxidized to aldehydes and ketones [30]. In the present study, we measured how the different acids used in the marinade affect the content of MDA in the Baltic herring fillets during four months of cold storage (Figure 5).

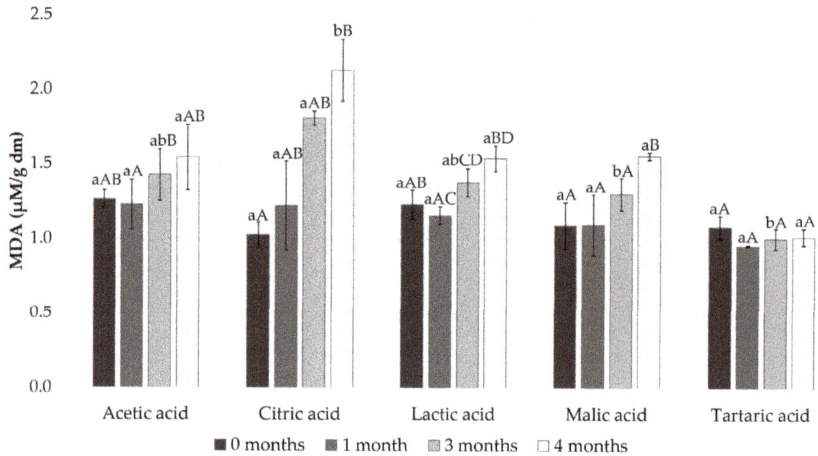

Figure 5. MDA content of the pickled and marinated Baltic herring fillets measured after 0, 1, 3 and 4 months in cold storage. Different small letters within the same month and capital letters within the same acid indicate a statistically significant difference on a 95% confidence level. Dm = dry matter.

The initial concentration of MDA in the Baltic herring fillets immediately after preparation varied from 1.0 ± 0.1 μM/g dry matter (dm) to 1.3 ± 0.1 μM/g dm; the lowest content was detected in the fillets in the citric acid marinade and the highest value in the fillets in the acetic acid marinade (Figure 5). The MDA content remained at the same level until after 1 month of cold storage, with all tested acids showing no statistically significant differences between the 0- and 1-month time points. However, the MDA levels increased significantly in all the tested acids when the storage period continued to 3 and 4 months except for tartaric acid (Figure 5, $p < 0.05$). After 3 and 4 months of cold storage, the citric acid marinated fillets were evidently the most oxidized sample. The highest MDA level (2.1 ± 0.2 μM/g dm) was observed after 4 months of cold storage in Baltic herring filets marinated with citric acid. After the 4 months of cold storage, Baltic herring fillets marinated with citric acid showed a significantly higher MDA level in comparison to all the other tested acids (Figure 5, $p < 0.05$). After 3 months of cold storage, the MDA content in the Baltic herring fillets marinated with citric acid was significantly higher in comparison to the filets marinated with tartaric and malic acid ($p < 0.05$). After 3 and 4 months of cold storage, Baltic herring fillets marinated with tartaric acid showed the lowest MDA content, but the difference in comparison with the other acids was not statistically significant ($p > 0.05$).

3.5. Sensory Profile

The sensory panel evaluated the sensory properties of the five Baltic herring products at three time points at 0, 1, and 4 months. Their odor, texture, taste, and flavor properties were evaluated using 20 sensory attributes. The panel recognized seven odor attributes, two texture attributes, six taste attributes, and five flavor attributes. The odor attributes were total intensity, sweetness, vinegar, fish, stuffy, rancid, and metallic. Texture attributes were crumbliness and moistness. Taste attributes were total intensity, sweetness, saltiness, umami, sourness, and bitterness. Flavor attributes were vinegar, fish, stuffy, rancid, and metallic. The differences between sensory characteristics of the samples can be seen in Table 5.

Table 5. The means and standard deviations of sensory attributes of Baltic herring pickled and marinated in different weak acids after 0, 1, and 4 months of storage evaluated on a scale from 0–10 (0 = not at all, 10 = very strong).

Attribute [1]	Months	Acetic Acid	Citric Acid	Lactic Acid	Malic Acid	Tartaric Acid
O_total	0	7.4 ± 1.8 ^{aA}	4.6 ± 1.8 ^{bA}	5.4 ± 1.5 ^{bA}	4.8 ± 1.5 ^{bA}	5.7 ± 1.7 ^{bA}
	1	8.4 ± 0.9 ^{aB}	4.7 ± 1.5 ^{bA}	5.5 ± 1.6 ^{bcA}	5.4 ± 1.1 ^{bcA}	6.0 ± 1.5 ^{cA}
	4	8.1 ± 1.4 ^{aAB}	4.9 ± 1.4 ^{bA}	5.5 ± 1.7 ^{bA}	5.0 ± 1.6 ^{bA}	5.7 ± 1.3 ^{bA}
O_sweet	0	2.4 ± 1.5 ^{aA}	1.9 ± 1.5 ^{aA}	1.7 ± 1.2 ^{aA}	1.6 ± 1.5 ^{aA}	1.8 ± 1.5 ^{aA}
	1	2.3 ± 1.2 ^{aA}	1.9 ± 1.3 ^{aA}	2.0 ± 1.6 ^{aA}	2.1 ± 1.6 ^{aA}	1.9 ± 1.8 ^{aA}
	4	2.9 ± 2.2 ^{aA}	2.2 ± 1.4 ^{aA}	2.4 ± 1.7 ^{aA}	2.1 ± 1.7 ^{aA}	2.7 ± 1.5 ^{aA}
O_vinegar	0	6.7 ± 2.8 ^{aA}	2.0 ± 2.0 ^{bA}	2.0 ± 2.4 ^{bA}	1.9 ± 1.9 ^{bA}	1.9 ± 1.9 ^{bA}
	1	8.1 ± 1.2 ^{aB}	1.6 ± 1.3 ^{bA}	2.0 ± 1.7 ^{bA}	2.4 ± 1.8 ^{bA}	2.5 ± 2.2 ^{bA}
	4	8.6 ± 1.0 ^{aAB}	3.0 ± 2.1 ^{bA}	3.7 ± 2.3 ^{bA}	2.7 ± 2.2 ^{bA}	3.2 ± 2.2 ^{bA}
O_fish	0	3.1 ± 2.0 ^{aA}	3.4 ± 2.2 ^{aA}	4.2 ± 2.2 ^{aA}	4.3 ± 2.2 ^{aA}	4.6 ± 2.5 ^{aA}
	1	2.7 ± 1.4 ^{aA}	3.5 ± 1.9 ^{abA}	4.3 ± 1.6 ^{bA}	4.0 ± 1.5 ^{bA}	4.3 ± 1.5 ^{bA}
	4	2.5 ± 1.6 ^{aA}	4.1 ± 1.7 ^{bA}	4.8 ± 1.8 ^{bA}	4.9 ± 1.6 ^{bA}	5.0 ± 1.6 ^{bA}
O_stuffy	0	1.3 ± 1.8 ^{aA}	2.0 ± 2.3 ^{aA}	1.7 ± 2.3 ^{aA}	1.7 ± 2.0 ^{aA}	2.0 ± 2.2 ^{aA}
	1	0.4 ± 0.6 ^{aA}	2.1 ± 1.9 ^{bA}	1.7 ± 1.5 ^{bA}	1.2 ± 1.1 ^{bA}	1.8 ± 1.6 ^{bA}
	4	1.2 ± 1.5 ^{aA}	2.1 ± 1.7 ^{aA}	2.2 ± 1.8 ^{aA}	1.8 ± 1.5 ^{aA}	2.0 ± 1.3 ^{aA}
O_rancid	0	0.9 ± 2.1 ^{aA}	1.7 ± 2.4 ^{aA}	1.7 ± 2.6 ^{aA}	2.1 ± 2.6 ^{aA}	2.2 ± 2.9 ^{aA}
	1	1.2 ± 2.3 ^{aA}	2.3 ± 2.5 ^{aA}	2.4 ± 2.4 ^{aA}	2.0 ± 2.2 ^{aA}	2.3 ± 2.3 ^{aA}
	4	0.5 ± 1.1 ^{aA}	2.0 ± 2.5 ^{aA}	1.7 ± 1.7 ^{aA}	1.7 ± 1.9 ^{aA}	2.0 ± 2.2 ^{aA}
O_metallic	1	0.8 ± 0.9 ^{aA}	1.3 ± 1.2 ^{aA}	1.1 ± 1.2 ^{aA}	1.6 ± 1.1 ^{aA}	1.8 ± 1.5 ^{aA}
	4	1.6 ± 1.8 ^{aA}	1.9 ± 1.7 ^{aA}	2.2 ± 2.0 ^{aB}	2.1 ± 1.9 ^{aA}	2.0 ± 1.1 ^{aA}
Tx_crumbly	0	8.0 ± 1.9 ^{aA}	5.4 ± 1.9 ^{bA}	6.2 ± 1.9 ^{bA}	6.3 ± 1.8 ^{bA}	6.4 ± 1.8 ^{bA}
	1	9.4 ± 0.9 ^{aB}	7.5 ± 1.7 ^{bcB}	7.5 ± 1.6 ^{bcB}	7.2 ± 1.6 ^{bA}	8.4 ± 1.1 ^{cB}
	4	9.9 ± 0.3 ^{aC}	9.3 ± 0.7 ^{bC}	9.4 ± 1.1 ^{abC}	8.8 ± 1.2 ^{bB}	9.2 ± 1.1 ^{abB}
Tx_moistness	0	7.0 ± 1.3 ^{aA}	5.4 ± 1.4 ^{bA}	5.1 ± 1.4 ^{bA}	5.7 ± 1.4 ^{bA}	5.4 ± 1.5 ^{bA}
	1	7.3 ± 1.8 ^{aA}	5.8 ± 1.5 ^{bA}	5.5 ± 1.5 ^{bAB}	5.8 ± 1.1 ^{bA}	6.4 ± 1.7 ^{abB}
	4	7.7 ± 2.0 ^{aA}	6.6 ± 2.0 ^{aA}	6.6 ± 1.9 ^{aB}	6.3 ± 2.0 ^{aA}	6.2 ± 1.9 ^{aAB}
T_total	0	8.7 ± 0.5 ^{aA}	5.7 ± 1.2 ^{bA}	5.7 ± 1.1 ^{bA}	6.1 ± 0.9 ^{bA}	6.2 ± 1.0 ^{bA}
	1	9.0 ± 0.5 ^{aAB}	5.8 ± 1.0 ^{bA}	6.0 ± 1.1 ^{bA}	5.6 ± 1.2 ^{bA}	6.2 ±1.4 ^{bAB}
	4	9.3 ± 0.7 ^{aB}	6.6 ± 1.2 ^{bB}	6.5 ± 1.2 ^{bA}	6.0 ± 1.6 ^{bA}	7.0 ± 1.3 ^{bB}
T_sweet	0	4.1 ± 2.3 ^{aA}	5.2 ± 2.1 ^{aA}	5.4 ± 2.0 ^{aA}	5.1 ± 1.9 ^{aA}	5.4 ± 2.3 ^{aA}
	1	3.2 ± 2.1 ^{aA}	5.0 ± 2.2 ^{aA}	4.4 ± 2.2 ^{aA}	4.9 ± 2.1 ^{aA}	4.9 ± 2.4 ^{aA}
	4	4.7 ± 2.3 ^{aA}	6.2 ± 2.1 ^{aA}	6.2 ± 2.0 ^{aA}	6.1 ± 2.0 ^{aA}	6.1 ± 2.4 ^{aA}
T_salt	0	1.9 ± 1.7 ^{aA}	1.9 ± 1.5 ^{aA}	1.9 ± 1.9 ^{aAB}	2.0 ± 1.6 ^{aA}	2.0 ± 1.7 ^{aA}
	1	1.9 ± 2.1 ^{aA}	1.6 ± 1.5 ^{aA}	1.5 ± 1.5 ^{aA}	1.9 ± 1.9 ^{aA}	1.8 ± 1.9 ^{aA}
	4	2.5 ± 2.6 ^{aA}	3.0 ± 2.5 ^{aA}	3.1 ± 2.3 ^{aB}	3.0 ± 2.3 ^{aA}	3.1 ± 2.3 ^{aA}
T_umami	0	0.8 ± 1.3 ^{aA}	1.4 ± 1.6 ^{aA}	1.4 ± 1.6 ^{aA}	1.4 ± 1.5 ^{aA}	1.1 ± 1.3 ^{aA}
	1	1.0 ± 1.4 ^{aA}	1.2 ± 1.4 ^{aA}	1.5 ± 1.7 ^{aA}	1.3 ± 1.3 ^{aA}	1.8 ± 1.8 ^{aA}
	4	0.7 ± 1.0 ^{aA}	1.0 ± 1.1 ^{aA}	1.4 ± 1.3 ^{aA}	1.3 ± 1.4 ^{aA}	1.1 ± 1.2 ^{aA}
T_sour	0	6.3 ± 1.3 ^{aA}	3.0 ±1.3 ^{bAB}	2.9 ± 1.3 ^{bA}	3.2 ± 1.6 ^{bA}	3.6 ± 1.5 ^{bA}
	1	6.9 ± 1.5 ^{aA}	2.7 ± 1.5 ^{bA}	2.8 ± 1.4 ^{bA}	3.0 ± 1.2 ^{bA}	3.5 ± 1.3 ^{bA}
	4	7.6 ± 1.1 ^{aB}	4.1 ± 2.1 ^{bB}	4.0 ± 2.1 ^{bA}	4.2 ± 2.1 ^{bA}	5.3 ± 2.1 ^{bB}

Table 5. *Cont.*

Attribute [1]	Months	Acetic Acid	Citric Acid	Lactic Acid	Malic Acid	Tartaric Acid
T_bitter	0	1.2 ± 1.3 ^{aA}	0.9 ± 2.1 ^{aA}	0.8 ± 1.9 ^{aA}	0.9 ± 1.9 ^{aA}	1.2 ± 1.9 ^{aA}
	1	1.8 ± 0.9 ^{aA}	1.6 ± 2.1 ^{aA}	2.0 ± 1.5 ^{aAB}	1.1 ± 1.7 ^{aA}	1.8 ± 2.1 ^{aA}
	4	1.5 ± 1.0 ^{aA}	1.9 ± 2.2 ^{aA}	2.6 ± 2.2 ^{aB}	1.7 ± 1.9 ^{aA}	2.5 ± 1.9 ^{aA}
F_vinegar	0	8.0 ± 1.1 ^{aA}	3.4 ± 1.1 ^{bA}	2.7 ± 1.5 ^{bAB}	3.9 ± 1.3 ^{bA}	3.8 ± 1.5 ^{bA}
	1	8.5 ± 1.7 ^{aA}	2.9 ± 1.6 ^{bA}	2.3 ± 1.8 ^{bA}	3.3 ± 1.1 ^{bA}	3.7 ± 1.5 ^{bA}
	4	9.0 ± 1.5 ^{aA}	3.9 ± 2.2 ^{bA}	4.2 ± 2.5 ^{bB}	3.6 ± 1.4 ^{bA}	5.1 ± 2.1 ^{bA}
F_fish	0	3.2 ± 1.5 ^{aA}	4.6 ± 1.7 ^{bA}	4.7 ± 1.7 ^{bA}	4.8 ± 1.8 ^{bA}	4.8 ± 2.0 ^{bA}
	1	2.8 ± 1.4 ^{aA}	4.2 ± 1.6 ^{bA}	4.2 ± 1.7 ^{bA}	4.3 ± 1.5 ^{bA}	4.3 ± 1.8 ^{bA}
	4	2.9 ± 1.4 ^{aA}	4.8 ± 1.3 ^{bA}	5.1 ± 1.6 ^{bA}	5.0 ± 1.3 ^{bA}	4.8 ± 1.4 ^{bA}
F_stuffy	0	0.6 ± 1.3 ^{aA}	1.0 ± 1.5 ^{aA}	0.9 ± 1.6 ^{aA}	1.4 ± 2.2 ^{aA}	1.1 ± 1.4 ^{aA}
	1	0.7 ± 1.2 ^{aA}	1.2 ± 1.8 ^{aA}	1.5 ± 1.2 ^{aA}	1.1 ± 1.8 ^{aA}	1.7 ± 2.4 ^{aAB}
	4	0.7 ± 1.2 ^{aA}	1.8 ± 1.8 ^{abA}	2.1 ± 2.0 ^{abA}	1.4 ± 1.4 ^{abA}	2.7 ± 2.1 ^{bB}
F_rancid	0	0.5 ± 1.5 ^{aA}	1.6 ± 2.3 ^{aA}	1.7 ± 2.5 ^{aA}	1.2 ± 2.1 ^{aA}	2.1 ± 3.0 ^{aA}
	1	1.2 ± 2.0 ^{aA}	1.5 ± 2.0 ^{aA}	1.9 ± 2.2 ^{aA}	1.9 ± 2.2 ^{aA}	2.1 ± 2.2 ^{aA}
	4	1.1 ± 1.3 ^{aA}	1.7 ± 1.8 ^{aA}	2.5 ± 2.0 ^{aA}	1.5 ± 1.8 ^{aA}	2.6 ± 2.2 ^{aA}
F_metallic	0	1.0 ± 1.3 ^{aA}	1.4 ± 1.4 ^{abA}	1.9 ± 1.9 ^{abA}	2.1 ± 2.0 ^{abA}	2.4 ± 1.8 ^{bA}
	1	1.4 ± 1.2 ^{aB}	1.9 ± 1.5 ^{abA}	2.7 ± 1.8 ^{bA}	1.9 ± 1.6 ^{abA}	2.7 ± 1.6 ^{bAB}
	4	2.7 ± 1.9 ^{abB}	2.9 ± 1.8 ^{abB}	3.8 ± 1.8 ^{abB}	2.6 ± 1.8 ^{aA}	4.1 ± 1.8 ^{bB}

[1] O = odor, Tx = texture, T = taste, F = flavor. $^{a\text{–}c}$ Different letter indicates statistically significant difference between samples at each time point's evaluation. $^{A\text{–}C}$ Different letter within an attribute indicates statistically significant difference between each time point's evaluation of each acid sample.

During the 4 months of preservation, there were statistically significant changes between the samples in 12 of the twenty sensory attributes: total intensity of odor and taste, vinegary and metallic odor and flavor, crumbly and moist texture, salty, sour, and bitter taste, and stuffy and metallic flavor. There were no differences between the samples at any point in eight of the sensory attributes: sweet, stuffy, and metallic odor, sweet, salty, umami, and bitter taste, and rancid flavor.

The acetic acid sample was clearly stronger than most of the other samples in total intensity of odor and taste, vinegary odor and flavor, sour taste, and crumbly texture. The differences were significant at all time points. The crumbly texture was more intense in the acetic acid sample than in the other samples after 0 and 1 months of storage. It was more intense in the acetic acid sample compared to the citric acid and malic acid samples after 4 months of storage. The intensity of the crumbly texture was above 9 in all samples except the malic acid sample after 4 months of storage. Although the fishy odor and flavor were recognized in all the samples, the acetic acid sample was noticeable as being the mildest in these attributes during the whole study. The tartaric acid sample was significantly stronger in total intensity of odor compared to the citric acid sample and in crumbliness compared to the malic acid sample after 1 month of storage. The tartaric acid sample was significantly stronger in metallic flavor compared to the malic acid sample after 4 months of storage.

The sensory quality changed during the preservation. The lactic acid sample had the greatest number of sensory attributes that changed during the 4-month storage period. In the lactic acid sample seven of twenty sensory attributes had statistically significant differences in intensities when comparing the sensory evaluations over the different months; metallic odor and flavor, crumbliness and moistness of texture, salty and bitter taste, and vinegary flavor and their intensity had increased over time. In contrast, in the malic acid sample the only sensory attribute that changed over time was its increasing crumbliness. The crumbly texture was the only attribute that changed in all the samples over time.

4. Discussion

The influence of acetic, citric, lactic, malic, and tartaric acids on the preservability and quality of pickled and marinated herring was studied by measuring the microbiological quality, pH, composition, and lipid oxidation and by sensory profiling.

Organic acids are weak acids and vary in strength (pKa). For instance, acetic acid has a pKa of 4.76, making it a weaker acid than most other organic acids, such as tartaric acid (pKa = 2.98) or lactic acid (pKa = 3.86). The pKa is determined by the degree of dissociation in water and therefore the pH for a certain concentration of an acid [31]. However, different muscles vary in their buffering ability, i.e., their ability to resist the change of pH upon the addition of acid (or base) [32]. Due to the different buffering abilities of the acids, the acid concentrations in the pickling and marinating solutions varied as pH level in this study was aimed at 4. Since the differences between the samples regarding pH at the end of the pickling and marinating were so small, it seems likely that the acid accounts for the differences in the results of chemical analyses rather than the pH and different acid concentrations.

Food intended for human consumption must be microbiologically safe. According to the results the microbial quality of all the tested products was excellent in respect of both food-borne pathogens *L. monocytogenes* and the sulfite-reducing Clostridia and the food spoilage microbes including yeasts, aerobic mesophilic and psychro-trophic microorganisms and hydrogen sulfide-producing bacteria. This was expected due to the low pH (pH < 4.2) of the samples. In the weak organic acids used in our study, pH is considered the determining factor in their antimicrobial effects, since it is connected to the degree of dissociation. The suggested mechanism behind this effect is that, in an undissociated form, acids can pass through the bacterial cell wall, and in the neutral cytoplasm of the cell dissociate into H^+ ions and anions, both of which can be detrimental to bacteria [31].

In addition to low pH, other environmental factors, including NaCl content, water activity, and storage temperature affect microbial growth. In this study, the effects of the aforementioned factors on microbial growth were not studied.

Pickling and storage in a marinade has an effect on the chemical composition of fish. In this study, the moisture content of the Baltic herring fillets increased, while both the protein and fat content decreased during the 4-month storage. This is in contrast to the results reported in previous studies [15–17]. This is probably due to two reasons: the low salinity of the pickling solution in the pickling procedure and the complete absence of salt in the marinade solution. The salinity of the water affects the solubility of the proteins. At low concentrations (<5.8%, here 5%) the proteins swell, and their solubility increases. [33–35] As a result, salt and water are absorbed into the muscle and soluble proteins and non-protein nitrogen compounds diffuse into the surrounding solution [12].

During the pickling and marinating processes, acid diffuses into the fish muscle, lowering the pH and causing protein denaturation and lower water absorption [17]. If the pH of the pickling solution is low and, thus, the pH of the muscle tissue is on the acidic side of the isoelectric point, it results in electrostatic repulsion of actin and myosin filaments, which causes an open structure in the muscle tissue and increases the water retention. However, when the salt content increases further, the repulsive charges are shielded, which lowers the water retention capacity. Therefore, salt is needed during the pickling and marinating process, as the use of too low salt content results in soft fillets [36]. In future studies and product development the optimal level of salt in the pickling and marinating solutions should be carefully adjusted.

Despite a direct comparison of the MDA levels of the marinated Baltic herring fillet with those in the literature being difficult, due to the different methods and units used, the levels measured in this study were in the same range as the literature. Halamickova and Malota [37] measured the level of TBARS in marinated Atlantic herring muscles collected in a market. The lowest TBARS levels were in warm baked marinades (1.17 ± 0.40 mg MDA/kg muscle) while the highest TBARS levels were observed in warm cooked marinades (16.48 ± 4.22 mg MDA/kg muscle). The MDA levels of Baltic herring fillets in our study after four months of storage varied from

1.0 ± 0.1 μM/g dm–2.1 ± 0.2 μM/g dm corresponding to approximately 4–8 mg MDA/kg muscle, which is the same level as the warm baked marinated Atlantic herring [37].

Baltic herring is rich in PUFAs which are susceptible to oxidation. The risk of oxidation is further increased due to the high content of pro-oxidative heme proteins found in the muscle of this species [38]. The results of our study indicate that all the acids tested are able to maintain the commercial quality of Baltic herring fillets for four months in terms of oxidation. However, the differences between the acids are significant, highlighting the need for careful selection of the acid to be used in preservation. The possible pro-oxidative effect of citric acid requires further studies.

Pickling and marinating with different weak acids resulted in different products in terms of the sensorial quality. The samples pickled and marinated with citric, lactic, malic, and tartaric acid did not have any differences in the intensities of the sensory attribute at the beginning of the storage, whereas the acetic acid sample was noticeably different from the other samples in nine sensorial properties. These were total intensity of odor and taste, intensity of vinegary odor and flavor, crumbly and moist texture, sour taste, fishy flavor, and metallic flavor. Later during the storage, the same attributes, apart from moist texture, continued to demonstrate differences between the samples.

The sensory attributes indicated that the strong and distinctive vinegary character of acetic acid dominated in the sample pickled and marinated in acetic acid. The strong vinegary odor and flavor as well as the sourness identified in this sample may have suppressed other sensory properties of the pickled fish. The dominating sensory attributes in the other samples were the total intensity of odor and flavor, a fishy odor and a flavor and sweet taste, as well as a crumbly and moist texture. However, the intensities of these attributes were not strong. These results clearly indicate that by replacing acetic acid with other weak acids frequently used in food industry, pickled and marinated fish products that have novel and milder sensory profiles can be produced.

In the sensory analysis, the crumbliness of the texture of the samples increased. The intensity of the crumbly texture was already significantly higher in the acetic acid treated sample at the beginning of the storage compared to the other samples. The intensity of crumbliness increased throughout the study in all the samples but at different rates: in acetic, citric, or lactic acid treated samples the crumbliness increased steadily throughout the study, whereas in the sample treated with malic acid the significant increase took place only between 1 and 4 months and in the sample treated with tartaric acid between 0 and 1 month. The development of crumbliness was most likely due to the changes in the protein structure and solubility. There was also a trend for the moist texture in all the samples to increase in intensity, although not all of them showed a significant increase. This, together with the chemical analysis, indicates that during marination several processes caused by the pH and the salt content affect the moisture of the meat. However, the current data is not sufficient enough to identify a causal relationship between chemical analysis and sensory data.

No statistically significant indication on the development of a rancid odor and flavor was detected in the sensory evaluations despite the chemically measured increase of MDA in the samples. This was the case especially in the citric acid sample. Refsgaard et al. [39] found that the metallic flavor of cooked salmon correlated with hexanal, a main oxidation product of linoleic acid. The sample pickled and marinated in citric acid showed a statistically significant increase in the intensity the metallic flavor. However, the intensity of the metallic flavor also increased in all the other samples except the sample pickled and marinated in malic acid, although it also showed oxidation over time. The sample pickled and marinated in tartaric acid did not show any oxidation over time. Therefore, it is likely that the levels of MDA formation measured by chemical analyses are not reflected in the sensory quality.

Sensory profile studies of pickled or marinated herring are very limited. However, Nielsen et al. [26,27] studied the sensory properties of Atlantic and Baltic herring pickled in acetic acid. The sensory attributes of the profiles are similar to those of this study. The attributes in the studies described the sweetness and saltiness of the marinade, the sour and vinegary properties of acetic acid, as well as the fishy and metallic odor and flavor of herring. Rancidity was also included in the sensory attributes. Texture properties were also evaluated with the following attributes: firmness, elasticity, fatty mouth feel, juicy, and gritty. The texture properties in this study were evaluated with only two attributes describing the fracturability and moistness of the samples. However, the previous studies also included samples with Atlantic herring which contains more fat and can be larger in size than Baltic herring. This affects the descriptors in the sensory profile. Furthermore, the pickling process was different between this and the previous studies; this study utilized a considerably lower level of salt which was most likely the reason for the crumbly texture.

5. Conclusions

In conclusion, pickling with citric, lactic, malic, and tartaric acids in addition to acetic acid with pH levels of 3.7–4.2 resulted in pickled Baltic herring products with high microbiological quality similar to the traditional pickling with acetic acid. The results of the chemical analysis of the samples indicate that pickling and storage in marinade influenced the chemical composition of fish, e.g., by increasing the moisture and decreasing the protein content of the samples. Fat content decreased during the storage period in acetic acid and malic acid samples, and minor lipid oxidation took place especially in the citric acid sample. The flavor profiles of the samples other than the acetic acid sample were mild, especially in sourness and vinegariness. The fishy odor and flavor were perceived more strongly in citric, lactic, malic, and tartaric acid samples compared to the acetic acid sample. Overall, the results indicate that Baltic herring pickled and marinated in other food grade organic acids than the traditionally used acetic acid resulted in preservable fish products with novel and milder sensory profiles.

Supplementary Materials: The following are available online at https://www.mdpi.com/article/10.3390/foods11121717/s1, Table S1: Concentration and pH values of pickling and marinade solutions of training samples of 0 months' sensory evaluation.

Author Contributions: Conceptualization, N.L., J.H., T.K., A.-L.V., S.M., N.R., P.M., B.Y., A.H.; methodology, N.L., J.H., T.K., A.-L.V., S.M., N.R., P.M., B.Y., A.H.; formal analysis, N.L., J.H., T.K., A.-L.V., S.M.; investigation, N.L., J.H., T.K., A.-L.V., S.M., N.R., P.M., B.Y., A.H.; writing—original draft preparation, N.L., J.H., T.K., A.-L.V., S.M., N.R., P.M., B.Y., A.H.; writing—review and editing, N.L., J.H., T.K., A.-L.V., S.M., N.R., P.M., B.Y., A.H.; visualization, N.L., J.H., T.K., A.-L.V., S.M., A.H.; supervision, S.M., B.Y., A.H.; project administration, A.H.; funding acquisition, J.H., S.M., P.M., B.Y., A.H. All authors have read and agreed to the published version of the manuscript.

Funding: This research was funded by Finnish Operational Programme of the European Maritime and Fisheries Fund (the Project "Blue Welfare Network-Blue Products", Project no. 33338).

Institutional Review Board Statement: The study was conducted in accordance with the Declaration of Helsinki, and approved by the Ethics Committee of the Hospital District of Southwest Finland (Sandell 145/1801/2014).

Informed Consent Statement: Informed consent was obtained from all subjects involved in the study.

Data Availability Statement: The data presented in this study are available on request from the corresponding author. The data are not publicly available due to privacy.

Acknowledgments: Our sincere thanks go to Tiina Väyrynen, Natural Resources Institute Finland (Luke), for performing the microbial analysis in Natural Resources Institute Finland (Luke), Oulu. Tiina Väyrynen and Jouni Karhu, Natural Resources Institute Finland (Luke), are acknowledged for carrying out statistical analysis. Mari Sandell, Department of Food and Nutrition, University of Helsinki and Functional Foods Forum, University of Turku, is acknowledged for her assistance in statistical analysis. Saila Mattila, Functional Foods Forum, University of Turku, is acknowledged for her assistance in sensory analysis. Satu Örling-Vigren, Natural Resources Institute Finland (Luke), is acknowledged for performing lipid oxidation analysis. We acknowledge Camillas Fiskdelikatess Ltd. (Molpe, Finland) for providing facilities and assistance in producing the research samples.

Conflicts of Interest: The authors declare no conflict of interest. The funders had no role in the design of the study; in the collection, analyses, or interpretation of data; in the writing of the manuscript, or in the decision to publish the results.

References

1. Belitz, H.D.; Grosch, W.; Schieberle, P. *Food Chemistry*, 4th ed.; Springer: Berlin/Heidelberg, Germany, 2009; ISBN 9783540699330.
2. *The European Union One Health 2019 Zoonoses Report*; European Food Safety Authority; European Centre for Disease Prevention and Control, 2021.
3. Jami, M.; Ghanbari, M.; Zunabovic, M.; Domig, K.J.; Kneifel, W. Listeria monocytogenes in Aquatic Food Products—A Review. *Compr. Rev. Food Sci. Food Saf.* **2014**, *13*, 798–813. [CrossRef]
4. Bergis, H.; Bonanno, L.; Asséré, A.; Lombard, B.; Polet, M.; Andersen, J.K. EURL Lm Technical Guidance Document on Challenge Tests and Durability Studies for Assessing Shelf-Life of Ready-to-Eat Foods Related to Listeria Monocytogenes. Maisons-Alfort, France, 24 August 2021.
5. Cheng, C.; Yang, Y.; Dong, Z.; Wang, X.; Fang, C.; Yang, M.; Sun, J.; Xiao, L.; Fang, W.; Song, H. Listeria monocytogenes Varies Among Strains to Maintain Intracellular pH Homeostasis Under Stresses by Different Acids as Analyzed by a High-Throughput Microplate-Based Fluorometr. *Front. Microbiol.* **2015**, *6*, 15. [CrossRef] [PubMed]
6. Andrés-Bello, A.; Barreto-Palacios, V.; García-Segovia, P.; Mir-Bel, J.; Martínez-Monzó, J. Effect of pH on Color and Texture of Food Products. *Food Eng. Rev.* **2013**, *5*, 158–170. [CrossRef]
7. Ertbjerg, P.; Mielche, M.M.; Larsen, L.M.; Møller, A.J. Relationship Between Proteolytic Changes and Tenderness in Prerigor Lactic Acid Marinated Beef. *J. Sci. Food Agric.* **1999**, *79*, 970–978. [CrossRef]
8. Damodaran, S.; Parkin, K.L. *Fennema's Food Chemistry*, 5th ed.; Kirk, L., Ed.; CRC Press: Boca Raton, FL, USA, 2017; ISBN 1-5231-1785-0.
9. Burke, R.M.; Monahan, F.J. The Tenderisation of Shin Beef Using a Citrus Juice Marinade. *Meat Sci.* **2003**, *63*, 161–168. [CrossRef]
10. Szymczak, M.; Kołakowski, E.; Felisiak, K. Effect of Addition of Different Acetic Acid Concentrations on the Quality of Marinated Herring. *J. Aquat. Food Prod. Technol.* **2015**, *24*, 566–581. [CrossRef]
11. Topuz, O.K. Effects of Marinating Time, Acetic Acid and Salt Concentrations on the Quality of Little Tunny Fish (Euthynnus alletteratus) Fillet. *J. Food Processing Preserv.* **2016**, *40*, 1154–1163. [CrossRef]
12. Szymczak, M.; Kołakowski, E. Losses of Nitrogen Fractions from Herring to Brine During Marinating. *Food Chem.* **2012**, *132*, 237–243. [CrossRef]
13. Ahmed, Z.; Donkor, O.; Street, W.A.; Vasiljevic, T. Calpains- and Cathepsins-Induced Myofibrillar Changes in Post-Mortem Fish: Impact on Structural Softening and Release of Bioactive Peptides. *Trends Food Sci. Technol.* **2015**, *45*, 130–146. [CrossRef]
14. Fusek, M.; Mares, M.; Vetvicka, V. *Cathepsin, D. Handbook of Proteolytic Enzymes*; Rawlings, N.D., Salvesen, G., Eds.; Academic Press: Cambridge, MA, USA, 2013; Volume 1, pp. 54–63; ISBN 9780123822192.
15. Wang, P.A.; Stenvik, J.; Larsen, R.; Maehre, H.; Olsen, R.L. Cathepsin D from Atlantic Cod (Gadus morhua L.) Liver. Isolation and Comparative Studies. *Comp. Biochem. Physiol.* **2007**, *147*, 504–511. [CrossRef]
16. Szymczak, M.; Lepczyński, A. Occurrence of Aspartyl Proteases in Brine After Herring Marinating. *Food Chem.* **2016**, *194*, 470–475. [CrossRef] [PubMed]
17. Richards, M.P.; Modra, A.M.; Li, R. Role of Deoxyhemoglobin in Lipid Oxidation of Washed Cod Muscle Mediated by Trout, Poultry and Beef Hemoglobins. *Meat Sci.* **2002**, *62*, 157–163. [CrossRef]
18. Damerau, A.; Kakko, T.; Tian, Y.; Tuomasjukka, S.; Sandell, M.; Hopia, A.; Yang, B. Effect of Supercritical CO_2 Plant Extract and Berry Press Cakes on Stability and Consumer Acceptance of Frozen Baltic Herring (Clupea harengus membras) Mince. *Food Chem.* **2020**, *332*, 127396. [CrossRef]
19. Aitta, E.; Marsol-Vall, A.; Damerau, A.; Yang, B. Enzyme-Assisted Extraction of Fish Oil from Whole Fish and By-Products of Baltic Herring (Clupea harengus membras). *Foods* **2021**, *10*, 1811. [CrossRef] [PubMed]
20. Ke, S.; Huang, Y.; Decker, E.A.; Hultin, H.O. Impact of Citric Acid on the Tenderness, Microstructure and Oxidative Stability of Beef Muscle. *Meat Sci.* **2009**, *82*, 113–118. [CrossRef]
21. Pourashouri, P.; Shabanpour, B.; Aubourg, S.P.; Rohi, J.D.; Shabani, A. An Investigation of Rancidity Inhibition During Frozen Storage of Wels Catfish (Silurus glanis) Fillets by Previous Ascorbic and Citric Acid Treatment. *Int. J. Food Sci. Technol.* **2009**, *44*, 1503–1509. [CrossRef]
22. Kawai, M.; Sekine-Hayakawa, Y.; Okiyama, A.; Ninomiya, Y. Gustatory Sensation of L- and D-Amino Acids in Humans. *Amino Acids* **2012**, *43*, 2349–2358. [CrossRef] [PubMed]
23. Demirok, E.; Kolsarici, N.; Çelik, S.; Doğan, Z.; Hamdan, S.; Öztürk, F. Proteolytic and Sensory Changes During Marination of Rainbow Trout (Oncorhynchus mykiss) Flesh in Pomegranate Juice. *J. Aquat. Food Prod. Technol.* **2014**, *23*, 621–632. [CrossRef]
24. Šimat, V.; Mićunović, A.; Bogdanović, T.; Listeš, I.; Generalić Mekinić, I.; Hamed, I.; Skroza, D. The Impact of Lemon Juice on the Marination of Anchovy (Engraulis encrasicolus): Chemical, Microbiological and Sensory Changes. *Ital. J. Food Sci.* **2019**, *31*, 604–617. [CrossRef]
25. Babikova, J.; Hoeche, U.; Boyd, J.; Noci, F. Nutritional, Physical, Microbiological, and Sensory Properties of Marinated Irish Sprat. *Int. J. Gastron. Food Sci.* **2020**, *22*, 100277. [CrossRef]
26. Nielsen, D.; Hyldig, G.; Nielsen, H.H.; Nielsen, J. Sensory Properties of Marinated Herring (Clupea harengus)—Influence of Fishing Ground and Season. *J. Aquat. Food Prod. Technol.* **2004**, *13*, 3–24. [CrossRef]

27. Nielsen, D.; Hyldig, G.; Nielsen, J.; Nielsen, H.H. Sensory Properties of Marinated Herring (Clupea harengus) Processed from Raw Material from Commercial Landings. *J. Sci. Food Agric.* **2005**, *85*, 127–134. [CrossRef]
28. SFS-EN ISO 11290-1:2017: En. Available online: https://sales.sfs.fi/fi/index/tuotteet/SFS/CENISO/ID2/1/515736.html.stx (accessed on 2 February 2022).
29. Mäkinen, S.; Hellström, J.; Mäki, M.; Korpinen, R.; Mattila, P.H. Bilberry and Sea Buckthorn Leaves and Their Subcritical Water Extracts Prevent Lipid Oxidation in Meat Products. *Foods* **2020**, *9*, 265. [CrossRef] [PubMed]
30. Parisi, S. Oxidative Stability and Shelf Life of Foods Containing Oils and Fats. *Anal. Bioanal. Chem.* **2016**, *408*, 7549–7550. [CrossRef]
31. Ricke, S.C. Perspectives on the Use of Organic Acids and Short Chain Fatty Acids as Antimicrobials. *Poult. Sci.* **2003**, *82*, 632–639. [CrossRef]
32. Castellini, M.A.; Somero, G.N. Buffering Capacity of Vertebrate Muscle: Correlations with Potentials for Anaerobic Function. *J. Comp. Physiol.* **1981**, *143*, 191–198. [CrossRef]
33. Gallart-Jornet, L.; Barat, J.M.; Rustad, T.; Erikson, U.; Escriche, I.; Fito, P. A Comparative Study of Brine Salting of Atlantic Cod (Gadus morhua) and Atlantic Salmon (Salmo salar). *J. Food Eng.* **2007**, *79*, 261–270. [CrossRef]
34. Thorarinsdottir, K.A.; Arason, S.; Sigurgisladottir, S.; Gunnlaugsson, V.N.; Johannsdottir, J.; Tornberg, E. The Effects of Salt-Curing and Salting Procedures on the Microstructure of Cod (Gadus morhua) Muscle. *Food Chem.* **2011**, *126*, 109–115. [CrossRef]
35. Erikson, U.; Veliyulin, E.; Singstad, T.E.; Aursand, M. Salting and Desalting of Fresh and Frozen-Thawed Cod (Gadus morhua) Fillets: A Comparative Study Using 23Na NMR, 23Na MRI, Low-Field 1H NMR, and Physicochemical Analytical Methods. *J. Food Sci.* **2004**, *69*, 107–114. [CrossRef]
36. McLay, R.; Pirie, R. Development of Marinated Herring. *Int. J. Food Sci. Technol.* **1971**, *6*, 29–38. [CrossRef]
37. Halamíčková, A.; Malota, L. Muscle Thiobarbituric Acid Reactive Substance of the Atlantic Herring (Clupea harengus) in Marinades Collected in the Market Network. *Acta Vet. Brno* **2010**, *79*, 329–333. [CrossRef]
38. Larsson, K.; Almgren, A.; Undeland, I. Hemoglobin-Mediated Lipid Oxidation and Compositional Characteristics of Washed Fish Mince Model Systems Made from Cod (Gadus morhua), Herring (Clupea harengus), and Salmon (Salmo salar) Muscle. *J. Agric. Food Chem.* **2007**, *55*, 9027–9035. [CrossRef] [PubMed]
39. Refsgaard, H.H.F.; Brockhoff, P.B.; Jensen, B. Sensory and Chemical Changes in Farmed Atlantic Salmon (Salmo solar) during Frozen Storage. *J. Agric. Food Chem.* **1998**, *46*, 3473–3479. [CrossRef]

foods

MDPI

Article

Effect of High Hydrostatic Pressure Combined with Sous-Vide Treatment on the Quality of Largemouth Bass during Storage

Mingzhu Zhou [1,2,†], Yuzhao Ling [1,3,†], Fangxue Chen [1,2], Chao Wang [2], Yu Qiao [1,*], Guangquan Xiong [1], Lan Wang [1], Wenjin Wu [1], Liu Shi [1] and Anzi Ding [1]

[1] Key Laboratory of Cold Chain Logistics Technology for Agro-Product, Ministry of Agriculture and Rural Affairs, Institute of Agro-Products Processing and Nuclear Agricultural Technology, Hubei Academy of Agricultural Sciences, Wuhan 430064, China; 101900476@hbut.edu.cn (M.Z.); lingyuzhao2012@163.com (Y.L.); aphroditesnow@hotmail.com (F.C.); xiongguangquan@163.com (G.X.); lilywang_2016@163.com (L.W.); wuwenjin@hbaas.com (W.W.); shiliu@hbaas.com (L.S.); anzi@hbaas.com (A.D.)

[2] Key Laboratory of Fermentation Engineering (Ministry of Education), Hubei Key Laboratory of Industrial Microbiology, National "111" Center for Cellular Regulation and Molecular Pharmaceutics, Cooperative Innovation Center of Industrial Fermentation (Ministry of Education & Hubei Province), Hubei Research Center of Food Fermentation Engineering and Technology, Hubei University of Technology, Wuhan 430068, China; wangchao@hbut.edu.cn

[3] School of Environmental Ecology and Biological Engineering, Wuhan Institute of Technology, Wuhan 430205, China

* Correspondence: qiaoyu@hbaas.com; Tel.: +86-13971371204

† These authors contributed equally to this work.

Abstract: In order to estimate the effects of high hydrostatic pressure treatment at 400 MPa for 0 min and 10 min (HHP-0, HHP-10) and high hydrostatic pressure in combination with sous-vide treatment (HHP-0+SV, HHP-10+SV) on the quality of largemouth bass stored at 4 °C for 30 days, the physicochemical changes were evaluated by microbiological determinations, pH, sensory evaluation and texture analysis, and the flavour changes were analysed by solid-phase microextraction–gas chromatography–mass spectrometry (SPME-GC-MS) and amino acid automatic analyser. The results show that HHP-0+SV and HHP-10+SV treatment effectively inhibited microbiological growth and attenuated physiochemical changes (pH, sensory evaluation, flesh and texture) of largemouth bass fillets. HHP+SV treatment prolonged the storage period of largemouth bass fillets for 24 days. The content of total free amino acids in control (CK) samples was high, but HHP+SV treatment caused the loss of free amino acid content. Especially when stored for 30 days, the total free amino acid content of HHP-0+SV and HHP-10+SV was only 14.67 mg/100 g and 18.98 mg/100 g, respectively. In addition, a total of 43 volatile compounds were detected and elucidated, among which hexanal, heptaldehyde, octanal and nonanal showed a decreasing tendency in HHP groups and an increasing trend in HHP+SV groups throughout the storage.

Keywords: high hydrostatic pressure; sous-vide; sensory; storage; largemouth

Citation: Zhou, M.; Ling, Y.; Chen, F.; Wang, C.; Qiao, Y.; Xiong, G.; Wang, L.; Wu, W.; Shi, L.; Ding, A. Effect of High Hydrostatic Pressure Combined with Sous-Vide Treatment on the Quality of Largemouth Bass during Storage. *Foods* **2022**, *11*, 1931. https://doi.org/10.3390/foods11131931

Academic Editor: Eduardo Puértolas

Received: 27 April 2022
Accepted: 21 June 2022
Published: 29 June 2022

1. Introduction

In many places, especially in coastal areas, fish is considered to be an excellent source of nutrition, containing more proteins than other animals [1]. Largemouth bass (Micropterus salmoides) is native to north America. Compared with other farmed fish, it is one of the most commercial fish [2]. Protein, vitamins (including vitamin A, B2, B6) and minerals (such as iron, calcium, potassium) are abundant in largemouth bass [3]. In recent years, the intake of fish has been linked with potential health benefits that can help people prevent a variety of diseases. Nowadays, there is a greater focus on safe pretreatment methods in order to keep the original flavour and nutrition of the food to the maximum extent. For example, pretreatment such as ultrasonic and HHP.

Since consumers prefer ready-to-eat products that are as natural as possible and have sufficient shelf life, it is important to consider appropriate minimal processing without altering the content and bioavailability of major nutrients [4]. In this sense, HHP, a non-thermal technology, seems to be a promising technique for preserving heat-sensitive compounds (for example, aldehydes, pyrazines and furans) and maintaining sensory quality such as colour, as well as inactivating microorganisms and enzymes [5,6]. HHP (100–800 MPa) has been shown to inactivate a wide range of microorganisms and endogenous enzymes while retaining the sensory properties and nutritional value of food [7,8]. Ye et al. [9] reported the effect of high-pressure treatment (300 MPa/20 min) on crab meat during storage at 4 °C, and indicated that the total viable counts (TVC) (5.71 log10 CFU/g) were still below the limit (6 log10 CFU/g) on day 8. Lin et al. [10] found that HHP could reduce the bacterial load and retard TVBN production, in addition, HHP conditions of ≥400 MPa for 3 min significantly extended the shelf life of hard clams during refrigerated storage. Kim et al.'s [11] results suggested that HHP treatment would be useful for inhibiting the activity of urease, thereby reducing the fishy smells from fish. However, expensive equipment and limited inactivation of spores are major restrictions to the widespread application of HHP [12].

SV temperature cooking is a food preparation process of heating samples for a long time at a lower temperature [13]. The food is heated to the central temperature of 58–95 °C in a vacuum sealing system. Compared with traditional cooking methods, this method is characterized by less efflux of soluble nutrients [14]. The sample is vacuum-packed in cling bags before heating, which reduces oxidation and prevents cross contamination. The reported advantages of using SV are the reduction of cooking losses and lipid oxidation while enhancing colour and flavour [15]. In addition, SV can improve meat tenderness [16]. Jiang et al. [17] showed that during long-term thermal processing, SV could significantly reduce the level of protein oxidation and fat oxidation, and effectively preserve the flavour and nutrition of braised meat. The cooking yield of poultry meat processed by the SV method (88.5%) was higher than that by the traditional steaming method (71.0%). The meat was found to be redder (a* values of 2.54 and 0.74, respectively) and less yellow (b* values of 15.12 and 16.49, respectively), as well as more tender [18].

Two emerging technologies, HHP and SV are regarded as effective and economical methods. The utilization of HHP or SV alone to effectively control food spoilage has been widely reported. However, reports are rare exploring the effects of HHP combined with SV on largemouth bass as far as we know. In this study, the effects of HHP and HHP+SV on the quality of largemouth bass were evaluated by measuring the indexes of microorganism, pH value, sensory, colour, texture, GC-MS and FAA.

2. Materials and Methods

2.1. Raw Materials

Cultured largemouth bass (weight of 0.38 ± 0.07 kg) were originated from Wuhan, and subsequently transported to the slaughter room alive in plastic bags filled with crushed ice within an hour. After stunning, the scales were removed. The whole bass were immediately washed with sterile water, and then cut into similar-sized fillets (about 7 cm × 5 cm × 2 cm). Fillets were randomly divided into five groups, and then fillets were subjected to HHP and SV pretreatments. Then, the fish fillets were drained, packaged separately in high-density polyethylene bags and stored at 4 °C. Three samples of each group were selected randomly for detection and analysis at day 0, 6, 12, 18, 24 and 30 of the storage.

2.2. SV Cooking

The vacuum-packed fish fillets were immersed in a constant temperature water bath (DF-101S, Zhengzhou Great Wall Technology Industry and Trade Co., Ltd. Zhengzhou, China), which contained 1.3 L distilled water that was pre-heated to 90 °C. The cooking process was ended until the core temperature of samples reached 90 ± 5 °C. In the light of pre-experiments, the cooking time was 15 min. Then the samples were immediately cooled to 7 ± 2 °C in an ice water bath.

2.3. HHP and SV Treatment

The treatment of the packaged fish fillets was implemented using high hydrostatic pressure equipment (HPPL2-600 MPa/2L, Tianjin Huatai SenMiao Biotechnology Engineering Limited by Share Ltd., TianJin, China). High hydrostatic pressure treatments were performed at 400 MPa for different times. High hydrostatic pressure treatment was to place the packaged bass slices in the high hydrostatic pressure chamber (1 L) with water as the medium (25 °C), and keep them for a certain time when the pressure reached 400 MPa (30 s). There were three replicates for each treatment, and 300 g bass slices were used in each replicate. After HHP treatment, SV treatment was carried out according to Section 2.2. Fish fillets were subjected to four kinds of treatment.

(1) High hydrostatic pressure treatment at 400 MPa for 0 min. (HHP-0)
(2) High hydrostatic pressure treatment at 400 MPa for 10 min. (HHP-10)
(3) High hydrostatic pressure treatment at 400 MPa for 0 min in combination with sous vide. (HHP-0+SV).
(4) High hydrostatic pressure treatment at 400 MPa for 10 min in combination with sous vide. (HHP-10+SV).

2.4. Microbiological Determinations

A total of 10 g sample was added into 90 mL 0.85% sterilized physiological solution and the mixture was homogenized for 60 s. The microbiological evaluations were performed based on the work of Rezaeigolestani et al. [19], where serial ten-fold dilutions were made with sterilized physiological saline solution and 1 mL of the appropriately diluted sample was poured onto total viable counts (TVC) (enumerated on plate count agar (PCA) and incubated for 72 h at 30 °C).

2.5. pH

The pH value was recorded on a pH meter (FiveEosy, Mettler-Toledo company, Shanghai, China). Fish muscle (10 g) was thoroughly homogenized with 100 mL distilled water and soaked for 30 min, then filtered to determine the pH.

2.6. Sensory Evaluation

According to the description of Liu et al. [20] with some modifications, sensory analysis of flesh colour, texture, taste and aroma was conducted on a ten-point scale, completed by eight experienced panellists (four females and four males, 20 to 35 years old), in which a score of 8.0–10.0 indicated good quality, 6.0–8.0 indicated acceptable quality, 4.0–6.0 indicated unacceptable quality, and 1.0–4.0 indicated an intense dislike.

2.7. Texture Analysis

Texture profile analysis was measured using a Texture Analyzer (Ta. Xt 2i/50, Stability Microsystems, Surrey, UK) equipped with a specific cylindrical probe placed horizontally on a heavy platform, according to the method reported by Yu et al. [21]. The device parameters were set to trigger: force 5 g and distance 15 mm, pretest speed 5 mm/s, test speed and post-test speed were 2 mm/s and 10 mm/s. Six replicates were made from each sample under different pretreatments.

2.8. Headspace SPME-GC/MS Analysis

According to Shi et al. [22]. Solid-phase microextraction–gas chromatography–mass spectrometry (SPME-GC-MS) was used to detect volatile compounds. Fillets of fish (2 g) were placed into a headspace vial containing 20 mL of water and equilibrated at 40 °C for 10 min. SPME fibre (50/30 μm DVB/CAR/PDMS extractor, Supelco, Bellefonte, PA, USA) was exposed to the vial headspace for another 40 min at 40 °C. Then, the fibre was inserted into the GC and desorbed at 250 °C for 5 min. Compounds were separated on a DB-1ms capillary column (30 m × 0.25 mm × 0.25 μm, Agilent J&W, Palo Alto, CA, USA) with helium as the carrier gas at a flow rate of 1 mL min^{-1}.

The temperature in the GC oven was held at 40 °C for 2 min, then increased at 2 °C/min to 90 °C, held for 5 min, then increased at 8 °C/min to a final temperature of 250 °C. The ion source was EI. The MS source and transfer line were kept at 230 °C and 280 °C, respectively. The mass spectrometry data were collected over the full scan range of 35 to 350 m/z using positive ionization with an electron energy of 70 eV. Volatile compounds can be identified by comparing the mass spectrometry results of volatile compounds with data stored in the National Institute of Standards and Technology (NIST08) spectrum database. Area normalization was used to quantify volatile compounds.

2.9. Free Amino Acids (FAA) Analysis

FAAs were measured according to Yu et al. [23]. Fish fillets (2 g) were homogenized with 15 mL of 5% trichloroacetic acid for 1 min. Then they were placed in a refrigerator at 4 °C for 2 h and centrifuged at 9000× g for 15 min with 10 mL supernatant. Taking 5 mL supernatant, pH was adjusted to 2.0 with 6 mol/L NaOH, and then volume was adjusted with distilled water to 10 mL. The extraction was filtered using a 0.22 μm filter membrane and applied to an automatic amino acid analyser (L-8900; Hitachi, Tokyo, Japan). Three parallels for each sample.

2.10. Statistical Analysis

All data are presented as mean ± standard deviation. The significant difference was at the level of $p < 0.05$ by DPS software analysis and Duncan's multiple range test. All figures were obtained by Origin 2017 (OriginLab Co., Northampton, MA, USA).

3. Results and Discussion

3.1. Microbiological Analysis

The main factor limiting the shelf life of largemouth bass fillets is the development of microbial groups during the preservation. Meat products are often easily degraded by microbial activities. Figure 1 describes the changing trend in the number of TVC of largemouth bass with different treatments during storage. The results show that the TVC of largemouth bass was significantly different with different storage times and sample treatments ($p < 0.05$). The higher initial TVC for fresh largemouth bass was 4.5 log CFU/g (Figure 1). Shuai Zhuang et al. [24] found that the initial TVC of fresh largemouth bass fillets was about 4.55 log CFU/g, and the initial value was high. The TVC value of largemouth bass fillets exceeded 7 log CFU/g, which is recognized to be the maximum acceptable limit of microorganisms [25]. The TVC of fresh largemouth bass fillets on the sixth day of storage was 7.41 log CFU/g. HHP-0- and HHP-10-treated fish fillets exceeded the maximum limit (7 log CFU/g) on day 18 and 24 of storage, respectively. HHP-0+SV- and HHP-10+SV-treated fish fillets exceeded 7 log CFU/g on day 30 of storage, as shown in Figure 1. The population of the TVC increasingly elevated with storage time. Throughout the same storage time, HHP and HHP combined with SV treatment significantly reduced the number of TVC ($p < 0.05$). Ekonomou et al. [26] found that high hydrostatic pressure (HHP; 200 MPa, 15 min) could eliminate Listeria monocytogenes in trout. At the same time, the synergistic effect of HHP, liquid smoke and freezing can reduce its bacteria. Except for 20 min SV treatment at 65 °C, all temperature/time combinations used to cook turkey slices could greatly inactivate pathogenic microorganisms [27]. Compared with the control group, the shelf life of largemouth bass fillets treated with HHP-0 and HHP-10 may be extended by 12 days and 18 days, respectively, whereas the combination of HHP with SV was substantially more effective in extending the microbiological shelf life of largemouth bass fillets by 24 days.

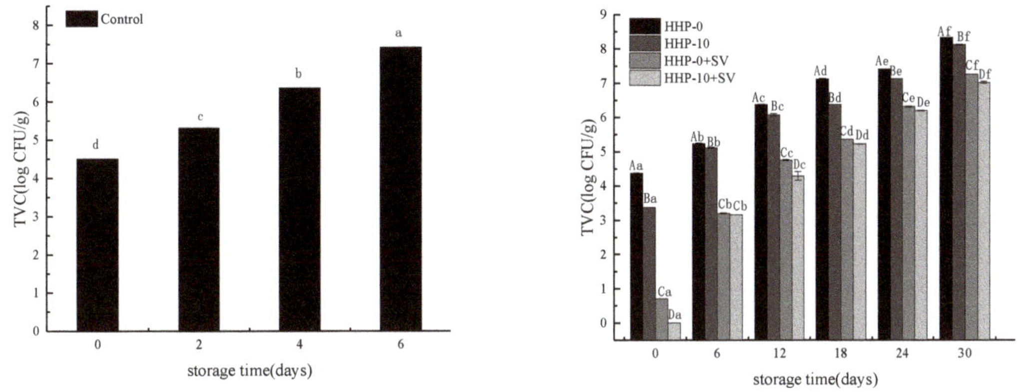

Figure 1. The effect of different treatments on the number of TVC of largemouth bass during storage. A–D: Differences between different treatment groups; a–f: Differences in treatment groups during storage.

3.2. *pH Value Analysis*

The changes in pH values for HHP, HHP+SV and the control are given in Table 1. The initial pH value of largemouth bass fillets was 6.98. Masniyom et al. [28] found that the initial pH value of fresh largemouth bass fillets was 6.80 and that the pH value was gradually increased (7.04 ± 0.01) during 2 days of storage. HHP-0, HHP-10, HHP-0+SV and HHP-10+SV showed a comparable level of pH on day 0, 6.97 ± 0.01, 6.92 ± 0.01, 6.89 ± 0.02 and 6.85 ± 0.04, respectively. However, the pH value started increasing after 6 days of storage. Presumably due to the volatile alkaline components produced by spoilage bacteria, such as alkaline amines (trimethylamine and other volatile amines) produced by fish-spoiling bacteria [29]. The pH of largemouth bass fillets was sharply decreased after one week of storage. The decrease in pH value may be due to the release of lactic acid produced by the anaerobic glycolysis of glucose and inorganic phosphate produced by ATP degradation [30]. On the 30th day of storage, HHP and HHP+SV treatment resulted in pH values of 6.4 and 6.8 for largemouth bass fillets, respectively. In this study, HHP+SV treatment contributed to keeping the pH level of largemouth bass fillets slowly changing since it may reduce the growth of bacteria that produce volatile basic components (for example, alcohols and ketones). The pH value of largemouth bass fillets treated with HHP+SV was higher than that of the control and HHP. The increase in the pH value of fish meat during cooking was owed to the formation of disulphide bonds during the cooking process.

Table 1. Effects of different treatments on the pH value of largemouth bass during storage.

Treatment	**Storage Time (Days)**					
	0	2	4	6	–	–
Control	6.98 ± 0.01 ^{b}	7.04 ± 0.01 ^{a}	6.74 ± 0.01 ^{c}	6.56 ± 0.01 ^{d}	–	–
Treatment	**Storage time (days)**					
	0	**6**	**12**	**18**	**24**	**30**
HHP-0	6.97 ± 0.01 ^{Ab}	7.02 ± 0.01 ^{Ca}	6.87 ± 0.01 ^{Dc}	6.60 ± 0.01 ^{Dd}	6.51 ± 0.01 ^{De}	6.40 ± 0.01 ^{Df}
HHP-10	6.92 ± 0.01 ^{Bc}	6.97 ± 0.01 ^{Da}	6.94 ± 0.01 ^{Cb}	6.86 ± 0.01 ^{Cd}	6.60 ± 0.01 ^{Ce}	6.43 ± 0.01 ^{Cf}
HHP-0+SV	6.89 ± 0.02 ^{Cc}	7.12 ± 0.01 ^{Aa}	7.07 ± 0.01 ^{Ab}	6.89 ± 0.01 ^{Bc}	6.81 ± 0.01 ^{Bd}	6.75 ± 0.01 ^{Be}
HHP-10+SV	6.85 ± 0.04 ^{De}	7.05 ± 0.01 ^{Ba}	7.01 ± 0.01 ^{Bb}	6.97 ± 0.01 ^{Ac}	6.89 ± 0.01 ^{Ad}	6.85 ± 0.01 ^{Ae}

A–D: Differences between different treatment groups; a–f: Differences in treatment groups during storage.

3.3. Sensory Evaluation and Texture Profile Analysis

Sensory characteristics of largemouth bass are shown in Figure 2. There were significant differences between HHP and HHP+SV in the sensory scores of colour, texture, taste and aroma, while the score of the taste attribute in HHP+SV was significantly higher than in HHP. The results of sensory analysis showed that HHP+SV would have a more intense taste attribute than HHP. The important quality characteristics of bass fillets include colour, texture, juiciness, aroma and taste. Processing technology and storage have important effects on sensory attributes of bass fillets. In the present study, aroma and taste were more significant than colour, texture and juiciness in discriminating the sensory characteristics of HHP and HHP+SV. It may be that SV treatment could help produce different aroma and taste components. Sensory characteristics tended to decrease as storage time increased, but sensory scores decreased more slowly in the HHP+SV treatment compared with the HHP group. In the later stages of storage, the fish in the HHP group became darker in colour, looser in texture and worse in smell.

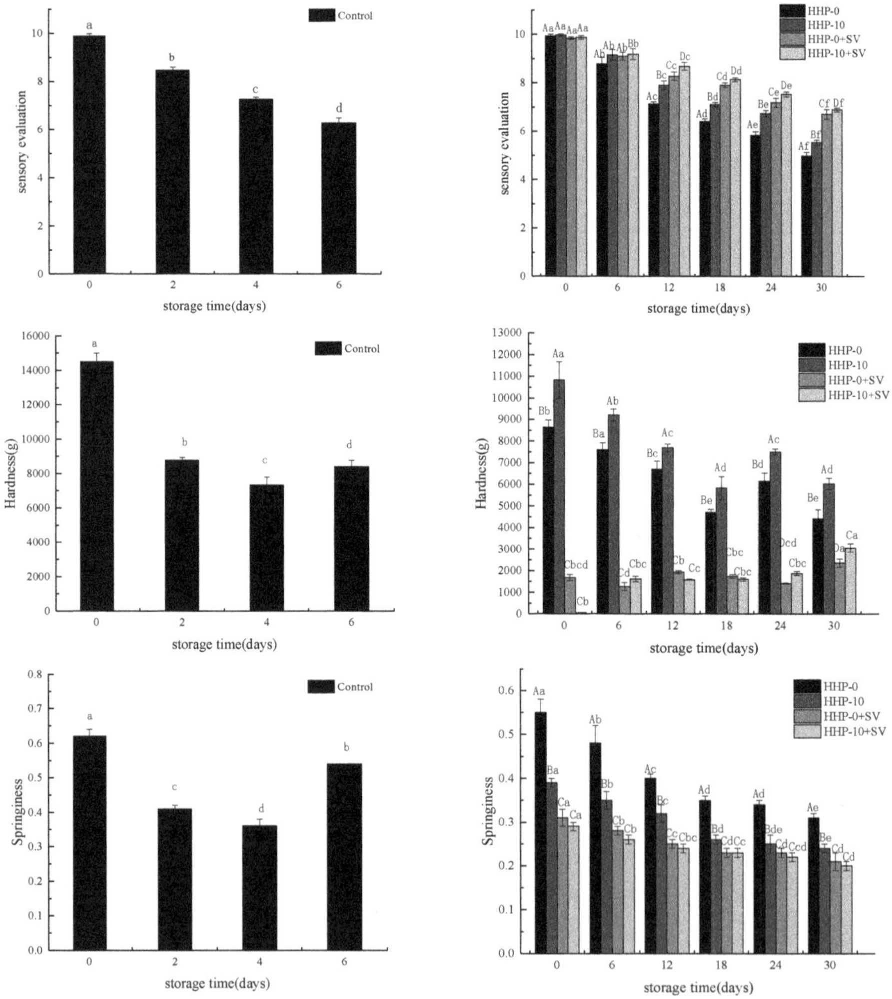

Figure 2. *Cont.*

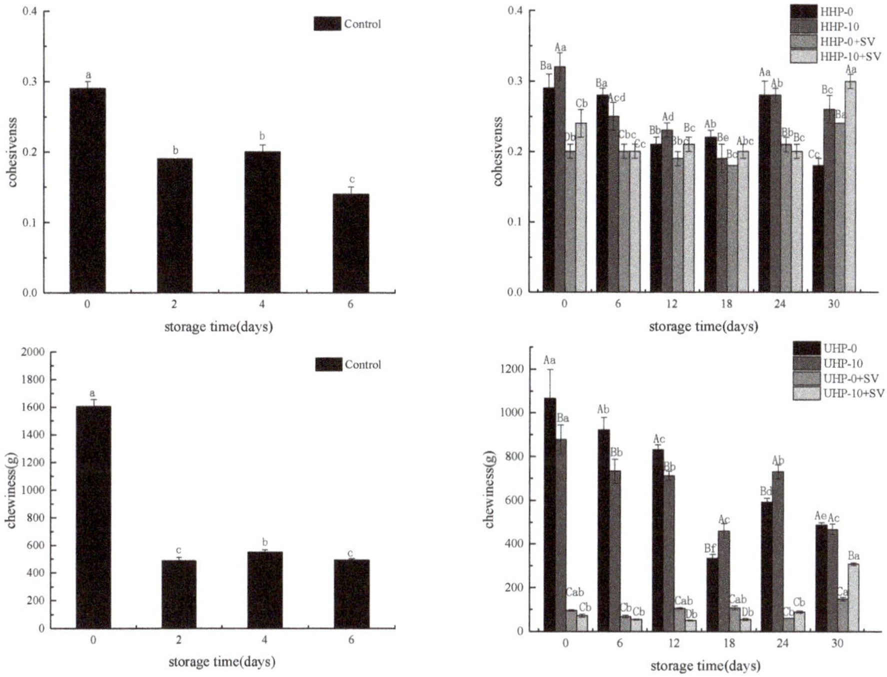

Figure 2. The effect of different treatments on sensory evaluations: springiness, hardness, cohesiveness and chewiness of largemouth bass during storage. A–D: Differences between different treatment groups; a–f: Differences in treatment groups during storage.

Texture analysis is displayed in Figure 2. The springiness, hardness, cohesiveness and chewiness of HHP+SV were significantly higher than those in HHP ($p < 0.05$). This showed a firm and elastic taste was obtained by SV pretreatment, which was popular with consumers. The initial hardness of the largemouth bass fillets from the CK group was 14,499 g. After treatment, the hardness showed a decreasing trend and gradually decreased as the storage time increased. The HHP+SV showed a lower elasticity value when compared with HHP. The springiness of bass fillets decreased with an increase in storage time and reached the minimum value at 30 days. Chewiness refers to the time required to chew food at a fixed speed to reduce it to a size suitable for swallowing [31]. Cohesiveness and chewiness exhibited the same change curves, which were complementary to hardness. This is consistent with the results of the sensory characteristics. HHP combined with SV treatment significantly reduced the hardness and chewability of fish and had a better texture, indicating that hardness and chewability are important components in this treatment.

3.4. Headspace SPME-GC/MS Analysis

The flavour changes of the pretreated bass fillets under storage conditions were investigated by HS-SPME GC/MS. A total of 43 volatile compounds including alcohol (10), aldehyde (10), ketone (3) hydrocarbons (17) and others (3) were detected in the sample.

As shown in Table 2, the most important compounds were alcohols and aldehydes, but alcohols were not the main contributors to the overall aroma because of their high odour threshold [32]. The odour threshold of nonanal, hexanal and other aldehydes was low. They come from the degradation of fatty acids and triglycerides, which can have a great impact on the overall aroma of marine products [33]. A few compounds such as valeraldehyde, trans-2-heptenal, cyclobutanol and 1-nonanol appeared occasionally during storage. According to the treatment and storage conditions of bass meat, most compounds can be divided into three categories. The first group included four compounds that were detected throughout the storage of bass meat. The second group included many compounds that were detected only in the middle and/or end stage, and at the end of treatment and storage; the initially detected compounds decreased or disappeared.

Table 2. Effects of different treatments on the volatile compounds of largemouth bass during storage.

Compound (×10⁶)	0 Day	2 Day	4 Day	6 Day	0 Day				30 Day			
	CK	CK	CK	CK	HHP-0	HHP-0+SV	HHP-10	HHP-10+SV	HHP-0	HHP-0+SV	HHP-10	HHP-10+SV
Aldehydes (10)												
Benzaldehyde	0.02	0.03	-	0.02	-	0.02	-	0.03	-	0.11	-	0.04
Valeraldehyde	0.28	-	0.19	0.36	-	1.23	0.23	3.42	-	-	-	-
Isovaleraldehyde	0.17	0.63	-	0.44	-	-	0.07	0.07	0.16	-	0.28	-
Hexanal	14.73	3.13	7.68	2.38	11.40	2.57	11.34	2.27	1.50	4.04	4.12	15.02
Heptaldehyde	2.19	2.35	0.41	1.43	0.70	2.05	1.09	1.86	0.32	3.16	0.48	2.28
trans-2-Heptenal	-	0.10	0.01	0.08	0.05	-	-	-	-	-	-	-
Octanal	2.50	2.52	0.50	1.66	1.12	1.38	1.00	1.38	0.26	3.54	0.37	1.04
(E)-2-Octenal	0.07	0.21	0.02	0.22	0.04	-	-	-	-	0.07	-	-
Nonanal	3.92	5.12	1.12	3.16	2.34	1.24	1.96	0.97	1.15	3.51	0.76	0.86
Decanal	0.11	0.15	0.04	0.12	0.11	-	0.05	-	0.06	0.05	-	-
Alcohols (10)												
Cyclobutanol	-	-	-	-	0.16	-	0.17	-	-	-	-	-
1-Pentanol	0.72	1.20	0.28	0.24	0.24	0.52	0.22	0.28	0.01	1.79	0.09	-
1-Penten-3-ol	-	-	-	-	-	-	-	-	-	1.66	-	3.11
3-Methyl-1-butanol	-	-	-	-	-	-	-	0.20	11.65	0.90	3.31	-
Hexyl alcohol	1.32	1.76	0.96	1.27	1.51	-	0.54	-	0.37	5.68	-	-
n-Heptanol	-	-	-	0.91	0.53	-	0.36	-	-	-	-	-
1-Octanol	0.50	0.87	0.15	0.54	0.48	0.04	0.21	0.09	0.08	1.01	0.06	-
1-Octen-3-ol	1.89	4.39	0.67	2.90	0.90	2.09	-	1.90	0.38	9.97	1.15	9.14
1-Nonanol	-	0.15	0.01	0.03	-	-	-	-	-	-	-	-
Decyl alcohol	0.03	-	-	-	-	-	-	-	-	-	-	-
Ketones (3)												
2-Heptanone	-	0.07	-	-	-	-	-	-	0.25	0.64	-	-
2,3-Octanedione	1.69	3.58	0.43	1.75	0.72	-	0.64	0.93	0.20	0.98	0.45	0.55
2-Nonanone	0.05	0.08	-	0.04	-	0.02	-	0.04	0.03	0.14	0.37	0.10
Hydrocarbons (17)												
Cycloheptane	-	0.09	-	-	0.04	-	-	-	-	-	-	-
1-Octane	0.33	0.39	0.03	0.06	0.08	-	0.05	0.13	0.04	2.21	-	-
n-Nonane	-	-	0.01	0.02	-	-	-	0.16	0.34	-	0.85	-
Decane	-	0.09	0.00	0.08	0.01	0.05	-	0.06	0.04	0.17	0.07	0.20
Undecane	0.02	0.06	0.03	0.03	0.02	0.04	0.04	-	0.04	0.20	-	-
Dodecane	0.02	0.02	0.09	0.04	-	0.03	0.01	0.06	0.06	-	0.06	0.33
Tridecane	0.01	0.03	0.04	0.03	0.01	0.06	0.02	0.08	0.04	0.16	0.04	0.25
Tetradecane	0.03	0.04	0.03	0.02	0.02	0.08	0.02	0.13	0.04	0.09	0.11	0.29
Pentadecane	0.15	0.19	0.18	0.11	0.14	2.26	0.11	3.46	0.25	0.49	3.69	7.34
Hexadecane	0.01	0.01	0.01	0.01	-	0.02	0.01	0.02	0.01	0.02	0.03	0.06
Phenylethylene	-	-	0.17	0.04	0.04	0.07	0.03	0.05	0.02	-	0.08	0.65
Benzocyclobutene	0.01	-	0.01	0.04	-	0.17	-	0.07	-	-	0.57	1.01
1,3-Cyclooctadiene	0.19	0.07	-	-	0.18	-	-	0.02	-	-	-	3.05
1,3,5,7-Cyclooctatetraene	-	-	0.12	0.02	0.01	0.16	0.02	0.09	-	-	0.04	-
1,3,6-Octatriene,3,7-dimethyl-	0.05	0.06	-	-	-	-	-	-	-	0.21	-	-
1-Tridecene	-	-	-	-	-	0.02	-	0.06	-	-	0.03	0.10
l-Caryophyllene	-	-	-	-	-	0.02	-	0.02	-	0.10	0.03	0.07
Others (3)												
2-Pentylfuran	0.03	0.01	-	-	0.02	0.02	0.02	0.02	-	0.50	-	0.10
3-Methylpyridazine	0.02	-	-	-	-	-	-	-	-	-	-	-
Dimethyl disulfide	-	-	-	-	-	-	-	-	-	0.19	-	-

In the first group of compounds, hexanal, heptaldehyde, octanal and nonanal exhibited a tendency to reduce gradually in HHP groups throughout the whole storage, whereas they tended to increase in HHP+SV groups. The relative content of the second group of compounds in fish fillets was low. The species and peak areas of aldehydes were significantly reduced after HHP and UHP+SV treatment. Only the HHP-treated samples did not detect benzaldehyde, indicating that the HHP-only treatment caused the samples to lose their nutty aroma, while the HHP+SV treatment avoided this phenomenon and maintained the original flavour characteristics. After treatment and storage of the samples, trans-2-heptenal and 1-nonanol compounds disappeared. In some cases, metabolites produced during corruption may be further metabolized by microorganisms [34]. These aspects probably explain the above VOC changes. At the same time, some VOC accumulation was caused by the metabolism of spoilage microorganisms (such as *Pseudomonas*,

Shewanella and *Enterobacteriaceae*) and chemical reactions mainly including the oxidation of polyunsaturated fatty acids [35].

3.5. Free Amino Acid (FAA) Analysis

Table 3 presents the effects of different treatments on the content of FAA (expressed in mg/100 g dry matter) of largemouth bass during storage. Statistical analysis indicates that HHP and HHP+SV treatment had significant effects on the total FAA content of bass ($p < 0.05$). There were significant differences among CK, HHP and HHP+SV treatments. The total amount of FAA in CK samples was high. This indicates that HHP and HHP+SV treatment would lead to the loss of free amino acid content. The total FAA content of the HHP group was higher than that of the HHP+SV group at early storage (0 day) and late storage (30 day), which indicates that SV treatment can slow down the production of amino acid molecules. In the latter stage of storage (30 day), the total FAA in the HHP+SV group decreased more than that in HHP group. In the composition of FAA, alanine was the only amino acid with the highest content in untreated samples. Higher levels of specific amino acids in HHP and HHP+SV samples compared with untreated samples may influence the perception of sweet (sweet amino acids include alanine, serine, threonine and glycine), bitter (bitter amino acids include phenylalanine, histidine, valine, methionine, isoleucine, leucine and arginine), umami (umami amino acids include glutamic and aspartic acid) and tasteless (tasteless amino acids include cysteine, lysine and tyrosine) attributes in comparison with untreated samples (Table 3). After HHP and HHP+SV treatments, the sweet and umami amino acid content of the sea bass fillets was reduced, whereas the bitter amino acid content was increased. On the 30th day of storage, the content of sweet and umami amino acids of largemouth bass decreased significantly, and the content of sweet and fresh amino acids of HHP-SV was lower than that of HHP.

Table 3. Effects of different treatments on the free amino acids in largemouth bass during storage.

FAA (mg/100 g)	0 Day	2 Day	4 Day	6 Day	0 Day				30 Day			
	CK				HHP-0	HHP-10	HHP-0+SV	HHP-10+SV	HHP-0	HHP-10	HHP-0+SV	HHP-10+SV
Thr	12.29 ^{b}	10.10 ^{e}	11.53 ^{c}	9.26 ^{g}	7.35 ^{j}	10.38 ^{d}	8.97 ^{h}	14.52 ^{a}	1.26 ^{l}	10.02 ^{f}	8.89 ^{i}	7.29 ^{k}
Ser	3.94 ^{g}	3.83 ^{h}	4.21 ^{e}	4.43 ^{d}	5.09 ^{b}	5.33 ^{a}	4.79 ^{c}	3.39 ^{j}	0.53 ^{l}	4.10 ^{f}	1.51 ^{k}	3.54 ^{i}
Ala	38.68 ^{g}	45.27 ^{b}	45.05 ^{c}	40.61 ^{f}	40.68 ^{e}	54.27 ^{a}	42.89 ^{d}	11.14 ^{k}	12.59 ^{j}	14.99 ^{i}	5.11 ^{l}	21.75 ^{h}
Gly	16.76 ^{a}	14.57 ^{c}	13.90 ^{d}	13.41 ^{e}	11.15 ^{h}	14.57 ^{c}	11.77 ^{g}	13.29 ^{f}	14.73 ^{b}	6.69 ^{j}	5.39 ^{k}	7.95 ^{i}
Sweet	71.67	73.77	74.69	67.71	64.26	84.55	68.42	42.33	29.11	35.80	20.90	40.54
Asp	0.48 ^{g}	0.27 ^{l}	0.32 ^{j}	0.51 ^{f}	0.61 ^{d}	0.45 ^{h}	0.77 ^{b}	0.29 ^{k}	0.66 ^{c}	0.87 ^{a}	0.55 ^{e}	0.36 ^{i}
Glu	5.96 ^{e}	6.05 ^{d}	6.88 ^{b}	6.63 ^{c}	4.95 ^{g}	5.69 ^{f}	4.28 ^{i}	4.76 ^{h}	11.68 ^{a}	5.69 ^{f}	2.37 ^{k}	2.87 ^{j}
Umami	6.44	6.32	7.20	7.14	5.56	6.13	5.05	5.05	12.34	6.56	2.92	3.23
Val	3.53 ^{h}	3.70 ^{e}	3.82 ^{c}	3.64 ^{f}	3.61 ^{g}	3.95 ^{b}	3.63 ^{f}	4.12 ^{a}	3.74 ^{d}	3.40 ^{i}	2.54 ^{j}	2.43 ^{k}
Met	2.15 ^{e}	2.38 ^{b}	2.43 ^{a}	2.37 ^{c}	1.82 ^{i}	2.31 ^{d}	2.03 ^{g}	2.13 ^{f}	2.31 ^{d}	2.00 ^{h}	1.35 ^{j}	1.28 ^{k}
Ile	2.43 ^{g}	2.53 ^{f}	2.74 ^{d}	2.58 ^{e}	2.38 ^{h}	3.22 ^{a}	2.98 ^{c}	3.00 ^{b}	1.73 ^{k}	2.21 ^{i}	1.51 ^{l}	1.86 ^{j}
Leu	4.69 ^{f}	4.70 ^{e}	5.16 ^{d}	4.69 ^{e}	4.25 ^{h}	5.85 ^{a}	5.38 ^{c}	5.56 ^{b}	3.56 ^{i}	4.27 ^{g}	3.13 ^{k}	3.53 ^{j}
Phe	3.39 ^{c}	3.29 ^{e}	3.42 ^{b}	3.14 ^{g}	3.12 ^{h}	3.58 ^{a}	3.33 ^{d}	3.23 ^{f}	3.13 ^{gh}	2.23 ^{i}	1.46 ^{k}	1.64 ^{j}
His	28.08 ^{k}	27.70 ^{l}	33.59 ^{f}	29.72 ^{j}	32.05 ^{h}	42.78 ^{b}	38.35 ^{c}	44.14 ^{a}	33.47 ^{g}	36.22 ^{e}	36.35 ^{d}	30.46 ^{i}
Arg	0.41 ^{h}	0.25 ^{k}	0.37 ^{i}	0.62 ^{a}	0.28 ^{j}	0.59 ^{b}	0.49 ^{d}	0.58 ^{c}	0.19 ^{l}	0.45 ^{f}	0.48 ^{e}	0.43 ^{g}
Bitter	44.68	44.55	51.53	46.76	47.51	62.29	56.19	62.76	48.14	50.78	46.82	41.64
Cys	0.00 ^{f}	0.00 ^{f}	0.00 ^{f}	0.00 ^{f}	0.00 ^{f}	0.00 ^{f}	0.00 ^{f}	0.11 ^{e}	3.53 ^{a}	0.89 ^{b}	0.84 ^{c}	0.34 ^{d}
Tyr	3.17 ^{b}	3.15 ^{c}	2.10 ^{l}	2.72 ^{h}	2.81 ^{g}	3.01 ^{e}	2.87 ^{f}	2.70 ^{i}	4.45 ^{a}	3.04 ^{d}	2.23 ^{j}	2.20 ^{k}
Lys	23.60 ^{d}	21.61 ^{f}	28.03 ^{a}	22.89 ^{e}	15.11 ^{j}	21.23 ^{g}	16.59 ^{h}	13.72 ^{k}	25.89 ^{b}	24.59 ^{c}	11.60 ^{l}	16.44 ^{i}
Tasteless	26.77	24.76	30.13	25.61	17.93	24.23	19.46	16.53	33.87	28.51	14.67	18.98
Total	149.56	149.40	163.55	147.22	135.26	177.21	149.11	126.67	123.47	121.66	85.31	104.38

Note: letters represent significant differences. $p < 0.05$.

4. Conclusions

The microbe, pH, sensory, colour, texture, GC-MS, FAA and other quality indicators of largemouth bass stored for 30 days after HHP and SV treatment were studied. The results indicated that along with the increase in storage time, the TVC of largemouth bass increased and the sensory evaluation decreased. HHP-0- and HHP-10-treated fish fillets exceeded the maximum limit (7 log CFU/g) on day 18 and 24 of storage, respectively. HHP-0+SV- and HHP-10+SV-treated fish fillets exceeded 7 log CFU/g on day 30 of storage. This showed that HHP+SV treatment can prolong the storage period of bass compared with HHP. The results of sensory evaluation, texture characteristics, GC-MS and FAA showed that HHP+SV had higher sensory and texture and good wind characteristics compared with HHP, but it would cause the loss of free amino acid content of bass. In general, HHP+SV processing is considered to be an effective technology, compared with HHP processing and SV processing alone.

Author Contributions: Conceptualization, M.Z., Y.Q. and G.X.; Data curation, M.Z., Y.L. and F.C.; Formal analysis, M.Z., Y.L. and F.C.; Investigation, M.Z. and Y.L.; Methodology, M.Z., Y.L. and F.C.; Visualization, M.Z., and Y.L.; Writing—original draft, M.Z. and Y.L.; Writing—review and editing, M.Z., F.C., C.W., Y.Q., L.W., W.W., L.S. and A.D.; Resources, C.W., Y.Q., G.X. and L.W.; Supervision, C.W., L.W., W.W., L.S. and A.D.; Funding acquisition, Y.Q. and G.X.; Validation, F.C. All authors have read and agreed to the published version of the manuscript.

Funding: This work was funded by the China Agriculture Research System (CARS-46) and the Maior project of the Scientific and Technological R&D of Hubei Agricultural Scientific and Technological Innovation Center (2020-620-000-002-03).

Institutional Review Board Statement: Not applicable.

Informed Consent Statement: Not applicable.

Data Availability Statement: Data is contained within the article.

Acknowledgments: We appreciate the research was funded by the China Agriculture Research System (CARS-46) and the Maior project of the Scientific and Technological R&D of Hubei Agricultural Scientific and Technological Innovation Center (2020-620-000-002-03).

Conflicts of Interest: The authors declare no conflict of interest.

References

1. Nieva-Echevarría, B.; Manzanos, M.J.; Goicoechea, E.; Guillén, M.D. Changes provoked by boiling, steaming and sous-vide cooking in the lipid and volatile profile of European sea bass. *Food Res. Int.* **2017**, *99*, 630–640. [CrossRef] [PubMed]
2. Nie, X.; Wang, L.; Wang, Q.; Lei, J.; Hong, W.; Huang, B.; Zhang, C. Effect of a sodium alginate coating infused with tea polyphenols on the quality of fresh Japanese sea bass (Lateolabrax japonicas) fillets. *J. Food Sci.* **2018**, *83*, 1695–1700. [CrossRef]
3. Hu, L.; Ren, S.; Shen, Q.; Chen, J.; Ye, X.; Ling, J. Proteomic study of the effect of different cooking methods on protein. oxidation in fish fillets. *RSC Adv.* **2017**, *7*, 27496–27505. [CrossRef]
4. Xu, B.; Chang, S.K.C. Effect of soaking, boiling, and steaming on total phenolic content and antioxidant activities of cool season food legumes. *Food Chem.* **2008**, *110*, 1–13. [CrossRef] [PubMed]
5. Bolumar, T.; Orlien, V.; Sikes, A.; Aganovic, K.; Bak, K.H.; Guyon, C.; Brüggemann, D.A. High-pressure processing of meat: Molecular impacts and industrial applications. *Compr. Rev. Food Sci. Food Saf.* **2021**, *20*, 332–368. [CrossRef] [PubMed]
6. Suemitsu, L.; Cristianini, M. Effects of high pressure processing (HPP) on quality attributes of tilapia (*Oreochromis niloticus*) fillets during refrigerated storage. *LWT* **2019**, *101*, 92–99. [CrossRef]
7. Deng, L.Z.; Mujumdar, A.S.; Pan, Z.; Vidyarthi, S.K.; Xu, J.; Zielinska, M.; Xiao, H.W. Emerging Chemical and Physical Disinfection Technologies of Fruits and Vegetables: A Comprehensive Review. *Crit. Rev. Food Sci. Nutr.* **2020**, *60*, 2481–2508. [CrossRef]
8. Aubourg, S.P. Impact of High-Pressure Processing on Chemical Constituents and Nutritional Properties in Aquatic Foods: A Review. *Int. J. Food Sci. Technol.* **2018**, *53*, 873–891. [CrossRef]
9. Ye, T.; Chen, X.; Chen, Z.; Yao, H.; Wang, Y.; Lin, L.; Lu, J. Quality and Microbial Community of High Pressure Shucked Crab (Eriocheir Sinensis) Meat Stored at 4 °C. *J. Food Process. Preserv.* **2021**, *45*, e15330. [CrossRef]
10. Lin, C.S.; Lee, Y.C.; Kung, H.F.; Cheng, Q.L.; Ou, T.Y.; Chang, S.K.; Tsai, Y.H. Inactivation of microbial loads and retardation of quality loss in Asian hard clam (*Meretrix lusoria*) using high-hydrostatic-pressure processing during refrigerated storage. *Food Control* **2022**, *133*, 108583. [CrossRef]
11. Kim, H.H.; Ryu, S.H.; Jeong, S.M.; Kang, W.S.; Lee, J.E.; Kim, S.R.; Ahn, D.H. Effect of High Hydrostatic Pressure Treatment on Urease Activity and Inhibition of Fishy Smell in Mackerel (*Scomber japonicus*) during Storage. *J. Microbiol. Biotechnol.* **2021**, *31*, 1684–1691. [CrossRef] [PubMed]
12. Zhao, Y.M.; de Alba, M.; Sun, D.W.; Tiwari, B. Principles and Recent Applications of Novel Non-Thermal Processing Technologies for the Fish Industry-a Review. *Crit. Rev. Food Sci. Nutr.* **2019**, *59*, 728–742. [CrossRef] [PubMed]
13. Dominguez-Hernandez, E.; Salaseviciene, A.; Ertbjerg, P. Low-temperature long-time cooking of meat: Eating quality and underlying mechanisms. *Meat Sci.* **2018**, *143*, 104–113. [CrossRef] [PubMed]
14. Saito, K.; Yoshinari, M.; Ishikawa, S.I. Effects of Low-temperature Long-time Sous-vide Cooking on the Physicochemical andSensory Characteristics of Beef and Pork Shank. *J. Culin. Sci. Technol.* **2020**, *20*, 165–179. [CrossRef]
15. García-Segovia, P.; Andr´es-Bello, A.; Martínez-Monz´o, J. Effect of cooking method on mechanical properties, color and structure of beef muscle (*M. pectoralis*). *J. Food Eng.* **2007**, *80*, 813–821. [CrossRef]

16. Park, C.H.; Lee, B.; Oh, E.; Kim, Y.S.; Choi, Y.M. Combined effects of sousvide cooking conditions on meat and sensory quality characteristics of chicken breast meat. *Poult. Sci.* **2020**, *99*, 3286–3291. [CrossRef]
17. Jiang, S.; Xue, D.; Zhang, Z.; Shan, K.; Ke, W.; Zhang, M.; Zhao, D.; Nian, Y.; Xu, X.; Zhou, G.; et al. Effect of Sous-vide cooking on the quality and digestion characteristics of braised pork. *Food Chem.* **2021**, *375*, 131683. [CrossRef]
18. Wiesław, P.; Danuta, J.; Katarzyna, K.; Piotr, S.; Kacper, P. Effect of Heat Treatment by the Sous-Vide Method on the Quality of Poultry Meat. *Foods* **2021**, *10*, 1610. [CrossRef]
19. Rezaeigolestani, M.; Misaghi, A.; Khanjari, A.; Basti, A.A.; Abdulkhani, A.; Fayazfar, S. Antimicrobial evaluation of novel poly-lactic acid based nanocomposites incorporated with bioactive compounds in-vitro and in refrigerated vacuum-packed cooked sausages. *Int. J. Food Microbiol.* **2017**, *260*, 1–10. [CrossRef]
20. Liu, X.; Huang, Z.; Jia, S.; Zhang, J.; Li, K.; Luo, Y. The roles of bacteria in the biochemical changes of chill-stored bighead carp (*Aristichthys nobilis*): Proteins degradation, biogenic amines accumulation, volatiles production, and nucleotides catabolism. *Food Chem.* **2018**, *255*, 174–181. [CrossRef]
21. Yu, D.; Jiang, Q.; Xu, Y.; Xia, W. The shelf life extension of refrigerated grass carp (*Ctenopharyngodon idellus*) fillets by chitosan coating combined with glycerol monolaurate. *Int. J. Biol. Macromol.* **2017**, *101*, 448–454. [CrossRef] [PubMed]
22. Shi, J.; Nian, Y.; Da, D.; Xu, X.; Zhou, G.; Zhao, D.; Li, C. Characterization of flavor volatile compounds in sauce spareribs by gas chromatography–mass spectrometry and electronic nose. *LWT-Food Sci. Technol.* **2020**, *124*, 109182. [CrossRef]
23. Yu, D.; Xu, Y.; Regenstein, J.M.; Xia, W.; Yang, F.; Jiang, Q.; Wang, B. The effects of edible chitosan-based coatings on flavor quality of raw grass carp (*Ctenopharyngodon idellus*) fillets during refrigerated storage. *Food Chem.* **2018**, *242*, 412. [CrossRef]
24. Zhuang, S.; Li, Y.; Hong, H.; Liu, Y.; Shu, R.; Luo, Y. Effects of ethyl lauroyl arginate hydrochloride on microbiota, quality and biochemical changes of container-cultured largemouth bass (*Micropterus salmonides*) fillets during storage at 4 °C. *Food Chem.* **2020**, *324*, 126886. [CrossRef] [PubMed]
25. ICMSF. *Microorganisms in Foods. Sampling for Microbiological Analysis: Principles and Scientific Applications*; University of Toronto Press: Toronto, ON, Canada, 1986; pp. 181–196.
26. Ekonomou, S.I.; Bulut, S.; Karatzas, K.A.G.; Boziaris, I.S. Inactivation of Listeria monocytogenes in raw and hot smoked trout fillets by high hydrostatic pressure processing combined with liquid smoke and freezing. *Innov. Food Sci. Emerg. Technol.* **2020**, *64*, 102427. [CrossRef]
27. Bıyıklı, M.; Akoğlu, A.; Kurhan, Ş.; Akoğlu, İ.T. Effect of different Sous Vide cooking temperature-time combinations on the physicochemical, microbiological, and sensory properties of Turkey cutlet. *Int. J. Gastron. Food Sci.* **2020**, *20*, 100204. [CrossRef]
28. Masniyom, P.; Benjakul, S.; Visessanguan, W. Shelf-life extension of refrigerated seabass slices under modified atmosphere. packaging. *J. Sci. Food Agric.* **2002**, *82*, 873–880. [CrossRef]
29. Ordonez, J.A.; Lopez-Galvez, D.E.; Fernandez, M.; Hierro, H.; Hoz, L. Microbial and physicochemical modifications of hake (*Merluccius merluccius*) steaks stored under carbon dioxide enriched atmospheres. *J. Sci. Food Agric.* **2000**, *80*, 1831–1840. [CrossRef]
30. Yu, D.; Regenstein, J.M.; Zang, J.; Xia, W.; Xu, Y.; Jiang, Q.; Yang, F. Inhibitory effects of chitosan-based coatings on endogenous enzyme activities, proteolytic degradation and texture softening of grass carp (*Ctenopharyngodon idellus*) fillets stored at 4 °C. *Food Chem.* **2018**, *262*, 1–6. [CrossRef]
31. Yu, X.; Li, L.; Xue, J.; Wang, J.; Song, G.; Zhang, Y.; Shen, Q. Effect of air-frying conditions on the quality attributes and lipidomic characteristics of surimi during processing. *Innov. Food Sci. Emerg. Technol.* **2020**, *60*, 102305. [CrossRef]
32. Enrique Cometto-Muñiz, J.; Cain, W.S. Perception of odor and nasal pungency from homologous series of volatile organic compounds. *Indoor Air* **2010**, *4*, 140–145. [CrossRef]
33. Hammer, M.; Schieberle, P. Model studies on the key aroma compounds formed by an oxidative degradation of ω-3 fatty acids initiated by either copper(ii) ions or lipoxygenase. *J. Agric. Food Chem.* **2013**, *61*, 10891–10900. [CrossRef] [PubMed]
34. Leduc, F.; Tournayre, P.; Kondjoyan, N.; Mercier, F.; Malle, P.; Kol, O.; Duflos, G. Evolution of volatile odorous compounds during the storage of European seabass (*Dicentrarchus labrax*). *Food Chem.* **2012**, *131*, 1304–1311. [CrossRef]
35. Parlapani, F.F.; Mallouchos, A.; Haroutounian, S.A.; Boziaris, I.S. Microbiological spoilage and investigation of volatile profile during storage of sea bream fillets under various conditions. *Int. J. Food Microbiol.* **2014**, *189*, 153–163. [CrossRef]

foods

Article

In-Vitro Study on the Antibacterial and Antioxidant Activity of Four Commercial Essential Oils and *In-Situ* Evaluation of Their Effect on Quality Deterioration of Pacific White Shrimp (*Litopenaeus vannamei*) during Cold Storage

Yun-Fang Qian [1,2,3], Ting Lin [1], Xiao Liu [4], Jiao Pan [1], Jing Xie [1,2] and Sheng-Ping Yang [1,2,*]

1 College of Food Science & Technology, Shanghai Ocean University, Shanghai 201306, China
2 Shanghai Engineering Research Center of Aquatic Product Processing & Preservation, Shanghai 201306, China
3 Department of Food and Nutrition, University of Helsinki, 00014 Helsinki, Finland
4 Henan Key Laboratory of Cold Chain Food Quality and Safety Control, Zhengzhou University of Light Industry, Zhengzhou 450001, China
* Correspondence: spyang@shou.edu.cn; Tel.: +86-21-61900400

Abstract: The antioxidant and antibacterial properties of four essential oils (oregano essential oil (OEO), tea tree essential oil (TTEO), wild orange essential oil (WOEO), and clove leaf essential oil (CLEO)) were determined. The *in-vitro* experiment indicated that CLEO had the highest total phenolic content and DPPH scavenging activity, and OEO displayed the highest antibacterial effect, so they were applied to maintain the quality of shrimp for further study. *In-situ* study, the total viable counts of shrimp were inhibited from 9.05 log CFU/g to 8.18 and 8.34 log CFU/g by 2% of OEO and CLEO treated alone on 10 d. The melanosis ratio was also retarded from 38.16% to 28.98% and 26.35% by the two essential oils. The inhibitory effects of OEO and CLEO on the increase of PPO activity, weight loss, and TCA-soluble peptides, and the decreasing tendency of whiteness, the contents of myofibrillar and sarcoplasmic proteins were also founded. The samples treated with 1% OEO + 1% CLEO had better quality than those treated alone. Therefore, the combination of OEO and CLEO had a synergistic effect, which displayed the highest efficiency to prevent the melanosis, bacterial growth, and protein hydrolysis of shrimp.

Keywords: oregano essential oil; clove leaf essential oil; melanosis ratio; protein degradation; TCA-soluble peptides; mechanism

Citation: Qian, Y.-F.; Lin, T.; Liu, X.; Pan, J.; Xie, J.; Yang, S.-P. *In-Vitro* Study on the Antibacterial and Antioxidant Activity of Four Commercial Essential Oils and *In-Situ* Evaluation of Their Effect on Quality Deterioration of Pacific White Shrimp (*Litopenaeus vannamei*) during Cold Storage. *Foods* **2022**, *11*, 2475. https://doi.org/10.3390/foods11162475

Academic Editors: Tao Yin and Liu Shi

Received: 19 July 2022
Accepted: 11 August 2022
Published: 17 August 2022

1. Introduction

Essential oils (EO) are important plant extracts and have attracted much interest from scientists for their various biological activities, including anticancer, anti-obesity, antispasmodic effects, etc. [1–3]. They are also potential to replace chemical preservatives in the field of food preservation due to their antibacterial, antiviral, antifungal, and antioxidant effects [4–6]. Studies have demonstrated the efficiency of EOs to inhibit the growth of pathogens and spoilers, and extend the shelf life of seafoods [7–9]. Furthermore, as they are derived from plants, they are generally recognized as safe (GRAS) by the United States Food and Drug Administration (USFDA) [4].

The effectiveness of EOs on inhibiting bacterial growth and prolonging the shelf-life of various foods are proven [10–12]. However, EOs are very complex, because they are composed of at least 50 components. It is difficult to predict the susceptibility of a microorganism to a kind of EOs [11]. Therefore, it is necessary to study the formula of the extracts on different food products individually.

Pacific white shrimp (*Litopenaeus vannamei*) is an important aquacultured products favored worldwide, whose global production reached 5.8 million in 2020 [13,14]. After harvesting, chilling or freezing is often applied to retard the metabolic activities of microorganisms and endogenous protease hydrolysis. However, the hydrolysis and bacterial spoilage progress still move on even in chilling environment, leading to a limited shelf-life of several days [15]. However, besides the adverse effect of microorganisms and protein hydrolysis, melanosis is another essential problem to cause rejection from consumers. Melanosis occurs firstly when black spots appear on the cephalothorax of the shrimp, and finally develops thoroughly, mainly triggered by tyrosinase, which is also known as polyphenoloxidase (PPO), whose activity is related with catch/harvest season, gender, and size/age of specimens [16,17]. Although the precipitated melanin in shrimp is non-toxic, it can easily reduce the nutrient level, resulting in quality and economic loss [16]. Therefore, to extend the shelf life of shrimp, it is important to inhibit both the development of melanosis and the growth of bacteria. The inhibitory effect of some kinds of EOs on bacteria growth in seafood has been reported [7,9,10,18,19]. However, the studies about EOs on protein degradation and melanosis development of Pacific white shrimp still remains unclear.

In this study, the antibacterial and anti-melanosis activity of several essential oils (oregano essential oil (*Origanum vulgare*), wild orange essential oil (*Citrus sinensis*), tea tree essential oil (*Melaleuca Alternifolia*), and clove essential oil (*Eugenia caryophyllata*)) were evaluated both in vitro and *in situ*, so as to provide a whole view about the efficiency of the essential oils on the preservation of shrimp and propose a possible fresh-maintaining mechanism.

2. Materials and Methods

2.1. Chemicals

Oregano essential oil (OEO) and wild orange essential oil (WOEO) were obtained from döTERRA Shanghai Trading Co., Ltd. (Shanghai, China). Tea tree essential oil (TTEO) was obtained from Camenae Botanical Technology Co., Ltd. (Guangzhou, China). Clove leaf essential oil (CLEO) was brought from Ecoarts Enterprise Co., Ltd. (Shanghai, China). Trichloroacetic acid (TCA), sodium chloride, DPPH, and other chemicals were all obtained from Sinopharm chemical reagent company (Shanghai, China). All reagents were of the highest grade level available commercially. All microbial media were procured from Qingdao Hope Biol-Technology Co., Ltd. (Qingdao, China).

2.2. Bacterial Strains

Staphylococcus aureus (ATCC14222), *Escherichia coli* (ATCC25922), and *Bacillus cereus* (ATCC14579) were obtained from American Type Culture Collection (ATCC). *Shewanella putrefaciens* QY38 (NCBI accession no: KX692894) was isolated from spoiled Pacific white shrimps and stored in the laboratory previously.

2.3. In-Vitro Analysis of the Antioxidant and Antibacterial Ability

2.3.1. Preparation of the Preservatives

The preservatives, including OEO, tea tree essential oil TTEO, wild orange essential oil WOEO, and clove leaf essential oil CLEO were diluted to 0.5%, 2%, and 10% in ethanol for antioxidant assays, respectively.

For the *in-vitro* antibacterial assay, the essential oils were diluted to 2.5%, 2%, 0.5%, 0.1%, 0.025%, and 0.005% (v/v), respectively, and sterilized by filtration through a 0.22 μm pore-size filter before use. The solvent ethanol was used as the control.

2.3.2. Total Phenolic Content

Total phenolic content (TPC) of all the preservatives was determined as described by Olatunde, et al. [20] using gallic acid as standard.

2.3.3. DPPH Scavenging Activity

DPPH scavenging activity of the preservatives was evaluated according to the method with some modifications [21]. The dilution of preservatives (25 μL) was mixed with 975 μL of DPPH ethanol solution (100 μmol/L), and then incubated at room temperature for 30 min in the dark. The absorbance of the solution was measured at 515 nm. The mixture of deionized water and DPPH ethanol solution was used as the blank, while the mixture of the preservatives and ethanol solvent was used as the control. The DPPH scavenging activity was calculated as follows:

$$\text{DPPH scavenging activity\%} = \left(1 - \frac{A_s - A_c}{A_b}\right) \times 100\% \quad (1)$$

where A_s represents the absorbance of the sample, A_b represents the absorbance of blank, and A_c represents the absorbance of the control.

2.3.4. Oxford Cup Assay

These strains were activated by cultivation in Brain Heart Infusion (BHI) broth twice at 30 °C until the concentration reached approximately 10^8 cells/mL. Then culture was diluted to about 10^6 cells/mL by sterilized Luria-Bertani (LB) broth. Then the Oxford cup assay was conducted according to Qian, Cheng, Ye, Zhao, Xie and Yang [18]. Finally, the diameter of the inhibition zone was measured after incubation at 30 °C for 24 h.

2.4. In-Situ Study on the Effect of the Selected Preservatives on the Qualities of Pacific White Shrimp

2.4.1. Sample Preparation

Pacific white shrimp samples (16–20 g for each) were purchased from local market near Luchaogang Port (Shanghai, China) alive in water (14–18 °C) within 30 min. Upon arrival at the laboratory, the shrimps were immersed into the ice-slurry and washed. Then the shrimp was drained and separated into 4 groups randomly.

2.4.2. Preparation and Application of Preservatives

Oregano essential oil (OEO) and clove leaf essential oil (CLEO) were used to preserve Pacific white shrimp during cold storage. According to the *in-vitro* antioxidant and antibacterial experiment, the two essential oils were diluted to 2% (v/v) in water. A compound preserving solution was prepared with 2% OEO and 2% CLEO at a ratio of 1:1. Then the shrimp were immersed into the essential oils for 5 min before draining. The four treatments are listed as following: (1) Con: treated by sterilized water; (2) OEO: treated by 2% OEO; (3) CLEO: treated by 2% CLEO; (4) OEO + CLEO: treated by 1% OEO + 1% CLEO.

2.4.3. Microbiological Analysis

The total viable counts, psychrotrophic bacteria, and H_2S-producing bacteria growth were measured according to Yu, et al. [22]. In brief, 25 g of shrimp was homogenized with 225 mL sterilized saline water (0.85%, w/v), and diluted in serial 10-fold solutions. One milliliter of the dilutions was mixed with Plate count agar (PCA, No.HB0101, Qingdao Hope Biol-Technology Co., Ltd., Qingdao, China) or Iron agar (IA, NO. HB8735, Qingdao Hope Biol-Technology Co., Ltd., Qingdao, China). For total viable counts and psychrotrophic bacteria counts, the plates with PCA were incubated at 30 °C for 48 h and 4 °C for 10 days, respectively. For H_2S-producting bacteria counts, the plates with IA were incubated at 25 °C for 72 h, and the black colonies were counted for calculation.

2.4.4. Total Volatile Basic Nitrogen (TVB-N)

The TVB-N content of shrimp was detected by using the steam-distillation procedure on an Automatic Kjeldahl Apparatus (Kjeltec Analyzer Unit, Foss Tecator AB; Hoganas, Sweden) [22]. The results were recorded as mg N/100 g. The procedure was repeated in triplicate.

2.4.5. Weight LOSS

Weight loss (%) was evaluated by weighing shrimp during storage and evaluated by the following equation according to the reference [23].

$$\text{Weight loss }(\%) = \frac{m_0 - m_n}{m_0} \times 100 \tag{2}$$

where m_0 was the weight of shrimp at the beginning of storage, and m_n was the weight of shrimp after n days of storage.

2.4.6. Determination of TCA-Soluble Peptides Content

Trichloroacetic acid (TCA)-soluble peptides contents were determined according to the method of Chen, et al. [24].

2.4.7. Determination of Sarcoplasmic and Myofibrillar Protein Contents

The protein was extracted according to the method of Lv, et al. [25]. About 2 g minced fish sample was homogenized with 20 mL Tris-maleate buffer A (20 mmol/L pH 7.0, 0.05 mol/L KCl) at 4 °C. Then the solution was centrifuged for 15 min at the speed of 10,000× g. The supernatant containing sarcoplasmic protein was collected. The precipitate was homogenized with 20 mL Tris-maleate buffer B (20 mmol/L, pH 7.0, containing 0.6 mol/L KCl). The solution was incubated for 1 h at 4 °C. After centrifugation at 10,000× g for 15 min, the supernatant containing myofibrillar protein was obtained. The protein content of the extracted solution was determined by the BCA Protein Assay Kit (Tiangen Biotech (Beijing) Co., Ltd., Beijing, China) and the results were expressed as mg/mL of the extraction.

2.4.8. Colorimetric Measurement

The surface color including L^* (lightness), a^* (redness and greenness), b^* (yellowness and blueness) of shrimp cephalothoraxes were measured via YS6060 Benchtop Grating Spectrophotometer (Shenzhen Threenh Technology Co., Ltd., Shenzhen, China) in accordance with the method described by Qian, et al. [26]. The whiteness value was calculated according to Equation (3), and then the mean values of samples per group were calculated.

$$\text{Whiteness} = 100 - \sqrt{(100 - L^*)^2 + (a^*)^2 + (b^*)^2} \tag{3}$$

2.4.9. Melanosis Ratio

The photos of the shrimp samples were loaded to Adobe Photoshop 2020 (Adobe Systems Incorporated, San Jose, CA, USA). After eliminating the background, the black area was identified using the Magic Wand Tool with a tolerance of 4%, and the black pixels and the pixels of the whole shrimp were calculated. The melanosis ratio of the black pixels to the total pixels of the whole shrimp was used to evaluate the development of melanosis (Equation (4)).

$$\text{Melanosis ratio}\% = \frac{\text{Pixels of black area}}{\text{Total pixels of the whole shrimp area}} \times 100\% \tag{4}$$

2.4.10. Polyphenol Oxidase Activity

The polyphenol oxidase activity was determined using the polyphenol oxidase kit (Nanjing Jiancheng Bioengineering Institute, Nanjing, China). The shrimp (0.1 g) was homogenized with 1 mL extraction buffer. Then the extraction solution was mixed with the substrate solution, and the absorbance was measured via a spectrophotometer (752 N, Shanghai INESA Scientific Instrument Co., Ltd., Shanghai, China) at 420 nm. One unit (1 U) of PPO activity was defined as an increment of 0.01 absorbance/min. The PPO activity of the sample was expressed as specific activity in units of enzyme activity per mg protein (U/mg protein).

2.5. Statistical Analysis

All experiments were conducted at least in triplicate. The one-way ANOVA procedure followed by Duncan's test was used by the SPSS software ($p < 0.05$, SPSS Version 20.0, Inc., Chicago, IL, USA), and the results were expressed as means ± SD. The diagrams were designed by the Origin2018 software (OriginLab, Northampton, MA, USA).

3. Results and Discussion

3.1. In-Vitro Analysis of the Antioxidant and Antibacterial Ability

3.1.1. Total Phenolic Content

The total phenolic contents of the preservatives are shown in Table 1. It is known that polyphenols are a kind of bioactive compounds, which are considered to be responsible for the antibacterial and antioxidant activity [27]. The results showed that the TPC of OEO, TTEO, and CLEO were close to each other, while the TPC of WOEO was much lower. This phenomenon should be related to the different predominant compounds in these essential oils. The predominant component in WOEO was limonene, belonging to terpene groups, and other essential oils had higher contents of phenol compounds.

Table 1. Total phenolic content (TPC) and DPPH scavenging activity (%) of different preservatives.

	Concentration (*v*/*v*)	OEO	TTEO	WOEO	CLEO
TPC (mg GAE/100 g)	10%	20.97 ± 0.22	20.29 ± 0.35	13.34 ± 0.51	21.12 ± 0.01
DPPH scavenging activity (%)	0.5%	42.28 ± 1.15	22.76 ± 1.15	27.64 ± 1.15	79.67 ± 1.15
	2%	73.11 ± 0.54	28.22 ± 0.27	54.55 ± 0.54	84.85 ± 0.54
	10%	84.55 ± 3.45	61.79 ± 2.30	65.85 ± 1.15	87.80 ± 2.30

Note: oregano essential oil (OEO), tea tree essential oil (TTEO), wild orange essential oil (WOEO), and clove leaf essential oil (CLEO).

3.1.2. DPPH Scavenging Activity

Melanosis begins from the oxidation of tyrosine triggered by tyrosinase [16], therefore an antioxidant is considered to have a potential ability to prevent melanosis. The antioxidative capacities of the preservatives were determined by the DPPH free radical scavenging assay, and the results are also displayed in Table 1. Generally, the DPPH scavenging activity of the essential oils exhibited a dose-dependent relationship at different extents. The DPPH scavenging activity of CLEO was significantly higher than other essential oils. The results of OEO were close to the study by Kosakowska et al. [28], which contributed the DPPH scavenging activity of OEO to carvacrol and/or thymol. It is reported that the free radical scavenging activity of CLEO is attributed to the phenylpropanoids such as eugenol and its derivatives [29]. The results confirmed that having the same level of phenolic content does not necessarily mean the same antioxidant response [30]. It is reported that the antioxidant effect relates with the hydrophilic-lipophilic balance of the extract and the chemistry of the surrounding environment [31]. Therefore, the antioxidant activity of EOs is also related to the solvent polarity [28]. To scavenge 50% DPPH, the concentrations of OEO and CLEO should not be less than 2% and 0.5%, respectively.

3.1.3. Oxford Cup Assay

The sensitivity of the four bacterial species to the four essential oils is shown in Table 2. In general, all preservatives showed promising antibacterial properties against the selected bacteria. The dimensions of the inhibition halos increased with the increasing concentrations of essential oils, but the sensitiveness differed from the bacterial species and the oils used. The minimum inhibitory concentration (MIC) is defined when the diameters of inhibition halos were higher than 10 mm. According to the results, the minimum inhibitory concentrations (MICs) of oregano essential oil against *S. aureus*, *B. subtilis*, *S. putrefaciens*, and *E. coli* were 0.1%, and the MIC of TTEO was 10%, and that of CLEO was about 0.5%. The MIC of OEO was close to the previous study [28]. WOEO had the highest MICs than other oils, indicating the lowest antibacterial activity. The antibacterial activity of these essential oils could be contributed to the polyphenolic compounds, which can interact with bacterial cell membrane due to the hydrophobic/lipophilic property, consequently leading to the destruction of membrane and leakage of cell components [32,33].

Table 2. The sensitivity of the four bacterial species to the four essential oils.

Bacterium	Presser-Vatives	Concentrations (%)						
		10	2.5	0.5	0.1	0.025	0.005	Negative Control
S. aureus (+)	OEO	++++	++++	++++	++	+	−	−
	TTEO	++	+	+	−	−	−	−
	WOEO	++	+	+	−	−	−	−
	CLEO	++++	+++	++	+	+	−	−
B. subtilis (+)	OEO	++++	++++	++++	++	+	−	−
	TTEO	++	++	+	−	−	−	−
	WOEO	+	+	−	−	−	−	−
	CLEO	++++	+++	++	+	+	−	−
S. putrefaciens (−)	OEO	++++	++++	++++	++	+	+	−
	TTEO	++	++	+	−	−	−	−
	WOEO	+++	++	+	−	−	−	−
	CLEO	++++	++++	++	+	+	+	−
E. coli (−)	OEO	++++	++++	++++	++	−	−	−
	TTEO	++	+	−	−	−	−	−
	WOEO	+	+	−	−	−	−	−
	CLEO	++++	+++	++	+	+	+	−

Note: "++++" means extremely sensitive (bacteriostatic diameter $\geq$ 20 mm); "+++" means quite sensitive (15 mm $\leq$ bacteriostatic diameter < 20 mm); "++" means sensitive (10 mm $\leq$ bacteriostatic diameter < 15 mm); "+" means slightly sensitive (8 mm < bacteriostatic diameter < 10 mm); "−" means not sensitive (bacteriostatic diameter $\leq$ 8 mm).

Among these selected plant essential oils, OEO and CLEO displayed a better antioxidant and antibacterial effect. Considering the low adhesion rate into shrimp, the 4MICs should be applied for further *in-situ* study. Therefore, the concentrations of OEO and CLEO applied in Pacific white shrimp for further study were 2%.

3.2. In-Situ Study on the Effect of the Selected Essential Oils on the Quality of Pacific White Shrimp

3.2.1. Microbiological Growth

The changes of total viable counts, psychrotrophic bacterial counts, and H_2S-producing bacterial counts of shrimp treated with essential oils are displayed in Figure 1A–C. The initial total viable counts (TVCs) of the four groups were close to each other, i.e., 4.37 log CFU/g, 4.16 log CFU/g, 4.50 log CFU/g, and 4.01 log CFU/g, respectively (Figure 1A). The total viable counts of all groups were all increased during storage. The essential oil treatment group was significantly lower than the control group ($p < 0.05$). It is illustrated that the essential oils had an obvious effect of bacteriostasis, which was consistent with the results of the *in-vitro* experiments mentioned above. The phenolic compounds such as carvacrol in OEO and eugenol in CLEO should be attributed to the antibacterial effect [34]. The TVC of samples treated with OEO or CLEO only was close to each other, indicating that they had a similar antibacterial effect, but was slightly higher than the combined treatment. The control exceeded the threshold of 7 log CFU/g [35] on day 6 (7.71 log CFU/g), while the sample treated by OEO + CLEO reached the limitation on day 10 (7.83 log CFU/g). Therefore, the shelf life of shrimp was prolonged by about 4 days by OEO + CLEO. It was demonstrated the combination of the two essential oils could enhance the effect of bacteriostasis.

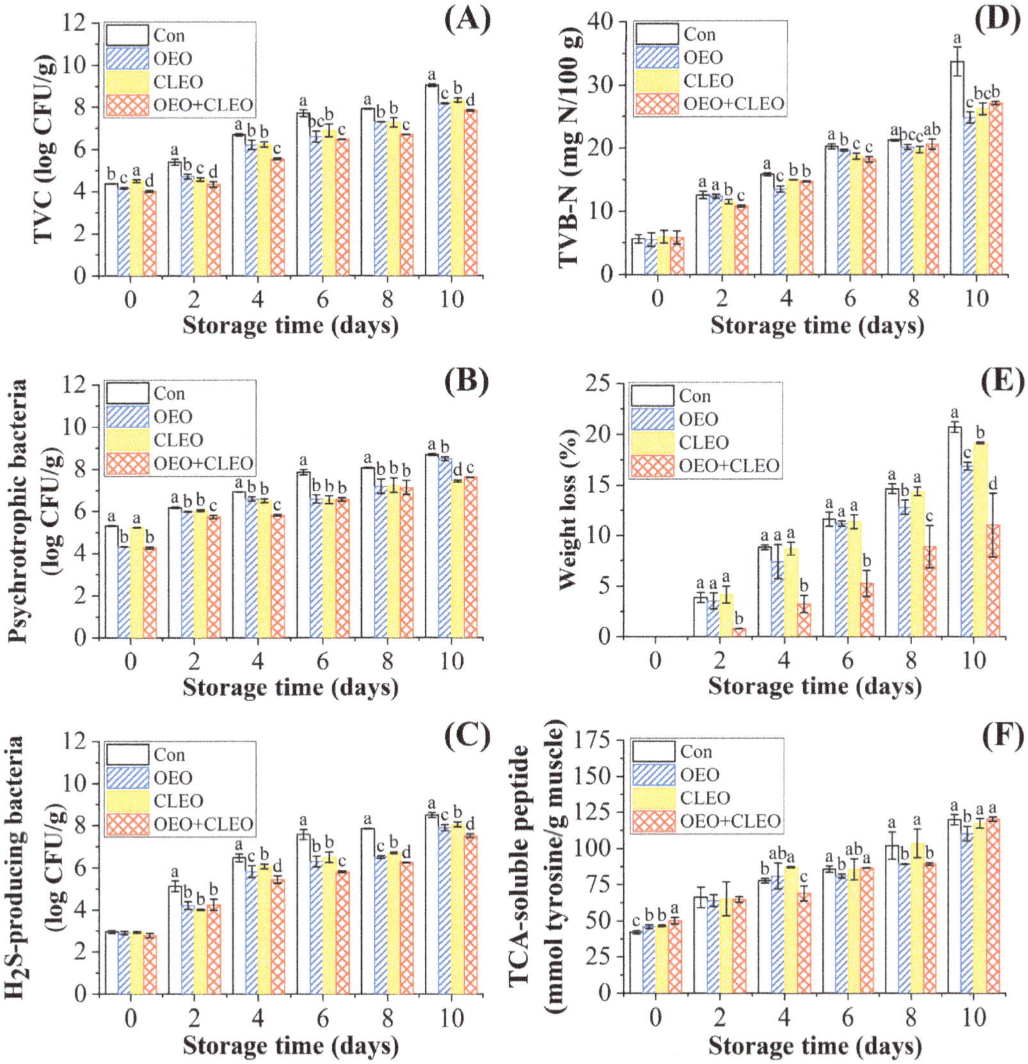

Figure 1. Changes of total viable counts (**A**), psychrotrophic bacterial counts (**B**), H_2S-producing bacterial counts (**C**), TVB-N (**D**), weight loss (**E**) and TCA-soluble peptide (**F**) of refrigerated Pacific white shrimp treated with OEO and CLEO at 4 °C. Data are shown as mean ± SD (n = 3). Note: Different lowercase letters indicate significant differences ($p < 0.05$) among different treatments within the same storage time.

The initial counts of psychrotrophic bacteria of shrimp were higher than TVC, and also increased during the whole storage period (Figure 1B). Therefore, the spoilage of shrimp during refrigerated storage was mainly attributed to the proliferation of psychrotrophic bacteria. During the storage period, the samples treated with essential oils were always lower than the control group, indicating its inhibitory effect on the growth of psychrotrophic bacteria.

The changes of H_2S-producing bacterial counts of Pacific white shrimp during storage are shown in Figure 1C. *Shewanella* species, including *S. putrefaciens* and *S. baltica* are reported to be the predominant H_2S-producing bacteria in shrimp, which has the capacity to produce trimethylamine, H_2S, putrescine, and other off-odor compounds [36]. The initial counts of H_2S-producing bacteria in the four groups were close to each other, but the increase in samples treated with essential oils was retarded. At the end of storage, the counts of the four groups increased to 8.48 log CFU/g, 7.89 log CFU/g, 8.04 log CFU/g, and 7.51 log CFU/g, respectively. CLEO displayed a slightly lower effect of bacteriostasis than OEO, in accordance with the *in-vitro* measurement. However, the combination of the two essential oils showed the highest antibacterial effect, indicating that there was a synergistic effect.

3.2.2. Total Volatile Basic Nitrogen (TVB-N)

The TVB-N value reflects the amount of trimethylamine, dimethylamine, ammonia and other amines produced by the decomposition of protein in aquatic products under the combined action of microorganisms and enzymes [37]. The changes of TVB-N value of Pacific white shrimp treated with essential oils during storage are shown in Figure 1D. The initial TVB-N value of shrimp was about 5.52~5.85 mg N/100 g, but increased to 33.73 mg N/100 g, 24.83 mg N/100 g, 26.21 mg N/100 g, and 27.15 mg N/100 g at the end of storage, respectively. The contents of TVB-N of the control during storage were close to the study previously reported [18]. The TVB-N values of shrimp treated with essential oils were significantly lower than the control after 10 days of storage.

3.2.3. Weight Loss

Figure 1E showed the weight loss of Pacific white shrimp treated with essential oils during storage. The weight loss is mainly related with the drip loss of muscle, due to the decomposition, denaturation, or oxidation of proteins [38,39]. The changes of weight loss were indeed positively correlated with changes of bacterial growth, TCA-soluble peptide, and negatively correlated with protein contents displayed as following in this study. Therefore, the capacity to prevent weight loss of products is an important indicator to evaluate the efficiency of preservatives. The results showed that all samples treated with essential oils had lower weight loss than the control. Among the three treated samples, the samples treated with OEO + CLEO had the lowest weight loss, indicating that OEO + CLEO could inhibit the changes of protein effectively.

3.2.4. Changes of TCA-Soluble Peptide Content

The content of TCA-soluble peptide in shrimp increased during the storage period (Figure 1F), which was negatively correlated with the changes of sarcoplasmic and myofibrillar proteins. The increase of TCA-soluble peptide should be attributed to the degradation of proteins triggered by endogenous enzymes and contaminated bacteria [40]. Therefore, the higher the amount of TCA-soluble peptide means the greater the degradation of the protein. Compared with the four groups, it was found that OEO and OEO + CLEO displayed a better effect on inhibiting the increase of TCA-soluble peptide.

3.2.5. Changes of Sarcoplasmic and Myofibrillar Contents

The muscle protein of shrimp mainly consists of the sarcoplasmic and myofibrillar proteins, which are water-soluble and salt-soluble protein, respectively [41]. As shown in Figure 2A,B, the contents of sarcoplasmic and myofibrillar protein of shrimp decreased during storage, which were negatively correlated with the changes of bacterial gowth, TCA-soluble peptide, and TVB-N. The initial contents of sarcoplasmic protein of shrimp were 12.56 mg/mL, 9.10 mg/mL, 8.82 mg/mL, and 10.56 mg/mL on 0 d, respectively. It is reported that the sarcoplasmic protein is mainly composed by proteases and other low-molecular proteins [42]. The control group was generally lower than the treatment group, especially after 8 days of storage ($p < 0.05$). It is illustrated that the treatment with essential oils could significantly delay the protein degradation of Pacific white shrimp.

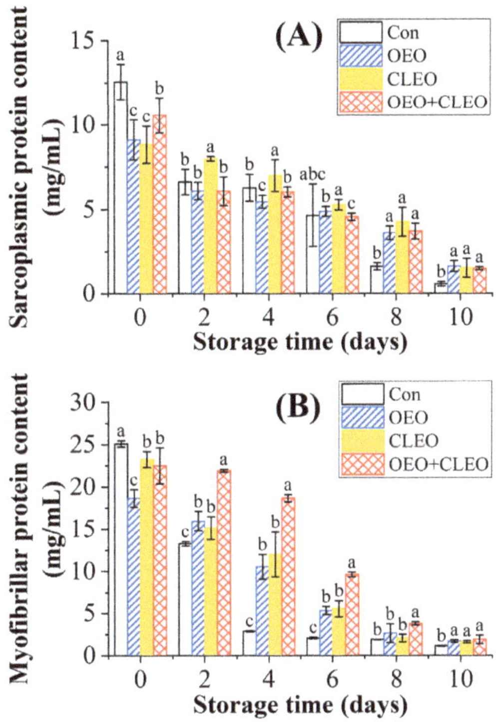

Figure 2. Changes of the content of sarcoplasmic protein (**A**) and myofibrillar protein (**B**) of refrigerated Pacific white shrimp treated with OEO and CLEO at 4 °C. Data are shown as mean ± SD (n = 3). Note: Different lowercase letters indicate significant differences ($p < 0.05$) among different treatments within the same storage time.

The myofibrillar protein is regarded as the main muscle protein in shrimp, which is composed of salt soluble proteins such as myosin, actin, and troponins [41]. The initial contents of myofibrillar protein in shrimp treated with essential oils were 18.63 mg/mL, 23.25 mg/mL, and 22.50 mg/mL, respectively, which were slightly lower than the control (25.10 mg/mL). The content of myofibrillar protein of the control decreased significantly in the first 4 days and then maintained at a low level till the end of storage. It was found that the essential oils could retard the decreasing rates of the myofibrillar protein, which should be attributed to their antibacterial activity. The samples with the combined treatment had the highest content of myofibrillar protein, indicating the efficiency to preventing protein degradation.

3.2.6. Colorimetric Measurement

The whiteness of cephalothorax and abdomen of Pacific white shrimp are observed in Figure 3A,B. The whiteness value of all groups showed a decreasing tendency during storage. The shrimp treated with essential oils had the higher whiteness values than the control, especially the samples treated with CLEO and OEO + CLEO. At the end of storage, the whiteness values of shrimp cephalothorax of the four groups were 21.19, 21.20, 26.48, and 23.25, respectively, and the whiteness values of the abdomen were 32.71, 38.67, 38.16, and 38.44. The anti-melanosis effect of essential oils was also confirmed by the results. The CLEO was more helpful to maintain the whiteness value of shrimp cephalothorax, due to its higher antioxidant activity.

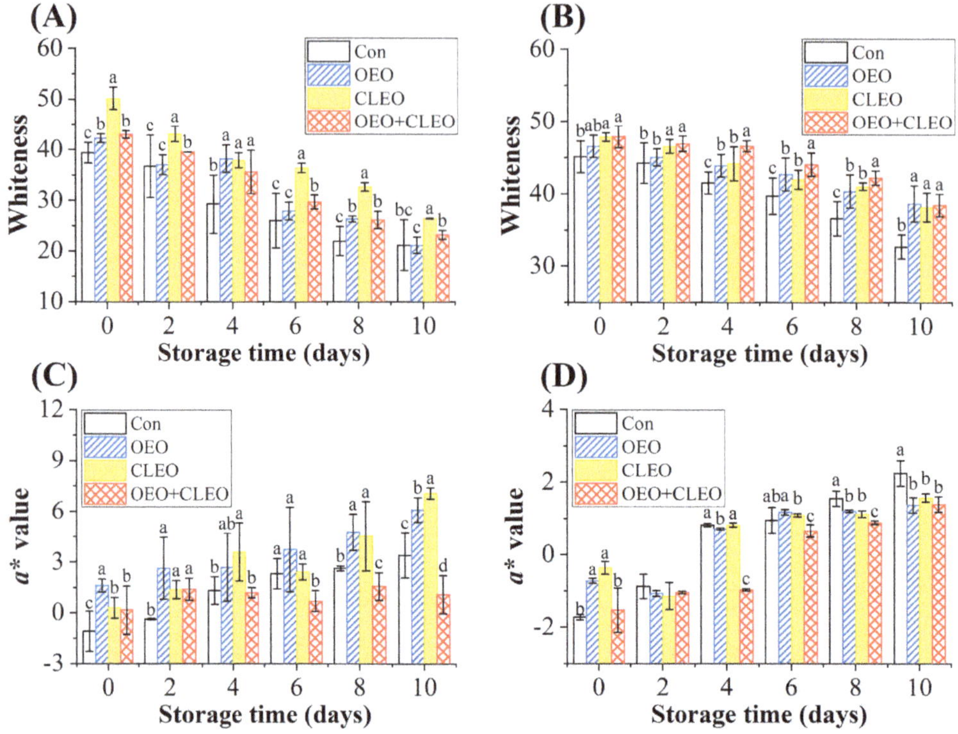

Figure 3. Changes of whiteness and a^* of cephalothorax (**A**,**C**) and abdomen (**B**,**D**) of refrigerated Pacific white shrimp treated with OEO and CLEO at 4 °C. Data are shown as mean ± SD (n = 3). Note: Different lowercase letters indicate significant differences ($p < 0.05$) among different treatments within the same storage time.

The redness (a^*) value of shrimp cephalothorax and abdomen both increased during the storage (Figure 3C,D). The increase of a^* value of shrimp cephalothorax was much faster than that of the shrimp abdominal section. The samples treated with OEO + CLEO were lower than other groups. The increase of redness value could be related with the decomposition of astaxanthin-protein complexes and the release of astaxanthin monomers [43].

3.2.7. Melanosis Ratio

The overall photographs of Pacific white shrimp during storage are shown in Figure 4A. The melanosis area ratio was calculated accordingly to evaluate the development of melanosis all over the shrimp sample, and the results are shown in Figure 4B. The melanosis usually starts from the cephalothorax and then spreads to the abdomen and the tail [44]. The melanosis ratio of the control was significantly higher than the treated samples throughout the storage period, indicating more black spots occurred on the whole shrimp. The melanosis ratio was negatively correlated with the changes of whiteness values. CLEO seemed to have a stronger inhibitory effect on melanosis than OEO, as the shrimp treated with OEO had a higher melanosis ratio. This might be contributed to the higher antioxidant activity of CLEO as demonstrated in the DPPH scavenging activity. The combined treatment of OEO and CLEO displayed the lowest melanosis ratio, indicating that the two essential oils had a synergetic effect on inhibiting the spread of black spots.

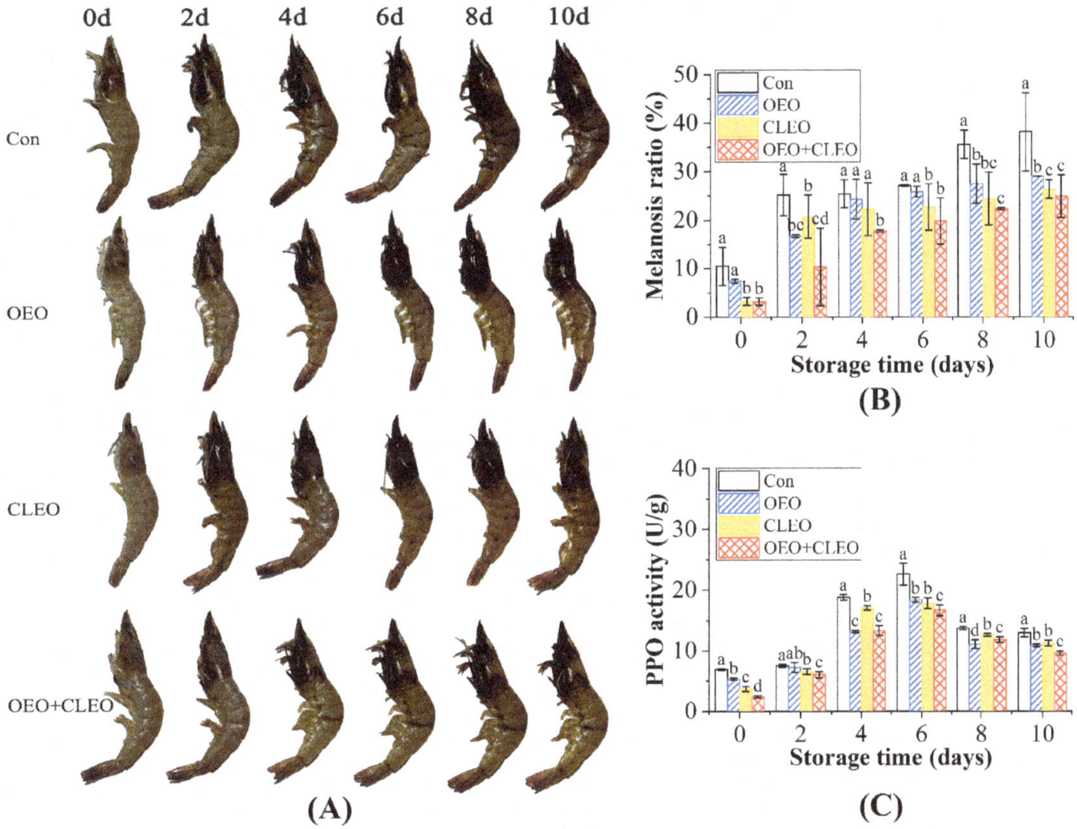

Figure 4. Photograph (**A**) and changes of melanosis ratio (**B**) and PPO activity (**C**) of refrigerated Pacific white shrimp treated by OEO and CLEO at 4 °C. Data are shown as mean ± SD (n = 3). Note: Different lowercase letters indicate significant differences ($p < 0.05$) among different treatments within the same storage time.

3.2.8. PPO Activity

PPO is an important enzyme involved in a variety of physiological activities in crustaceans, which catalyzes the oxidation of tyrosine and undergoes several reactions to form dark pigments [44]. The changes in the PPO activities of shrimp during storage are shown in Figure 4C. The initial PPO activity value of the samples treated with essential oils were 5.30 U/g, 3.69 U/g, and 2.40 U/g, which were lower than the control group (6.87 U/g). The result indicated that essential oils could inhibit the PPO activity immediately after treatment. The PPO activity of all groups went up to the peak on 6 d, and then decreased to some extent. The decreasing tendency of PPO activity should be contributed to proteolytic degradation by endogenous and bacteria protease. Similar phenomenon was also illustrated in other literatures [45]. The PPO activities of treated samples were 18.37 U/g, 17.79 U/g, and 16.64 U/g on day 6, respectively, while the control was about 22.57 U/g. The CLEO displayed a higher inhibitory effect on PPO activity than OEO, which was in accordance with their DPPH scavenging activity. It is hypothesized that phenolic compounds in essential oils inhibit the activity of PPO by competing with the substrates [46]. Moreover, the combination of the two essential oils had the highest efficiency. The increasing tendency of PPO activity in the first period was corresponded with the changes of the melanosis ratio and whiteness value.

3.3. Schematic Illustration

A schematic illustration of the effect of essential oils on melanosis and quality changes of Pacific white shrimp was proposed (Figure 5). The effect of OEO + CLEO to maintain the quality of Pacific white shrimp should basically contributed to their antibacterial and antioxidant activity as illustrated in the *in-vitro* study, and consequently inhibiting the quality deterioration related with bacterial growth and oxidation reactions. During cold storage, the bacteria proliferated and induced the degradation of muscle protein, leading to the decrease of protein content and increase of TCA-soluble peptides and TVB-N. In this study, the effect of OEO + CLEO on retarding the deterioration process due to antibacterial process was proved. Regarding to the melanosis development, it is known that the melanin is produced from tyrosine triggered by PPO in the presence of oxygen [47]. Therefore, inhibitors which can inhibit the activity of PPO or scavenge free oxygen can be used to as anti-melanosis agents [48]. OEO and CLEO deprived from plants, are known to be rich in phenolic compounds, and they could probably inhibit melanosis by inhibit PPO activity and scavenging free oxygen. On the other hand, PPO is known to exist in crustaceans as zymogens and plays a key role in the primary immune response system, and it can be activated because of the serine proteinase cascade triggered by microbial compounds (carbohydrates and lipopolysaccharides) and a series of other proteins [45,47,49]. Therefore, there could be another possible way for the essential oils to inhibit melanosis by retarding the proliferation of bacteria. A relationship between bacteria and melanosis was also displayed in the diagram.

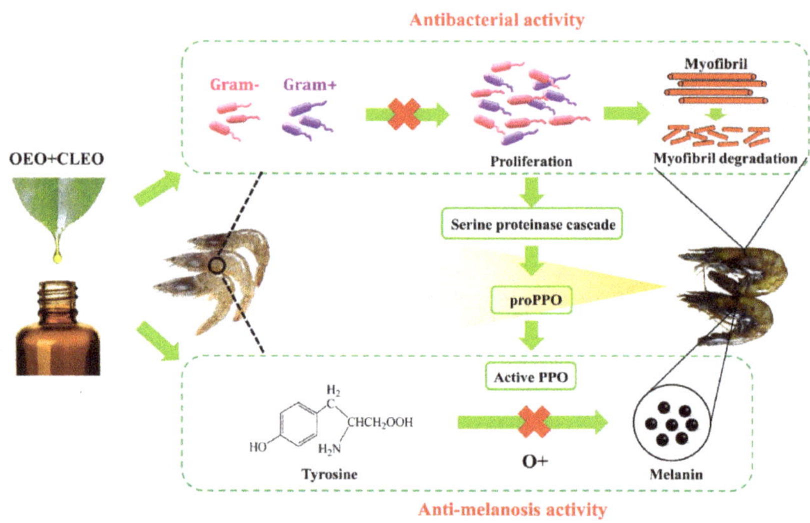

Figure 5. Schematic diagram of the antibacterial and anti-melanosis activity of OEO + CLEO on Pacific white shrimp during cold storage.

4. Conclusions

This work describes the ability of essential oil treatment to extend the shelf life of Pacific white shrimp. The antioxidant and antibacterial efficiency of the four essential oils were screening via *in-vitro* experiment. It was found that CLEO and OEO had higher antioxidant and antibacterial effect than WOEO and TTEO. Therefore, CLEO and OEO were applied to preserve shrimp for further *in-situ* study. The results indicated that both CLEO and OEO could inhibit the quality deterioration of shrimp effectively. The CLEO displayed higher antioxidant activity, as well as a higher inhibitory effect on PPO activity and melanosis, while OEO had higher effect on preventing the increase of bacterial growth, TVB-N value, weight loss, and TCA-soluble peptides. This study highlighted that the combination of

CLEO and OEO enhanced their antioxidant and antibacterial effect, and could maintain the quality of shrimp most effectively, consequently prolonging the microbiological shelf-life by about 4 days. Therefore, this study expanded the usage of EOs in Pacific white shrimp, and the convenience of this method made it to be a potential use for industrial application.

Author Contributions: Y.-F.Q., conceptualization, writing—review and editing, funding acquisition; T.L., investigation, data curation, writing—original draft; X.L., writing—review and editing; J.P., investigation; J.X., funding acquisition, project administration, supervision; S.-P.Y., methodology, investigation, project administration. All authors have read and agreed to the published version of the manuscript.

Funding: This work was financially supported by National Natural Science Foundation of China (No: 31972142, 31501551), China Scholarship Council (No: 202008310018), Special Fund for the Development of Science and Technology of Shanghai Ocean University (No: A2-2006-20-200203), and Henan Key Laboratory of Cold Chain Food Quality and Safety Control (No: CCFQ2021).

Institutional Review Board Statement: Not applicable.

Informed Consent Statement: Not applicable.

Data Availability Statement: The datasets for this study can be found in https://figshare.com/s/4a4358daa3054f71b4ed (accessed on 1 February 2022).

Conflicts of Interest: The authors declare no conflict of interest.

References

1. Ju, J.; Chen, X.; Xie, Y.; Yu, H.; Guo, Y.; Cheng, Y.; Qian, H.; Yao, W. Application of essential oil as a sustained release preparation in food packaging. *Trends Food Sci. Technol.* **2019**, *92*, 22–32. [CrossRef]
2. Mancianti, F.; Ebani, V.V. Biological activity of essential oils. *Molecules* **2020**, *25*, 678. [CrossRef] [PubMed]
3. Duque-Soto, C.; Borrás-Linares, I.; Quirantes-Piné, R.; Falcó, I.; Sánchez, G.; Segura-Carretero, A.; Lozano-Sánchez, J. Potential antioxidant and antiviral activities of hydroethanolic extracts of selected Lamiaceae species. *Foods* **2022**, *11*, 1862. [CrossRef] [PubMed]
4. Ju, J.; Xie, Y.; Guo, Y.; Cheng, Y.; Qian, H.; Yao, W. Application of edible coating with essential oil in food preservation. *Crit. Rev. Food Sci. Nutr.* **2019**, *59*, 2467–2480. [CrossRef]
5. Perricone, M.; Arace, E.; Corbo, M.R.; Sinigaglia, M.; Bevilacqua, A. Bioactivity of essential oils: A review on their interaction with food components. *Front. Microb.* **2015**, *6*, 76. [CrossRef]
6. Bakkali, F.; Averbeck, S.; Averbeck, D.; Idaomar, M. Biological effects of essential oils—A review. *Food Chem. Toxicol.* **2008**, *46*, 446–475. [CrossRef]
7. Chuesiang, P.; Sanguandeekul, R.; Siripatrawan, U. Enhancing effect of nanoemulsion on antimicrobial activity of cinnamon essential oil against foodborne pathogens in refrigerated Asian seabass (*Lates calcarifer*) fillets. *Food Control* **2021**, *122*, 107782. [CrossRef]
8. Oğuzhan Yildiz, P. Effect of essential oils and packaging on hot smoked rainbow trout during storage. *J. Food Process. Preserv.* **2015**, *39*, 806–815. [CrossRef]
9. Huang, Z.; Liu, X.; Jia, S.; Zhang, L.; Luo, Y. The effect of essential oils on microbial composition and quality of grass carp (*Ctenopharyngodon idellus*) fillets during chilled storage. *Int. J. Food Microbiol.* **2018**, *266*, 52–59. [CrossRef]
10. Zheng, X.; Han, B.; Kumar, V.; Feyaerts, A.F.; Van Dijck, P.; Bossier, P. Essential oils improve the survival of Gnotobiotic brine shrimp (*Artemia franciscana*) challenged with *Vibrio campbellii*. *Front. Immunol.* **2021**, *12*, 693932. [CrossRef]
11. Hyldgaard, M.; Mygind, T.; Meyer, R. Essential oils in food preservation: Mode of action, synergies, and interactions with food matrix components. *Front. Microb.* **2012**, *3*, 12. [CrossRef]
12. Bhavaniramya, S.; Vishnupriya, S.; Al-Aboody, M.S.; Vijayakumar, R.; Baskaran, D. Role of essential oils in food safety: Antimicrobial and antioxidant applications. *Grain Oil Sci. Technol.* **2019**, *2*, 49–55. [CrossRef]
13. Okpala, C.O.R.; Choo, W.S.; Dykes, G.A. Quality and shelf life assessment of Pacific white shrimp (*Litopenaeus vannamei*) freshly harvested and stored on ice. *LWT-Food Sci. Technol.* **2014**, *55*, 110–116. [CrossRef]
14. FAO. *The State of World Fisheries and Aquaculture*; Food and Agriculture Organization: Rome, Italy, 2022.
15. Qian, Y.F.; Xie, J.; Yang, S.-P.; Huang, S.; Wu, W.H.; Li, L. Inhibitory effect of a quercetin-based soaking formulation and modified atmospheric packaging (MAP) on muscle degradation of Pacific white shrimp (*Litopenaeus vannamei*). *LWT-Food Sci. Technol.* **2015**, *63*, 1339–1346. [CrossRef]
16. Sae-Leaw, T.; Benjakul, S. Prevention of quality loss and melanosis of Pacific white shrimp by cashew leaf extracts. *Food Control* **2019**, *95*, 257–266. [CrossRef]

17. Bono, G.; Badalucco, C.; Corrao, A.; Cusumano, S.; Mammina, L.; Palmegiano, G.B. Effect of temporal variation, gender and size on cuticle polyphenol oxidase activity in deep-water rose shrimp (*Parapenaeus longirostris*). *Food Chem.* **2010**, *123*, 489–493. [CrossRef]
18. Qian, Y.F.; Cheng, Y.; Ye, J.-X.; Zhao, Y.; Xie, J.; Yang, S.P. Targeting shrimp spoiler *Shewanella putrefaciens*: Application of ε-polylysine and oregano essential oil in Pacific white shrimp preservation. *Food Control* **2021**, *123*, 107702. [CrossRef]
19. Alparslan, Y.; Baygar, T. Effect of chitosan film coating combined with Orange peel essential oil on the shelf life of Deepwater pink shrimp. *Food Bioprocess Technol.* **2017**, *10*, 842–853. [CrossRef]
20. Olatunde, O.O.; Benjakul, S.; Vongkamjan, K. Antioxidant and antibacterial properties of guava leaf extracts as affected by solvents used for prior dechlorophyllization. *J. Food Biochem.* **2018**, *42*, e12600. [CrossRef]
21. Lu-Martínez, A.A.; Báez-González, J.G.; Castillo-Hernández, S.; Amaya-Guerra, C.; Rodríguez-Rodríguez, J.; García-Márquez, E. Studied of *Prunus serotine* oil extracted by cold pressing and antioxidant effect of *P. longiflora* essential oil. *J. Food Sci. Technol.* **2021**, *58*, 1420–1429. [CrossRef]
22. Yu, Y.J.; Yang, S.P.; Lin, T.; Qian, Y.-F.; Xie, J.; Hu, C. Effect of cold chain logistic interruptions on lipid oxidation and volatile organic compounds of salmon (*Salmo salar*) and their correlations with water dynamics. *Front. Nutr.* **2020**, *7*, 155. [CrossRef] [PubMed]
23. Wang, S.; Xiang, W.; Fan, H.; Xie, J.; Qian, Y.F. Study on the mobility of water and its correlation with the spoilage process of salmon (*Salmo solar*) stored at 0 and 4 °C by low-field nuclear magnetic resonance (LF NMR ^{1}H). *J. Food Sci. Technol.* **2018**, *55*, 173–182. [CrossRef] [PubMed]
24. Chen, L.; Jiao, D.; Liu, H.; Zhu, C.; Sun, Y.; Wu, J.; Zheng, M.; Zhang, D. Effects of water distribution and protein degradation on the texture of high pressure-treated shrimp (*Penaeus monodon*) during chilled storage. *Food Control* **2022**, *132*, 108555. [CrossRef]
25. Lv, M.; Mei, K.; Zhang, H.; Xu, D.; Yang, W. Effects of electron beam irradiation on the biochemical properties and structure of myofibrillar protein from Tegillarca granosa meat. *Food Chem.* **2018**, *254*, 64–69. [CrossRef]
26. Qian, Y.F.; Xie, J.; Yang, S.-P.; Wu, W.H. Study of the quality changes and myofibrillar proteins of white shrimp (*Litopenaeus vannamei*) under modified atmosphere packaging with varying CO_2 levels. *Eur. Food Res. Technol.* **2013**, *236*, 629–635. [CrossRef]
27. Maqsood, S.; Benjakul, S.; Abushelaibi, A.; Alam, A. Phenolic compounds and plant phenolic extracts as natural antioxidants in prevention of lipid oxidation in seafood: A detailed review. *Compr. Rev. Food Sci. Food Saf.* **2014**, *13*, 1125–1140. [CrossRef]
28. Kosakowska, O.; Węglarz, Z.; Pióro-Jabrucka, E.; Przybył, J.L.; Kraśniewska, K.; Gniewosz, M.; Bączek, K. Antioxidant and antibacterial activity of essential oils and hydroethanolic extracts of Greek oregano (*O. vulgare* L. subsp. *hirtum* (Link) Ietswaart) and common oregano (*O. vulgare* L. subsp. *vulgare*). *Molecules* **2021**, *26*, 988. [CrossRef]
29. Hadidi, M.; Pouramin, S.; Adinepour, F.; Haghani, S.; Jafari, S.M. Chitosan nanoparticles loaded with clove essential oil: Characterization, antioxidant and antibacterial activities. *Carbohydr. Polym.* **2020**, *236*, 116075. [CrossRef]
30. Rocha-Guzmán, N.E.; Gallegos-Infante, J.A.; González-Laredo, R.F.; Ramos-Gómez, M.; Rodríguez-Muñoz, M.E.; Reynoso-Camacho, R.; Rocha-Uribe, A.; Roque-Rosales, M.R. Antioxidant effect of oregano (*Lippia berlandieri* v. Shauer) essential oil and mother liquors. *Food Chem.* **2007**, *102*, 330–335. [CrossRef]
31. Chang, Y.C.; Almy, E.A.; Blamer, G.A.; Gray, J.I.; Frost, J.W.; Strasburg, G.M. Antioxidant activity of 3-dehydroshikimic acid in liposomes, emulsions, and bulk oil. *J. Agric. Food Chem.* **2003**, *51*, 2753–2757. [CrossRef]
32. Badola, R.; Panjagari, N.R.; Singh, R.R.B.; Singh, A.K.; Prasad, W.G. Effect of clove bud and curry leaf essential oils on the anti-oxidative and anti-microbial activity of *burfi*, a milk-based confection. *J. Food Sci. Technol.* **2018**, *55*, 4802–4810. [CrossRef] [PubMed]
33. Cui, H.; Zhang, C.; Li, C.; Lin, L. Antibacterial mechanism of oregano essential oil. *Ind. Crops Prod.* **2019**, *139*, 111498. [CrossRef]
34. Hać-Szymańczuk, E.; Cegiełka, A.; Karkos, M.; Gniewosz, M.; Piwowarek, K. Evaluation of antioxidant and antimicrobial activity of oregano (*Origanum vulgare* L.) preparations during storage of low-pressure mechanically separated meat (BAADER meat) from chickens. *Food Sci. Biotechnol.* **2019**, *28*, 449–457. [CrossRef] [PubMed]
35. International Commission on Microbiological Specifications for Foods. Microorganisms in Foods. *Microbiol. Test. Food Saf. Manag.* **2002**, *7*, 362.
36. Yang, S.P.; Xie, J.; Qian, Y.F. Determination of spoilage microbiota of Pacific white shrimp during ambient and cold storage using next-generation sequencing and culture-dependent method. *J. Food Sci.* **2017**, *82*, 1178–1183. [CrossRef]
37. Annamalai, J.; Sivam, V.; Unnikrishnan, P.; Kuppa Sivasankara, S.; Kaushlesh Pansingh, R.; Shaik Abdul, K.; Lakshmi, N.M.; Chandragiri Nagarajarao, R. Effect of electron beam irradiation on the biochemical, microbiological and sensory quality of *Litopenaeus vannamei* during chilled storage. *J. Food Sci. Technol.* **2020**, *57*, 2150–2158. [CrossRef]
38. Luan, L.; Wu, C.; Wang, L.; Li, Y.; Ishimura, G.; Yuan, C.; Ding, T.; Hu, Y. Protein denaturation and oxidation in chilled hairtail (*Trichiutus haumela*) as affected by electrolyzed oxidizing water and chitosan treatment. *Int. J. Food Prop.* **2018**, *20*, S2696–S2707. [CrossRef]
39. Zhang, X.; Pan, C.; Chen, S.; Xue, Y.; Wang, Y.; Wu, Y. Effects of modified atmosphere packaging with different gas ratios on the quality changes of golden pompano (*Trachinotus ovatus*) fillets during superchilling storage. *Foods* **2022**, *11*, 1943. [CrossRef]
40. Benjakul, S.; Visessanguan, W.; Tueksuban, J. Changes in physico-chemical properties and gel-forming ability of lizardfish (*Saurida tumbil*) during post-mortem storage in ice. *Food Chem.* **2003**, *80*, 535–544. [CrossRef]
41. Ertbjerg, P.; Puolanne, E. Muscle structure, sarcomere length and influences on meat quality: A review. *Meat Sci.* **2017**, *132*, 139–152. [CrossRef]

42. Díaz-Tenorio, L.M.; García-Carreño, F.L.; Pacheco-Aguilar, R. Comparison of freezing and thawing treatments on muscle properties of whiteleg shrimp (*Litopenaeus vannamei*). *J. Food Biochem.* **2007**, *31*, 563–576. [CrossRef]
43. Castañeda-López, G.G.; Ulloa, J.A.; Rosas-Ulloa, P.; Ramírez-Ramírez, J.C.; Gutiérrez-Leyva, R.; Silva-Carrillo, Y.; Ulloa-Rangel, B.E. Ultrasound use as a pretreatment for shrimp (*Litopenaeus vannamei*) dehydration and its effect on physicochemical, microbiological, structural, and rehydration properties. *J. Food Process. Preserv.* **2021**, *45*, e15366. [CrossRef]
44. Sae-Leaw, T.; Benjakul, S. Distribution and characteristics of polyphenoloxidase from Pacific white shrimp (*Litopenaeus vannamei*). *J. Food Sci.* **2019**, *84*, 1078–1086. [CrossRef]
45. Qian, Y.-F.; Xie, J.; Yang, S.-P.; Wu, W.-H.; Xiong, Q.; Gao, Z.-L. In vivo study of spoilage bacteria on polyphenoloxidase activity and melanosis of modified atmosphere packaged Pacific white shrimp. *Food Chem.* **2014**, *155*, 126–131. [CrossRef] [PubMed]
46. Ibrahim, A.; Ibrahim, M.S.C.; Bakar, K.; Bakar, J.; Ikhwanuddin, M.; Karim, N.U. Effects of *Annona muricata* extraction on inhibition of polyphenoloxidase and microbiology quality of *Macrobrachium rosenbergii*. *J. Food Sci. Technol.* **2022**, *59*, 859–868. [CrossRef]
47. Goncalves, A.A.; de Oliveira, A.R.M. Melanosis in crustaceans: A review. *LWT-Food Sci. Technol.* **2016**, *65*, 791–799. [CrossRef]
48. Sae-leaw, T.; Benjakul, S. Prevention of melanosis in crustaceans by plant polyphenols: A review. *Trends Food Sci. Technol.* **2019**, *85*, 1–9. [CrossRef]
49. Xu, D.; Yang, X.; Wang, Y.; Sun, L. Cascading mechanism triggering the activation of polyphenol oxidase zymogen in shrimp *Litopenaeus vannamei* after postmortem and the correlation with melanosis development. *Food Bioprocess Technol.* **2020**, *13*, 1131–1145. [CrossRef]

foods

MDPI

Article

Preparation, Characterization, and Application of Modified Starch/Chitosan/Sweet Orange Oil Microcapsules

Liang Qiu [1,2], Hui Ma [3], Qinghua Luo [3], Chan Bai [1,2], Guangquan Xiong [1,2], Shiwei Jin [3], Juguang Wang [1,2], Xiaoyan Zu [1,2], Hailan Li [1,2] and Tao Liao [1,2,*]

1 Hubei Engineering Research Center for Agricultural Products Irradiation, Institute of Agro-Products Processing and Nuclear Agricultural Technology, Hubei Academy of Agricultural Sciences, Wuhan 430064, China; qiuliang@hbas.com (L.Q.); baichan09@hbas.com (C.B.); xiongguangquan@hbaas.com (G.X.); juguangw@hbas.com (J.W.); zuxiaoyan@hbaas.com (X.Z.); hl.li@hbaas.com (H.L.)

2 Key Laboratory of Cold Chain Logistics Technology for Agro-Product, Ministry of Agriculture and Rural Affairs, Wuhan 430064, China

3 Key Laboratory of Catalysis and Energy Materials Chemistry of Education, South-Central University for Nationalities, Wuhan 430074, China; 2020120311@scuec.edu.cn (H.M.); luoqinghua@scuec.edu.cn (Q.L.); jinsw@mail.scuec.edu.cn (S.J.)

* Correspondence: liaotao0728@hbaas.com; Tel.: +86-87-389-705

Citation: Qiu, L.; Ma, H.; Luo, Q.; Bai, C.; Xiong, G.; Jin, S.; Wang, J.; Zu, X.; Li, H.; Liao, T. Preparation, Characterization, and Application of Modified Starch/Chitosan/Sweet Orange Oil Microcapsules. *Foods* **2022**, *11*, 2306. https://doi.org/10.3390/foods11152306

Academic Editor: Alejandra Acevedo-Fani

Received: 7 July 2022
Accepted: 27 July 2022
Published: 2 August 2022

Abstract: Aquatic products have an important role in global agriculture, but the challenges associated with preservation have limited their marketability. Essential oil (EO), such as sweet orange oil (SOEO), has been widely used for preservation due to its excellent antibacterial ability. However, the volatilization of EO limits its application in food preservation. In this study, SOEO was extracted from sweet orange peel by steam distillation and then stored in microcapsules. The components of the microcapsules were as follows: the porous starch was chosen as an adsorbed substrate to store SOEO (PS/SOEO), and sodium alginate (SA) and chitosan (CMCS) were used as shell material to delay the volatilization of SOEO using the sharp pore coagulation method. Our results showed that the main antibacterial ingredients in SOEO were aldehydes (33.93%) and d-limonene (15.38%). The microcapsules were of an irregular shape (oval), and the size of the microcapsules was 1.2 $\pm$ 0.1 cm as measured by a digital micrometer. Scanning electron microscopy (SEM) results showed that there were a lot of pores on the surface of the starch after modification, but sodium alginate and chitosan could well encapsulate these pores. The results of Fourier transform infrared (FTIR) spectroscopy and X-ray diffraction (XRD) analysis also showed that SOEO was successful encapsulated into the porous starch. The results of compression test and releasing kinetics studies suggested that CMCS and SA improved the mechanical and slow-releasing ability of SOEO microcapsules. The best antibacterial performance was obtained when 0.8 g of SOEO microcapsules was added. Finally, the shelf life of crawfish could be extended to 6 days by SOEO microcapsule (1/10 g, SOEO microcapsule/crawfish) under room temperature. These results provide a systematic understanding of the antibacterial capabilities of sweet orange essential oil microcapsules, which can contribute to the development of preservation methods for aquatic products.

Keywords: sweet orange essential oil; antibacterial activity; porous starch; microcapsule; crawfish

1. Introduction

Aquaculture is an essential economic pillar industry in many countries. However, due to its high nutrition, high moisture, and low connective tissue content, aquatic products easily decay, leading to severe economic loss. In recent years, plant essential oil (EO) has been considered a potential food preservative because of its antioxidant, biophilic, and antibacterial properties [1–3]. It is relatively simple to extract using the leaves, roots, and fruits of plants as raw materials through simple squeezing and distillation [4]. However, the direct use of EO has many limitations, such as lower water solubility, high volatility, high sensitivity to environment, and strong odor. These limitations inhibit its use as a preservation material in food packaging engineering [5]. How to improve the stability of EO has become a key issue for researchers.

At present, microencapsulation technology is widely used in the food, biological, and pharmaceutical fields [6]. Studies have shown that microencapsulation could improve the stability of bioactive compounds and endow EO with sustained-release characteristics that could extend the actuation duration [7]. Besides playing an antibacterial role, EO can also protect or improve the physicochemical properties of food [8]. Sotelo et al. used thyme essential oil as the core material to prepare microcapsules by coprecipitation and found that EO microcapsules had the same antibacterial ability as thyme essential oil [9]. This suggests that microencapsulation can not only effectively prevent the instability of EO and increase its contact area with food but also be easily dispersed in the food surface to inhibit microbial growth and proliferation [10]. Microcapsules can be prepared by interfacial polymerization [11], spray drying [12], sharp pore-coagulation bath [13], etc. Among these methods, pore-coagulation bath has the advantages of simple operation and low cost. Moreover, it can be operated at low temperature, which can reduce loss of EO during the embedding process [14]. Many researchers have used pore-coagulation bath to produce EO microcapsules, such as star anise oil, clove oil, etc. [15]. The components of microcapsules are substrate and shell. The substrate determines the amount of EO in microcapsules, while the shell can enhance the stability of EO and prolong its releasing time. The substrate is mainly produced by adsorption materials as it stores essential oils through adsorption [16]. As a renewable natural polymer material, starch is a natural, cheap, and biodegradable polymer. It is also the main source of carbohydrates in the human diet [17]. Therefore, starch products have been widely used to carry active ingredients, such as antiseptic, antioxidants, colorants, spices, and nutrients, which can be used as a kind of conveying system [18]. Unlike other active ingredients, EO is more sensitive to the environment. The core needs higher mechanical strength and good qualitative property [19]. Therefore, modification is needed for starch to embed EO. According to the structure and physicochemical properties of starch, the main modification methods are physical and chemical [20]. For example, Wang et al. used porous starch that was produced by enzyme as a core of microcapsule to carry clove essential oil and found that the modified microcapsule had strong heat resistance and significant inhibitory effect on the mold spore [21]. Dons et al. found that EO-modified starch microcapsule could minimize the effects on food taste and act as a substance to increase bioactivity [22]. Thus, modified starch is an effective and innovative substrate to maintain the effectiveness of EO.

The shell of microcapsules can separate the substrate from the external environment, reducing the impact of oxygen, heat, lighting, and pH on the EO and enhancing its stability [23]. Many researchers have used chitosan to embed EO microcapsules because of its biodegradability and safety. For example, Sotelo et al. used chitosan to prepare thyme essential oil microcapsule and found that the embedding rate of the microcapsules was more than 68% and had a strong inhibitory effect on *Bacillus cereus* [9].

However, the forming capability of chitosan is very poor, so it needs to be combined with other materials. Many studies have shown that the combined use of two or more shell materials is beneficial to improve the loading capacity and stability of microcapsules [24]. Sodium alginate (SA) is a linear polysaccharide extracted from seaweed and can be cross-linked to chitosan to form hydrogels or insoluble polymers [25]. Han et al. used chitosan and SA as substrates to store thyme essential oil by the layer self-assembly method. The results showed that the embedding rate of microcapsule was up to 80.23% [26].

In the above-described context, the objectives of this study were to (1) prepare a porous corn starch using the enzymolysis method; (2) use porous corn starch as a substance to adsorb sweet orange essential oil and sodium alginate (SA) and modified carboxymethyl chitosan (CMCS) as shell to protect the SOEO; (3) characterize the resulting materials by Fourier transform infrared spectroscopy (FT-IR), pressure resistance analysis, X-ray diffraction (XRD), and scanning electron microscopy (SEM); (4) explore the effective antibacterial ingredients of sweet orange oil and its release kinetics; (5) test the antibacterial properties of SOEO and compare it with natural starch; and (6) determine the preservation of crawfish by SOEO microcapsules.

2. Materials and Methods

2.1. Materials

The main materials used were natural starch (food grade), ethanol, n-hexane, acetic acid, disodium hydrogen phosphate, citric acid, calcium chloride anhydrous, sodium hydroxide, potassium bromide (KBr), magnesium oxide (MgO), nutrient agar, and carboxymethyl chitosan (CMCS); all were analytically pure materials and purchased from Sinopharm Chemical Reagent Co. (Shanghai, China). Sodium alginate (SA) was purchased from Shanghai Maclean Biochemical Technology Co. (Shanghai, China). BCP (bromocresol purple plate count agar) medium with 0.2% soluble starch was purchased from Beijing Wokai Biotechnology Co. (Beijing, China). Two strains, *Aeromonas sobria* (ATCC43979) and *Shwanella putrefaciens* (ATCCBAA-1097), were purchased from Shanghai Preservation Biotechnology Centre (Shanghai, China). Sweet orange essential oil (SOEO) was sourced from sweet orange essential peel. The chitosan used in this experiment was treated with 10 KGY irradiation (Hubei Irradiation Experimental Center (Beijing, China), ^{60}Co-ray; the irradiation dose rate was 6.25 Gy/min, and the irradiation doses were set at 20, 40, 60, 80, and 100 KGy.

2.2. Preparation of Modified Porous Starch

According to the method of Fu et al. [27], the orthogonal test was used to find the best conditions of porous modified starch by enzymatic digestion, with the ratio of enzyme to starch, the pH of the buffer solution, and the reaction temperature set as three factors and three levels (Table 1) [28,29]. Natural corn starch was mixed with citric acid/disodium hydrogen phosphate buffer solution at a mass ratio of 1:8, placed in a magnetic heating stirrer, and stirred in a water bath at 25 °C for 0.5 h. Next, α-amylase and glucoamylase were added, and the reaction was terminated by the addition of sodium hydroxide at a concentration of 0.1 mol/L after 8 h in the water bath. The product was centrifuged at 3000 r/min for 10 min and then washed 3–5 times with deionized water. Finally, the product was dried in an oven at 45 °C for 3 h, ground, and prepared.

Table 1. Levels and factors of orthogonal experiment.

Serial Number	Influencing Factors		
	Enzyme/Starch Ratio	pH	Temperature (°C)
1	0.015:1	4.5	45
2	0.020:1	5.0	50
3	0.025:1	5.5	55

The adsorption properties of porous starch on SOEO (PS/SOEO microspheres) were analyzed by the differential weight method. The porous starch was stirred at high speed with SOEO and then kept at room temperature to remove the surface residue. The adsorption rate of SOEO was then calculated by the following equation:

$$\text{Adsorption rate (\%)} = (m_2 - m_1)/m_1 \times 100\% \quad (1)$$

where m_1 (g) is the weight of the initial porous starch, and m_2 (g) is the final weight of the porous starch.

2.3. Composition Analysis of Sweet Orange Essential Oil (GC–MS)

The sweet orange peel was placed into a thermostatic drying chamber for drying and crushed. Then, 100 g treated sweet orange peel was put into 1000 mL distillation flask with a certain amount of water. The upper oil liquid was sweet orange oil (SOEO) after steam distillation.

The essential oil of sweet orange was analyzed by 7890A-5975C gas chromatography–mass spectrometry (GC–MS). First, 10 μL of SOEO was diluted 1000-fold with n-hexane, and 1.5 μL of diluted solution was placed into the HS-SPME sample flask. The chromatographic conditions were as follows: DB-WAX model chromatographic column (30 × 0.25 mm); inlet temperature of 240 °C; pressure of 100 kPa; and high-purity helium as carrier gas. The initial temperature of 60 °C was held for 4 min, then ramped up to 240 °C at 4 °C/min and held for 2 min. The mass spectrometry conditions were as follows: ion source temperature of 200 °C and interface temperature of 220 °C. The solvent delay time was 5 min [30].

2.4. Determination of the Antibacterial Properties of Sweet Orange Essential Oil

Referring to the relevant literature, the Oxford cup method was used to determine the diameter of the essential oil inhibition zone (IZ) [31]. First, 100 μL each of the already cultured *Aeromonas sobria* and *Shwanella putrefaciens* were applied to a suitable solid medium by the spread plate method. Three Oxford cups were placed on the surface of the plates, and 200 μL of SOEO was added to the Oxford cups. An equal volume of sterile water was used as control. After about 24 h, a vernier caliper was used to measure the diameter of the IZ.

2.5. Preparation of Sweet Orange Essential Oil Microcapsules

According to the method of Fu et al. [27], 1 g of PS/SOEO microspheres was weighed and 3 g of SA was added and stirred at high speed for 1 h, followed by drop-by-drop addition of 1 g of 10KGY irradiated modified CMCS; the mass ratio of the three materials was 1:3:1. The resulting emulsion was dripped into calcium chloride solution using a syringe and cured in a refrigerator for 2 h. The residue was then dried in an oven to obtain CMCS-SA-PS/SOEO microcapsules. The rates of CMCS and SA are shown in Table 2.

Table 2. Material ratio.

Sample Number	CMCS/%	SA/%
A	0.8	2.4
B	0.8	3.2
C	1.2	2.4
D	1.2	3.2

2.6. Characterisation of Materials

2.6.1. Morphology of Samples by Scanning Electron Microscopy Analysis (SEM)

The surface morphology of modified starch, natural starch, and SOEO microcapsules was studied by SEM (Hitachi SU8010). The images obtained were 50, 20, and 1 μm.

2.6.2. Fourier Infrared Spectroscopy (FTIR)

Fourier infrared spectroscopy (FTIR, NICOLE 6700) was used to analyze the chemical composition of the samples (modified starch, natural starch, and SOEO microcapsules). First, 1–2 mg of powder sample and 200 mg of pure KBr were finely ground, homogenized, and pressed into a transparent sheet on an oil press. The samples were then put into an infrared spectrometer for testing. The wavenumber range was 4000–400 cm^{-1}, and 32 scans at 4 cm^{-1} resolution were achieved for each spectrum.

2.6.3. X-ray Diffraction Analysis (XRD)

To study the packing characteristics and crystallinity of the material, the diffraction intensity of the samples (modified starch, natural starch, and SOEO microcapsules) were recorded every 5° by XRD (Bruker D8 Advance 25) over a 2θ range of 5° to 60° [31]. The results were analyzed by Jade program (9.0) and Origin (9.0).

2.6.4. Pressure Resistance Analysis

The pressure resistance of SOEO microcapsules (Groups A, B, C, and D) was analyzed by a universal mechanical testing machine (CMT6103). The range of the load sensor was 10 kN, and the test speed was 1.5 mm/min. The compression deformation was obtained by displacement of the crosshead of the testing machine, so the compression elastic modulus calculated in this study is the apparent elastic modulus. Each sample was conducted in triplicate.

2.7. Study of the Release Kinetics of SOEO Microcapsules

Ethanol is soluble in any proportion with water. It reduces the interfacial tension between EO and water, accelerates the dissolution of EO on the surface of microcapsules, accelerates the release of the core material, and shortens the period of the experiment [18]. For this reason, 50% ethanol was used as a solvent for the release kinetics of the microcapsules. Referring to the method of Liu et al. [32], SOEO was diluted using 50% anhydrous ethanol, and the absorbance of SOEO at a wavelength of 671 nm was tested using a UV spectrophotometer [33]. The linear standard equation was $y = 0.1203x - 0.0078$, $R^2 = 0.9941$.

Next, 0.2 g of SOEO microcapsules was added to 30 mL of 50% ethanol solution in a closed system at room temperature (25 °C). At each time point (5, 10, 15, 45, 55, 75, 100, 150, 180, and 300 min), an amount of the solution was removed and the absorbance at 670 nm (Figure S1) was measured using a UV spectrophotometer.

2.8. Determination of the Antibacterial Potential of SOEO Microcapsules

According to the method of Tu et al. [33], 100 μL of *Shewanella putrefaciens* and *Aeromonas sobria* were spread on the medium, and microcapsules of SOEO (0.2, 0.4, 0.6, 0.8, and 1.0 g) were fixed with double adhesive on the lids of the dishes. The culture dishes were then inverted and incubated in a constant temperature (37 °C) incubator for 48 h. The optimum amount of microcapsules was selected by the spread plate count method. The antibacterial properties against composite strains were also examined, and the proportion of composite strains was as follows: *Aeromonas sobria*/*Shwanella putrefaciens* ratio = 1:1, 50 μL.

2.9. Packaging of Crawfish with SOEO Microcapsules

Alive crawfish with an average weight of 30 ± 2.4 g and average length of 20 ± 1.8 cm was purchased in June from the Qiyimeng fresh market, Wuhan. Upon arrival to the laboratory, crawfish was cleaned in water, the head and sand vein were removed, and the tail was rinsed. The crawfish was cooked at 100 °C for 7 min and then vacuum packed quickly with SOEO microcapsules (1/10 g) before storing at room temperature (25 °C) in a biochemical incubator (BPN-40CRH, blue pard, Shanghai Bluepard Instruments Co., Ltd., Shanghai, China).

2.9.1. Determination of the Total Volatile Base Nitrogen (TVB-N)

TVB-N value was estimated by the FOSS method. Crawfish was hacked through the packing, cleaned and drained in sterile environment (5.0 g), and homogenized with MgO (0.5 g) using a mixer. Finally, the mixture was placed into a FOSS digestion tube. TVB-N value was measured by an automatic Kjeldahl apparatus (8400 Kjeltec Distillation, FOSS Analytical Instrument Co. Ltd., Shanghai, China), and the level of TVB-N was expressed as mg N/100 g of crawfish.

2.9.2. Determination of the Total Plate Count (TPC)

The TPC value was estimated as follows. First, the crawfish was hacked through the packing, cleaned and drained in sterile environment (10 g), and homogenized. Then, it was placed into a sterile bag with 90 mL aseptic sterile saline and beaten using a beating homogenizer for 3 min to obtain uniform bacterial suspension. The bacterial suspension was diluted by 10× dilution method. The appropriate bacterial solution (0.1 mL) with three gradients was coated on the surface of the plate. The TPC of the dilution was measured using the standard plate count method using nutrient agar after incubating at 30 °C for 48 h, and the value of TPC was expressed as log CFU/g of crawfish.

2.10. Statistical Analysis

One-way analysis of variance (ANOVA) was utilized for a completely randomized statistical analysis; $p < 0.05$ was considered statistically significant. The SPSS software (version 26.0) was used for all statistical analyses.

3. Results and Discussion

3.1. Modified Starch Adsorption Performance Analysis

Table 3 shows the results of orthogonal optimization experiments for the modified starches. The optimum conditions for preparation of porous modified starch were as follows: enzyme/starch ratio of 0.020:1, buffer solution pH of 5.5, and 45 °C water bath temperature. The adsorption of SOEO by the modified starch could reach 17.2%. The impact of the factors was pH > temperature > enzyme/starch ratio. This may be because the pH is the main factor in enzyme activity.

Table 3. Results of the orthogonal experiment.

Serial Number	Influencing Factors			Adsorption Rate (%)
	Enzyme/Starch Ratio	pH	Temperature (°C)	
1	0.015:1	4.5	45	13.07
2	0.015:1	5.0	50	14.28
3	0.015:1	5.5	55	13.39
4	0.020:1	4.5	50	14.28
5	0.020:1	5.0	55	13.69
6	0.020:1	5.5	45	17.22
7	0.025:1	4.5	55	13.94
8	0.025:1	5.0	45	12.60
9	0.025:1	5.5	50	11.36
K_1	13.58	13.76	14.30	
K_2	15.06	13.52	13.31	
K_3	12.63	13.99	13.67	
Signif	0.021	0.02	0.022	
Quadratic sum	124.92	125.58	120.89	

Kn is sum of experimental results at all levels. Signif < 0.05 means this factor has an important influence on the subject. Quadratic sum: the degree of influence of each level on the adsorption.

3.2. Composition Analysis of SOEO

The chromatograms (shown in Figure S2 and Table 4) revealed the main compounds of SOEO as follows: β-myrcene (35.94%), n-octanal (31.91%), d-limonene (15.38%), 3-carene (3.83%), and pentadecane (1.59%). The total proportion of the five ingredients reached 88.65%, with a very small proportion of other ingredients (<2%).

Other researchers have proven that aldehydes (33.93%) [14] and d-limonene (15.38%) [13] are the main antibacterial agents among these compounds. It has been shown that aldehydes cause bacterial death by inhibiting the activity of enzymes in the cell membrane and causing extravasation of intracellular material [19]. Limonene preserves food by disrupting bacterial cell membranes and cell proteins [20].

Table 4. Analysis of components of SOEO.

Serial Number	Retention Time	Components	Relative Content (%)
1	23.556	Tridecanal	0.91
2	27.391	D-Limonene	15.38
3	29.675	Octacosane	1.50
4	32.048	Eicosane	1.51
5	32.527	1-Hexanol, 2-ethyl-	0.12
6	32.527	Decanal	0.65
7	35.198	Pentadecane	1.59
8	35.198	Octanal	31.91
9	35.198	Nonanal	0.27
10	35.198	Cyclohexane, 1-ethenyl-1-methyl-2,	0.46
11	35.198	Caryophyllene	0.41
12	35.198	Octadecane	0.48
13	35.198	Dodecanal	0.19
14	35.198	Heneicosane	0.19
15	35.198	Toluene	0.17
16	35.949	β-Myrcene	35.94
17	36.142	β-Pinene	0.68
18	37.604	Naphthalene, 1,2,3,5,6,8a-hexahydr	0.55
19	37.698	Camphene	0.74
20	37.793	3-Carene	3.83

3.3. Antibacterial Properties of SOEO

In this study, we used *Shewanella putrefaciens* and *Aeromonas sobria* to carry out the bacteriostasis test because our previous studies found these bacteria to be the dominant spoilage bacteria for crawfish stored at room temperature (unpublished data). As shown in Figure 1, obvious inhibition circles were observed in the experimental groups of *Shewanella putrefaciens* and *Aeromonas sobria*. Measurements showed that the inhibition zone size was 1.1 ± 0.1 cm for *Shewanella putrefaciens* and 0.9 ± 0.1 cm for *Aeromonas sobria*. No inhibition zone was observed in the control group of both strains, indicating that SOEO has relatively high inhibition potential against *Shewanella putrefaciens* and *Aeromonas sobria*. Regarding the mechanism of inhibition, He et al. studied the inhibitory effect of eucalyptus essential oil on four strains of bacteria, including *Escherichia coli* and *Shewanella putrefaciens*, and showed that the use of eucalyptus essential oil resulted in increased permeability of bacterial cell membranes and impaired cellular integrity and cellular physiological functions, demonstrating the inhibitory effect of plant essential oil on specific spoilage bacteria in aquatic products [21].

3.4. Preparation of SOEO Microcapsules

The modified starch was used as an adsorbent to adsorb SOEO, and SA and CMCS were used as wall materials to encapsulate the modified starch with adsorbed SOEO to form a microcapsule structure. The carboxyl group (−COOH) of SA and the amino group ($-NH_2$) of CMCS wrapped the starch between them due to electrostatic attraction to form an electron layer, and white emulsion-like CMCS-SA-PS/SOEO was finally obtained [21]. The SOEO microcapsules were cured in 1.5% calcium chloride solution, and the hardness index became high.

As shown in Table 2, different ratios of CMCS-SA-PS/SOEO microcapsules were prepared by using different mass fractions of SA and CMCS solutions. Microcapsules with the best ratios were selected according to some characterization analysis.

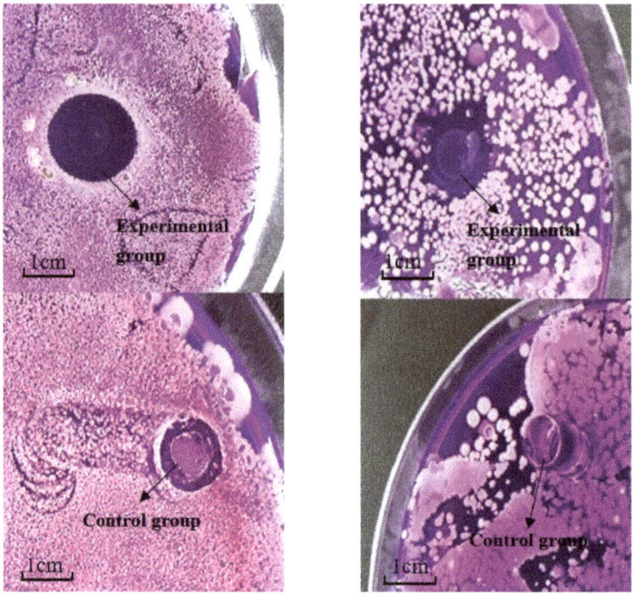

Figure 1. Size of the inhibition zone of the two strains.

3.5. Characterisation of SOEO Microcapsule

3.5.1. SEM Analysis

The morphology of the natural starch and modified porous starch were observed under 2000× magnification (Figure 2a,b). The natural starch (Figure 2a) was an irregular granule with a smooth surface. The modified starch (Figure 2b) had not significantly changed in shape but had an irregular appearance overall. Moreover, the surface had become rough, and a large number of microporous structures appeared (Figure 2g and Figure S3). This is very similar to the modified starch prepared using enzymatic digestion by other results. Both had many microporous structures on their surfaces [31].

Figure 2c–f shows the A, B, C, and D microcapsules (Table 2) obtained from different material ratios and observed under 1000× magnification. As can be seen, the surface space of microcapsules in group C was smaller than that in the other three groups, proving that group C microcapsules were more tightly wrapped around SOEO. Figure 2g shows the surface morphology of group C microcapsules as observed under 50× SEM. As can be seen, an encapsulated structure was formed on the surface of the microcapsules and no oil-like material was visible on the surface. Figure 2h shows the morphology of group C microcapsules observed under 2000× magnification. The oil-like material was faintly visible, further demonstrating the successful encapsulation of SOEO by the microcapsules.

3.5.2. FTIR Analysis

Figure 3 illustrates the FTIR spectra of modified-starch-adsorbed SOEO, modified starch, natural starch, starch SOEO microcapsules, and microcapsules unloaded with SOEO. The absorption peaks of modified starch (B in Figure 3) and natural starch (C in Figure 3) did not change significantly, indicating that enzymatic reaction did not change the functional groups and the molecular structure of the starch remained essentially the same. The PS/SOEO (A in Figure 3) and SOEO microcapsules (D in Figure 3) had similar characteristic peaks because they both contained SOEO. The characteristic peaks of SOEO were at 1465, 1733, and 2929 cm^{-1}. This could be attributed to the presence of alkanes in SOEO with C–H_2 and C–H_3, the aldehydes in SOEO with high C–H and C=C stretching intensity and thus a sharper characteristic peak, and O–H stretching, respectively. The peaks at 1159 and 997 cm^{-1} in the microcapsules without SOEO were due to the N–H stretching of the group in CMCS, and the peak at 1666 cm^{-1} was the stretching vibration of the double bond of the carboxyl group in SA [34,35].

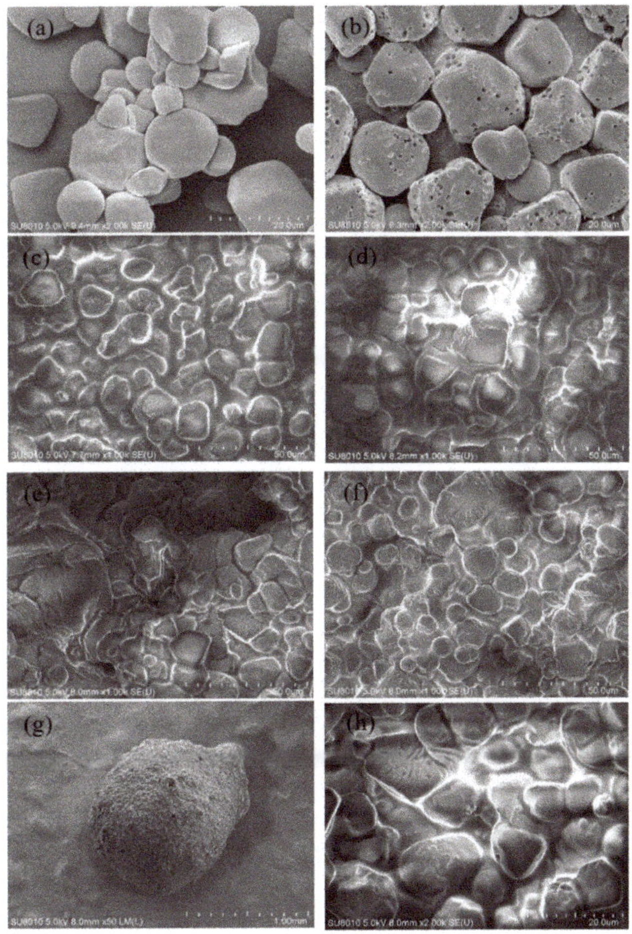

Figure 2. Scanning electron microscopy of starch and microcapsules: (**a**) natural starch; (**b**) modified starch; (**c**) group A microcapsules, magnification ratio: 1000×; (**d**) group B microcapsules, magnification ratio: 1000×; (**e**) group C microcapsules, magnification ratio: 1000×; (**f**) group D microcapsules, magnification ratio: 1000×; (**g**) group C microcapsules, magnification ratio: 50×; and (**h**) group C microcapsules, magnification ratio: 2000×.

3.5.3. XRD Analysis

The crystallinity of natural starch, enzymatically modified starch, and SOEO microcapsules are shown in Figure 4. The results showed that the natural maize starch had an A-shaped structure, and the A-pattern was still retained after modification. Previous studies have pointed out that starch granules are amorphous structure as the starch consists of straight-chain starch and twisted or unstable branched chains [36,37]. From the XRD pattern, we can see that the natural (Figure 4, Natural starch) and modified (Figure 4, Modified starch) starches showed strong diffraction peaks at 15° and 23°2θ, double peaks at 17° and 18°2θ, and some smaller diffraction peaks at around 20°2θ Researchers have generally associated imperfections in starch granules with amorphous structural domains, such as broken chains of straight-chain starch and disordered or unstable branched starch. The same properties of starch, such as symmetry, order, and stability, are associated with branched starch [38,39].

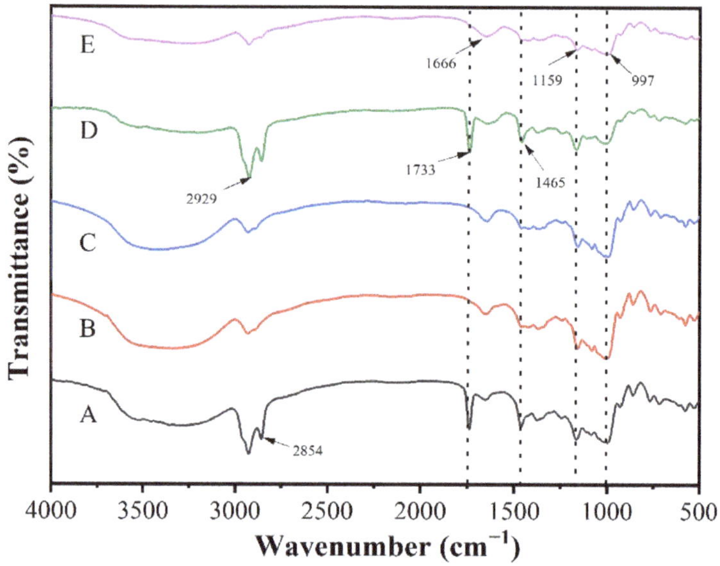

Figure 3. FTIR spectra of (A) PS/SOEO; (B) modified starch; (C) natural starch; (D) SOEO microcapsules; and (E) microcapsules without SOEO.

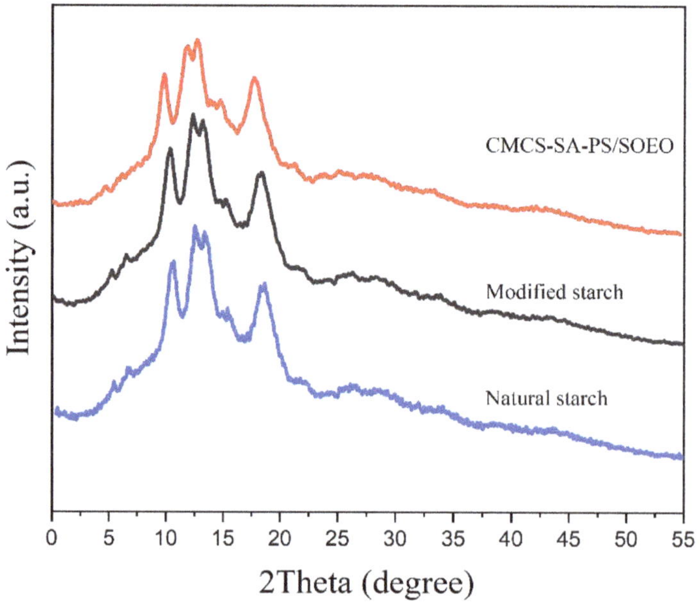

Figure 4. XRD pattern of natural starch; modified starch; and CMCS-SA-PS/EO.

In contrast, the peaks in SOEO microcapsules (Figure 4, CMCS-SA-PS/EO) were not very sharp, and the position of the peaks were the same as natural starches. This may be because the compounds of SOEO were very volatile, and its crystallinity was poor.

3.5.4. Pressure Resistance Analysis

The pressure resistance analysis of SOEO microcapsules is shown in Figure 5. It can be concluded that the higher the rate of CMCS and SA, the higher was the pressure resistance of microcapsules. This might be attributed to the mechanical properties that come from CMCS and SA, which improved the resilient ability of SOEO microcapsules. However, the microcapsules were wrapped with SOEO. The shell would be thicker with increasing rate of CMCS and SA, and a thicker shell is not conducive to the release of SOEO.

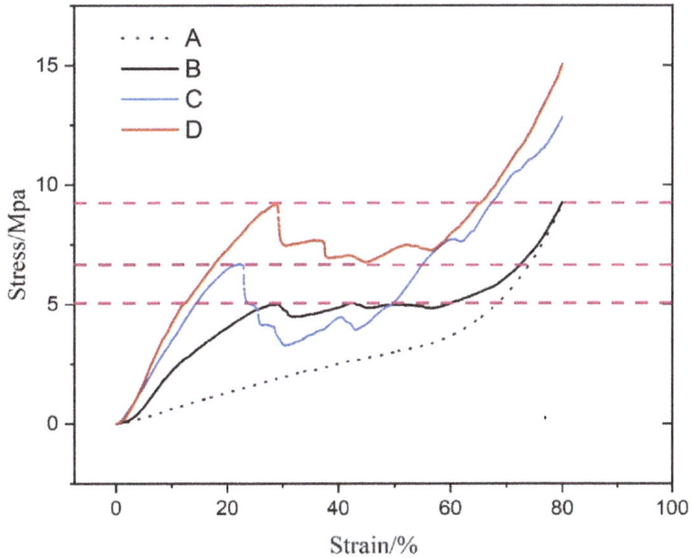

Figure 5. Pressure resistance analysis of (A) group A microcapsules; (B) group B microcapsules; (C) group C microcapsules; and (D) group D microcapsules.

3.6. *Study of the Release Kinetics of SOEO Microcapsules*

Figure 6 shows the release of sweet orange essential oil microcapsules in 50% anhydrous ethanol. In the range of 0–200 min, the release curves generally conformed to the mathematical model of y = a × b [40]. After fitting the C capsules by the origin, the resulting release kinetic equation was y = 3.6108 × 1.7226 with a well-fitted correlation coefficient of R^2 = 0.9716. It can be seen from the graph that SOEO in the four groups was released rapidly in the anhydrous ethanol solution during the period of 0–50 min; after 50 min, the release rate of the four groups of sweet orange essential oil microcapsules tended to stabilize. It can be speculated that some essential oils may have been attached to the surface of the microcapsules at the beginning and when entering the ethanol solution, these surface oils were released into the ethanol solution in a shorter period of time, thus resulting in a faster release rate at the initial stage [28,39].

Although the releasing curves of the four groups of microcapsules were similar, the final release rate of SOEO encapsulated in the microcapsules of group C was about 38%, which was higher than the other groups of microcapsules. On the other hand, it also proved that the best microcapsules were prepared by the material ratio of group C.

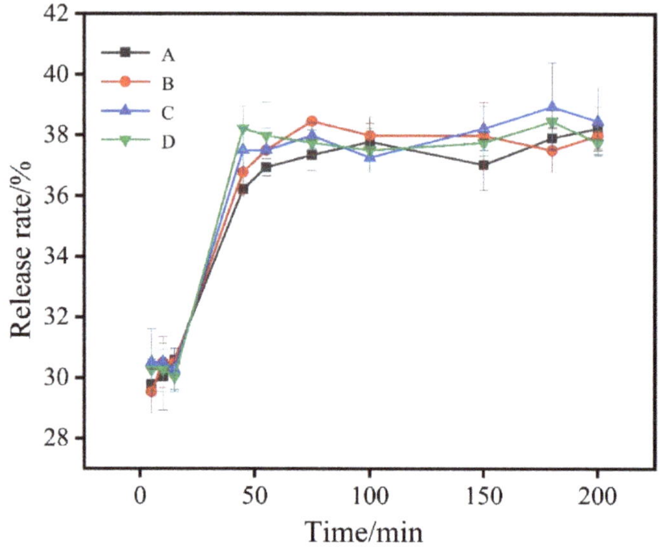

Figure 6. Release curves of (A) group A microcapsules; (B) group B microcapsules; (C) group C microcapsules; and (D) group D microcapsules in 50% ethanol solution.

3.7. Antibacterial Potential of SOEO Microcapsules

The results of the inhibition of *Shewanella putrefaciens*, *Aeromonas sobria*, and the complex bacterium by SOEO microcapsules are given in Table 5. In our experiments, 0.8 g of SOEO microcapsules showed the best antibacterial ability, suggesting that the antibacterial effect of SOEO does not increase with the increase in dose because of its volatility. Compared to natural starch, the ability of inhibition by 0.8 g SOEO microcapsules against *Shewanella putrefaciens*, *Aeromonas sobria*, and the complex bacterium reached 56.43, 61.27, and 62.91%, respectively, at 48 h.

Table 5. Bacterial colony count for microencapsulation inhibition test.

Weight/g	CMCS-SA-PS/EO/								
	12 h			24 h			48 h		
	Shewanella putrefaciens	*Aeromonas sobria*	Composite Strains	*Shewanella putrefaciens*	*Aeromonas sobria*	Composite Strains	*Shewanella putrefaciens*	*Aeromonas sobria*	Composite Strains
0.2	42 ± 3	35 ± 2	67 ± 2	71 ± 8	113 ± 12	122 ± 10	73 ± 7	137 ± 12	138 ± 10
0.4	38 ± 1	29 ± 3	61 ± 9	70 ± 5	106 ± 13	117 ± 11	72 ± 4	125 ± 13	127 ± 11
0.6	30 ± 3	26 ± 4	63 ± 7	66 ± 3	111 ± 11	109 ± 12	68 ± 4	121 ± 11	119 ± 12
0.8	24 ± 4	11 ± 1	47 ± 3	55 ± 2	103 ± 14	99 ± 8	57 ± 3	106 ± 14	102 ± 8
1.0	27 ± 5	22 ± 5	55 ± 3	61 ± 7	136 ± 11	103 ± 11	64 ± 8	129 ± 11	124 ± 11
CK_2	20 ± 6	28 ± 8	59 ± 4	81 ± 4	160 ± 15	141 ± 15	89 ± 10	162 ± 15	154 ± 15
CK_1	61 ± 3	52 ± 4	109 ± 12	99 ± 7	171 ± 18	148 ± 14	101 ± 11	173 ± 18	162 ± 14

CK_2 represents SOEO with modified starch, and CK_1 represents the culture condition.

With the extension of time, the antibacterial effect of sweet orange essential oil and SOEO microcapsules was weakened, but SOEO microcapsules showed more persistent antibacterial properties than essential oil on its own. In the initial 12 h, pure essential oil showed excellent antibacterial properties because a large amount of essential oil was released. When the incubation time was extended to 24–48 h, the modified porous starch and chitosan in the microcapsule protected the orange peel essential oil, reduced its instability, and prolonged the antibacterial effect of the essential oil. Moreover, our results showed that the antibacterial effect of microcapsules on *Aeromonas thermophila* was better than that of *Shewanella*. This might be because *Aeromonas thermophila* can use starch substrate, so the EO has better contact with the microbial cell membrane. In the complex flora, microcapsules also showed certain antibacterial ability due to the competitive growth relationship between the two bacteria.

Unlike free essential oils, EO starch microcapsules can be divided into three steps. First, the wall of the microcapsules is diffused (because the compounds of shells are mainly from macromolecule) by degradation and released from the microcapsules; second, the substrate starts to swell and the degradation or corrosion of the shells leads to the release of the EO; third, the released EO works on the bacteria [40]. In this study, we used chitosan to protect essential oil. Compared to other studies that used, for example, the combination of sodium caseinate and baili EO or the combination of sodium alginate and citronella EO, our SOEO microcapsules showed better antibacterial ability because of synergistic bacteriostasis [41].

3.8. Research on the Preservation of Crawfish

3.8.1. Total Volatile Base Nitrogen (TVB-N)

TVB-N is one of the main indicators used to judge whether aquatic products have decayed. According to hygienic standards for marine products, the acceptable TVB-N values for marine fish and shrimp should be no more than 20 mg N/100 g [42]. As shown in Figure 7a, the TVB-N value of crawfish was 4.21 ± 0.26 mg N/100 g, and it increased with longer time. The TVB-N value of the 25 °C group increased faster and exceeded 20 mg N/100 g on the second day. However, the TVB-N value of SOEO microcapsules exceeded 20 mg N/100 g on the sixth day.

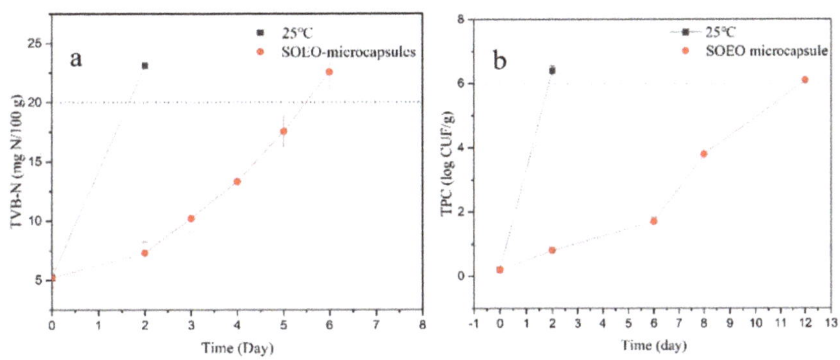

Figure 7. The TVB-N (**a**) and TPC (**b**) of crawfish under 25 °C and SOEO microcapsules.

3.8.2. Total Plate Count (TPC)

Figure 7b shows changes in TPC in all crawfish during storage. Crawfish is considered fresh if the TPC is lower than 6 log CFU/g; TPC higher than this means the crawfish is spoilt [42]. On the second day, the TPC value of the 25 °C group was already over 6 log CFU/g. On the sixth day, the TPC value of crawfish stored at 4 °C was 6.17 ± 0.17 log CFU/g.

Overall, SOEO microcapsule showed a good ability in preserving crawfish, and the shelf life of crawfish could be extended to 6 days. Compared to other studies, our SOEO microcapsules were more stable. For example, Alotaibi used sweet potato starch and thyme essential oil to produce microcapsules and found that the rate of lipid oxidation in shrimp meat was significantly inhibited during storage. However, the shelf life of crawfish was only extended to 4 days because the essential oil was not protected in the form of a shell, which led to instability of the essential oil.

4. Conclusions

In this study, SOEO microcapsule were prepared by SOEO, modified starch, CMCS, and SA. The modified starch was used as a substance to adsorb SOEO, and the best condition to modify starch was as follows: enzyme/starch ratio of 0.020:1, buffer solution pH of 5.5, and 45 °C water bath temperature. The main antibacterial agents were aldehydes and d-limonene. The results of characterization revealed that (1) SOEO had successfully been adsorbed by modified starch; (2) CMCS and SA could not only wrap the starch well but also improved the pressure resistance of the microcapsule; (3) combined with the results of release kinetics, the best rates of CMCS and SA were 1.2 and 2.4%, respectively. The results of antibacterial test showed that the overall cost-effectiveness of the film was the highest when 0.8 g of SOEO microcapsules was added. In the preservation test of crawfish, it was found that the shelf life of crawfish was greatly extended compared to other treatments.

Supplementary Materials: The following supporting information can be downloaded at: https://www.mdpi.com/article/10.3390/foods11152306/s1, Figure S1: The Full scanning wavelength of SOEO; Figure S2: The chromatograms of SOEO; Figure S3: The Scanning electron microscopy of SOEO microcapsules: a: Group A microcapsules; b: Group B microcapsules; c: Group C microcapsules

Author Contributions: Formal analysis, H.M.; investigation, Q.L.; data curation, L.Q.; writing—original draft preparation, L.Q.; writing—review and editing, H.L. and T.L.; project administration, J.W. and X.Z.; funding acquisition, C.B., G.X., S.J., L.Q. and H.M. did the same contribution to this work. All authors have read and agreed to the published version of the manuscript.

Funding: This research was funded by National Key R&D Program of China, grant number 2019YFD0902000; Key R&D Program of Guangzhou City, grant number 202103000087 and Guangzhou people's Livelihood Science and technology project, grant number 202002020087.

Institutional Review Board Statement: Not applicable.

Informed Consent Statement: Not applicable.

Data Availability Statement: The data presented in this study are available on request from the corresponding author.

Conflicts of Interest: The authors declare no conflict of interest.

References

1. Ramos da Silva, L.R.; Oliveira Ferreira, O.; Nevez Cruz, J.; de Jesus Pereira Franco, C.; dos Anjos, T.O.; Moraes Cascaes, M.; Almeida da Costa, W.; de Aguiar Andrade, E.H.; Santana de Oliveira, M. *Lamiaceae* Essential Oils, Phytochemical Profile, Antioxidant, and Biological Activities. *Evid. Based Complement. Altern. Med.* **2021**, *2021*, 6748052. [CrossRef] [PubMed]
2. Gormez, A.; Bozari, S.; Yanmis, D.; Gulluce, M.; Agar, G.; Sahin, F. The Use of Essential Oils of Origanum rotundifolium as Antimicrobial Agent Against Plant Pathogenic Bacteria. *J. Essent. Oil Bear. Plants* **2016**, *19*, 656–663. [CrossRef]
3. Uma, K.; Huang, X.; Kumar, B.A. Antifungal effect of plant extract and essential oil. *Chin. J. Integr. Med.* **2017**, *23*, 233–239. [CrossRef] [PubMed]
4. El kharraf, S.; Farah, A.; Miguel, M.G.; El-Guendouz, S.; El Hadrami, E.M. Two Extraction Methods of Essential Oils: Conventional and Non-conventional Hydrodistillation. *J. Essent. Oil Bear. Plants* **2020**, *23*, 870–889. [CrossRef]
5. Sanches Abelan, U.; de Oliveira, A.C.; Cacoci, R.S.P.; Azevedo Martins, T.E.; Giacon, V.M.; Robles Velasco, M.V.; Ribeiro de Castro, C.R. Potential use of essential oils in cosmetic and dermatological hair products: A review. *J. Cosmet. Dermatol.* **2021**, *21*, 1407–1418. [CrossRef]
6. Li, L.; Song, W.; Shen, C.; Dong, Q.; Wang, Y.; Zuo, S. Active packaging film containing oregano essential oil microcapsules and their application for strawberry preservation. *J. Food Process. Preserv.* **2020**, *44*, e14799. [CrossRef]
7. Kujur, A.; Kiran, S.; Dubey, N.K.; Prakash, B. Microencapsulation of Gaultheria procumbens essential oil using chitosan-cinnamic acid microgel: Improvement of antimicrobial activity, stability and mode of action. *LWT* **2017**, *86*, 132–138. [CrossRef]
8. Li, X.C.; Wu, Y.T.; Zhang, D.F.; Xing, Y.K.; Meng, Y.; Ma, L. Influence of microcapsule preservatives containing cinnamon oil on soybean quality during storage. *Food Ind.* **2016**, *9*, 4–7.
9. Sotelo-Boyás, M.; Correa-Pacheco, Z.; Bautista-Baños, S.; Gómez, Y.G. Release study and inhibitory activity of thyme essential oil-loaded chitosan nanoparticles and nanocapsules against foodborne bacteria. *Int. J. Biol. Macromol.* **2017**, *103*, 409–414. [CrossRef]

10. Tian, Y.Q.; Li, Y.X.; Zhang, W.; Wang, Y.P. Oregano essential oilmicrocapsule was buried process and antibacterial effect. *Packag. Eng.* **2016**, *17*, 22.
11. Wang, Y.F.; Shao, J.J.; Wang, Z.L.; Lu, Z.X. Study of allicin microcapsules in β-cyclodextrin and porous starch mixture. *Food Res. Int.* **2012**, *49*, 641–647. [CrossRef]
12. Wang, F.; Dan, X.Y.; Wan, G.D. Study on the preparation and stability of orange peel essential oil microcapsule. *J. Chin. Cereals Oils Assoc.* **2012**, *27*, 78–83.
13. Wang, Y.; Ye, H.; Zhou, C.; Lv, F.; Bie, X.; Lu, Z. Study on the spray-drying encapsulation of lutein in the porous starch and gelatin mixture. *Eur. Food Res. Technol.* **2012**, *234*, 157–163. [CrossRef]
14. Tajkarimi, M.M.; Ibrahim, S.A.; Cliver, D.O. Antimicrobial herb and spice compounds in food. *Food Control* **2010**, *21*, 1199–1218. [CrossRef]
15. Wang, Y.Y.; Ren, G.Y.; Yang, L. Optimization and Quality of Onion Essential Oil Microencapsulation Technology. *Food Sci.* **2017**, *113*, 10–12.
16. Yao, W.R.; Yao, H.Y. Adsorption characteristics of porous starch and its application. *Grain Feed. Ind.* **2005**, *1*, 22–24.
17. Ye, F.; Miao, M.; Lu, K.; Jiang, B.; Li, X.; Cui, S.W. Structure and physicochemical properties for modified starch-based nanoparticle from different maize varieties. *Food Hydrocoll.* **2017**, *67*, 37–44. [CrossRef]
18. Zhang, B.; Cui, D.; Liu, M.; Gong, H.; Huang, Y.; Han, F. Corn porous starch: Preparation, characterization and adsorption property. *Int. J. Biol. Macromol.* **2012**, *50*, 250–256. [CrossRef]
19. Zhang, L.; Yang, Z.; Wei, J.; Su, P.; Chen, D.; Pan, W.; Tang, J. Contrastive analysis of chemical composition of essential oil from twelve Curcuma species distributed in China. *Ind. Crops Prod.* **2017**, *108*, 17–25. [CrossRef]
20. Zhu, H.; Zhang, Y.; Tian, J.; Chu, Z. Effect of a new shell material—Jackfruit seed starch on novel flavor microcapsules containing vanilla oil. *Ind. Crops Prod.* **2018**, *112*, 47–52. [CrossRef]
21. Wang, Y.F.; Jia, J.X.; Tian, Y.Q.; Shu, X.; Ren, X.J.; Guan, Y.; Yan, Z.Y. Antifungal effects of clove oil microcapsule on meat products. *LWT* **2017**, *89*, 604–609. [CrossRef]
22. Donsì, F.; Annunziata, M.; Sessa, M.; Ferrari, G. Nanoencapsulation of essential oils to enhance their antimicrobial activity in foods. *LWT* **2011**, *44*, 1908–1914. [CrossRef]
23. Chen, M.; Yan, X.; Cheng, M.; Zhao, P.; Wang, Y.; Zhang, R.; Wang, X.; Wang, J.; Chen, M. Preparation, characterization and application of poly(lactic acid)/corn starch/eucalyptus leaf essential oil microencapsulated active bilayer degradable film. *Int. J. Biol. Macromol.* **2021**, *41*, 108–117. [CrossRef]
24. De Sousa, D.P.; Silva, R.H.N.; Silva, E.F.d.; Gavioli, E.C. Essential Oils and Their Constituents: An Alternative Source for Novel Antidepressants. *Molecules* **2017**, *22*, 1290. [CrossRef]
25. Mehran, M.; Masoum, S.; Memarzadeh, M. Microencapsulation of Mentha spicata essential oil by spray drying: Optimization, characterization, release kinetics of essential oil from microcapsules in food models. *Ind. Crops Prod.* **2020**, *154*, 112694. [CrossRef]
26. Han, X.; Wen, H.; Luo, Y.; Yang, J.; Xiao, W.; Ji, X.; Xie, J. Effects of α-amylase and glucoamylase on the characterization and function of maize porous starches. *Food Hydrocoll.* **2021**, *116*, 106661–106667. [CrossRef]
27. Fu, Q.; Song, H. Preparation Properties of Thyme Essential Oil/Porous Starch Microcapsules. *Packag. Eng.* **2020**, *41*, 77–82.
28. Guo, L.; Li, J.; Gui, Y.; Zhu, Y.; Yu, B.; Tan, C.; Fang, Y.; Cui, B. Porous starches modified with double enzymes: Structure and adsorption properties. *Int. J. Biol. Macromol.* **2020**, *164*, 1758–1765. [CrossRef]
29. Lito, M.J.; Guiomar, H.M.; Filomena, M.; Camões, G.F.C.; Covington, A.K. Effect of citrate impurities on the reference pH value of potassium dihydrogen buffer solution. *Anal. Chim. Acta* **2003**, *482*, 137–146. [CrossRef]
30. Tangpao, T.; Chung, H.-H.; Sommano, S.R. Aromatic Profiles of Essential Oils from Five Commonly Used Thai Basils. *Foods* **2018**, *7*, 175. [CrossRef]
31. Zhang, J.; Cui, X.; Zhang, M.; Bai, B.; Yang, Y.; Fan, S. The antibacterial mechanism of perilla rosmarinic acid. *Biotechnol. Appl. Biochem.* **2021**, *11*, 2248–2256. [CrossRef] [PubMed]
32. Pozo, C.; Rodríguez-Llamazares, S.; Bouza, R.; Barral, L.; Castaño, J.; Müller, N.; Restrepo, I. Study of the structural order of native starch granules using combined FTIR and XRD analysis. *J. Polym. Res.* **2018**, *25*, 266–270. [CrossRef]
33. Tu, Q.B.; Wang, P.-Y.; Sheng, S.; Xu, Y.; Wang, J.-Z.; You, S.; Zhu, A.; Wang, J.; Wu, F.-Y. Microencapsulation and Antimicrobial Activity of Plant Essential Oil Against Ralstonia solanacearum. *Waste Biomass Valorization* **2020**, *11*, 5273–5282. [CrossRef]
34. Hadidi, M.; Pouraminb, S.; Adinepourc, F.; Haghanib, S.; Jafaric, S.M. Chitosan nanoparticles loaded with clove essential oil: Characterization, antioxidant and antibacterial activities. *Carbohydr. Polym.* **2020**, *236*, 116075. [CrossRef] [PubMed]
35. Chen, M.; Hu, Y.; Zhou, J.; Xie, Y.; Wu, H.; Yuan, T.; Yang, Z. Facile fabrication of tea tree oil-loaded antibacterial microcapsules by complex coacervation of sodium alginate/quaternary ammonium salt of chitosan. *RSC Adv.* **2016**, *6*, 13032–13039. [CrossRef]
36. Zhang, H.; Wang, R.; Chen, Z.; Zhong, Q. Enzymatically modified starch with low digestibility produced from amylopectin by sequential amylosucrase and pullulanase treatments. *Food Hydrocoll.* **2019**, *95*, 195–202. [CrossRef]
37. Punia Bangar, S.; Ashogbon, A.O.; Singh, A.; Chaudhary, V.; Whiteside, W.S. Enzymatic modification of starch: A green approach for starch applications. *Carbohydr. Polym.* **2022**, *287*, 119265. [CrossRef]
38. Gonzalez, A.; Wang, Y.J. Enhancing the Formation of Porous Potato Starch by Combining α-Amylase or Glucoamylase Digestion with Acid Hydrolysis. *StarchStärke* **2020**, *72*, 1900269. [CrossRef]
39. Chong, Y.-B.; Zhang, H.; Yue, C.Y.; Yang, J. Fabrication and Release Behavior of Microcapsules with Double-Layered Shell Containing Clove Oil for Antibacterial Applications. *ACS Appl. Mater. Interfaces* **2018**, *10*, 15532–15541. [CrossRef]

40. Ozogul, Y.; Boğa, E.K.; Akyol, I.; Durmus, M.; Ucar, Y.; Regenstein, J.M.; Köşker, A.R. Antimicrobial activity of thyme essential oil nanoemulsions on spoilage bacteria of fish and food-borne pathogens. *Food Biosci.* **2020**, *36*, 100635. [CrossRef]
41. Qian, J.; Chen, Y.; Wang, Q.; Zhao, X.; Yang, H.; Gong, F.; Guo, H. Preparation and antimicrobial activity of pectin-chitosan embedding nisin microcapsules. *Eur. Polym. J.* **2021**, *157*, 110676. [CrossRef]
42. Valizadeh, S.; Naseri, M.; Babaei, S.; Hosseini, S.M.H. Shelf-life extension of fish patty using biopolymer-coated active paper sheets. *Food Packag. Shelf Life* **2020**, *47*, 133–140. [CrossRef]

foods

MDPI

Article

Assessing the Gel Quality and Storage Properties of *Hypophythalmalmichthys molitrix* Surimi Gel Prepared with Epigallocatechin Gallate Subject to Multiple Freeze-Thaw Cycles

Zhihang Tian [1], Xin Jiang [1], Naiyong Xiao [1], Qiang Zhang [1], Wenzheng Shi [1,2,*] and Quanyou Guo [3,*]

[1] College of Food Sciences & Technology, Shanghai Ocean University, Shanghai 201306, China; 17865561866@163.com (Z.T.); 13053523375@163.com (X.J.); xny931215@163.com (N.X.); qzzz1274@163.com (Q.Z.)

[2] National R & D Branch Center for Freshwater Aquatic Products Processing Technology (Shanghai), Shanghai 201306, China

[3] East China Sea Fisheries Research Institute, Chinese Academy of Fishery Sciences, Shanghai 200090, China

* Correspondence: wzshi@shou.edu.cn (W.S.); dhsguoqy@163.com (Q.G.); Tel.: +86-156-9216-5859 (W.S.)

Abstract: Epigallocatechin gallate (EGCG) with concentrations of 0–0.03% was added to manufacture surimi gels, respectively, while effects on gel quality and storage properties indicators during freeze-thaw (F-T) cycles were investigated. The results implied that the gel quality and storage properties of surimi gels added without EGCG were seriously destroyed during F-T cycles. The addition of EGCG could inhibit the decline of texture and gel strength. Moreover, EGCG has effect on inhibiting the microbial growth and the formation of off-odor compounds such as total volatile basic nitrogen (TVB-N) and malondialdehyde (MDA). Low-field nuclear magnetic resonance (LF-NMR) and water-holding capacity (WHC) results showed that immobilized water migrated to free water with the extension of F-T cycles. The scanning electron microscope (SEM) observed denser protein networks and smaller holes from the surimi gels added with EGCG. However, excessive (0.03%) EGCG showed the loose network structure and moisture loss. Overall, EGCG in 0.01–0.02% addition was good for resisting damage of surimi gels during F-T cycles.

Keywords: surimi gels; EGCG; gel quality; storage properties; freeze-thaw cycle

Citation: Tian, Z.; Jiang, X.; Xiao, N.; Zhang, Q.; Shi, W.; Guo, Q. Assessing the Gel Quality and Storage Properties of *Hypophythalmalmichthys molitrix* Surimi Gel Prepared with Epigallocatechin Gallate Subject to Multiple Freeze-Thaw Cycles. *Foods* **2022**, *11*, 1612. https://doi.org/10.3390/foods11111612

Academic Editors: Tao Yin and Liu Shi

Received: 6 May 2022
Accepted: 26 May 2022
Published: 30 May 2022

1. Introduction

Surimi products, a kind of elastic and nutritious food with high protein and low fat, are made from frozen surimi by suwari and kamaboko. Frozen storage is a common way to limit protein oxidation and microbial growth, thus prolonging the shelf life of surimi products [1]. However, repeated freeze-thawing occurs inevitably in surimi products during processing, storage, transportation, sale, and consumption. Due to the temperature fluctuation, the gel quality of surimi products has been destructed by ice crystals, such as hardness reduction, moisture loss, and gel network fracture. The adverse biochemical reaction of surimi is occurred swiftly by ice crystals which leave the hole [2]. Besides, myofibrillar protein is the main part of surimi, which plays a key role in gel formation. With the increase of F-T cycles, the structural integrity of the surimi myofibrillar protein network gradually decreased by extrusion of ice crystals, thus affecting the texture, whiteness, and WHC of surimi products [3]. Microbial growth is also an important cause of quality changes in surimi products during frozen storage. How to effectively control the quality deterioration of surimi during freezing has always been a hot research topic. However, to our best knowledge, few studies on delaying quality deterioration of surimi products during freeze-thaw cycle.

Tea polyphenols, as the emerging food additive for aquatic products, have played a vital role in antioxidant and bacteriostatic properties [4]. The main component of tea polyphenols is catechins. There are four main catechins, including epicatechin (EC), epicatechin gallate (ECG), epigallocatechin (EGC), and EGCG [5]. Due to the ability of antiseptic and sterilization, EGCG was widely applied in the medical field to opposite cancer, virus, and obesity [6]. In the field of food processing, transglutaminase and EGCG had double cross-linking impact on myofibrillar proteins modification, and that EGCG could further improve the gel properties under the induction of transglutaminase [7]. The adducts reacted with EGCG and malondialdehyde could enhanced gel quality and fixed the microstructure of myofibrillar proteins [8]. Besides, EGCG plays an essential role in preserving the freshness of food. EGCG combined with squid ring shell oligosaccharides was used to extend the product life of yellowfin tuna fillets by 12 days at 4 °C [9]. Furthermore, the gum tragacanth-sodium alginate active coating containing EGCG and lysozyme could prolong the shelf life of large yellow croaker [10]. As a natural additive for use in frozen food, EGCG has been proved to be viable in aquatic product preservation. However, few studies are focused on the addition of EGCG to surimi gels.

The study aimed to sight how EGCG affected surimi products during F-T cycles. Aerobic plate count (APC), total volatile basic nitrogen (TVB-N), thiobarbituric acid reactive substance (TBARS), and pH were used to identify the state of corruption. Moreover, the texture properties, gel strength, whiteness, WHC, and moisture distribution were investigated to study the gel properties and storage properties of surimi gels. Collectively, the microstructure was analyzed the changes in the myofibrillar proteins network. This research can serve as an experimental foundation for the production and quality control of high-quality surimi products during frozen storage.

2. Materials and Methods

2.1. Materials and Reagents

Silver carp (*Hypophythalmalmichthys molitrix*) frozen surimi (AAA grade) was obtained from Jingli Aquatic Food Co., Ltd. (Honghu, China) and then cut into approximately 1 kg blocks. These surimi blocks were kept at −20 °C until further need. EGCG with a purity of 98% was provided from Shanghai Meryer Co., Ltd. (Shanghai, China). Other chemicals were of analytical grade and supplied from Sinopharm Chemical Reagent Co., Ltd. (Shanghai, China).

2.2. Preparation of Heat-Induced Gel and F-T Cycles

Raw surimi blocks were defrosted at 4 °C overnight until a semi-thawed state and cut into about 3 cm cubes. Then, the surimi cubes were blended using AM-CG108 food processor at 5000 rpm for 2 min. NaCl (2.5%) was added and the blending was continued for 2 min. Subsequently, different content EGCG (0%, 0.01%, 0.02%, and 0.03%) was sprinkled into the surimi and added ice water to adjust the moisture content of surimi to 78%, during which time the mixture was blended at 5000 rpm for 6 min. The entire procedure was kept at a temperature of less than 10 °C. Next, the surimi paste was filled into polyvinylidene chloride casings (25 mm diameter) to make surimi sausages (20 cm long) and make sure that both ends were sealed tightly to prevent water from immigrating. Finally, the two-step heating method was used, with 60 min at 40 °C followed by 30 min at 90 °C. After sterilization, the sausages were promptly chilled in ice water for 30 min before being frozen at −20 °C.

The F-T cycles were designed as follows. The one F-T cycle was defined that surimi gels were prepared by freezing at −20 °C for 72 h after thawing at 4 °C for 12 h. Repeated 6 cycles and selected 0, 2, 4, 6 cycles to measure relevant indicators.

2.3. Texture Profile Analysis (TPA)

Surimi gels were sliced into the cylinders with a height of 20 mm, and then equilibrated to room temperature for 30 min. Hardness, springiness, cohesiveness, and chewiness were estimated by TA-XT plus texture analyzer (Stable Micro Systems, Vienna, UK) equipped with probe P/50. The parameters were as follows: 3 mm/s pre-test speed, 1 mm/s test speed, 3 mm/s post-test speed, 50% compression strain, and 10 g trigger force.

2.4. Gel Strength Measurement

Gel samples were balanced at room temperature for about 30 min before testing. Breaking force (g) and deformation (cm) were measured in gel cylinders cut into 25 mm lengths, using TA-XT plus texture analyzer (Stable Micro Systems, Vienna, UK) equipped with a P/5S probe attachment. The gels were examined at a press down distance of 15 mm with a pretest speed of 2 mm/s, test speed of 1 mm/s, and post-test speed of 10 mm/s. Gel strength was calculated by the following Equation (1):

$$\text{Gel strength } (\text{g}\cdot\text{cm}) = \text{Breaking force}(\text{g}) \times \text{Deformation}(\text{cm}) \quad (1)$$

2.5. Color Measurement

The apparent color of the gel was measured by CR-400 colorimeter (Konica Minolta, Osaka, Japan) to analyze L* (lightness), a* (redness-greenness), and b* (yellowness-blueness). The whiteness (W) was computed in the following Equation (2):

$$W = 100 - \left[(100 - L^*)^2 + a^{*2} + b^{*2}\right]^{1/2} \quad (2)$$

2.6. Storage Properties Measurement

2.6.1. Aerobic Plate Count (APC) Measurement

Samples were prepared using the approach of Xu et al. [11] with some modifications. Accurately 2 g of surimi gel was weighed, and 18 mL of normal saline (0.85%, *w/w*) was added. After aseptic homogenization for 3 min, 10-fold dilution was continued. Standard plate count agar was selected for APC. Sample liquid (1 mL) which needed 3 dilution gradients was injected into plate count agar medium and kept at 30 °C for 72 h. 2 parallels were set up per dilution gradient. Results of APC were expressed as log colony forming units (CFU)/g sample.

2.6.2. Total Volatile Basic Nitrogen (TVB-N) Measurement

The TVB-N value was determined according to Wang et al. [12] using FOSS 8400 Kjeldahl Nitrogen apparatus (FOSS Corporation, Denmark) with slight modification. The samples of surimi gels (10.0 g) were minced and mixed with 10.0 g magnesium oxide slightly in a digestive tube. Following distillation, the volatile nitrogen was recovered and titrated with 0.1 M hydrochloric acid in a 1% boric acid solution (*w/v*).

2.6.3. pH Value Measurement

The surimi gel was equilibrated at normal temperature for 30 min. Then, 2 g of gel sample was homogenized with 18 mL of deionized water for 1 min by AD200L-P high-speed dispersing homogenizer. The homogenate was centrifuged at 10,000 r/min for 10 min, followed by filtering and retaining the supernatant. The pH was measured with an FE28 pH meter and performed at least in triplicate.

2.6.4. TBA Measurement

First, 1 g of sample was weighed and added to 10 mL of 7.5% trichloroacetic acid (containing 0.1% of EDTA). After being homogenized for 1 min at 15,000 rpm, samples were left in the ice bath for 10 min and then filtered. To 3 mL of filtrate in test tubes, 3 mL of 0.02 mol/L 2-thiobarbituric acid in distilled water was added. The sample solution was reacted at boiling bath for 40 min. Finally, the absorbance was measured at 532 nm after cooling with ice water. A standard curve was built by 1,1,3,3-Tetraethoxypropane (TEP).

2.7. Water-Holding Capacity (WHC) Measurement

Surimi gels were sliced into 5 mm cylindrical slices and weighed as W_1, which closed to 3 g. Next, the gel sample was wrapped by a piece of double-layer filter, and centrifuged at 5000 r for 15 min. Finally, the filter paper was removed and the weight of samples was measured as W_2. The Equations (3) and (4) were used to determine WHC.

$$\text{Centrifugal loss } (\%) = \frac{W_2}{W_1} \times 100 \quad (3)$$

$$\text{WHC } (\%) = \frac{G-CL}{G} \times 100 \quad (4)$$

Thereinto, G represented the moisture content determined by drying the sample at 105 °C until constant weight and CL was the centrifugal loss.

2.8. Low-Field Nuclear Magnetic Resonance (LF-NMR) Spin-Spin Relaxation (T_2) Measurement

The moisture distribution and relaxation time were determined using the approach of Jiang et al. [13] by a Niumag Pulsed NMR analyzer (MesoMR23-060H-I, Niumag Electric Co., Shanghai, China). Surimi gels were cut into cylinders (20 mm × 20 mm), and transverse relaxation (T_2) was determined using the Carr-Purcell-Meiboom-Gill (CPMG) pause sequence. The data from 8000 echoes were collected throughout the course of 8 scan repeats. All measurements were performed at 32 °C in triplicate.

2.9. Magnetic Resonance Imaging (MRI) Measurement

The measurement of MRI was described by Cheng et al. [14] with slight modifications. Samples were wrapped in plastic wrap and put into an MRI tube. Proton density imaging was obtained by MSE imaging sequence with the main following parameters: Repetition Time = 500 ms, Echo Time = 18.2 ms, and Average = 6. Finally, the condition of mapping and pseudocolor was provided by Shanghai Niumag Electric Co., Ltd.

2.10. Microstructure Measurement

The samples were sliced into 3 mm × 3 mm × 1 mm pieces by surgical knife and fixed with glutaraldehyde (2.5%, *v*/*v*) for 12 h at 4 °C. Then, 0.1 M phosphoric acid buffer (pH 7.2–7.4) was used to rinse the gel samples three times. Afterwards, a serious density of ethanol (30%, 50%, 70%, 80%, 90%, and 100%) was added in gel samples for gradient elution. The dehydrated samples were immersed in a mixture of ethanol and tert butanol (3:1, 1:1, 1:3, 0:1) to remove the ethanol. The processed pieces that sputter-coated with gold after freeze-drying were observed by a Scanning Electron Microscope instrument (Hitachi SU5000, Hitachi High-Tech Co., Ltd., Shanghai, China) at an acceleration voltage of 5 kV.

2.11. Statistical Analysis

Each experiment was repeated three times. Statistical Package for Social Science 26.0 software (SPSS Inc., Chicago, IL, USA) was used to analyze the experimental data, and the mean ± standard deviation (SD) was provided. Duncan's multiple range test method was used to compare significant differences using one-way analysis of variance (ANOVA) at the 0.05 level.

3. Results and Discussion

3.1. TPA

TPA is widely applied to research the physical characteristics of surimi gels, which become an effective method to investigate texture [15]. The change in hardness, springiness, cohesiveness, and chewiness of gel matrix with different content EGCG during F-T cycles were shown in Figure 1. Samples were compressed twice to provide texture test curve and grasp the characteristic parameters of the gel matrix [16]. Texture-related properties were overall decreased with increasing F-T cycles. The phenomenon might be attributed to the disruption of the F-T cycles to the formation, melting, and regeneration of ice crystals, resulting in larger and more irregular ice crystals that physically disrupt the protein network than ice crystals before F-T cycles [17]. Besides, the growth of ice crystals affected the tightness of the myofibrillar network and part of the water was migrated outside the gel, which might be associated with the gel strength of surimi gel [18]. Thereinto, Figure 1a,d showed that F-T cycles had a substantial impact on hardness and chewiness of the 0% EGCG group. The hardness and chewiness changed the most between the zeroth and second F-T cycles, demonstrating that the quality of surimi gel was seriously degraded during this period. Besides, there were significant differences in chewiness of the sample with 0.01% EGCG and three other samples after sixth F-T cycles ($p < 0.05$). Figure 1b,c revealed that the springiness and cohesiveness reduced considerably after fourth F-T cycle ($p < 0.05$). It was indicating that negative impact of the springiness and cohesiveness was mainly occurred in the late F-T period. Moreover, the 0.01% EGCG group had better springiness and cohesiveness, illustrating the effective preservation of texture. Moreover, the downward trend of hardness, springiness, cohesiveness, and chewiness of gels without EGCG was larger than those in the other added amount groups during the F-T cycles. These findings showed that EGCG had a beneficial influence on surimi gel F-T stability.

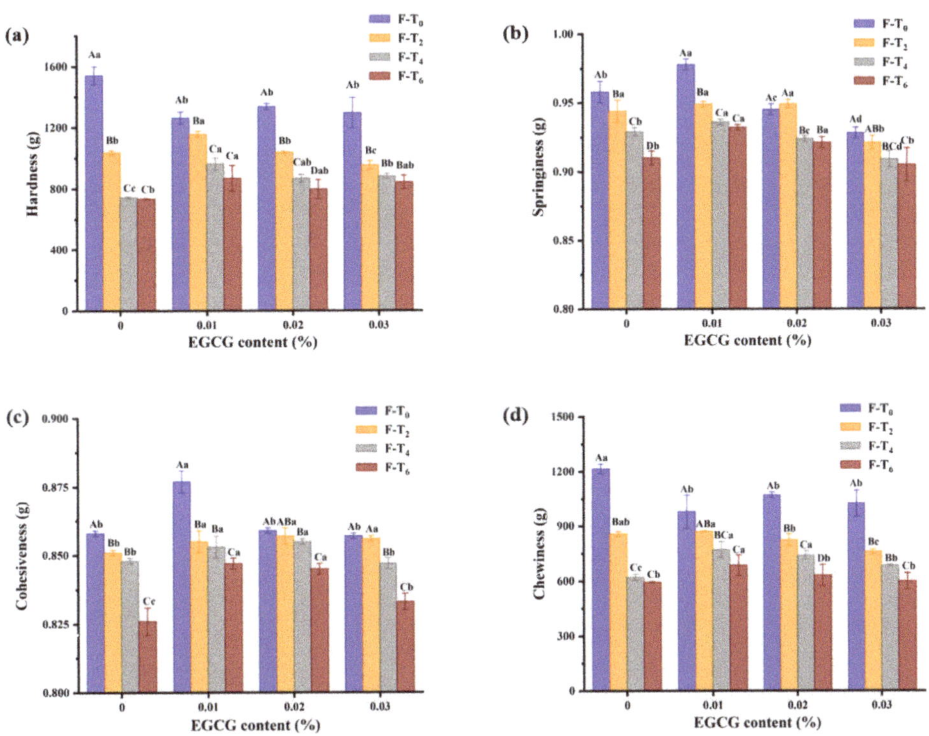

Figure 1. Changes in hardness (**a**), springiness (**b**), cohesiveness (**c**), and chewiness (**d**) of surimi gels treated with different content EGCG during F-T cycles. Uppercase letters indicate significant difference ($p < 0.05$) between different F-T cycle, lowercase letters indicate the difference between gels with different EGCG content ($p < 0.05$). F-T_0: the zeroth freeze-thaw cycle; F-T_2: the second freeze-thaw cycle; F-T_4: the fourth freeze-thaw cycle; F-T_6: the sixth freeze-thaw cycle.

3.2. Gel Strength

Figure 2 showed that breaking force, deformation, and gel strength of samples were all decreased due to the F-T damage. Generally, the breaking force could characterize the hardness of gels and the deformation could represent the gel flexibility and elasticity [19].

Correspondingly, breaking force of samples rose followed by a reduction as the EGCG content increased after each F-T cycle, and the gels added with 0.01% EGCG had the maximum breaking force. The phenomenon was consistent with the result of TPA. As the main functional protein, the higher-order structure of myosin unfolds after heating to form a stable network structure, which is intertwined with each other through hydrogen bonding to form fibrous macromolecules [18]. It was hypothesized that the gel strength treated with EGCG was connected to the changes in protein structure of surimi gels. Li et al. [7] reported that EGCG could promote the cross-linking of myofibrillar proteins and increase the gel strength of surimi gels with the non-covalent interactions such as hydrophobic interaction and hydrogen bond. However, the group which treated high content of EGCG exhibited poor gel strength in Figure 2c. A possible explanation for this might be the self-aggregation of EGCG, which caused the loss in the ability of myofibrillar proteins cross-linking [20].

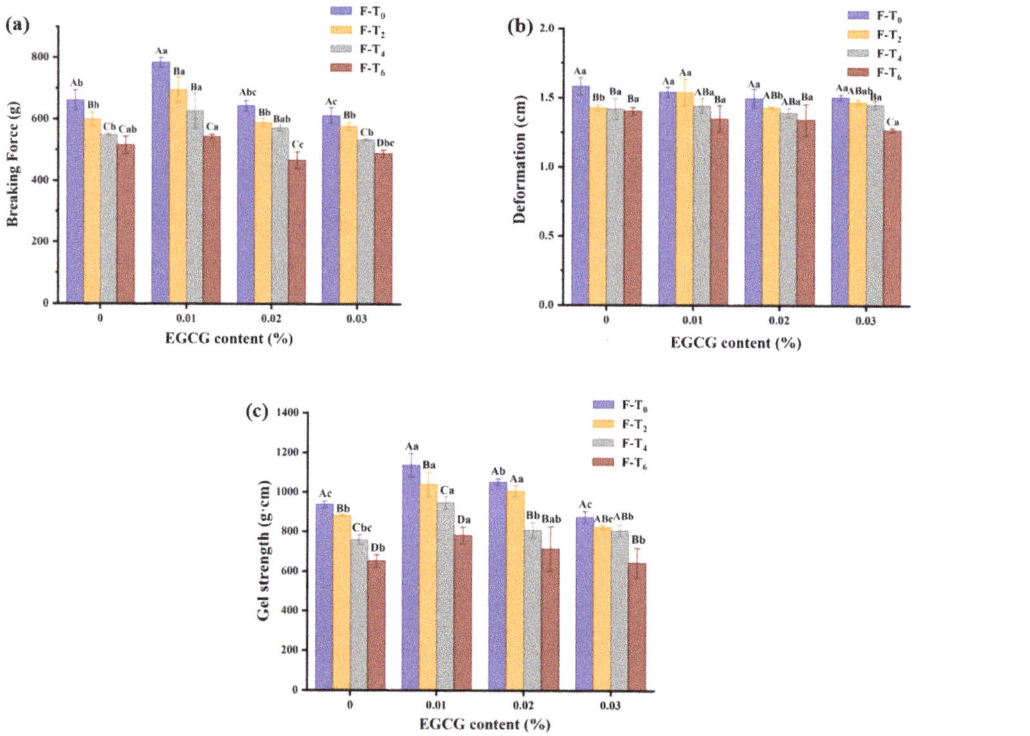

Figure 2. Changes in breaking force (**a**), deformation (**b**), and gel strength (**c**) of surimi gels treated with different content EGCG during F-T cycles. Uppercase letters indicate significant difference ($p < 0.05$) between different F-T cycle, lowercase letters indicate the difference between gels with different EGCG content ($p < 0.05$). F-T_0: the zeroth freeze-thaw cycle; F-T_2: the second freeze-thaw cycle; F-T_4: the fourth freeze-thaw cycle; F-T_6: the sixth freeze-thaw cycle.

3.3. Color

Whiteness is the most intuitive indication for consumers to identify surimi quality. L*, a*, b* values, and whiteness of surimi gels, with and without EGCG, as affected by F-T cycles were shown in Table 1. As the cycles of F-T increased, the lightness decreased significantly ($p < 0.05$). The more the F-T cycle is, the worse quality the gels exhibit, the dimmer the lightness of the gel surface is. Compared with the 0% EGCG group, the group treated with EGCG exhibited low whiteness. There was a view that the tight gel network has positive effect on light absorption, which caused the deep color of gels [21]. Combined with the previous analysis, the greater the gel strength is, the lower the whiteness of surimi gel is. Besides, the quinones obtained by EGCG oxidation were brown, which resulted in the decline of whiteness. The redness-greenness of aquatic products might be related to the oxidation of myoglobin and the reduction of methemoglobin [22]. In the whole F-T cycles, the a* value of each group had no significant change. It was supposed that the myoglobin was removed when the surimi was rinsed, resulting in the stability of a* value. In addition, the b* value of each EGCG group was significantly lower than the control group. It was reported that a small amount of fat oxidation and pigment degradation in surimi could affect the color of surimi gels, especially the yellowness-blueness [23]. It might be that EGCG could better inhibit fat oxidation.

Table 1. Changes in whiteness of surimi gels with difference content EGCG during F-T cycles.

Color	EGCG Content (%)	F-T Cycles			
		0	2	4	6
L*	0	72.80 ± 0.40^{Aa}	71.30 ± 0.13^{Ba}	70.90 ± 0.07^{Ba}	70.2 ± 0.29^{Ca}
	2	71.00 ± 0.59^{Ab}	70.10 ± 0.11^{Bb}	69.40 ± 0.47^{BCb}	69.0 ± 0.12^{Cb}
	4	70.80 ± 0.11^{Ab}	69.20 ± 0.36^{Bc}	68.90 ± 0.14^{Bc}	68.1 ± 0.25^{Cbc}
	6	70.40 ± 0.38^{Ab}	69.20 ± 0.40^{Bc}	68.60 ± 0.47^{BCc}	68.0 ± 0.43^{Cc}
a*	0	-1.60 ± 0.08^{Ac}	-1.70 ± 0.03^{Ac}	-1.60 ± 0.01^{Ad}	-1.70 ± 0.10^{Ac}
	2	-0.50 ± 0.05^{Ab}	-0.70 ± 0.04^{Bb}	-0.80 ± 0.05^{Bc}	-0.60 ± 0.10^{ABb}
	4	-0.30 ± 0.04^{Aa}	-0.60 ± 0.02^{Bab}	-0.20 ± 0.06^{Ab}	-0.30 ± 0.08^{Aa}
	6	-0.30 ± 0.08^{Ba}	-0.50 ± 0.09^{Ca}	0.10 ± 0.03^{Aa}	-0.20 ± 0.04^{Ba}
b*	0	4.60 ± 0.09^{Aa}	3.50 ± 0.16^{Ca}	3.60 ± 0.21^{Ca}	4.00 ± 0.26^{Ba}
	2	2.40 ± 0.01^{Ab}	2.30 ± 0.13^{ABb}	2.30 ± 0.46^{ABb}	2.20 ± 0.25^{Bb}
	4	2.50 ± 0.08^{Ab}	2.10 ± 0.03^{Bc}	2.00 ± 0.31^{BCbc}	1.70 ± 0.18^{Cc}
	6	2.50 ± 0.09^{Ab}	1.80 ± 0.06^{Bd}	1.80 ± 0.08^{Bc}	1.70 ± 0.13^{Bc}
W	0	72.40 ± 0.40^{Aa}	71.00 ± 0.04^{Ba}	70.70 ± 0.09^{Ba}	69.9 ± 0.30^{Ca}
	2	70.90 ± 0.60^{Ab}	70.00 ± 0.11^{Bb}	69.30 ± 0.50^{BCb}	69.0 ± 0.11^{Cab}
	4	70.30 ± 0.40^{Ab}	69.10 ± 0.36^{Bc}	68.90 ± 0.10^{Bc}	68.0 ± 0.23^{Cbc}
	6	70.70 ± 0.10^{Ab}	69.20 ± 0.40^{Bc}	68.50 ± 0.50^{BCc}	68.0 ± 0.43^{Cc}

Uppercase letters indicate significant difference ($p < 0.05$) between different F-T cycle, lowercase letters indicate the difference between gels with different EGCG content ($p < 0.05$), and the values are expressed as mean ± SD. L*: lightness, a*: redness-greenness, b*: yellowness-blueness, and W: whiteness.

3.4. Storage Properties

3.4.1. Aerobic Plate Count (APC)

One of the major contributors to the spoiling of surimi products during storage is microbial development, which has an impact on shelf life [24]. Throughout the F-T cycles, the APC of surimi gels showed an upward trend, especially from the zeroth to second F-T cycles. It might be that fat oxidation was accelerated, which obtained foul-smelling rancid products due to the F-T cycles. Therefore, the conditions for microbial contamination were created.

As the EGCG content increased, the APC of surimi gels decreased after each F-T cycle, confirming the antibacterial effect of EGCG. Some assumptions could be explained to describe the results of APC: (1) EGCG destroyed the structural integrity of individual microorganisms, resulting in the release of intracellular components and functional impairment. (2) Respiration of microorganisms was influenced. (3) Proteins and enzymes of microorganisms were destroyed. (4) EGCG disturbed the metabolism of microorganisms with chelating metal ions [25]. EGCG has strong antioxidant properties [26]. Klancnik et al. [27] discovered that the growth of *L. monocytogenes*, *E. coli*, and *C. jejuni* was prevented by the role of AlpE (ethanolic extract of *A. katsumadai* seeds) and EGCG in jointly protecting the surimi gels. EGCG has been shown to have regulatory activity and strong antibacterial activity against Gram-positive and Gram-negative bacteria [28].

3.4.2. Total Volatile Basic Nitrogen (TVB-N)

TVB-N is another important indicator for evaluating the storage properties of surimi gels [29]. In the process of corruption, volatile ammonia, and biogenic amines were obtained from proteins, amino acids, and nitrogen-containing compounds, which decomposed by microorganisms [30]. As can be seen from Figure 3b, the TVB-N values of the control group, 0.01%, 0.02%, and 0.03% EGCG group in the zeroth F-T cycle were respectively 2.77, 2.50, 2.61, and 2.61 mg/100 g. Subsequently, the value of the control group increased rapidly and reached 7.69 mg/100 g on the sixth F-T cycle, which was higher than other experimental groups. It was illustrated that EGCG treatment helped to suppress the development of TVB-N and remain the storage properties of surimi gels. Correspondingly, the curve trend of TVB-N value was similar to the APC.

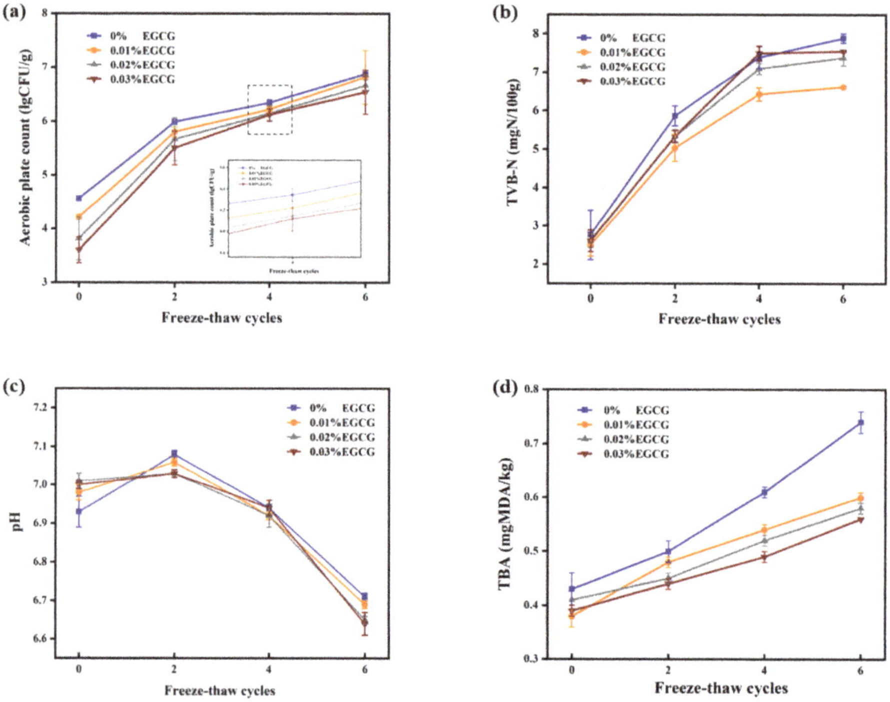

Figure 3. Changes in APC (**a**), TVB-N (**b**), pH (**c**), and TBA (**d**) of surimi gels treated with different content EGCG during F-T cycles.

3.4.3. pH Value

The changes in pH of surimi gels with different EGCG content during F-T cycles were shown in Figure 3c. With the increase in the times of F-T cycle, the pH value of surimi gels was first increased, and then decreased. Various basic nitrogenous compounds such as amines and trimethylamine were generated due to the decomposition of protein, which induced the increase in pH [31]. During 0–2 F-T cycles, the pH of the control group rose faster than other groups, indicating that the protein decomposition of the control group was more serious. The changes in pH correspond to previous TVB-N results. Thereafter, depending on more suitable growth conditions supplied by F-T cycles, microorganisms decomposed small organic molecules to generate acidic substances [32].

3.4.4. TBA

TBA value reflects the degree to which lipids in surimi products are oxidized to malonaldehyde [33]. The occurrence of lipid deterioration is inevitable in F-T meat [34]. It could be seen in Figure 3d that the TBA value of surimi gels increased continuously during the F-T cycles. After the fourth F-T cycle, the TBA value of the control group increased rapidly and was significantly higher than EGCG group ($p < 0.05$). In comparison to the control group, the TBA value of surimi gels was dramatically lowered when EGCG was added. The reason was that EGCG contains eight phenolic hydroxyl groups, which possess the ability to scavenge free radical and slow down or terminate free radicals chain reactions effectively as hydrogen donors [35,36].

3.5. Water-Holding Capacity (WHC)

WHC, defined as the ability to retain water in samples, is reflected by centrifugal loss after external influence [37]. The changes in WHC of surimi gels treated with different EGCG content during F-T cycles were shown in Figure 4. As the F-T cycles increased, the WHC of each group decreased. During freezing, the protein network was disrupted by growing ice crystals, leading to tissue deformation and moisture loss [38]. Therefore, WHC is directly affected by the density of the gel network. When the F-T cycles were not started, there were substantial variations between the samples with and without EGCG. Between the fourth and sixth F-T cycles, the WHC of surimi gel added without EGCG declined seriously from 86.06% to 83.34%. Among the samples added with EGCG, the 0.02% group maintained a high WHC in the whole F-T cycle. It represented that EGCG could effectively suppress the decline of WHC. However, the WHC of 0.03% EGCG group was significantly lower than the group added with 0.01% and 0.02% EGCG after the fourth F-T cycle ($p < 0.05$). It was assumed that a high dose of EGCG led to excessive crosslink of myofibrillar protein that induced negative aggregation. The loose gel network structure was obtained due to the decreasing uniformity and density. It resulted that surimi gels could not effectively intercept water.

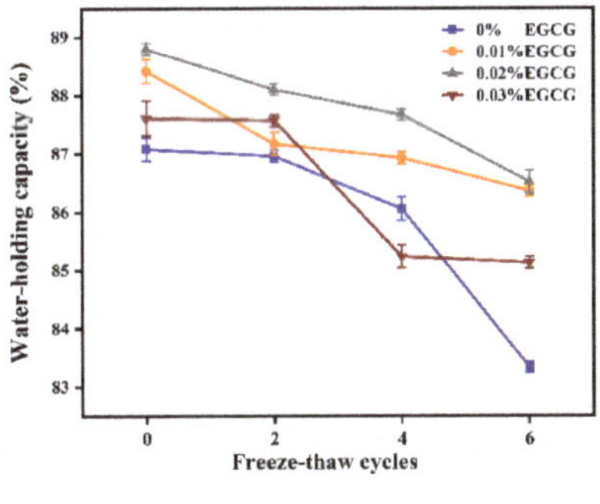

Figure 4. Changes in WHC of surimi gels treated with different content EGCG during F-T cycles.

3.6. Moisture Mobility and Distribution Analysis

3.6.1. Moisture Mobility

Water mobility and distribution in the gel matrix may be explored using LF-NMR, which is a fast and non-destructive approach [39]. T_2 represents the degree of water freedom. According to Figure 5a, three kinds of peaks appeared at approximately 0.1~10 ms (T_{2b}), 20~200 ms (T_{21}), and 300~1500 ms (T_{22}), which represented bound water, immobilized water, and free water, respectively.

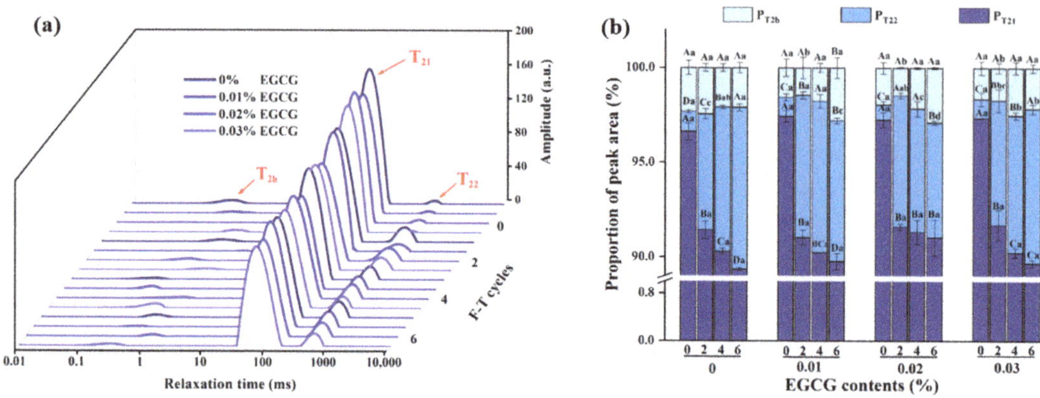

Figure 5. Changes in T_2 relaxation times (**a**) and peak area proportion (P_{T2b}, P_{T21}, P_{T22}) (**b**) of surimi gels treated with different content EGCG during F-T cycles. Uppercase letters indicate significant difference ($p < 0.05$) between different F-T cycle, lowercase letters indicate the difference between gels with different EGCG content ($p < 0.05$).

The T_{21} and T_{22} relaxation periods of the surimi gels were dramatically lengthened when the F-T cycles were extended. Due to the crystallization and recrystallization of water, the moisture migrated outside the protein networks and the degree of water freedom inside the gel was changed [40]. The shorter the relaxation time is, the higher the binding capability between protein molecule and water is, and the stronger the water stability is. The relaxation time of the other groups differs from that of the control group after the addition of EGCG. Thereinto, surimi gels treated with 0.02% EGCG exhibited shorter T_{21} and T_{22}. In addition, the peak corresponding to the relaxation time T_{21} and T_{22} of 0.02% EGCG group shifted to the left in each F-T cycle and it was found that the value of the second peak decreased remarkably. It implied that EGCG can inhibit the fluidity of moisture and control the loss of water, which was in accordance with the WHC.

The percentage of peak area (P_{T2b}, P_{T21}, and P_{T22}) can objectively reflect the relative moisture content of surimi gels in Figure 5b. EGCG had no influence on the alterations in PT_{2b} of the samples. It might be that the bound water was closely linked with proteins and had low fluidity during the F-T cycles. Furthermore, F-T cycles caused P_{T21} to decrease, while P_{T22} increased. When thawing, the internal moisture migrated to large network gaps. After freezing, bigger ice crystals were formed to extrude the protein network, resulting in the reduction of reabsorbed water while thawing again. There has been researching pointed out that myofibrillar protein denaturation reduces protein binding capacity for water during the F-T cycles, thus inducing the conversion of immobilized water to free water [41]. Similar observations were also reported in previous studies [42]. In the fourth F-T cycle, the P_{T22} of 0.02% EGCG group was 6.08%, which was lower than the control group (7.62%), indicating that EGCG could effectively inhibit the growth of free water. The result of the sixth F-T cycle was also in agreement with above phenomenon. EGCG hastened the formation of protein gel and increased the degree of cross-linking of myofibrillar proteins. Therefore, dense protein networks could limit the conversion of immobilized water to free water. However, a too dense protein network caused by excessive cross-linking (0.03% EGCG) could not maintain the state of moisture, which was similar to the result of WHC.

3.6.2. Moisture Distribution

MRI is also an auxiliary way to evaluate moisture distribution in surimi gels during F-T cycles [43]. Pseudo color image (value range: 0~255) produced by nuclear magnetic resonance could directly observe the situation of moisture content and distribution in Figure 6. The more the hydrogen protons exist, the redder the image exhibits, and the higher water content the gel has. On the contrary, the blue area represents low moisture. During the F-T cycles, the signal strength decreased gradually, illustrating that the partial moisture of the surimi gels was lost. Gels added with 0.01% and 0.02% showed stronger signals, which implied that they had higher water content, which supported the previous results. Li et al. [7] reported that the fraction of immobilized water rose with the addition of EGCG and transglutaminase. Overall, the dense gel network induced by EGCG is the primary cause for preserving moisture and making the water distribution uniform.

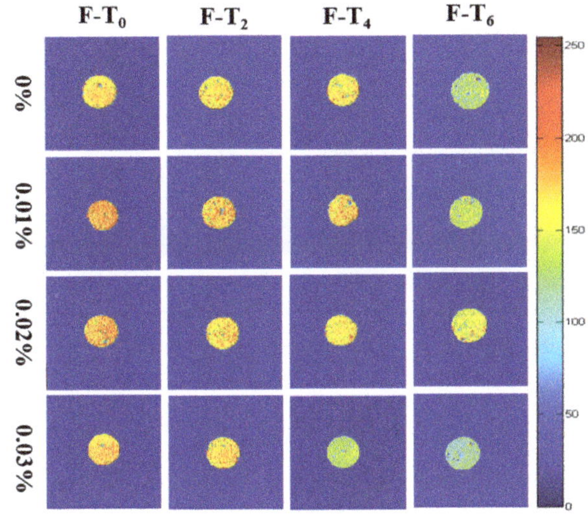

Figure 6. Changes in NMR pseudo color images of surimi gels treated with different content EGCG during F-T cycles. 0%, 0.01%, 0.02%, and 0.03% represented 0%, 0.01%, 0.02%, and 0.03% EGCG content, respectively. Caption: see Figure 1.

3.7. Microstructure

The changes in microstructure inside the gel system observed by scanning electron microscope can intuitively explain the close relationship between 3D network structure and gel properties [44]. Figure 7 showed the pore size and compactness of the protein network structure at 10,000× magnification. After two F-T cycles, large holes were obviously observed in the control group and there were traces of breakage in the protein network. However, samples treated with EGCG presented less ice crystal damage and better protein structure. In the initial stage of F-T cycle, 0.01% and 0.02% EGCG group have already shown a dense and regular gel network. Changes in microstructural compactness may account for the reduced whiteness of surimi gels with EGCG. It was reported that the addition of EGCG could expand myofibrillar protein structure and expose reactive groups in proteins, inducing massive protein aggregation relies on chemical bond action [45]. Furthermore, the declined trend of gel strength is related to the integral and dense extent of the gel microstructure, which is consistent with previously reported results [2].

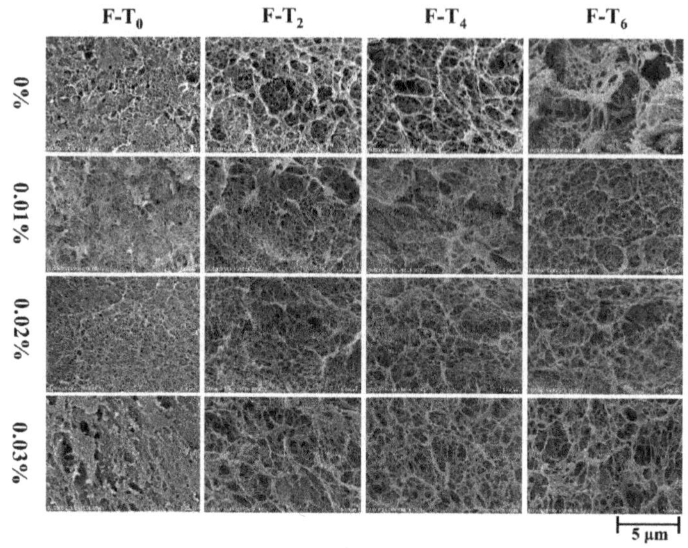

Figure 7. Changes in microstructure micrographs (magnification ×10,000) of surimi gels treated with different content EGCG during F-T cycles. 0%, 0.01%, 0.02%, and 0.03% represented 0%, 0.01%, 0.02%, and 0.03% EGCG content, respectively. Caption: see Figure 1.

4. Conclusions

Changes in the gel properties and storage properties of surimi gels treated with four different dosages of EGCG during F-T cycles were investigated. The addition of EGCG could delay the decrease of hardness, springiness, chewiness, and gel strength. By the antioxidant and bacteriostatic properties of EGCG, the indicators of storage properties such as APC, TVB-N, pH, and TBA were also significantly inhibited. Meanwhile, the higher the EGCG content is, the greater the inhibitory effect is. In addition, LF-NMR and WHC indicated that immobilized water migrated to free water, leading to severe water loss during F-T cycles. Compared with the control group, 0.02% EGCG group could better maintain the moisture of gels. This is attributed to the cross-linking of myofibrillar proteins induced by EGCG, which enhances the binding ability of the gel to water. Through the microstructure, EGCG groups represented denser protein networks and smaller holes. Namely, samples treated with EGCG suffered less damage during F-T cycles. However, overdose (0.03%) of EGCG exhibited fragile and loose network structure, which caused poor gel strength and WHC. In conclusion, this study found that 0.01–0.02% EGCG is an ideal added amount for surimi gels, which provides a data basis for resisting damage during F-T cycles.

Author Contributions: Conceptualization, Z.T.; methodology, Z.T. and N.X.; validation, N.X.; formal analysis, Z.T., X.J. and. N.X.; investigation, X.J. and N.X.; resources, Z.T. and Q.Z.; data curation, Z.T.; writing—original draft, Z.T.; writing—review & editing, X.J. and N.X.; visualization, Z.T.; supervision, W.S.; project administration, W.S. and Q.G.; funding acquisition, W.S. All authors have read and agreed to the published version of the manuscript.

Funding: This research was funded by the National Key R&D Program of China, grant number 2019YFD0902003.

Informed Consent Statement: Informed consent was obtained from all subjects involved in the study.

Data Availability Statement: Not applicable.

Conflicts of Interest: The authors declare no conflict of interest.

References

1. Wang, B.; Du, X.; Kong, B.; Liu, Q.; Li, F.; Pan, N.; Xia, X.; Zhang, D. Effect of ultrasound thawing, vacuum thawing, and microwave thawing on gelling properties of protein from porcine *longissimus dorsi*. *Ultrason. Sonochem.* **2020**, *64*, 104860. [CrossRef] [PubMed]
2. Leygonie, C.; Britz, T.J.; Hoffman, L.C. Impact of freezing and thawing on the quality of meat: Review. *Meat Sci.* **2012**, *91*, 93–98. [CrossRef] [PubMed]
3. Kingwascharapong, P.; Benjakul, S. Effect of phosphate and bicarbonate replacers on quality changes of raw and cooked Pacific white shrimp as influenced by the repeated freeze–thawing. *Int. J. Refrig.* **2016**, *67*, 345–354. [CrossRef]
4. Higdon, J.V.; Frei, B. Tea catechins and polyphenols: Health effects, metabolism, and antioxidant functions. *Crit. Rev. Food Sci. Nutr.* **2003**, *43*, 89–143. [CrossRef]
5. Khan, N.; Mukhtar, H. Tea Polyphenols in Promotion of Human Health. *Nutrients* **2019**, *11*, 39. [CrossRef]
6. Ediriweera, M.K.; Tennekoon, K.H.; Samarakoon, S.R.; Thabrew, I.; de Silva, E.D. Protective Effects of Six Selected Dietary Compounds against Leptin-Induced Proliferation of Oestrogen Receptor Positive (MCF-7) Breast Cancer Cells. *Medicines* **2017**, *4*, 56. [CrossRef]
7. Li, J.; Munir, S.; Yu, X.; Yin, T.; You, J.; Liu, R.; Xiong, S.; Hu, Y. Double-crosslinked effect of TGase and EGCG on myofibrillar proteins gel based on physicochemical properties and molecular docking. *Food Chem.* **2021**, *345*, 128655. [CrossRef]
8. Lv, Y.; Feng, X.; Wang, Y.; Guan, Q.; Qian, S.; Xu, X.; Zhou, G.; Ullah, N.; Chen, L. The gelation properties of myofibrillar proteins prepared with malondialdehyde and (-)-epigallocatechin-3-gallate. *Food Chem.* **2021**, *340*, 10. [CrossRef]
9. Singh, A.; Benjakul, S.; Zhou, P.; Zhang, B.; Deng, S. Effect of squid pen chitooligosaccharide and epigallocatechin gallate on discoloration and shelf-life of yellowfin tuna slices during refrigerated storage. *Food Chem.* **2021**, *351*, 129296. [CrossRef]
10. Pei, J.; Mei, J.; Yu, H.; Qiu, W.; Xie, J. Effect of Gum Tragacanth-Sodium Alginate Active Coatings Incorporated With Epigallocatechin Gallate and Lysozyme on the Quality of Large Yellow Croaker at Superchilling Condition. *Front. Nutr.* **2022**, *8*. [CrossRef]
11. Xu, Y.; Li, L.; Xia, W.; Zang, J.; Gao, P. The role of microbes in free fatty acids release and oxidation in fermented fish paste. *LWT* **2019**, *101*, 323–330. [CrossRef]
12. Wang, Y.; Huang, H.; Shi, W. Effect of different drying time on physicochemical properties of black carp (*Mylopharyngodon piceus*) by hot air. *J. Food Process. Preserv.* **2021**, *46*, e16217. [CrossRef]
13. Jiang, X.; Chen, Q.; Xiao, N.; Du, Y.; Feng, Q.; Shi, W. Changes in Gel Structure and Chemical Interactions of *Hypophthalmichthys molitrix* Surimi Gels: Effect of Setting Process and Different Starch Addition. *Foods* **2021**, *11*, 9. [CrossRef] [PubMed]
14. Cheng, X.F.; Zhang, M.; Adhikari, B.; Islam, M.N. Effect of Power Ultrasound and Pulsed Vacuum Treatments on the Dehydration Kinetics, Distribution, and Status of Water in Osmotically Dehydrated Strawberry: A Combined NMR and DSC Study. *Food Bioprocess Technol.* **2014**, *7*, 2782–2792. [CrossRef]
15. Alakhrash, F.; Anyanwu, U.; Tahergorabi, R. Physicochemical properties of Alaska pollock (*Theragra chalcograma*) surimi gels with oat bran. *LWT Food Sci. Technol.* **2016**, *66*, 41–47. [CrossRef]
16. Zhang, H.; Zhu, Y.; Chen, S.; Xu, C.; Yu, Y.; Wang, X.; Shi, W. Determination of the effects of different high-temperature treatments on texture and aroma characteristics in Alaska pollock surimi. *Food Sci. Nutr.* **2018**, *6*, 2079–2091. [CrossRef] [PubMed]
17. Kong, C.H.; Hamid, N.; Liu, T.; Sarojini, V. Effect of Antifreeze Peptide Pretreatment on Ice Crystal Size, Drip Loss, Texture, and Volatile Compounds of Frozen Carrots. *J Agric Food Chem.* **2016**, *64*, 4327–4335. [CrossRef]
18. Sun, X.D.; Holley, R.A. Factors Influencing Gel Formation by Myofibrillar Proteins in Muscle Foods. *Compr. Rev. Food Sci. Food Saf.* **2011**, *10*, 33–51. [CrossRef]
19. Sakamoto, H.; Kumazawa, Y.; Toiguchi, S.; Seguro, K.; Soeda, T.; Motoki, M. Gel strength enhancement by addition of microbial transglutaminase during onshore surimi manufacture. *J. Food Sci.* **1995**, *60*, 300–304. [CrossRef]
20. Balange, A.; Benjakul, S. Enhancement of gel strength of bigeye snapper (*Priacanthus tayenus*) surimi using oxidised phenolic compounds. *Food Chem.* **2009**, *113*, 61–70. [CrossRef]
21. Sochaya, C.; Soottawat, B. Effect of formaldehyde on protein cross-linking and gel forming ability of surimi from lizardfish induced by microbial transglutaminase. *Food Hydrocoll.* **2013**, *30*, 704–711. [CrossRef]
22. Medic, H.; Djurkin Kusec, I.; Pleadin, J.; Kozacinski, L.; Njari, B.; Hengl, B.; Kusec, G. The impact of frozen storage duration on physical, chemical and microbiological properties of pork. *Meat Sci.* **2018**, *140*, 119–127. [CrossRef] [PubMed]
23. Thanonkaew, A.; Benjakul, S.; Visessanguan, W.; Decker, E.A. The effect of metal ions on lipid oxidation, colour and physicochemical properties of cuttlefish (*Sepia pharaonis*) subjected to multiple freeze–thaw cycles. *Food Chem.* **2006**, *95*, 591–599. [CrossRef]
24. Aguirrezabal, M.M.; Mateo, J.; Dominguez, M.C.; Zumalacarregui, J.M. The effect of paprika, garlic and salt on rancidity in dry sausages. *Meat Sci.* **2000**, *54*, 77–81. [CrossRef]
25. Klancnik, A.; Mozina, S.S.; Zhang, Q. Anti-Campylobacter activities and resistance mechanisms of natural phenolic compounds in Campylobacter. *PLoS ONE* **2012**, *7*, e51800. [CrossRef] [PubMed]
26. Nagle, D.G.; Ferreira, D.; Zhou, Y.D. Epigallocatechin-3-gallate (EGCG): Chemical and biomedical perspectives. *Phytochemistry* **2006**, *67*, 1849–1855. [CrossRef]
27. Klancnik, A.; Piskernik, S.; Bucar, F.; Vuckovic, D.; Mozina, S.S.; Jersek, B. Reduction of microbiological risk in minced meat by a combination of natural antimicrobials. *J. Sci. Food Agric.* **2014**, *94*, 2758–2765. [CrossRef]

28. Friedman, M. Overview of antibacterial, antitoxin, antiviral, and antifungal activities of tea flavonoids and teas. *Mol. Nutr. Food Res.* **2007**, *51*, 116–134. [CrossRef]
29. Goulas, A.E.; Kontominas, M.G. Combined effect of light salting, modified atmosphere packaging and oregano essential oil on the shelf-life of sea bream (*Sparus aurata*): Biochemical and sensory attributes. *Food Chem.* **2007**, *100*, 287–296. [CrossRef]
30. Zhou, Y.; Yang, H. Effects of calcium ion on gel properties and gelation of tilapia (*Oreochromis niloticus*) protein isolates processed with pH shift method. *Food Chem.* **2019**, *277*, 327–335. [CrossRef]
31. Benjakul, S.; Visessanguan, W.; Thongkaew, C.; Tanaka, M. Comparative study on physicochemical changes of muscle proteins from some tropical fish during frozen storage. *Food Res. Int.* **2003**, *36*, 787–795. [CrossRef]
32. Suvanich, V.; Jahncke, M.L.; Marshall, D.L. Changes in selected chemical quality characteristics of channel catfish frame mince during chill and frozen storage. *J. Food Sci.* **2000**, *65*, 24–29. [CrossRef]
33. Ozpolat, E.; Patir, B. Determination of Shelf Life for Sausages Produced From Some Freshwater Fish Using Two Different Smoking Methods. *J. Food Saf.* **2016**, *36*, 69–76. [CrossRef]
34. Li, F.; Zhong, Q.; Kong, B.; Wang, B.; Pan, N.; Xia, X. Deterioration in quality of quick-frozen pork patties induced by changes in protein structure and lipid and protein oxidation during frozen storage. *Food Res. Int.* **2020**, *133*, 109142. [CrossRef] [PubMed]
35. Nikoo, M.; Regenstein, J.M.; Ahmadi Gavlighi, H. Antioxidant and Antimicrobial Activities of (-)-Epigallocatechin-3-gallate (EGCG) and its Potential to Preserve the Quality and Safety of Foods. *Compr. Rev. Food Sci. Food Saf.* **2018**, *17*, 732–753. [CrossRef]
36. Quan, T.H.; Benjakul, S.; Sae-leaw, T.; Balange, A.K.; Maqsood, S. Protein–polyphenol conjugates: Antioxidant property, functionalities and their applications. *Trends Food Sci. Technol.* **2019**, *91*, 507–517. [CrossRef]
37. Zhou, F.; Wang, X. Assessing the gelling properties of the silver carp surimi gel prepared with large yellow croaker processing by-product in freeze-thaw cycles. *J. Food Process. Preserv.* **2021**, *45*, e15479. [CrossRef]
38. Jiang, Q.; Nakazawa, N.; Hu, Y.; Osako, K.; Okazaki, E. Changes in quality properties and tissue histology of lightly salted tuna meat subjected to multiple freeze-thaw cycles. *Food Chem.* **2019**, *293*, 178–186. [CrossRef]
39. Zhang, Z.Y.; Regenstein, J.M.; Zhou, P.; Yang, Y.L. Effects of high intensity ultrasound modification on physicochemical property and water in myofibrillar protein gel. *Ultrason. Sonochem.* **2017**, *34*, 960–967. [CrossRef]
40. Provesi, J.G.; Valentim Neto, P.A.; Arisi, A.C.M.; Amante, E.R. Extraction of antifreeze proteins from cold acclimated leaves of Drimys angustifolia and their application to star fruit (*Averrhoa carambola*) freezing. *Food Chem.* **2019**, *289*, 65–73. [CrossRef]
41. Zhang, M.; Li, F.; Diao, X.; Kong, B.; Xia, X. Moisture migration, microstructure damage and protein structure changes in porcine longissimus muscle as influenced by multiple freeze-thaw cycles. *Meat Sci.* **2017**, *133*, 10–18. [CrossRef] [PubMed]
42. Wang, B.; Li, F.; Pan, N.; Kong, B.; Xia, X. Effect of ice structuring protein on the quality of quick-frozen patties subjected to multiple freeze-thaw cycles. *Meat Sci.* **2021**, *172*, 108335. [CrossRef] [PubMed]
43. Warschawski, D.E.; Arnold, A.A.; Marcotte, I. A New Method of Assessing Lipid Mixtures by P-31 Magic-Angle Spinning NMR. *Biophys. J.* **2018**, *114*, 1368–1376. [CrossRef] [PubMed]
44. Nunez-Flores, R.; Cando, D.; Javier Borderias, A.; Moreno, H.M. Importance of salt and temperature in myosin polymerization during surimi gelation. *Food Chem.* **2018**, *239*, 1226–1234. [CrossRef] [PubMed]
45. Huang, Y.; Hua, Y.; Qiu, A. Soybean protein aggregation induced by lipoxygenase catalyzed linoleic acid oxidation. *Food Res. Int.* **2006**, *39*, 240–249. [CrossRef]

foods MDPI

Article

Microbial Evaluation of Ozone Water Combined with Ultrasound Cleaning on Crayfish (*Procambarus clarkii*)

Yuzhao Ling [1,2], Hongyuan Tan [1,2], Lingwei Shen [1,3], Lingyun Wei [2], Guangquan Xiong [1], Lan Wang [1], Wenjin Wu [1] and Yu Qiao [1,*]

[1] Key Laboratory of Cold Chain Logistics Technology for Agro-Product, Ministry of Agriculture and Rural Affairs, Institute of Agricultural Products Processing and Nuclear Agricultural Technology, Hubei Academy of Agricultural Sciences, Wuhan 430064, China; lingyuzhao2012@163.com (Y.L.); tanhongyuan1206@163.com (H.T.); shenlingwei2021@163.com (L.S.); xiongguangquan@163.com (G.X.); lilywang_2016@163.com (L.W.); 272081603@163.com (W.W.)

[2] School of Environmental Ecology and Biological Engineering, Wuhan Institute of Technology, Wuhan 430205, China; lingyun.wei@outlook.com

[3] School of Bioengineering and Food, Hubei University of Technology, Wuhan 430068, China

* Correspondence: qiaoyu@hbaas.com

Abstract: The effects of ozone water (OW) and ultrasound cleaning (UL) on microbial community diversity of crayfish were studied through microbial viable count and 16S rRNA gene sequencing. The results showed that compared with the control (CK), the ozone water combined with ultrasound cleaning (OCU) showed a significant reduction ($p < 0.05$) in total viable count (TVC), psychrophilic viable count (PVC), mesophilic viable count (MVC), Pseudomonas, hydrogen sulfide-producing bacteria (HSPB), molds and yeasts. Concretely, the TVC of the CK, OW, UL and OCU were 5.09, 4.55, 4.32 and 4.06 log CFU/g, respectively. The dominant bacterium in untreated crayfish was *Chryseobacterium*, and its relative abundance was reduced by combined treatment. Color measurement and sensory evaluation suggested that a satisfactory sensory experience could be obtained on the crayfish applied with OCU. In brief, OCU could be used as a cleaning strategy to control the microbial quality of crayfish and have no influence on its quality.

Keywords: crayfish; ozone water; ultrasound; 16S rRNA gene sequencing

Citation: Ling, Y.; Tan, H.; Shen, L.; Wei, L.; Xiong, G.; Wang, L.; Wu, W.; Qiao, Y. Microbial Evaluation of Ozone Water Combined with Ultrasound Cleaning on Crayfish (*Procambarus clarkii*). *Foods* **2022**, *11*, 2314. https://doi.org/10.3390/foods11152314

Academic Editor: Paula Bourke

Received: 21 June 2022
Accepted: 27 July 2022
Published: 3 August 2022

1. Introduction

Crayfish (*Procambarus clarkii*) is an important part of the fishery trade, and it is related to the outbreak of foodborne diseases [1], as well as crayfish plague [2]. The food-borne pathogens associated with crayfish mainly include *Vibrio*, *Listeria*, *Salmonella*, and *Shigella*. In order to solve the hidden dangers of food-borne pathogenic bacteria, cleaning and disinfection are considered essential to control the microbiological quality of crayfish.

Considering live crayfish, non-thermal sterilization is the first choice. Common physical food processing technologies have cold plasma, irradiation, high pressure, ozonation, ultrasound, ultraviolet light, pulsed light, and pulsed electric fields [3]. Ultrasound and ozone are considered harmless, effective and economical methods, and are widely used in food technology. As an emerging green technique, ultrasound is considered as a safe and efficient tool in food processing industry [4–6]. The wide application of ultrasound treatment was attributed to the cavitation effect caused by ultrasonic waves propagating in liquid systems, as well as thermal effect, shear force, micro-jet, and shock wave [7–9]. Ultrasonic inactivation of microbes is mainly due to cell rupture, localized high temperature and pressures, and the generation of several free radicals [9–13]. Kordowska–Wiater et al. reported that sonication of chicken wings in water for 3 min resulted in a reduction of bacteria on the skin surface, and the most sensitive was *Escherichia coli* (*E. coli*) [14]. However, ultrasound alone cannot significantly reduce microbial contamination [3]. The study suggested that bacteria were mostly washed away rather than destroyed by ultrasound in water [14]. Hurdle technology is feasible, which is defined as cleaning of ultrasound combined with other detergents. As a highly effective strong oxidant [15], ozone could inactivate a variety of fungi, Gram-positive bacteria, Gram-negative bacteria, and viruses [16,17]. Ozone use in the food industry has been reported. Rodrigues et al. indicated that ozone water immersion treatment could remove some pesticides (azoxystrobin, chlorothalonil, and difenoconazole) remaining in tomatoes [18]. Disinfection of red-meat-processing wastewater using ozone showed that 99% of TVC, total coliforms, and *E. coli* were inactivated [19]. Ozone spray effectively reduced the initial load of aerobic bacteria in refrigerated salmon fillets, and did not significantly increase the level of lipid oxidation [20]. Therefore, ozone water could be used as a disinfectant and washing liquid to facilitate subsequent food processing [21]. Studies have shown that ultrasound combined with ozone could be used to increase oxidative capacity, thereby providing a faster degradation rate of organic pollutants [22,23].

In recent years, the use of 16S rRNA gene sequencing has become popular in the analysis of bacterial community diversity; it is a highly sensitive and relatively quantitative technique [24]. Cleaning is one of the pre-treatments in the processing of crayfish, and it is the basic step to ensure the safety and quality of crayfish. Therefore, studying the bacterial spectrum in crayfish after pre-cleaning is integral to quality control. According to our current knowledge, ozone combined with ultrasound has been expanded in various areas, including device descaling [25], water sterilization [26], fruit preservation [27], biosludge reduction [28], and degradation of antibiotic drugs [29], but rarely in the cleaning and disinfection of crayfish. In this study, the effectiveness of ozone water combined with ultrasound cleaning in crayfish via microbiological (traditional plate count and 16S rRNA gene sequencing) and physicochemical parameters (color test, electronic nose, and sensory evaluation) were studied.

2. Materials and Methods

2.1. Materials

Crayfish were purchased from the Baili Supermarket, Hongshan District, Wuhan, China. The mean length and weight of crayfish were 13.0 ± 0.5 cm and 13.2 ± 0.4 g, respectively. The purchased crayfish were immediately transported to the laboratory in no more than one hour. The obtained crayfish were not pretreated before cleaning.

2.2. Treatment of Crayfish

Live crayfish were divided into four treatments, with 30 in each. Ozonizer (GCQJ-1-3, Wuhan, China), air-liquid mixer (HPSJ-25, Wuhan, China) and ultrasonic bath (KQ-500VDV, Kunshan, China) were applied: the concentration of ozone water and the intensity of ultrasound were 26.60 mg/L and 200 W, respectively. The batches were respectively washed by different cleaning methods with treatment groups as follows: (a) OW, immersed in ozone water for 20 min, (b) UL, ultrasound cleaning with ultra-pure water for 10 min, and (c) OCU, at first immersed in ozone water for 20 min followed by ultrasound cleaning with ultra-pure water for 10 min. In addition, crayfish immersed in ultra-pure water for 20 min set as the control (CK). The whole crayfish is covered with the soaking solution. The temperature of the ultrasound cleaning solution was kept at about 25 °C by means of an ice bath. After that, treated crayfish were used for microbiological and physicochemical analysis.

2.3. Microbial Count

Cleaned crayfish head and shell were removed. Precisely 5.0 g of shredded crayfish tail muscle with 45 mL of sterilized 0.85% (w/v) NaCl solution was stirred for 2 min. Sterilized glass beads were added to thoroughly contact the mixture. The suspensions were serially diluted in tubes with 9 mL of sterilized NaCl solution, then 1 mL suitable dilution was poured into petri dishes with agar medium, making the suspension and medium mix well by rotating the petri dish slightly. Total viable count (TVC) was cultivated by plate count agar (Hopebio, Qingdao, China) at 30 °C for 72 h; similarly, psychrophilic viable count

(PVC) and mesophilic viable count (MVC) were incubated at 4 °C for 7 d and at 37 °C for 48 h, respectively. *Pseudomonas* were cultivated with cetrimide fucidin cephaloridine agar (Hopebio, Qingdao, China) with selected medium additives at 30 °C for 72 h. Hydrogen sulfide-producing bacteria (HSPB) were cultivated with triple sugar iron agar (Hopebio, Qingdao, China) at 30 °C for 72 h (black colonies formed due to the precipitation of iron sulfide). Molds and yeasts were incubated on rose bengal medium (Hopebio, Qingdao, China) at 28 °C for 5 d. Each crayfish sample was used for determination in duplicate. The number of cultured colonies was converted to colony-forming units per gram of sample (CFU/g).

2.4. DNA Extraction and Amplification

Twenty-four swabs were prepared containing surface inclusions of crayfish muscle from four treatments. E.Z.N.A @ Soil DNA kit (Omega Bio-Tek, Norcross, GA, USA) was used to extract the total DNA of crayfish samples, followed by using 1.2% agarose gel electrophoresis to determine DNA quality. The DNA concentration was measured by using BioSpec-nano (Shimadzu, Japan). The PCR mix is as described above, containing PCR ExTaq Buffer, DNA template, dNTP, ExTaq, primer1, and primer2. Universal bacterial primers are 338-F (5′-ACTCCTACGGGAGGCAGCA-3′) and 806-R (5′-GGACTACHVGGGTWTCTAAT-3′), which are used to amplify the V3-V4 region of the bacterial 16S rRNA gene. The PCR amplified conditions were: 95 °C for 2 min, and followed by 30 cycles: degeneration at 95 °C for 1 min, renaturation at 60 °C for 40 s, and elongation at 70 °C for 40 s, and finally extension at 70 °C for 10 min. The PCR products were identified by 1.2% agarose gel electrophoresis, and purified through the AxyPrep DNA Gel Extraction Kit (Axygen Biosciences, Union City, CA, USA). The PCR-Free Sample Preparation Kit (Illumina, San Diego, CA, USA) was used for the generation of amplicon library, and the Illumina HiSeq platform (Beijing Novogene Bioinformation Science and Technology Co. Ltd., China) was used for high throughput sequencing. This information was collected on the online platform from Shanghai Majorbio Bio-pharm Technology Co. Ltd. (Shanghai, China)

High-quality sequences were stitched by Flash software (version 1.2.11, https://ccb.jhu.edu/software/FLASH/index.shtml accessed on 20 June 2022), using a similarity level of 97%, and using Uparse software (version 7.0.1090, http://www.drive5.com/uparse/ accessed on 20 June 2022) to classify all sequences into operational tax units (OTU). OTU can be a genus or a phylum. In order to obtain the species classification information corresponding to each OTU, the silva132/16S_bacteria database was used for taxonomic comparison of species classification, together with the Ribosomal Database Project (RDP) classifier (version 2.11, https://sourceforge.net/projects/rdp-classifier/ accessed on 20 June 2022). This process was carried out on the platform for quantitative insights into microbial ecology (Qiime) (http://qiime.org/scripts/assign_taxonomy.html, accessed on 6 November 2019). The confidence of species classification was set as 70%. Mothur software (version 1.30.2, https://www.mothur.org/wiki/Download_mothur accessed on 20 June 2022) was used to evaluate alpha diversity of each sample, namely Coverage, Shannon index, Simpson index, Ace index, and Chao index, and the significance level was set by the Duncan method. Mapping with the online platform from Shanghai Majorbio Bio-pharm Technology Co. Ltd. (Shanghai, China), included Venn plot, bar-plot, heat-map, and principal co-ordinates analysis (PCoA) at bacteria community level.

2.5. Examination of Chromaticity

The L* (brightness value), a* (red or green value), and b* (yellow or blue value) of crayfish were measured by a portable colorimeter (CR-400, Konica Minolta Inc., Japan), and the colorimeter was corrected by a white standard board (L* = 85.6, a* = 0.3162, b* = 0.3238) before testing. Whiteness was used to distinguish the color difference between crayfish samples. The muscle chromaticity of four crayfish was measured, and the muscle test parts of each crayfish were randomly selected.

$$\text{Whiteness} = 100 - \sqrt{(100 - \text{L})^2 + \text{a}^2 + \text{b}^2}$$

2.6. E-Nose Analysis

The PEN3 electronic nose (AIRSENSE, Schwerin, Germany) was used to analyze the odor difference of crayfish samples. 3 g of shredded crayfish muscle and 6 mL of saturated saline were put into the 20 mL headspace bottle, and then sealed with an E-Z Crimper (Huifen CO. Ltd., Zaozhuang, China). The headspace bottle was placed in a 45 °C water bath to equilibrate for 3 min, and then analyzed in the E-nose system. The test parameters were as follows: flush time and measurement time were set as 100 and 120 s, respectively, and chamber flow was set as 600 mL/min. The E-nose system consists of 10 metal oxide sensors, which are sensitive to different volatile components. As a result of previous experiments, the response values of the sensor from 90 to 94 s were used to visualize the odor difference between the samples, and this process was completed in the software WinMuster (Version 1.6.2, Schwerin, Germany) built into the E-nose system. Each sample was tested in triplicate.

2.7. Sensory Evaluation

Sensory evaluation was determined from three indexes: smell, color and muscle tissue, which were slightly modified [30]. Each indicator included three summary descriptive words based on sensory responses. The descriptors respectively represented three quality levels, which were excellent (representing 8 to 9 scores), good (representing 6 to 7 scores), and general (representing 5 to 6 scores). Seven experienced food professionals (4 males and 3 females) aged about 24 in the laboratory scored the descriptors, and the results were averaged.

2.8. Statistical Analysis

The results were expressed as means ± standard deviations derived from replicates. Origin 9.4 (Origin Lab, Northampton, MA, USA), Microsoft Excel 2010 and SPSS 18.0 (SPSS Inc., Chicago, IL, USA) were used for data analysis. One-way analysis of variance was carried out by Tukey multiple comparison method to evaluate the significance of differences between the data, and the value of $p < 0.05$ was considered as significant.

3. Results

3.1. Microbial Count

The MVC of OW was equivalent to that of UL (4.77 and 4.69 log CFU/g, respectively), but it decreased significantly ($p < 0.05$) in the combined treatment (4.24 log CFU/g). TVC (4.32 log CFU/g) was less than MVC (4.69 log CFU/g) in UL, which may be because the local high temperature generated by ultrasonic cavitation was not conducive to TVC growth. Compared with the control, using three cleaning methods (OW, UL, and OCU) significantly reduced ($p < 0.05$) the amount of TVC, MVC, PVC, *Pseudomonas*, HSPB, molds and yeasts (Figure 1). Most of the selective count in OCU showed a synergistic decline ($p < 0.05$), such as TVC, MVC, and PVC. Similarly, reduction of total coliform count can be observed in single treatments (ozone or ultrasound), but the combination showed a synergistic effect [31]. These counts were in agreement with previous studies. Compared with the control, ozone water reduced the TVC of tilapia fillets by 79.49% [32]. The microbial load of crayfish treated with ozone was significantly reduced [33,34]. Similarly, on day 0, the MVC of samples treated with ozone water was less than 1.40 log CFU/g, which was significantly lower than that of the control (2.97 log CFU/g) [35]. Ozone water spraying significantly reduced TVC of salmon fillets [20]. In contrast, the growth of psychrophilic bacteria was not observed in shrimp treated with ozone water on day 0 [35].

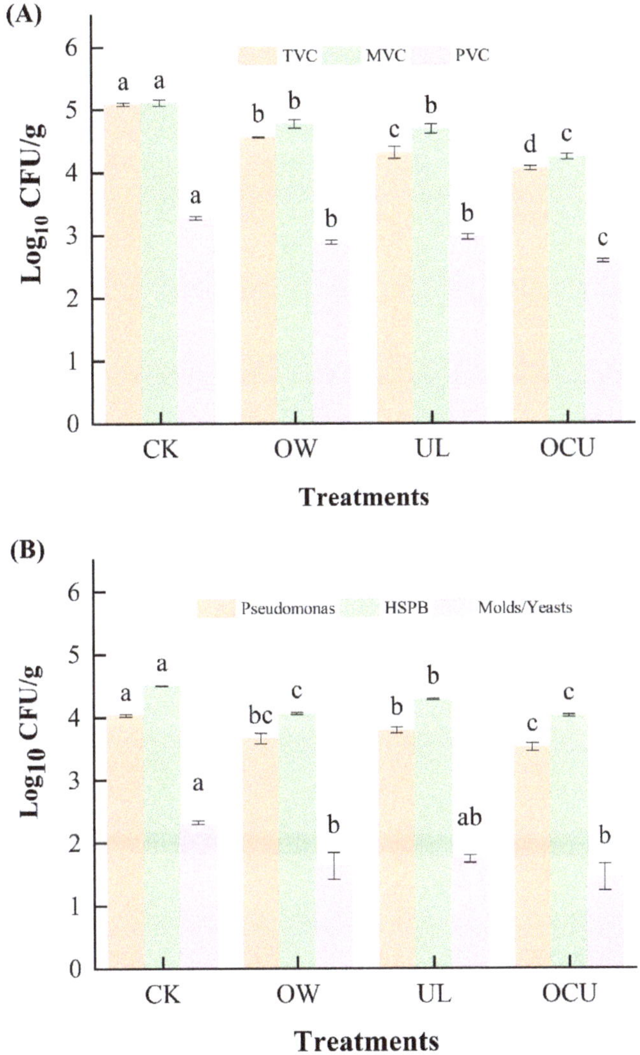

Figure 1. Panels (**A**,**B**) show the microbial count of crayfish washed by different treatments; mean values are expressed as log CFU/g. Error bars were derived from standard deviation of means. The significant differences ($p < 0.05$) between treatments for the same index are expressed alphabetically (a, b, c, and d). Notes: CK, the control; OW, ozone water cleaning; UL, ultrasound cleaning; OCU, ozone water combined with ultrasound cleaning; TVC: total viable count; MVC: mesophilic viable count; PVC: psychrophilic viable count; HSPB, hydrogen sulfide-producing bacteria.

For both *Pseudomonas* and HSPB, the trends were similar. *Pseudomonas* and HSPB in OCU (3.53 and 4.02 log CFU/g, respectively) were significantly ($p < 0.05$) lower than levels in the control (4.03 and 4.50 log CFU/g, respectively). Similar treatment showed that *Pseudomonas* did not show resistance to the combined treatment of ultrasound with lactic acid solution [14]. On the whole, the level of molds and yeasts was lower by 2 to 3 orders of magnitude than that of bacteria. However, compared with the control, UL did not significantly reduce ($p > 0.05$) the amount of molds and yeasts, while a significant reduction occurred in OCU, which showed that the ability of ozone water to inactivate molds and yeasts was better than that of ultrasound. Both gaseous and aqueous ozone reduced the amount of yeast by about 0.5 log CFU/mL [36]. Some non-thermal technologies, such as high pressure and pulsed electric fields cannot effectively kill spores [3]. It was found, however, that ozone treatment could completely eliminate yeast and bacteria [37]. For the reduction of yeast, the effect of ozone was not affected by its form or action time. There was no significant difference in total yeast count between use of ozonated water and gaseous ozone, and the count was also not affected by the extension of reaction time [36]. Previous studies suggested that ultrasound has no direct effect on spores or Gram-positive bacteria [12], and that Gram-negative bacteria are more sensitive to ozone than Gram-positive bacteria [38], and finally yeast [39]. In addition, vacuum packaging after ozonation seems to be effective in reducing yeasts and molds [40], which may be due to the inability of most aerobic yeasts and molds to grow normally.

3.2. Bacterial Diversity

3.2.1. Alpha Diversity

Illumina MiSeq high-throughput sequencing was used to evaluate the bacterial community diversity in crayfish samples. 1224285 effective 16S rRNA gene sequences (the reads from CK, OW, UL, and OCU were 293,960, 327,539, 302,607, and 300,179, respectively) were obtained from 24 crayfish samples in sextuplicate. Alpha diversity analysis was used to evaluate richness and diversity in the dominant microbiota of crayfish samples in this study, including Shannon index, Simpson index, Ace index, and Chao index. The average coverage from crayfish samples was higher 99% (data not shown), which suggested that the sequence reads almost characterized the microbial community of crayfish samples. As shown in Table 1, compared with CK, the Ace index and Chao index of both OW and OCU were significantly lower ($p < 0.05$), while UL decreased insignificantly ($p > 0.05$), which indicated that ozone water causes a more effective reduction of microbiota richness than ultrasound. In addition, the Simpson index of OCU (0.08) was the highest among the treatments, which illustrated that ozone water combined with ultrasound cleaning is the most beneficial for reducing the microbial community of crayfish.

Table 1. The alpha diversity indexes of crayfish samples.

Samples	Shannon	Simpson	Ace	Chao
CK	4.46 ± 0.22^{a}	0.03 ± 0.01^{a}	1230.25 ± 133.13^{a}	1235.84 ± 141.03^{a}
OW	4.09 ± 0.10^{ab}	0.05 ± 0.01^{ab}	996.00 ± 126.95^{bc}	980.03 ± 121.72^{bc}
UL	4.20 ± 0.34^{ab}	0.06 ± 0.03^{ab}	1190.31 ± 128.74^{ab}	1173.43 ± 123.24^{ab}
OCU	3.91 ± 0.26^{b}	0.08 ± 0.03^{b}	943.25 ± 91.79^{c}	930.18 ± 90.89^{c}

Notes: The significant difference ($p < 0.05$) between treatments for the same index was expressed alphabetically (a, b, and c). CK, the control; OW, ozone water cleaning; UL, ultrasound cleaning; OCU, ozone water combined with ultrasound cleaning.

3.2.2. Bacterial Community Composition

Most OTU in crayfish belonged to five bacteria phyla, including *Proteobacteria*, *Bacteroidetes*, *Deinococcus-Thermus*, *Actinobacteria*, and *Firmicutes*. *Bacteroidetes*, *Actinobacteria*, and *Firmicutes* were reduced to varying degrees by ozone water, ultrasound, and the union of both (Figure 2A). Compared with the control, the *Proteobacteria* from three treatments increased markedly. *Deinococcus-Thermus* from both OW and OCU was also significantly decreased (4.23 and 4.27%, respectively) compared with the control (5.05%), while UL (5.90%) was increased. These phyla include *Proteobacteria*, *Bacteroidetes*, *Actinobacteria*, and *Firmicutes*, which usually appear alone or in combination in the microbiota of crayfish and shrimp [41–45]. In fact, *Proteobacteria*, *Firmicutes*, *Actinobacteria*, and *Bacteroidetes* were also dominant phyla in other food samples and meat processing rooms [37,38,46].

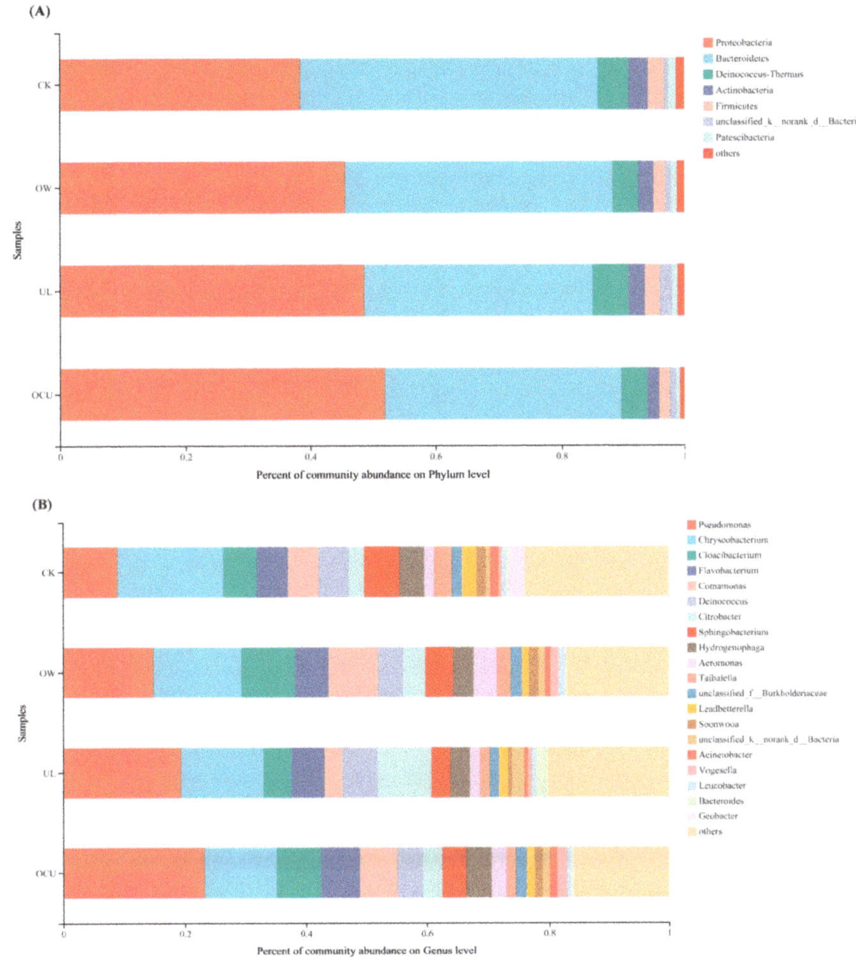

Figure 2. The relative abundance of crayfish washed by different treatments at the phylum (**A**) and genus (**B**) level. All genera with relative abundance less than 1% were classified as "other". Notes: CK, the control; OW, ozone water cleaning; UL, ultrasound cleaning; OCU, ozone water combined with ultrasound cleaning.

As shown in Figure 2B, twenty genera of total relative abundance value > 1% were identified at the genus level. The relative abundance of *Cloacibacterium*, *Soonwooa*, *Leucobacter*, and *Comamonas* was decreased after ultrasound cleaning alone; in detail, the values were 4.58, < 1, 0.68, and 2.94%, respectively, lower than that of the control (5.51, 1.56, 0.72, and 5.11%, respectively). As previously mentioned, the sterilization mechanism of ultrasound was mainly generated by free radicals that damage cell membranes and heat-inactivate enzymes [11]. However, *Escherichia-Shigella* was an exception, and its sequence reads increased rapidly in UL, while OTU level in CK was lower. Studies reported that *Escherichia-Shigella*, which belongs to the opportunistic pathogens [47], was significantly enriched after ultrasound treatment [48,49]. *Escherichia-Shigella* has a tenacious ability to survive, and its relative abundance was increased after the action of disinfectants such as ozone, acid solution, chlorine, and antibiotics [50–52].

The relative abundance of genera in OW decreased from *Chryseobacterium*, *Deinococcus*, *Sphingobacterium*, *Hydrogenophaga*, *Taibaiella*, *Leadbetterella*, and *Acinetobacter*. Furthermore, these genera showed synergistic effects in OCU. *Chryseobacterium* derived from the phylum *Bacteroidetes* was the second largest dominant genus in abundance level. Its abundance in CK, OW, UL, and OCU were 17.42, 14.49, 13.59, and 11.93%, respectively. This result indicated that the best reduction effect occurred in ozone water combined with ultrasound, followed by ozone water, and finally ultrasound. Similarly, cleaning treatments were performed on strawberries containing artificially inoculated *E. coli*, and it was found that ultrasound treatment could not completely remove the inoculum, but ozone could, and the combination of the two showed a synergistic reduction effect [27]. For the osmotic lysis of bacteria, although ultrasound has a mechanical effect, O_3 has a chemical effect that causes a significant increase in the production of reactive oxygen species [53]. Hydroxyl radicals have a powerful oxidizing capacity, can instantaneously penetrate into the cell, and then act on the components of cytoplasmic membrane and intracellular enzymes (such as superoxide dismutase and catalase) [54]. Likewise, Walker et al. used a presoak solution before ultrasound cleaning, producing the most effective instrument cleaning effect on the whole, compared to presoak cleaning only and ultrasound cleaning only [55]. For the two cleaning treatments (ultrasound and plasma liquids), synchronous treatment reduced the amount of *E. coli* and *Shewanella putrefaciens* more than that of separate treatment [56]. Ozone water combined with ultrasound enhanced the antibacterial ability of ozone and resulted in >3 log reduction of microorganisms in cherry tomato [57]. Due to the local high temperature and high pressure generated by ultrasonic cavitation, the bacterial cell membrane becomes fragile and hydroxyl radicals generated by ozone decomposition can penetrate into the cell more easily [58]. In addition, studies showed that ozone could react with biofilm components [59] and ultrasound could enhance the capability of ozone to reduce the thickness of biofilm [25].

Compared with CK (8.96%), the relative abundance of *Pseudomonas* in treatments increased. Concretely, the abundance in OW, UL, and OCU were 14.83, 19.41, and 23.22%, respectively. Generally, *Pseudomonas* was considered to be the dominant microbiota in crayfish [45,60]. However, because the total microbial load was unknown, although the relative abundance of *Pseudomonas* increased, it was not known whether this phenomenon was due to an increase in its absolute quantity or a decrease in the abundance of other microbiota. This phenomenon has occurred in other species. *Staphylococcus* counted by microbial plate was not affected by ozone treatment, but its relative abundance was decreased by high-throughput sequencing [38]. Taking *Geobacter* as an example, its detected level was extremely low (<1%) after ozone water or ultrasound cleaning compared with the control (2.59%), which indirectly increased the relative abundance of *Pseudomonas*. Another fact referred to the "other" category: the relative abundance of CK, OW, UL, and OCU were 24.03, 17.14, 20.23, and 15.93%, respectively, which indicated that some less abundant and unclassified genera in the "others" category were more sensitive to ozone water and ultrasound treatment.

3.2.3. Sample Difference

There were significant defferences in OTU levels in the microbiota of crayfish treated with ultrasound, ozone water, and a combination of the two compared to the control (Figure 3A). In a Venn plot (Figure 3B), the microbial species of CK, OW, UL, and OCU were 1917, 1658, 2072, and 1684, respectively. The shared species in two comparisons, OW&OCU and UL&OCU were 37 and 108, respectively. Furthermore, the unique species in OW and UL were 159 and 309, respectively, which showed that reduced species occurred more in the ozone water than ultrasound. Consistent with the preceding results, the sterilization effect of ozone water is better than that of ultrasound. The principal co-ordinates analysis (PCoA) reflected the differences in microbial community between different cleaning methods (Figure 3C). The first principal component (PC1) and the second principal component (PC2) were 31.03% and 19.04%, respectively, which represent the dominant microbial species in crayfish samples. Cluster analysis showed that the species distribution between CK and treatments showed great isolation, which indicated that ozone water and ultrasound treatment could significantly reduce the microbial community of crayfish. The species distribution between OW and OCU did not show great isolation, which suggested that the inactivation capacity of the combined cleaning mainly owed to ozone water. There was a potential synergistic effect on bacteria reduction between ozone water and ultrasound, and further research is urgently needed to understand the detailed mechanism of the combined treatment.

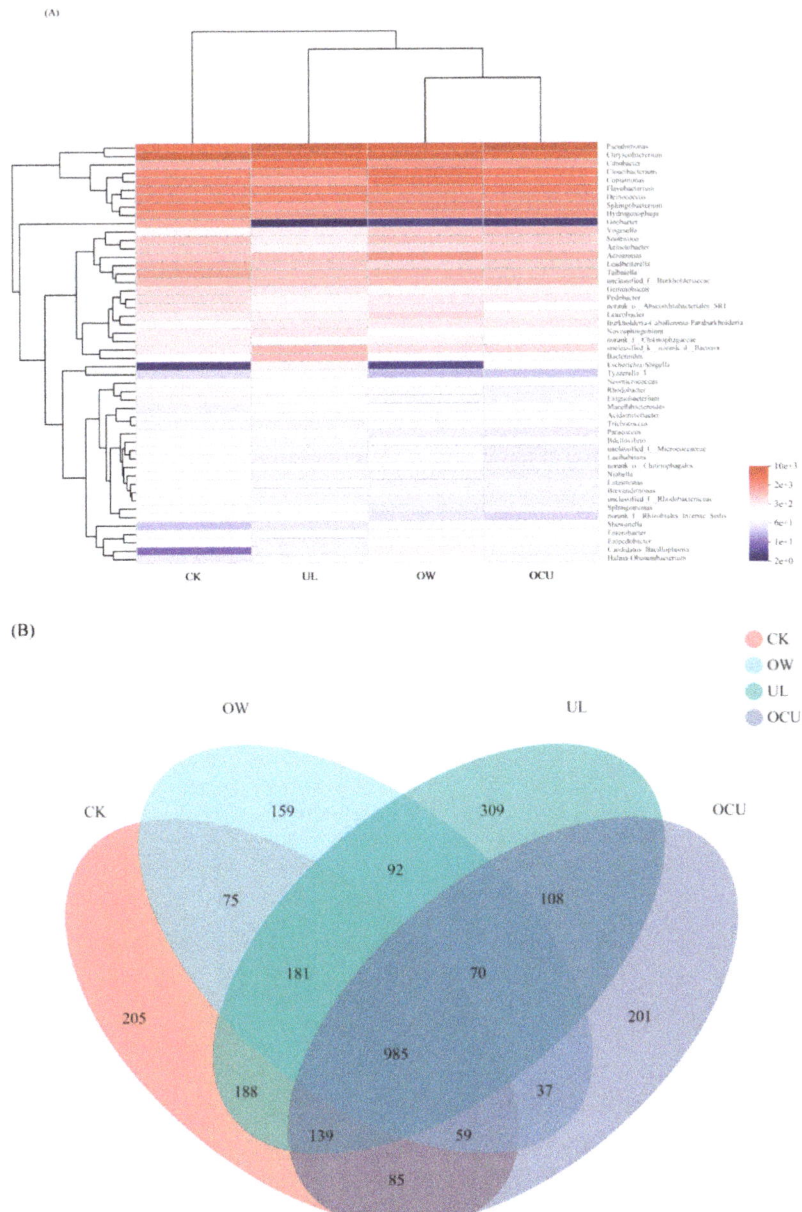

Figure 3. *Cont.*

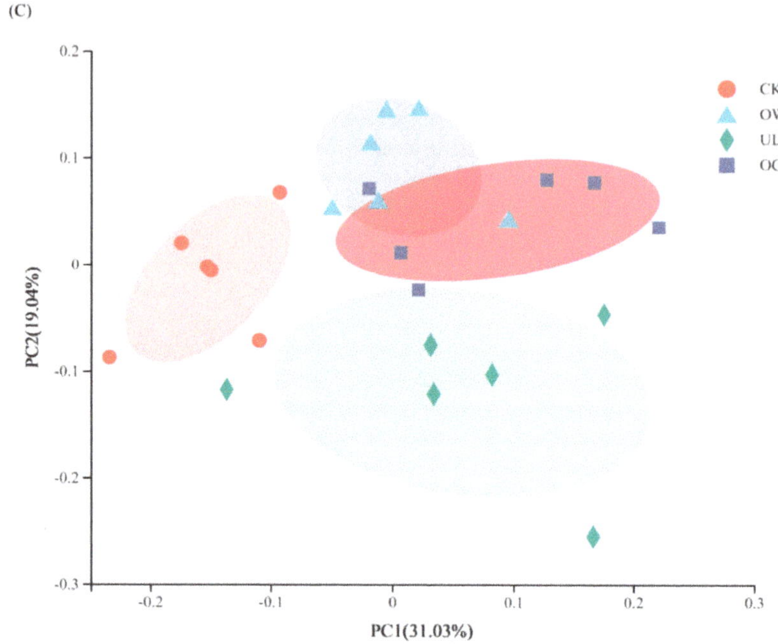

Figure 3. Panels (**A–C**) were respectively the heat-map, Venn plot, and principal co-ordinates analysis (PCoA) of crayfish (readers can choose the online version of this article to refer to the colors in the heat-map). Notes: CK, the control; OW, ozone water cleaning; UL, ultrasound cleaning; OCU, ozone water combined with ultrasound cleaning.

3.3. *Chromaticity*

The color of food is an important parameter of consumer acceptance. As shown in Figure 4A, compared to the control, the a* values (redness) of OW and OCU decreased significantly ($p < 0.05$), which may be attributed to excessive protein oxidation caused by ozone water. UL contributed the largest a* value (4.55), which indicated that the ruptured cells formed by ultrasound released more haemoglobin and myoglobin pigments. The increased a* value of tilapia fillets was attributed to the mechanical effect of ultrasound to induce cytochrome release [61]. Although OCU reduced redness, its yellowness is the lowest among all groups (Figure 4B). Generally speaking, low yellowness is a desirable result. Ozone water bleached the crayfish, but ultrasound treatment made up for this defect by increasing redness. Therefore, the lightly increased whiteness (43.96) caused by combined treatment was not significant ($p > 0.05$) compared with the control (43.30) (Figure 4C). Similarly, Esua et al. found that plasma functionalized liquid combined with ultrasound treatment of grass carp led to an increase in whiteness. In brief, crayfish cleaned by ozone water combined with ultrasound have obtained an acceptable appearance [56].

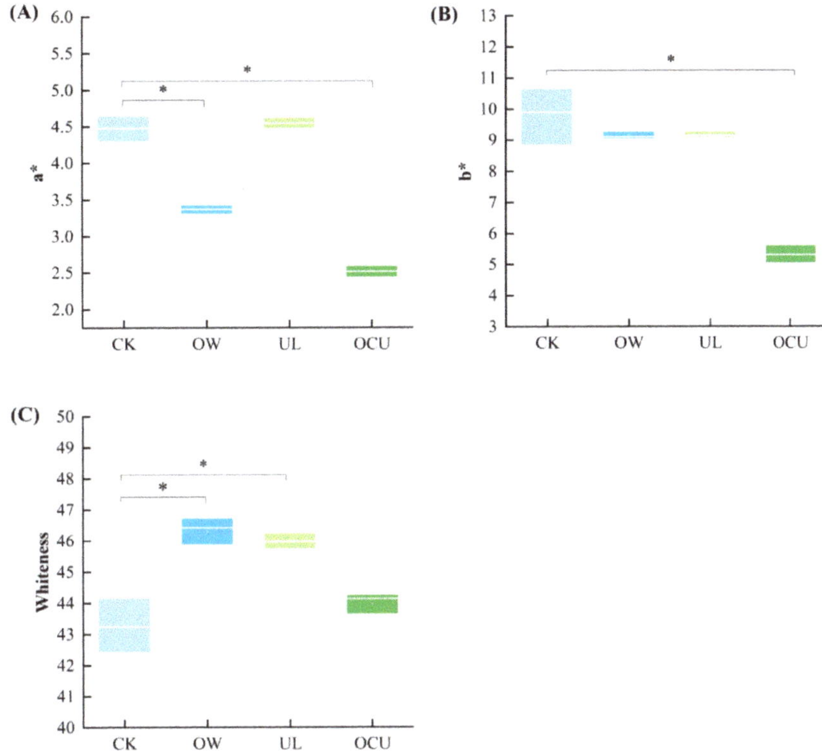

Figure 4. Panels (**A–C**) were respectively a* value, b* value, and whiteness, and expressed as boxplots. Compared with the control, the significant difference was expressed as * ($p < 0.05$). Notes: CK, the control; OW, ozone water cleaning; UL, ultrasound cleaning; OCU, ozone water combined with ultrasound cleaning.

3.4. E-Nose

Electronic nose technology is considered as a simple and fast parameter for characterizing complex odors in food samples [62]. Figure 5 shows the odor clustering of crayfish samples after different cleaning methods. The sample variances of PC1 and PC2 were 75.30% and 20.40%, respectively, which almost characterized the volatile odor components of all crayfish samples. The relatively larger isolation areas showed that the odor components of three treatments (OW, UL, and OCU) were significantly different from the control. The overlapping areas were shared between the ozone water alone, the ultrasound alone, and their combination. Wenzheng et al. reported that the content of odor compounds (for example geosmin), which contributed a lot to the fishy smell in aquatic products, was significantly reduced after ozonation [63]. As mentioned earlier, the cavitation effect of ultrasound could synergistically degrade and release contaminants attached to the surface of crayfish [8]. Ozone water combined with ultrasound cleaning changed the distribution of odor compounds in crayfish, so it is necessary to carry out research on the effect of combined treatment on fishy components.

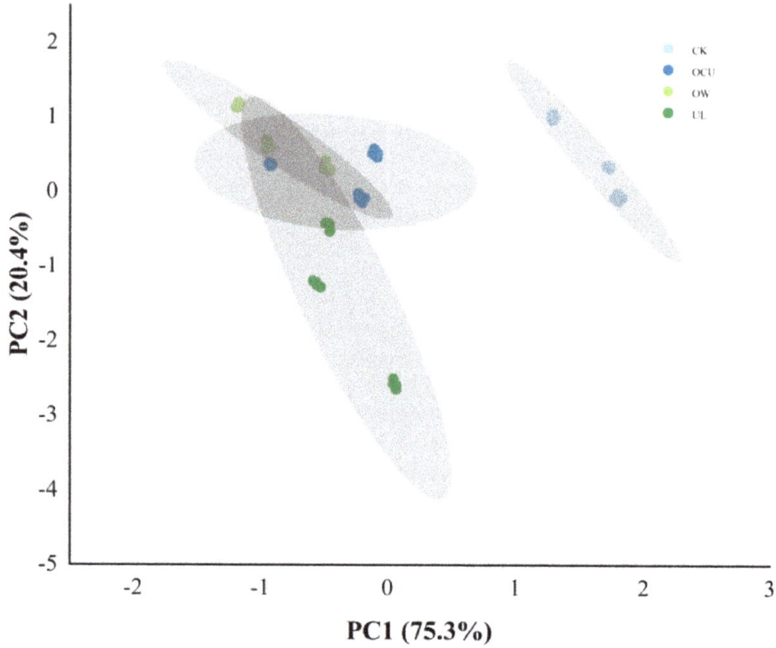

Figure 5. Crayfish odor clustering based on principal component analysis. Notes: CK, the control; OW, ozone water cleaning; UL, ultrasound cleaning; OCU, ozone water combined with ultrasound cleaning.

3.5. Sensory Evaluation

As shown in Figure 6, UL has the highest color score (8.14), which may be due to ultrasound destroying muscle cells and freeing myoglobin. The light red color (6.14) of OW was attributed to the strong oxidation of ozone water. In terms of texture score, there was no significant difference ($p > 0.05$) between treatments. Concretely, CK, OW, UL, and OCU scored 7.43, 6.57, 6.57, and 6.71, respectively. The intrinsic unpleasant odor of ozone water resulted in a lower odor score (6.71) in OW, while a higher odor score (7.57) in OCU may be attributed to the cavitation effect of ultrasound. In a word, crayfish cleaned by ozone water combined with ultrasound have obtained satisfactory sensory experiences, including color, texture, and smell.

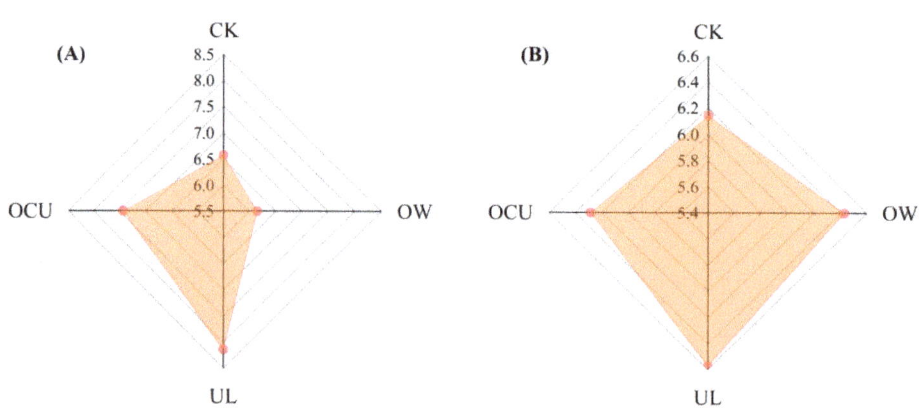

Figure 6. *Cont.*

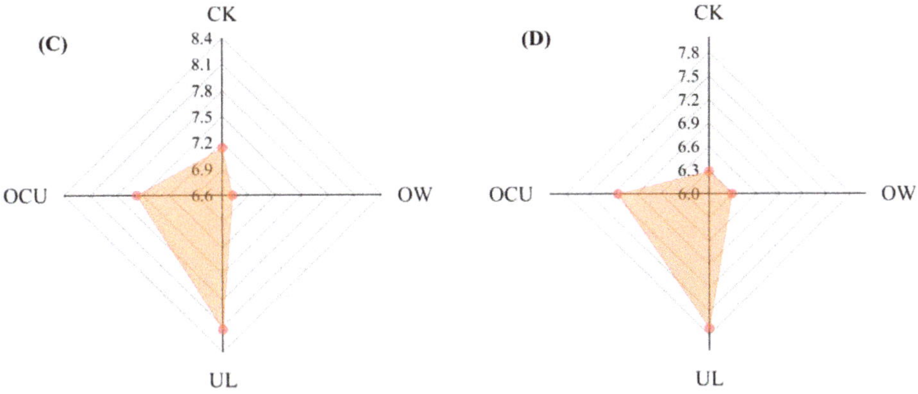

Figure 6. Panels (**A–D**) were respectively color, texture, odor, and overall evaluation of crayfish samples, and expressed by radar map. Notes: CK, the control; OW, ozone water cleaning; UL, ultrasound cleaning; OCU, ozone water combined with ultrasound cleaning.

4. Conclusions

Ozone water combined with ultrasound cleaning reduced the load of TVC, PVC, MVC, *Pseudomonas*, HSPB, molds and yeasts in crayfish. Color test, E-nose analysis, and sensory evaluation also showed that the combined treatment would not adversely affect the quality of crayfish. In brief, the combined cleaning can be used as a pretreatment method for crayfish products to reduce bacterial contamination and improve the quality of crayfish. In addition, there was a potential synergistic effect between ozone water and ultrasound in bacteria reduction. Thus, further research is urgently needed to understand the detailed mechanism of the combined treatment.

Author Contributions: Experiments, statistical analysis of data, writing of manuscripts and revised manuscripts, Y.L.; experimental design, conceptualization, visualization, Y.Q.; revised manuscripts, H.T.; revised manuscripts, L.S.; project proposal, research methodology, L.W. (Lingyun Wei).; funding acquisition, G.X.; supervision, L.W. (Lan Wang); review and editing, W.W. All authors have read and agreed to the published version of the manuscript.

Funding: The sponsors are the Ministry of Science and Technology of the People's Republic of China and the Department of Science and Technology of Hubei Province. This research was funded by the National Key R&D Program of China (2019YFD0902000) and the Major Program of Technical Innovation of Hubei Province (2019ABA087).

Institutional Review Board Statement: Not applicable.

Informed Consent Statement: Not applicable.

Data Availability Statement: The datasets generated during and/or analyzed during the current study are available from the corresponding author on reasonable request.

Acknowledgments: The authors sincerely acknowledge the support of the staff at the Key Laboratory of Cold Chain Logistics Technology for Agro-Product, Ministry of Agriculture and Rural Affairs, Institute of Agro-Products Processing and Nuclear agricultural Technology, Hubei Academy of Agricultural Sciences.

Conflicts of Interest: The authors declare no conflict of interest with the contents of this article.

Abbreviations

OW, ozone water cleaning; UL, ultrasound cleaning; OCU, ozone water combined with ultrasound cleaning; TVC, total viable count; MVC, mesophilic viable count; PVC, psychrophilic viable count; HSPB, hydrogen sulfide-producing bacteria; OTU, operational taxon units; PCoA, principal co-ordinates analysis; E-nose, electronic nose; PCA, principal components analysis.

References

1. Guo, M.; Jin, T.Z.; Yang, R.; Antenucci, R.; Mills, B.; Cassidy, J.; Scullen, O.J.; Sites, J.E.; Rajkowski, K.T.; Sommers, C.H. Inactivation of natural microflora and inoculated Listeria innocua on whole raw shrimp by ozonated water, antimicrobial coatings, and cryogenic freezing. *Food Control* **2013**, *34*, 24–30. [CrossRef]
2. Herbert, B. Emergency animal diseases bulletin: Crayfish plague. *Aust. Vet. J.* **2014**, *92*, N8–N9. [PubMed]
3. Zhao, Y.M.; de Alba, M.; Sun, D.W.; Tiwari, B. Principles and recent applications of novel non-thermal processing technologies for the fish industry—A review. *Crit. Rev. Food Sci. Nutr.* **2019**, *59*, 728–742. [CrossRef]
4. Zhang, J.; Zhang, Y.; Wang, Y.; Xing, L.; Zhang, W. Influences of ultrasonic-assisted frying on the flavor characteristics of fried meatballs. *Innov. Food Sci. Emerg. Technol.* **2020**, *62*, 102365. [CrossRef]
5. Ercan, S.Ş.; Soysal, Ç. Use of ultrasound in food preservation. *Nat. Sci.* **2013**, *05*, 5–13. [CrossRef]
6. Chandrapala, J.; Oliver, C.; Kentish, S.; Ashokkumar, M. Ultrasonics in food processing. *Ultrason. Sonochem.* **2012**, *19*, 975–983. [CrossRef] [PubMed]
7. Li, X.X.; Sun, P.; Ma, Y.; Cai, L.; Li, J.R. Effect of ultrasonic thawing on the water-holding capacity, physicochemical properties and structure of frozen tuna (Thunnus tonggol) myofibrillar proteins. *J. Sci. Food Agric.* **2019**, *99*, 5083–5091. [CrossRef]
8. Mason, T.J. Ultrasonic cleaning: An historical perspective. *Ultrason. Sonochem.* **2016**, *29*, 519–523. [CrossRef]
9. Ashokkumar, M. Applications of ultrasound in food and bioprocessing. *Ultrason. Sonochem.* **2015**, *25*, 17–23. [CrossRef]
10. Vetrimurugan, R.; Hooi, B. Study of ultrasonic parameters on removal of contamination from slider surface by using various cleaning chemistry. *Int. J. Chem. Environ. Eng.* **2012**, *3*, 392–396.
11. Chemat, F.; Zille, H.; Khan, M.K. Applications of ultrasound in food technology: Processing, preservation and extraction. *Ultrason. Sonochem.* **2011**, *18*, 813–835. [CrossRef] [PubMed]
12. Leong, T.; Ashokkumar, M.; Kentish, S. The Fundamentals of Power Ultrasound—A Review. *Acoust. Aust.* **2011**, *39*, 54–63.
13. Spiteri, D.; Chot-Plassot, C.; Sclear, J.; Karatzas, K.A.; Scerri, C.; Valdramidis, V.P. Ultrasound processing of liquid system(s) and its antimicrobial mechanism of action. *Lett. Appl. Microbiol.* **2017**, *65*, 313–318. [CrossRef]
14. Kordowska-Wiater, M.; Stasiak, D. Effect of ultrasound on survival of gram-negative bacteria on chicken skin surface. *Bull. Vet. Inst. Pulawy* **2011**, *55*, 207–210.
15. da Silva, S.; Luvielmo, M.; Geyer, M.; Pra, I. Potencialidades do uso do ozônio no processamento de alimentos Potential use of ozone in the food processing. *Semin. Ciênc. Agrárias* **2011**, *32*, 659–682. [CrossRef]
16. Zhang, L.; Luo, Y.; Wang, W.; Sun, Y.; Zhang, J.; Fatima, M.; Jia, X.; Qiu, H.J. Efficient inactivation of African swine fever virus by ozonized water. *Vet. Microbiol.* **2020**, *247*, 108796. [CrossRef]
17. Fonseca, P.M.M.; De Sa, P.L.; Miyakawa, W.; Damiao, A.J.; Melo, L.; Zangaro, R.A.; Fernandes, A.B.; De Lima, C.J. Analysis of Damage on the Streptococcus mutans Immersed in Ozonated Water: Preliminary Study for Application as Mouth Rinse. *Ozone-Sci. Eng.* **2019**, *41*, 242–249. [CrossRef]
18. Rodrigues, A.A.Z.; Queiroz, M.; Neves, A.A.; Oliveira, A.F.; Prates, L.H.F.; Freitas, J.F.; Heleno, F.F.; Faroni, L.R.D. Use of ozone and detergent for removal of pesticides and improving storage quality of tomato. *Food Res. Int.* **2019**, *125*, 108626. [CrossRef] [PubMed]
19. Wu, J.; Doan, H. Disinfection of recycled red-meat-processing wastewater by ozone. *J. Chem. Technol. Biotechnol.* **2005**, *80*, 828–833. [CrossRef]
20. Crowe, K.M.; Skonberg, D.; Bushway, A.; Baxter, S. Application of ozone sprays as a strategy to improve the microbial safety and quality of salmon fillets. *Food Control* **2012**, *25*, 464–468. [CrossRef]
21. Jurado-Alameda, E.; García-Román, M.; Altmajer-Vaz, D.; Jiménez-Pérez, J.L. Assessment of the use of ozone for cleaning fatty soils in the food industry. *J. Food Eng.* **2012**, *110*, 44–52. [CrossRef]
22. Tran, N.; Drogui, P.; Brar, S.K. Sonochemical techniques to degrade pharmaceutical organic pollutants. *Environ. Chem. Lett.* **2015**, *13*, 251–268. [CrossRef]
23. Abdurahman, M.H.; Abdullah, A.Z. Mechanism and reaction kinetic of hybrid ozonation-ultrasonication treatment for intensified degradation of emerging organic contaminants in water: A critical review. *Chem. Eng. Process.-Process Intensif.* **2020**, *154*, 108047. [CrossRef]
24. Duan, S.; Zhou, X.; Xiao, H.; Miao, J.; Zhao, L. Characterization of Bacterial Microbiota in Tilapia Fillets Under Different Storage Temperatures. *J. Food Sci.* **2019**, *84*, 1487–1493. [CrossRef]
25. Bott, T.R.; Tianqing, L. Ultrasound enhancement of biocide efficiency. *Ultrason. Sonochem.* **2004**, *11*, 323–326. [CrossRef]

26. Al-Hashimi, A.M.; Mason, T.J.; Joyce, E.M. Combined Effect of Ultrasound and Ozone on Bacteria in Water. *Environ. Sci. Technol.* **2015**, *49*, 11697–11702. [CrossRef]
27. Aday, M.S.; Caner, C. Individual and combined effects of ultrasound, ozone and chlorine dioxide on strawberry storage life. *LWT-Food Sci. Technol.* **2014**, *57*, 344–351. [CrossRef]
28. Yang, S.S.; Guo, W.Q.; Chen, Y.D.; Wu, Q.L.; Luo, H.C.; Peng, S.M.; Zheng, H.S.; Feng, X.C.; Zhou, X.; Ren, N.Q. Economical evaluation of sludge reduction and characterization of effluent organic matter in an alternating aeration activated sludge system combining ozone/ultrasound pretreatment. *Bioresour. Technol.* **2015**, *177*, 194–203. [CrossRef]
29. Zhao, Q.; Li, M.; Zhang, K.F.; Wang, N.; Wang, K.K.; Wang, H.B.; Meng, S.J.; Mu, R.M. Effect of ultrasound irradiation combined with ozone pretreatment on the anaerobic digestion for the biosludge exposed to trace-level levofloxacin: Degradation, microbial community and ARGs analysis. *J. Environ. Manag.* **2020**, *262*, 9. [CrossRef]
30. Li, P.; Peng, Y.; Mei, J.; Xie, J. Effects of microencapsulated eugenol emulsions on microbiological, chemical and organoleptic qualities of farmed Japanese sea bass (*Lateolabrax japonicus*) during cold storage. *LWT* **2020**, *118*, 108831. [CrossRef]
31. Chen, X.J.; Tang, R.; Wang, Y.L.; Yuan, S.J.; Wang, W.; Ali, I.M.; Hu, Z.H. Effect of ultrasonic and ozone pretreatment on the fate of enteric indicator bacteria and antibiotic resistance genes, and anaerobic digestion of dairy wastewater. *Bioresour. Technol.* **2021**, *320*, 9. [CrossRef] [PubMed]
32. de Mendonça Silva, A.M.; Gonçalves, A.A. Effect of aqueous ozone on microbial and physicochemical quality of Nile tilapia processing. *J. Food Process. Preserv.* **2017**, *41*, e13298. [CrossRef]
33. Okpala, C.O.R. Investigation of quality attributes of ice-stored Pacific white shrimp (*Litopenaeus vannamei*) as affected by sequential minimal ozone treatment. *LWT-Food Sci. Technol.* **2014**, *57*, 538–547. [CrossRef]
34. Wang, L.; Shi, L.; Jiao, C.; Qiao, Y.; Wu, W.; Li, X.; Wang, J.; Ding, A.; Liao, L.; Xiong, G. Effect of Ultrasound Combined with Ozone Water Pretreatment on the Bacterial Communities and the Physicochemical Properties of Red Swamp Crayfish Meat (*Procambarus clarkii*). *Food Bioprocess Technol.* **2020**, *13*, 1778–1790. [CrossRef]
35. Gonçalves, A.A.; Lira Santos, T.C. Improving quality and shelf-life of whole chilled Pacific white shrimp (*Litopenaeus vannamei*) by ozone technology combined with modified atmosphere packaging. *LWT* **2019**, *99*, 568–575. [CrossRef]
36. Cravero, F.; Englezos, V.; Rantsiou, K.; Torchio, F.; Giacosa, S.; Segade, S.R.; Gerbi, V.; Rolle, L.; Cocolin, L. Ozone treatments of post harvested wine grapes: Impact on fermentative yeasts and wine chemical properties. *Food Res. Int.* **2016**, *87*, 134–141. [CrossRef]
37. Guzzon, R.; Carafa, I.; Tuohy, K.; Cervantes, G.; Vernetti, L.; Barmaz, A.; Larcher, R.; Franciosi, E. Exploring the microbiota of the red-brown defect in smear-ripened cheese by 454-pyrosequencing and its prevention using different cleaning systems. *Food Microbiol.* **2017**, *62*, 160–168. [CrossRef]
38. Botta, C.; Ferrocino, I.; Pessione, A.; Cocolin, L.; Rantsiou, K. Spatiotemporal Distribution of the Environmental Microbiota in Food Processing Plants as Impacted by Cleaning and Sanitizing Procedures: The Case of Slaughterhouses and Gaseous Ozone. *Appl. Environ. Microbiol.* **2020**, *86*, e01861-20. [CrossRef]
39. Moore, G.; Griffith, C.; Peters, A. Bactericidal properties of ozone and its potential application as a terminal disinfectant. *J. Food Prot.* **2000**, *63*, 1100–1106. [CrossRef]
40. Gertzou, I.N.; Drosos, P.E.; Karabagias, I.K.; Riganakos, K.A. Combined effect of ozonation and packaging on shelf life extension of fresh chicken legs during storage under refrigeration. *J. Food Sci. Technol.-Mysore* **2016**, *53*, 4270–4277. [CrossRef]
41. Wu, Z.; Zhang, Q.; Zhang, T.; Chen, J.; Wang, S.; Hao, J.; Lin, Y.; Li, A. Association of the microbiota dysbiosis in the hepatopancreas of farmed crayfish (Procambarus clarkii) with disease outbreaks. *Aquaculture* **2021**, *536*, 736492. [CrossRef]
42. Zeng, S.; Khoruamkid, S.; Kongpakdee, W.; Wei, D.; Yu, L.; Wang, H.; Deng, Z.; Weng, S.; Huang, Z.; He, J.; et al. Dissimilarity of microbial diversity of pond water, shrimp intestine and sediment in Aquamimicry system. *AMB Express* **2020**, *10*, 180. [CrossRef] [PubMed]
43. Hou, D.; Huang, Z.; Zeng, S.; Liu, J.; Wei, D.; Deng, X.; Weng, S.; Yan, Q.; He, J. Intestinal bacterial signatures of white feces syndrome in shrimp. *Appl. Microbiol. Biotechnol.* **2018**, *102*, 3701–3709. [CrossRef] [PubMed]
44. Xavier, R.; Soares, M.C.; Silva, S.M.; Banha, F.; Gama, M.; Ribeiro, L.; Anastácio, P.; Cardoso, S.C. Environment and host-related factors modulate gut and carapace bacterial diversity of the invasive red swamp crayfish (*Procambarus clarkii*). *Hydrobiologia* **2021**, *848*, 4045–4057. [CrossRef]
45. Orlić, K.; Šver, L.; Burić, L.; Kazazić, S.; Grbin, D.; Maguire, I.; Pavić, D.; Hraščan, R.; Vladušić, T.; Hudina, S.; et al. Cuticle-associated bacteria can inhibit crayfish pathogen Aphanomyces astaci: Opening the perspective of biocontrol in astaciculture. *Aquaculture* **2021**, *533*, 736112. [CrossRef]
46. Bassey, A.P.; Chen, Y.; Zhu, Z.; Odeyemi, O.A.; Frimpong, E.B.; Ye, K.; Li, C.; Zhou, G. Assessment of quality characteristics and bacterial community of modified atmosphere packaged chilled pork loins using 16S rRNA amplicon sequencing analysis. *Food Res. Int.* **2021**, *145*, 110412. [CrossRef]
47. Jiang, L.; Li, M.; Tang, J.; Zhao, X.; Zhang, J.; Zhu, H.; Yu, X.; Li, Y.; Feng, T.; Zhang, X. Effect of Different Disinfectants on Bacterial Aerosol Diversity in Poultry Houses. *Front. Microbiol.* **2018**, *9*, 2113. [CrossRef]
48. Hu, Y.; Shen, Y.; Wang, J. Pretreatment of antibiotic fermentation residues by combined ultrasound and alkali for enhancing biohydrogen production. *J. Clean Prod.* **2020**, *268*, 122190. [CrossRef]
49. Yang, G.; Wang, J. Biohydrogen production from waste activated sludge pretreated by combining sodium citrate with ultrasonic: Energy conversion and microbial community. *Energy Convers. Manag.* **2020**, *225*, 113436. [CrossRef]

50. Lin, Y.; Li, D.; Zeng, S.; He, M. Changes of microbial composition during wastewater reclamation and distribution systems revealed by high-throughput sequencing analyses. *Front. Environ. Sci. Eng.* **2016**, *10*, 539–547. [CrossRef]
51. Tao, C.; Zhang, Q.; Zeng, W.; Liu, G.; Shao, H. The effect of antibiotic cocktails on host immune status is dynamic and does not always correspond to changes in gut microbiota. *Appl. Microbiol. Biotechnol.* **2020**, *104*, 4995–5009. [CrossRef] [PubMed]
52. Wang, J.; Sun, Y.; Tao, D.; Wang, S.; Li, C.; Zheng, F.; Wu, Z. Reduction of Escherichia coli O157:H7, Listeria monocytogenes, and Naturally Present Microbe Counts on Lettuce using an Acid Mixture of Acetic and Lactic Acid. *Microorganisms* **2019**, *7*, 373. [CrossRef] [PubMed]
53. Fonseca, P.M.M.; Palacios, D.A.B.; de Sa, P.L.; Miyakawa, W.; Damiao, A.J.; Fernandes, A.B.; de Lima, C.J. Preliminary Study: Comparative Analysis of the Effects of Ozone and Ultrasound onStreptococcus Mutans. *Ozone Sci. Eng.* **2021**, *43*, 263–275. [CrossRef]
54. Pandiselvam, R.; Sunoj, S.; Manikantan, M.R.; Kothakota, A.; Hebbar, K.B. Application and Kinetics of Ozone in Food Preservation. *Ozone Sci. Eng.* **2016**, *39*, 115–126. [CrossRef]
55. Walker, N.; Burke, F.J.T.; Palenik, C.J. Comparison of ultrasonic cleaning schemes: A pilot study. *Prim. Dent. Care J. Fac. Gen. Dent. Pract.* **2006**, *13*, 51–56. [CrossRef] [PubMed]
56. Esua, O.J.; Cheng, J.H.; Sun, D.W. Novel technique for treating grass carp (*Ctenopharyngodon idella*) by combining plasma functionalized liquids and Ultrasound: Effects on bacterial inactivation and quality attributes. *Ultrason. Sonochem.* **2021**, *76*, 14. [CrossRef]
57. Taiye Mustapha, A.; Zhou, C.; Wahia, H.; Amanor-Atiemoh, R.; Otu, P.; Qudus, A.; Abiola Fakayode, O.; Ma, H. Sonozonation: Enhancing the antimicrobial efficiency of aqueous ozone washing techniques on cherry tomato. *Ultrason. Sonochem.* **2020**, *64*, 105059. [CrossRef]
58. Pandiselvam, R.; Kaavya, R.; Jayanath, Y.; Veenuttranon, K.; Lueprasitsakul, P.; Divya, V.; Kothakota, A.; Ramesh, S.V. Ozone as a novel emerging technology for the dissipation of pesticide residues in foods–a review. *Trends Food Sci. Technol.* **2020**, *97*, 38–54. [CrossRef]
59. Tachikawa, M.; Yamanaka, K.; Nakamuro, K. Studies on the Disinfection and Removal of Biofilms by Ozone Water Using an Artificial Microbial Biofilm System. *Ozone Sci. Eng.* **2009**, *31*, 3–9. [CrossRef]
60. Topić Popović, N.; Sauerborn Klobučar, R.; Maguire, I.; Strunjak-Perović, I.; Kazazić, S.; Barišić, J.; Jadan, M.; Klobučar, G.; Čož-Rakovac, R. High-throughput discrimination of bacteria isolated fromAstacus astacusandA. leptodactylus. *Knowl. Manag. Aquat. Ecosyst.* **2014**, *413*, 4–16. [CrossRef]
61. Zhang, C. Effect of low power ultrasound on enhancing water-holding capacity of tilapia fillets. *Food Sci. Technol.* **2019**, *44*, 161–166. [CrossRef]
62. Zhao, D.; Hu, J.; Chen, W. Analysis of the relationship between microorganisms and flavour development in dry-cured grass carp by high-throughput sequencing, volatile flavour analysis and metabolomics. *Food Chem.* **2022**, *368*, 130889. [CrossRef] [PubMed]
63. Wenzheng, S.; Ying, L. Recent advances on deodorization technology of fishy odors. *Food Ferment. Ind.* **2021**, *47*, 282–287. [CrossRef]

foods

MDPI

Article

Effects of High-Voltage Atmospheric Cold Plasma Treatment on Microbiological and Quality Characters of Tilapia Fillets

Jiamei Wang [1,*], Tengfei Fu [1,2], Yuanyuan Wang [1] and Jianhao Zhang [3]

[1] College of Food Science and Engineering, Hainan University, Haikou 570228, China
[2] Agricultural Products Processing Research Institute, Chinese Academy of Tropical Agricultural Sciences, Zhanjiang 524000, China
[3] College of Food Science and Technology, Nanjing Agricultural University, Nanjing 210095, China
* Correspondence: 992918@hainanu.edu.cn

Abstract: Cold plasma (CP) has become an alternative to conventional thermal processing of food products. In this study, the effect of cold plasma treatment time on the inactivation and quality of tilapia fillets was investigated. The surfaces of tilapia fillets were inoculated with *Salmonella enteritis (S. enteritis)*, *Listeria monocytogenes* (*L. monocytogenes*), and a mixture of both before being treated with cold plasma at 70 kV for 0, 60, 120, 180, 240, and 300 s. With the extension of treatment time, the number of colonies on the surface of the fillets decreased gradually; after 300 s of cold plasma treatment, *S. enteritis* and *L. monocytogenes* populations were reduced by 2.34 log CFU/g and 1.69 log CFU/g, respectively, and the a* value and immobile water content decreased significantly ($p < 0.05$), while the free water content increased significantly ($p < 0.05$). TBARS value increased significantly ($p < 0.05$) to 1.83 mg MDA/kg for 300 s treatment. The carbonyl value and sulfhydryl value of sarcoplasmic protein significantly ($p < 0.05$) increased and decreased, respectively, as treatment time extension, while no significant changes were found in myofibrillar protein. No significant differences were observed in pH, b* value, elasticity, chewiness, thiol value, and TVB-N value. The results showed that cold plasma had an inactivation effect on tilapia fillets and could preserve their original safety indicators. It was concluded that CP treatment could be used as an effective non-thermal method to maintain the quality of tilapia fillets and extend their shelf-life.

Keywords: cold plasma; tilapia fillet; microbiological; quality

Citation: Wang, J.; Fu, T.; Wang, Y.; Zhang, J. Effects of High-Voltage Atmospheric Cold Plasma Treatment on Microbiological and Quality Characters of Tilapia Fillets. *Foods* **2022**, *11*, 2398. https://doi.org/10.3390/foods11162398

Academic Editors: Tao Yin and Liu Shi

Received: 7 July 2022
Accepted: 6 August 2022
Published: 10 August 2022

1. Introduction

Tilapia is popular with its high protein and low fat. China is the largest producer of farmed tilapia in the world, harvesting roughly 1.64 million tons of tilapia in 2019. At present, the main method of storing tilapia fillets is frozen storage (at 18 °C), which is highly energy-consuming and risks water loss during storage [1]. Cold storage is also commonly used for preserving aquatic products where the microbes are relatively active and have a negative impact on the quality of fish fillets [2]. While traditional heat processing methods (cooking, frying, drying, etc.) could sterilize the fish fillet successfully, they caused significant changes in the physical characters and chemical compositions of fish, including the denaturation of protein, decrease in water content, and changes in flavor, thus not suitable for the preservation of aquatic products. Therefore, developing a non-thermal inactivation process is particularly necessary for fish fillets preservation. Over the past decades, several non-thermal technologies have been studied, such as high hydrostatic pressure technology [3], pulsed electric field processing [4], and irradiation [5], which showed high inactivation efficacy with a minor negative influence on the natural quality of fish. However, some of them failed to be widely used in commercial applications due to the high price of equipment or the strict operating conditions [6]. Cold plasma, as an innovative non-thermal treatment, has advantages in terms of low temperature inactivation and operational ease [7]. Although products with high-fat content are easily oxidized by the active substances produced during cold plasma treatment [8], cold plasma treatment has a relatively broader prospect in researching tilapia fillets with low-fat content.

The research on cold plasma treatment of food has been widely reported [9]. Various reactive species such as reactive nitrogen species (RNS), reactive oxygen species (ROS), energetic ions, ultraviolet (UV) radiation, and charged particles generated during cold plasma have been reported as antibacterial agents [10]. It has been proved that with active substances that can destroy the cell membrane, DNA, and proteins of the bacteria, cold plasma can achieve high efficiency of inactivation [11,12]. In the study of herring (*Clupea harengus*) treated with a DBD cold plasma system, compared to the control samples, there were significant decreases in total aerobic mesophilic bacteria, total aerobic psychrotrophic bacteria, lactic acid bacteria, *Pseudomonas,* and *Enterobacteriaceae* [13]. When mackerel was treated with DBD cold plasma at 70 kV and 80 kV for 1, 3 and 5 min, respectively, the spoilage bacteria (total aerobic psychrotrophic bacteria, *Pseudomonas,* and lactic acid bacteria) on the surface of mackerel were significantly reduced [14]. The total aerobic psychrotrophic bacteria in the Asian sea bass decreased significantly after cold plasma treatment, which effectively extended the shelf life of the fish [15]. In addition, in the study of semi-dried Pacific saury treated by corona discharge plasma jet, cold plasma not only had a significant inactivation effect on total aerobic psychrotrophic bacteria, *Staphylococcus aureus,* and other bacteria, but also caused the number of mold and yeast to decrease significantly [16]. In terms of fish quality, lipids oxidation [14] and protein oxidation [15] of fish have been found to be the results of cold plasma treatment. Interestingly, it was reported that the cold plasma treatment had no significant effect on lipids oxidation of Atlantic mackerel, but it accelerated the oxidation of protein [17]. Additionally, the influence of cold plasma on *Scomber Japonicus* quality was demonstrated from the determination of TVB-N value, TBARS value, and PV value, respectively [18]. Similarly, in the study of Pacific White Shrimp (Litopenaeus Vannamei), the content of chemical indexes, including pH, TVB-N, TBARS, and PV, reflected the influence of cold plasma on the quality of Shrimp [19].

A search of the literature revealed few studies had been conducted on cold plasma treatment of tilapia at the present stage. Therefore, in this study, the material was the peeled tilapia fillet which was treated with cold plasma excited by air. The inactivation effect of cold plasma on the fish fillet inoculated with *Salmonella enteritis (S. enteritis)* and *Listeria monocytogenes (L. monocytogenes)* was evaluated. The quality indicators of fish fillet treated by cold plasma were evaluated by measuring the color, pH, TBARS level, carbonyl level, sulfhydryl level, total volatile basic nitrogen, texture, and low field NMR analysis.

2. Materials and Methods

The study was conducted from October 2020 to August 2021.

2.1. Sample Preparation

The fresh tilapia fillets were purchased from Hainan Quanyi Foods Co., Ltd. (Haikou, China). The fresh tilapias were put on ice and transported to the laboratory as soon as possible after being peeled, boned, and cleaned at the factory. The tilapia fillets were irradiated by an ultraviolet lamp before inoculation.

S. enteritis (21635) and *L. monocytogenes* (24119) were purchased from CICC (China Center of Industrial Culture Collection, China). *S. enteritis* and *L. monocytogenes* were cultivated in 100 mL nutrient broth and brain heart infusion broth at 37 °C for 10 h, respectively. Bacterial cells were harvested by repeating the centrifugation (6000 r/min for 15 min) at 4 °C and rinsing twice with sterile phosphate-buffered saline (PBS, pH 7.4). The cells were resuspended in sterile PBS to achieve a concentration of about 7 log CFU/mL.

For single bacterium inoculation, the fillets were soaked in *S. enteritis* and *L. monocytogenes* suspension for 1 min, respectively. After inoculation, the tilapia fillets were dried in a sterile and ventilated environment (30 min) and then placed in PP food packaging boxes sealed by a modified atmosphere packaging machine (MAP-H360, Senrui Fresh-keeping Equipment Ltd., Suzhou, China).

For mixture bacteria inoculation, the irradiated fillets were soaked in a mixed suspension for 1 min (*S. enteritis*: *L. monocytogenes* ratio of 1:1 (*v/v*)). After inoculation, the tilapia fillets were packaged using the same method as mentioned above.

2.2. Cold Plasma Treatment

In this study, A dielectric barrier discharge (DBD) cold plasma system was used for treatment as described by Wang, Zhuang, Lawrence, and Zhang [20]. The packaged tilapia fillets were treated by a DBD system at 70 kV voltage for 60 s, 120 s, 180 s, 240 s, and 300 s. Control groups without treatment were carried out at the same time.

2.3. Microbiological Analysis

The samples (5 g) were put into a sterile homogenizer containing 45 mL of 0.85% sterile saline and then flapped for 2 min using a slap homogenizer to make 1:10 sample diluents. After homogeneous mixing, the suspension was diluted serially with 0.85% sterile saline, and then appropriate dilutions of 0.1 mL were inoculated on solid media plates. The PCA (plate count agar) (Hopebio, Qingdao, China) was used for single bacterial inoculation samples. GNBSMs (Gram-Negative Bacteria Selective Medium) (Hopebio, Qingdao, China) and MMAs (Modified Mcbride Agar Base) (Hopebio, Qingdao, China) were used for mixed bacterial inoculation samples. GNBSMs were used as a selective media for *S. enteritidis*. Based on the introductions of media, 1 mL sterile penicillin solution (including penicillin G96 unit) (Hopebio, Qingdao, China) was added to a 200 mL sterile medium (45–50 °C) before the agar culture cooled. MMAs were used as the selective culture of *L. monocytogenes*; 3 mg of Fudaxin (Hopebio, Qingdao, China) was added to 100 mL sterile medium (50–55 °C) before the agar culture cooled. The agar plates were incubated at 37 °C for 48 h, and the colonies were counted and reported as log CFU/g. The plates without visible growth were incubated for 72 h to confirm the absence of colonies.

2.4. pH Analysis

The samples (5 g) were homogenized using 45 mL saturated KCl solution for 1 min, and then, the pH value of the resulting homogenate was determined using a pH meter (REX, pHS-3C, Shanghai, China).

2.5. Color Analysis

The color of fillets was determined using a color reader (CR-10, Konica Minolta, Japan). The manufacturer's standard white plate was used for calibration. The lightness (L*), red/green (a*), and yellow/blue (b*) were measured.

The color change (ΔE) was calculated as:

$$\Delta E = \sqrt{(L_t - L_0)^2 + (a_t - a_0)^2 + (b_t - b_0)^2}$$

The chromaticity was calculated as:

$$\Delta C = \sqrt{(a_t - a_0)^2 + (b_t - b_0)^2}$$

where subscript 0 denotes the initial color value of the fillet, and subscript t denotes the color value of the fillet after cold plasma treatment.

2.6. Low Field NMR (Nuclear Magnetic Resonance) Analysis

The samples (15 g) were placed in a 40 mm MRI tube and tested at 25 °C. The Low Field Magnetic Resonance Imaging Analyzer (NM120-040H-I, NIUMAG, Suzhou, China) with a resonance frequency of 21.3 MHz was used to generate NMR data. The Carr–Purcell–Meiboom–Gill (CPMG) pulse sequence was used for lateral measurement (T_2). The instrument was based on the SIRT algorithm with its own software version 4.0 (NIUMAG INSTRUMENT), and the T_2 spectrum was obtained using inversion when the iteration time was 100,000.

2.7. Texture Profile Analysis (TPA)

The TPA of fish fillets was determined using a texture analyzer (CT3, AMETEK. Inc., Brookfield, WI, USA) with a cylindrical probe (6 mm in diameter) at a test speed of 4 mm/s and a test distance of 2 mm. Each slice of fish was measured at three points, and the results were averaged.

2.8. Thiobarbituric Acid Reactive Substances (TBARS) Analysis

Lipids oxidation was measured by the method described by Huang, Wang, Zhuang, Yan, and Zhang [21] with little modification. Five grams of fish fillet were homogenized with 15 mL of 7.5% trichloroacetic acid (TCA) solution at 10,000 rpm for 30 s. The mixture was filtered, and 10 mL filtrate was mixed with an equivalent 20 mM TCA solution in a tube. Then, the tube was kept in a boiling bath for 30 min. The absorbance of the solution was read in a spectrophotometer (UV-2450, Shimadzu Co., Kyoto, Japan) after cooling. The concentration of TBARS was expressed as mg MDA (malondialdehyde) per kg of fish fillets.

2.9. Protein Carbonyls and Sulfhydryls Analysis

Sarcoplasmic protein and myofibrillar protein of tilapia fillets were extracted according to the method proposed by Toldra, Rico, and Flores [22]. Total carbonyls were measured by the method described by Mesquita et al. [23]. The 0.4 mL of protein solution was mixed with a 0.4 mL 10 mmol/L 2,4-dini-rophenylhydrazine (DNPH) solution (containing 0.5 mol/L H_3PO_4). The mixed solution was reacted in the dark at 25 °C for 10 min. Then, 0.2 mL 6 mol/L NaOH solution was added to the mixture and kept at 25 °C for 10 min. The absorbance of the mixture solution was measured at 450 nm in a spectrophotometer (JC-721, Qdjuchuang, Qingdao, China). Protein concentration was determined with a protein kit (Nanjing Jiancheng Biological Engineering Institute, Nanjing, China). Carbonyl concentration was expressed as nmol per mg protein.

Protein sulfhydryls were measured using Ellman's reagent in accordance with the method described by Srinivasan and Hultin [24]. The protein concentration was adjusted to 2 mg/mL with PBS buffer, 0.5 mL of the diluted protein solution mixed with 2 mL urea-sodium dodecyl sulfate (SDS) solution (containing 8.0 mol/L urea, 30 g/L SDS, 0.1 mol/L sodium phosphate buffer, pH 7.4) and 0.5 mL 10 mmol/L 2-nitrobenzoic acid (DTNB) reagent (dissolved in 0.1 mol/L sodium phosphate buffer, pH 7.4), The mixture was kept at room temperature for 15 min, then the absorbance value of supernatant was measured at 412 nm in a spectrophotometer (JC-721, Qdjuchuang, Qingdao, China). The results were expressed as mol per g protein.

2.10. Total Volatile Basic Nitrogen (TVB-N) Analysis

TVB-N value was determined using the method proposed by Cao et al. [25]. The fish sample was left to spread in 10 times of water for 30 min. The content of TVB-N was determined using a Kjeldahl apparatus (KDy-9820, Beijing, China) and titrated by 0.01 M HCl.

2.11. Statistical Analysis

The SPSS software (SPSS 20.0 for Windows, SPSS Inc., Chicago, IL, USA) was employed for data analysis. Data obtained from the experiment run in triplicates were subjected to one-way analysis of variance (ANOVA) and Duncan's multiple range test, and p-values less than 0.05 were considered statistically significant.

3. Results and Discussion

3.1. Bacterial Analysis

The effects of cold plasma on microorganisms are affected by the surface characteristics of the product. Potential irregularities on the surface of fish, including cracks, grooves, and pits, may have a blocking effect on the active material derived from cold plasma. The effect of cold plasma treatment time on the inactivation of tilapia fillet was investigated using 70 kV as the treatment voltage. When *S. enteritis* and *L. monocytogenes* were separately inoculated on fillets, after 300 s of cold plasma treatment, *S. enteritis* and *L. monocytogenes* populations were reduced by 2.34 log CFU/g and 1.69 log CFU/g, respectively (Figure 1a). The population of surviving bacteria of both strains decreased significantly ($p < 0.05$) with the extension of treatment time. Regardless of whether the fillet was inoculated with Gram-negative bacteria (*S. enteritis*) or Gram-positive bacteria (*L. monocytogenes*), the decrease in the population of bacteria was time-dependent after cold plasma treatment. In a mixed inoculation test, under the same cold plasma treatment conditions, the populations of *S. enteritis* and *L. monocytogenes* in tilapia fillet decreased by 1.84 log CFU/g and 1.36 log CFU/g, respectively (Figure 1b). The antibacterial efficiency of cold plasma against the bacteria in the mixture was weaker in comparison with the single bacterial inoculation test, probably because the mixed bacteria had a synergistic resistance reaction to cold plasma. In the treatment of beef jerkies inoculated with *L. monocytogenes* and *Staphylococcus aureus*, cold plasma had an excellent inactivation effect [26].

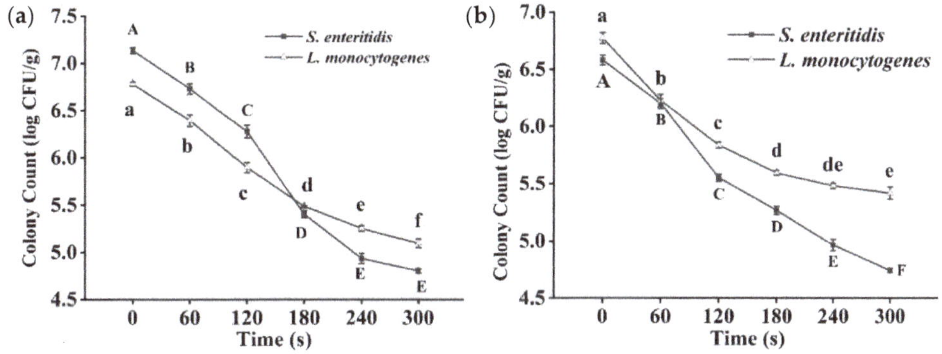

Figure 1. (**a**) Changes in bacteria numbers (individual inoculation) of tilapia fillets after cold plasma treatment. (**b**) Changes in bacteria numbers (mixed inoculation) of tilapia fillets after cold plasma treatment. Different letters indicate significant changes with changes in treatment time.

It is noteworthy that cold plasma has a better inactivation effect on G^- than G^+ bacteria inoculated separately or in combination, and many studies have reported similar results [12]. This may be due to the inactivation mechanism and the differences between the structures of G^- and G^+ bacteria; G^- bacteria have thinner cell walls, which are similar to cell membranes, with a double-membrane composed of unique lipopolysaccharides, phospholipids, and proteins. G^+ bacteria cell walls contain a peptidoglycan layer, which is much thinner than that in G^- bacteria [27]. It was found that due to the difference in cell walls, the leakage content of intracellular substances from G^- bacteria was more serious than that of G^+ bacteria after being treated with cold plasma [28]. On the other hand, charged particles, reactive oxygen species, reactive nitrogen radicals, and other active substances that are produced by cold plasma using air as the reaction medium are determining factors of inactivation efficiency [29]. They react directly with the outer membrane of G^- bacteria, while G^+ bacteria need to enter the cell before they can react with intracellular substances [12].

3.2. pH Analysis

The pH value is a reliable indicator of fish quality. The influence of cold plasma treatment on the pH value of tilapia fillet is shown in Figure 2. No significant difference in pH values was observed between tilapia fillet samples until they were treated for 300 s. Compared with the control group, the pH value of fish fillet treated for 300 s decreased from 6.97 to 6.85 ($p < 0.05$), similar results were reported in the studies of herring [13] and pork loin [30]. It was found that the decrease in pH value in cold plasma treatment was mainly due to the formation of acidic substances, such as NO_X, which mainly appeared in oxygen-containing multi-reaction systems [31].

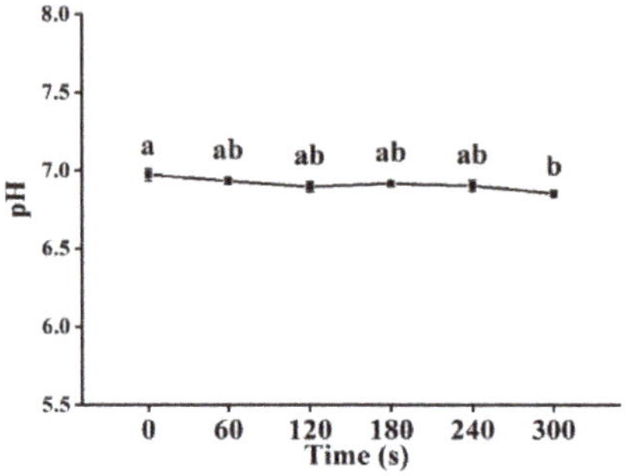

Figure 2. Changes in pH value of tilapia fillets after cold plasma treatment. Different letters indicate significant changes with changes in treatment time.

3.3. Color Analysis

Color is a prominent feature of fish, which directly affects consumers' acceptance. The influence of cold plasma treatment time on tilapia fillet color is shown in Table 1. The ΔE increased gradually after cold plasma treatment, which proves that the color change increased with the increase in treatment time. The chromatic value of tilapia fillet was related to a* value and b* value, increasing significantly after treatment. With the increase in cold plasma treatment time, the L* value of tilapia fillet increased, and a* value decreased (Table 1). The control group had the lowest L* value and the highest a* value, while the values of samples treated with cold plasma for 300 s showed the opposite trend. No significant difference in b* value was observed between the treatment group and the control group. It is possible that lipids oxidation resulted in changes in color parameters.

Table 1. Effect of cold plasma treatment on color of tilapia fillets.

Time (s)	ΔL	Δa	Δb	ΔE	ΔC
60	2.02 ± 0.43^{ab}	-1.03 ± 0.15^{b}	0.07 ± 0.18^{ab}	2.30 ± 0.41^{a}	1.06 ± 0.16^{a}
120	2.13 ± 0.57^{b}	-1.93 ± 0.09^{bc}	0.90 ± 0.14^{ab}	3.08 ± 0.37^{a}	2.14 ± 0.14^{ab}
180	3.85 ± 0.09^{c}	-3.07 ± 0.53^{c}	0.88 ± 0.60^{ab}	5.08 ± 0.47^{b}	3.25 ± 0.67^{bc}
240	5.07 ± 0.20^{d}	-3.86 ± 0.56^{d}	0.34 ± 0.18^{b}	6.38 ± 0.17^{c}	3.87 ± 0.56^{c}
300	5.37 ± 0.27^{d}	-4.04 ± 0.32^{d}	0.11 ± 0.22^{ab}	6.72 ± 0.41^{c}	4.04 ± 0.32^{c}

Different letters in same column indicate significant changes with changes in treatment time.

L* value of cold plasma treated chicken breast increased with the extension of treatment time [32]. It has been reported that there was no significant change in b* value after plasma treatment, regardless of the type of meat, including fish [14], chicken [33], and pork [34], which was similar to the results of this experiment. The a* value, which has the greatest effect on sensory perception, has been reported to decrease significantly in fresh pork after cold plasma treatment [35]. The color change of red meat is related to the oxidation of myoglobin (Mb); Mb is a binding protein composed of a peptide chain and a heme prosthetic group. Oxymyoglobin gives meat its bright red [36]. Our results of tilapia fillet research also confirmed this opinion. After exposure to strong oxides produced by cold plasma, the fish meat oxidized, the metmyoglobin formed, and the color became darker.

3.4. Low Field NMR Analysis

Moisture content and moisture distribution change dynamically during the whole process of meat processing and storage, which are important factors that determine fish meat quality and shelf life. Useful information on the interaction between water and myofibril was provided by low field NMR (Figure 3). Three peaks appearing in the spectrum were considered to be directly related to the three water components in muscle tissue, i.e., bound water, immobilized water, and free water. Bound water is very closely associated with biomacromolecules in tilapia fillets and is almost impossible to remove. T21 represented immobilized water located in a protein-dense myofibril network. Free water is also known as myofibril external water (T22). Most of the water detected in all the treatment conditions was immobilized water with a content of 95% and above. Low field NMR was performed for tilapia fillets treated with cold plasma in groups for 0 s, 180 s, and 300 s (Table 2). After cold plasma treatment, the content of immobilized water decreased significantly ($p < 0.05$), while the content of free water increased significantly ($p < 0.05$). Interestingly, from the relaxation time of low field NMR analysis in Figure 3, it can be observed that the relaxation time of free water moved forward slightly after cold plasma treatment. In the experiment of cold plasma treatment of mackerel, the results of low-field NMR analysis of moisture distribution in fish showed that the content of T21 immobilized water decreased significantly [14], which was consistent with the results of this experiment. It was speculated that the change in muscle fiber structure was related to immobilized water. At the same time, water holding capacity was also an important index for observing the change in water content. It was reported that the water holding capacity of chicken breast meat decreased significantly after cold plasma treatment at 100 kV for 5 min [37]. Studies have shown that cell membranes are damaged after being treated with cold plasma, which may be one of the causes of the loss of immobilized water in fish [38]. In addition, the denaturation of protein on the surface of fish must not be overlooked. Due to the changes in protein structure as a result of cold plasma treatment, the content of moisture bound to the protein will not remain the same.

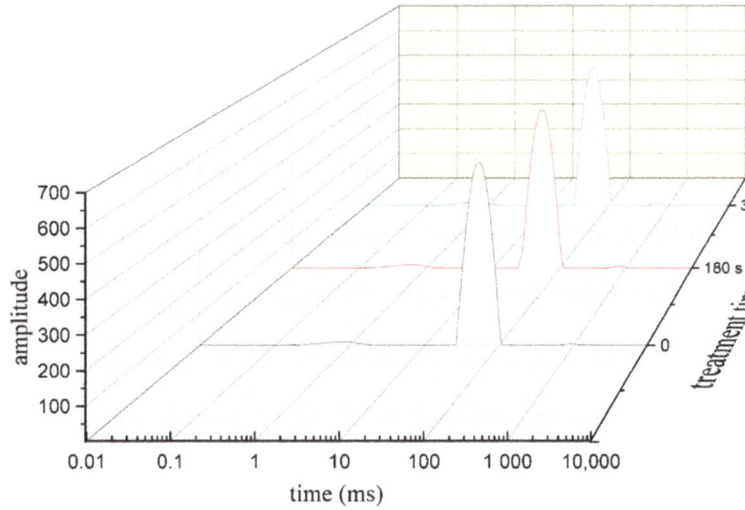

Figure 3. Low field NMR distribution of relaxation times of tilapia fillets after cold plasma treatment.

Table 2. Effect of cold plasma treatment on water content in different states of tilapia fillets.

Time (s)	T_{2b}	T_{21}	T_{22}
0	62.95 ± 3.42 [a]	2002.15 ± 27.43 [a]	8.47 ± 1.97 [a]
180	64.55 ± 3.88 [a]	1946.55 ± 18.97 [ab]	11.72 ± 1.39 [a]
300	64.26 ± 4.28 [a]	1925.85 ± 10.41 [b]	15.23 ± 1.84 [b]

Different letters in same column indicate significant changes with changes in treatment time.

3.5. Texture Analysis

The texture is one of the most important sensory factors for consumers of meat products, which can be used to evaluate the quality of the product. The hardness of the fish fillet decreased gradually after cold plasma treatment (Table 3). When samples were treated for 120 s, the hardness decreased much faster, and it decreased to 224.20 g after 300 s treatment. There was a significant difference in the chewiness of fish fillets between treatment and control groups, but no significant difference was observed between treatment groups. Nor was there any significant difference in the elasticity of cold plasma treated tilapia fillet. In terms of the viscosity and cohesion of tilapia, only the fish treated for 300 s showed significant differences from other factors.

Table 3. Changes in texture of tilapia fillets after cold plasma treatment.

Time (s)	Hardness (g)	Elasticity (mm)	Viscous (g)	Chewiness (mJ)	Cohesion
0	594.50 ± 21.50 [a]	2.96 ± 0.09 [a]	2.14 ± 0.34 [a]	6.77 ± 0.41 [a]	0.44 ± 0.02 [a]
60	373.80 ± 19.70 [ab]	2.89 ± 0.05 [a]	2.43 ± 0.37 [a]	5.08 ± 0.27 [b]	0.47 ± 0.01 [ab]
120	327.80 ± 18.90 [b]	2.96 ± 0.05 [a]	2.57 ± 0.29 [ab]	4.87 ± 0.19 [b]	0.49 ± 0.02 [ab]
180	283.70 ± 10.50 [b]	2.93 ± 0.02 [a]	2.71 ± 0.30 [ab]	4.75 ± 0.46 [b]	0.49 ± 0.01 [ab]
240	257.20 ± 13.40 [c]	2.89 ± 0.05 [a]	3.14 ± 0.51 [ab]	4.25 ± 0.44 [b]	0.51 ± 0.02 [b]
300	224.20 ± 10.20 [d]	2.99 ± 0.08 [a]	3.86 ± 0.59 [b]	4.13 ± 0.31 [b]	0.51 ± 0.01 [b]

Different letters in same column indicate significant changes with changes in treatment time.

A recent study [39] showed that cold plasma treatment had no significant effect on the texture of Asian bass. In the study of pork and beef treated with cold plasma, with the extension of treatment time, there was no significant difference in the hardness, elasticity, and chewability of pork and beef compared with the control group [35]. A similar result of elasticity was found in this study. The decrease in the hardness of fish may be related to the loss of immobilized water. In addition, only a slight difference between the fish before and after cold plasma treatment was observed in the elasticity, chewiness, viscosity, and cohesion, which means the whole structure of tilapia fillet was not significantly damaged by cold plasma.

3.6. TBARS Analysis

After cold plasma treatment, the TBARS value of tilapia fillet increased significantly ($p < 0.05$), especially after 180 s of treatment (Figure 4). When samples were treated for 300 s, the TBARS value increased to 1.83 mg MDA/kg but far below 8.0 mg MDA/kg, which was considered the acceptable limit for aquatic food [40,41].

The effect of cold plasma treatment on the TBARS level of meat has been demonstrated in a number of studies. In the studies of Bae, Park, Choe, and Ha [42], with the extension of exposure to atmospheric jet plasma, the TBARS levels of fresh beef loin, pork shoulder, and chicken breast increased significantly, and all of them were lower than 1.0 mg MDA/kg, which was within the acceptable range. A large number of prooxidants (RNS and ROS) were produced by atmospheric cold plasma, and the production of lipid oxidation by-products such as hexanal and malondialdehyde were formed during the reaction of ROS and fatty acids [43]. However, in the cold plasma treatment using argon as the reaction gas, the TBARS level did not change significantly in meat after cold plasma treatment, which indicated that the TBARS value was affected by the reaction medium gas [26].

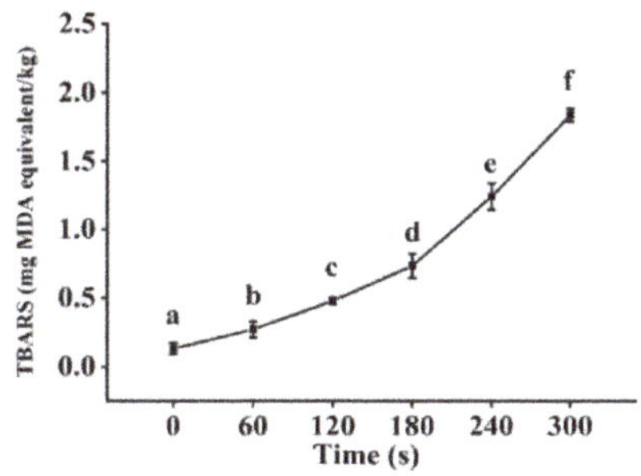

Figure 4. Changes in TBARS value of tilapia fillets after cold plasma treatment. Different letters indicate significant changes with changes in treatment time.

3.7. Carbonyl and Sulfhydryl Level Analysis

As shown in Figure 5a, the level of protein carbonyl in tilapia fillets varied as cold plasma treatment time changed. The carbonyl level of the sarcoplasmic protein did not change significantly in 60 s of treatment, and it showed an increasing trend accompanied by a slight fluctuation until it finally reached 4.2 nmol/mg prot when the treatment time was longer than 120 s. However, the carbonyl of myofibrillar protein was maintained at a low level, and no significant change in carbonyl level was observed in treated samples.

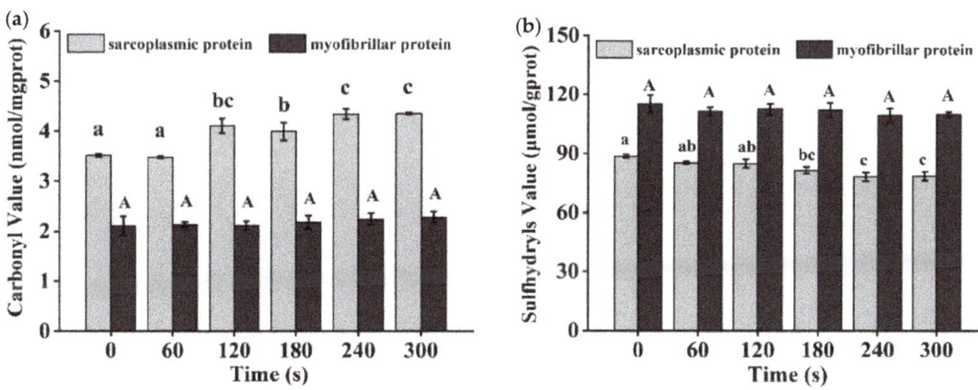

Figure 5. (**a**) Changes in carbonyl values of tilapia fillets after cold plasma treatment. (**b**) Changes in Sulfhydryl values of tilapia fillets after cold plasma treatment. Different letters indicate significant changes with changes in treatment time.

It was reported that the total carbonyl content in cold-plasma-treated Asian bass was significantly higher than that in the control group [44]. The oxidation of protein side chains is an important factor in the formation of carbonyl groups [45]. In addition, protein structures and peptide chains are vulnerable to cold plasma treatment, and peptide bond cleavage may result in the exposure of amino acid groups [46]. There are so many kinds of strong oxidizing substances in atmospheric cold plasma that can fully react with protein. Slight oxidation occurred in whey protein isolate (WPI) after 15 min of atmospheric cold plasma treatment, which changed the side chain of the amino acid residues and increased the carbonyl content [47].

The sulfhydryl groups of cysteine residues in proteins are extremely susceptible to oxidation induced by most forms of ROS, which provides an additional indicator of the level of protein oxidation. In all of the samples treated with cold plasma, the sulfhydryl content of sarcoplasmic protein decreased slightly to 78.51 mol/gprot after 300 s of treatment (Figure 5b). The sulfhydryl content of myofibril protein decreased slightly after cold plasma treatment.

After 300 s of cold plasma treatment with crude protease extracted from hairtail fish, sulfhydryl content decreased significantly ($p < 0.05$) compared with the control group [48]. Miao et al. [49] found that the sulfhydryl levels of myofibril extract from Alaskan cod decreased significantly after treatment with dielectric barrier discharge cold plasma. They considered that the reduction of sulfhydryl content was the result of their reaction with reactive oxygen species. Since sulfhydryl is involved in the formation of disulfide bonds, the reduction of disulfide bond formation has a significant impact on protein folding and stability. In the study of shrimp treated with plasma-activated water, no significant change in myosin sulfhydryl of shrimp was found [50], which was similar to the result of the tilapia research in this study. Protein oxidation induced by cold plasma treatment occurs easily in the extracted protein solution or protein model systems, and significant differences only occur after the directly treated foods are stored for a long period [51].

3.8. TVB-N Analysis

The TVB-N value of meat is usually the biomarker of bacterial and enzymatic degradation of protein and non-protein nitrogenide, including volatile ammonia, methylamine, dimethylamine, trimethylamine, and other low-level amines. The content of TVB-N is usually used as an indicator of fish freshness. In this study, no significant changes in TVB-N values of tilapia fillets were observed after cold plasma treatment (Figure 6).

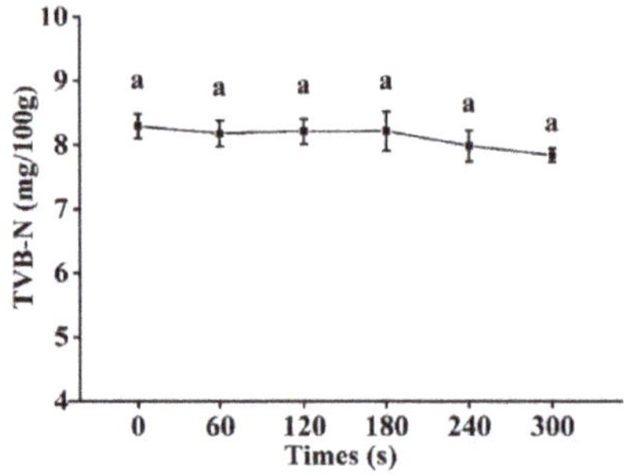

Figure 6. Changes in TVB-N value of tilapia fillets after cold plasma treatment. Different letters indicate significant changes with changes in treatment time.

In this study, the TVB-N values varied almost identically to the pH values. Zouelm, Abhari, Hosseini, and Khani [19] found that the TVB-N values in cold plasma treated samples kept to similar levels throughout the storage period. The acidic components produced during the plasma treatment can react with the alkaline components, resulting in a reduction of measurable alkaline substances and a lower TVB-N value. Meanwhile, due to its antibacterial effect, the cold plasma treatment could prevent protein breakdown in the meat and hinder the production of alkaline compounds, thus reducing the TVB-N value [18].

4. Conclusions

The inactivation effect of cold plasma on inoculated tilapia fillets was increased with the extension of treatment time and a better bactericidal effect of single bacteria strain than a mixture. As the treatment extended, the TBARS increased significantly ($p < 0.05$), which means that the longer time treatment enhanced the lipids oxidation. Sarcoplasmic protein was also oxidation during treatment processing. Therefore, the lipids and sarcoplasmic protein in tilapia fillets were sensitive to oxidation during treatment. Controlling treatment conditions would be helpful in limiting the oxidation degree of lipids and protein. In conclusion, cold plasma treatment could inhibit the growth of bacteria on fish surfaces without a significant negative influence on the organoleptic quality of tilapia fillets. Therefore, cold plasma is a promising alternative to traditional antibacterial treatments used in perishable seafood products.

Author Contributions: Writing—review and editing, project administration, and resources: J.W.; conducting experiments and writing—original draft preparation: T.F.; writing—review and editing: Y.W.; reviewing: J.Z. All authors have read and agreed to the published version of the manuscript.

Funding: This research was funded by the National Natural Science Foundation of China, grant number "32060568".

Data Availability Statement: Data is contained within the article.

Acknowledgments: The authors greatly acknowledge the financial support provided by the National Natural Science Foundation of China.

Conflicts of Interest: The authors declare no conflict of interest.

References

1. Samples, S. The effects of storage and preservation technologies on the quality of fish products: A review. *J. Food Process. Preserv.* **2015**, *39*, 1206–1215. [CrossRef]
2. Sivertsvik, M.; Jeksrud, W.K.; Rosnes, J.T. A review of modified atmosphere packaging of fish and fishery prod-ucts—Significance of microbial growth, activities and safety. *Int. J. Food Sci. Technol.* **2002**, *37*, 107–127. [CrossRef]
3. Vázquez, M.; Torres, J.A.; Gallardo, J.M.; Saraiva, J.; Aubourg, S.P. Lipid hydrolysis and oxidation development in frozen mackerel (Scomber scombrus): Effect of a high hydrostatic pressure pre-treatment. *Innov. Food Sci. Emerg. Technol.* **2013**, *18*, 24–30. [CrossRef]
4. Gómez, B.; Munekata, P.E.S.; Gavahian, M.; Barba, F.J.; Martí-Quijal, F.J.; Bolumar, T.; Campagnol, P.C.B.; Tomasevic, I.; Lo-renzo, J.M. Application of pulsed electric fields in meat and fish processing industries: An overview. *Food Res. Int.* **2019**, *123*, 95–105. [CrossRef]
5. Yang, Z.; Wang, H.; Wang, W.; Qi, W.; Yue, L.; Ye, Q. Effect of 10 MeV E-beam irradiation combined with vacuum-packaging on the shelf life of Atlantic salmon fillets during storage at 4 °C. *Food Chem.* **2014**, *145*, 535–541. [CrossRef]
6. Li, X.; Farid, M. A review on recent development in non-conventional food sterilization technologies. *J. Food Eng.* **2016**, *182*, 33–45. [CrossRef]
7. Ronit, M.; Anika, S.; Pratap, S.A. Recent developments in cold plasma decontamination technology in the food industry. *Trends Food Sci. Technol.* **2018**, *80*, 93–103. [CrossRef]
8. Gordillo-Vázquez, F.J. Air plasma kinetics under the influence of sprites. *J. Phys. D Appl. Phys.* **2008**, *41*, 234016. [CrossRef]
9. Misra, N.N.; Jo, C. Applications of cold plasma technology for microbiological safety in meat industry. *Trends Food Sci. Technol.* **2017**, *64*, 74–86. [CrossRef]
10. Scholtz, V.; Pazlarova, J.; Souskova, H.; Khun, J.; Julak, J. Nonthermal plasma—A tool for decontamination and disinfection. *Biotechnol. Adv.* **2015**, *33*, 1108–1119. [CrossRef]
11. Han, L.; Patil, S.; Boehm, D.; Milosavljević, V.; Cullen, P.J.; Bourke, P. Mechanisms of inactivation by high-voltage atmos-pheric cold plasma differ for Escherichia coli and *Staphylococcus aureus*. *Appl. Environ. Microbiol.* **2016**, *82*, 450–458. [CrossRef]
12. Huang, M.; Zhuang, H.; Zhao, J.; Wang, J.; Yan, W.; Zhang, J. Differences in cellular damage induced by dielectric barrier discharge plasma between Salmonella Typhimurium and *Staphylococcus aureus*. *Bioelectrochemistry* **2020**, *132*, 107445. [CrossRef] [PubMed]
13. Albertos, I.; Martin-Diana, A.B.; Cullen, P.J.; Tiwari, B.K.; Ojha, K.S.; Bourke, P.; Rico, D. Shelf-life extension of herring (Clu-pea harengus) using in-package atmospheric plasma technology. *Innov. Food Sci. Emerg. Technol.* **2019**, *53*, 85–91. [CrossRef]
14. Albertos, I.; Martín-Diana, A.; Cullen, P.J.; Tiwari, B.K.; Ojha, S.K.; Bourke, P.; Álvarez, C.; Rico, D. Effects of dielectric barri-er discharge (DBD) generated plasma on microbial reduction and quality parameters of fresh mackerel (Scomber scombrus) fillets. *Innov. Food Sci. Emerg. Technol.* **2017**, *44*, 117–122. [CrossRef]
15. Olatunde, O.O.; Benjakul, S.; Vongkamjan, K. Shelf-life of refrigerated Asian sea bass slices treated with cold plasma as affected by gas composition in packaging. *Int. J. Food Microbiol.* **2020**, *324*, 108612. [CrossRef]
16. Puligundla, P.; Choi, S.; Mok, C. Microbial decontamination of Gwamegi (semi-dried pacific saury) using corona discharge plasma jet, including physicochemical and sensory evaluation. *J. Aquat. Food Prod. Technol.* **2018**, *27*, 274–283. [CrossRef]
17. Pérez-Andrés, J.M.; de Alba, M.; Harrison, S.M.; Brunton, N.P.; Cullen, P.J.; Tiwari, B.K. Effects of cold atmospheric plasma on mackerel lipid and protein oxidation during storage. *LWT-Food Sci. Technol.* **2020**, *118*, 108697. [CrossRef]
18. Chen, J.; Wang, S.Z.; Chen, J.Y.; Chen, D.Z.; Deng, S.G.; Xu, B. Effect of cold plasma on maintaining the quality of chub mackerel (Scomber japonicus): Biochemical and sensory attributes. *J. Sci. Food Agric.* **2019**, *99*, 39–46. [CrossRef]
19. Zouelm, F.; Abhari, K.; Hosseini, H.; Khani, M. The effects of cold plasma application on quality and chemical spoilage of Pacific White Shrimp (Litopenaeus vannamei) during refrigerated storage. *J. Aquat. Food Prod. Technol.* **2019**, *28*, 624–636. [CrossRef]
20. Wang, J.M.; Zhuang, H.; Lawrence, K.; Zhang, J.H. Disinfection of chicken fillets in packages with atmospheric cold plasma: Effects of treatment voltage and time. *J. Appl. Microbiol.* **2018**, *124*, 1212–1219. [CrossRef]
21. Huang, M.; Wang, J.; Zhuang, H.; Yan, W.; Zhang, J. Effect of in-package high voltage dielectric barrier discharge on micro-biological, color and oxidation properties of pork in modified atmosphere packaging during storage. *Meat Sci.* **2019**, *149*, 107–113. [CrossRef] [PubMed]

22. Toldrá, F.; Rico, E.; Flores, J. Cathepsin B, D, H and L activities in the processing of dry-cured ham. *J. Sci. Food Agric.* **1993**, *62*, 157–161. [CrossRef]
23. Mesquita, C.S.; Oliveira, R.; Bento, F.; Geraldo, D.; Rodrigues, J.V.; Marcos, J.C. Simplified 2,4-dinitrophenylhydrazine spectrophotometric assay for quantification of carbonyls in oxidized proteins. *Anal. Biochem.* **2014**, *458*, 69–71. [CrossRef] [PubMed]
24. Srinivasan, S.; Hultin, H.O. Chemical, physical, and functional properties of cod proteins modified by a nonenzymic free-radical-generating system. *J. Agric. Food Chem.* **1997**, *45*, 310–320. [CrossRef]
25. Cao, J.; Wang, Q.; Ma, T.; Bao, K.; Yu, X.; Duan, Z.; Shen, X.; Li, C. Effect of EGCG-gelatin biofilm on the quality and micro-bial composition of tilapia fillets during chilled storage. *Food Chem.* **2020**, *305*, 125454. [CrossRef] [PubMed]
26. Gök, V.; Aktop, S.; Özkan, M.; Tomar, O. The effects of atmospheric cold plasma on inactivation of Listeria monocytogenes and Staphylococcus aureus and some quality characteristics of pastırma—A dry-cured beef product. *Innov. Food Sci. Emerg. Technol.* **2019**, *56*, 102188. [CrossRef]
27. Mai-Prochnow, A.; Clauson, M.; Hong, J.; Murphy, A.B. Gram positive and Gram negative bacteria differ in their sensitivity to cold plasma. *Sci. Rep.* **2016**, *6*, 38610. [CrossRef]
28. Qian, J.; Ma, L.; Yan, W.; Zhuang, H.; Huang, M.; Zhang, J.; Wang, J. Inactivation kinetics and cell envelope damages of foodborne pathogens Listeria monocytogenes and Salmonella Enteritidis treated with cold plasma. *Food Microbiol.* **2022**, *101*, 103891. [CrossRef]
29. Misra, N.N.; Yepez, X.; Xu, L.; Keener, K. In-package cold plasma technologies. *J. Food Eng.* **2019**, *244*, 21–31. [CrossRef]
30. Kim, H.J.; Yong, H.I.; Park, S.; Choe, W.; Jo, C. Effects of dielectric barrier discharge plasma on pathogen inactivation and the physicochemical and sensory characteristics of pork loin. *Curr. Appl. Phys.* **2013**, *13*, 1420–1425. [CrossRef]
31. Girard, F.; Badets, V.; Blanc, S.; Gazeli, K.; Marlin, L.; Authier, L.; Svarnas, P.; Sojic, N.; Clement, F.; Arbault, S. Formation of reactive nitrogen species including peroxynitrite in physiological buffer exposed to cold atmospheric plasma. *RSC Adv.* **2016**, *6*, 78457–78467. [CrossRef]
32. Lee, H.; Yong, H.I.; Kim, H.-J.; Choe, W.; Yoo, S.J.; Jang, E.J.; Jo, C. Evaluation of the microbiological safety, quality changes, and genotoxicity of chicken breast treated with flexible thin-layer dielectric barrier discharge plasma. *Food Sci. Biotechnol.* **2016**, *25*, 1189–1195. [CrossRef] [PubMed]
33. Wang, J.; Zhuang, H.; Hinton, A., Jr.; Zhang, J. Influence of in-package cold plasma treatment on microbiological shelf life and appearance of fresh chicken breast fillets. *Food Microbiol.* **2016**, *60*, 142–146. [CrossRef]
34. Ulbin-Figlewicz, N.; Brychcy, E.; Jarmoluk, A. Effect of low-pressure cold plasma on surface microflora of meat and quality attributes. *J. Food Sci. Technol.* **2015**, *52*, 1228–1232. [CrossRef] [PubMed]
35. Jayasena, D.D.; Kim, H.J.; Yong, H.I.; Park, S.; Kim, K.; Choe, W.; Jo, C. Flexible thin-layer dielectric barrier discharge plas-ma treatment of pork butt and beef loin: Effects on pathogen inactivation and meat-quality attributes. *Food Microbiol.* **2015**, *46*, 51–57. [CrossRef]
36. Bekhit, A.E.D.A.; Hopkins, D.L.; Fahri, F.T.; Ponnampalam, E.N. Oxidative processes in muscle systems and fresh meat: Sources, markers, and remedies. *Compr. Rev. Food Sci. Food Saf.* **2013**, *12*, 565–597. [CrossRef]
37. Moutiq, R.; Misra, N.N.; Mendonca, A.; Keener, K. In-package decontamination of chicken breast using cold plasma tech-nology: Microbial, quality and storage studies. *Meat Sci.* **2020**, *159*, 107942. [CrossRef]
38. Thirumdas, R.; Sarangapani, C.; Annapure, U.S. Cold plasma: A novel non-thermal technology for food processing. *Food Biophys.* **2015**, *10*, 1–11. [CrossRef]
39. Olatunde, O.O.; Benjakul, S.; Vongkamjan, K. High voltage cold atmospheric plasma: Antibacterial properties and its effect on quality of Asian sea bass slices. *Innov. Food Sci. Emerg. Technol.* **2019**, *52*, 305–312. [CrossRef]
40. Okpala, C.O.R.; Choo, W.S.; Dykes, G.A. Quality and shelf life assessment of Pacific white shrimp (Litopenaeus vannamei) freshly harvested and stored on ice. *LWT-Food Sci. Technol.* **2014**, *55*, 110–116. [CrossRef]
41. Zhao, Y.; Lan, W.; Shen, J.; Xu, Z.; Xie, J. Combining ozone and slurry ice treatment to prolong the shelf-life and quality of large yellow croaker (Pseudosciaena crocea). *LWT-Food Sci. Technol.* **2022**, *154*, 112615. [CrossRef]
42. Bae, S.C.; Park, S.Y.; Choe, W.; Ha, S.D. Inactivation of murine norovirus-1 and hepatitis A virus on fresh meats by atmos-pheric pressure plasma jets. *Food Res. Int.* **2015**, *76*, 342–347. [CrossRef]
43. Wang, J.; Zhuang, H.; Zhang, J. Inactivation of spoilage bacteria in package by dielectric barrier discharge atmospheric cold plasma—Treatment time effects. *Food Bioprocess Technol.* **2016**, *9*, 1648–1652. [CrossRef]
44. Olatunde, O.O.; Benjakul, S.; Vongkamjan, K. Cold plasma combined with liposomal ethanolic coconut husk extract: A po-tential hurdle technology for shelf-life extension of Asian sea bass slices packaged under modified atmosphere. *Innov. Food Sci. Emerg. Technol.* **2020**, *65*, 102448. [CrossRef]
45. Estevez, M. Protein carbonyls in meat systems: A review. *Meat Sci.* **2011**, *89*, 259–279. [CrossRef]
46. Bahrami, N.; Bayliss, D.; Chope, G.; Penson, S.; Perehinec, T.; Fisk, I.D. Cold plasma: A new technology to modify wheat flour functionality. *Food Chem.* **2016**, *202*, 247–253. [CrossRef]
47. Segat, A.; Misra, N.N.; Cullen, P.J.; Innocente, N. Effect of atmospheric pressure cold plasma (ACP) on activity and structure of alkaline phosphatase. *Food Bioprod. Process.* **2016**, *98*, 181–188. [CrossRef]
48. Koddy, J.K.; Miao, W.; Hatab, S.; Tang, L.; Deng, S. Understanding the role of atmospheric cold plasma (ACP) in maintain-ing the quality of hairtail (*Trichiurus Lepturus*). *Food Chem.* **2020**, *343*, 128418. [CrossRef]

49. Miao, W.; Nyaisaba, B.M.; Koddy, J.K.; Chen, M.; Hatab, S.; Deng, S. Effect of cold atmospheric plasma on the physicochem-ical and functional properties of myofibrillar protein from Alaska pollock (Theragra chalcogramma). *Int. J. Food Sci. Technol.* **2020**, *55*, 517–525. [CrossRef]
50. Liao, X.; Su, Y.; Liu, D.; Chen, S.; Hu, Y.; Ye, X.; Wang, J.; Ding, T. Application of atmospheric cold plasma-activated water (PAW) ice for preservation of shrimps (Metapenaeus ensis). *Food Control* **2018**, *94*, 307–314. [CrossRef]
51. Tolouie, H.; Mohammadifar, M.A.; Ghomi, H.; Hashemi, M. Cold atmospheric plasma manipulation of proteins in food systems. *Crit. Rev. Food Sci. Nutr.* **2018**, *58*, 2583–2597. [CrossRef] [PubMed]

foods

Article

Analysis of Water Distribution and Muscle Quality of Silver Carp (*Hypophthalmichthys molitrix*) Chunks Based on Electron-Beam Irradiation

Hai-Lan Li [1,2,†], Mei-Jin Li [1,3,†], Qing Zhao [1,3], Jia-Jun Huang [1,3] and Xiao-Yan Zu [1,2,*]

1 Institute of Agricultural Products Processing and Nuclear Agricultural Technology, Hubei Academy of Agricultural Sciences, Wuhan 430064, China
2 Key Laboratory of Cold Chain Logistics Technology for Agro-Product, Ministry of Agriculture and Rural Affairs, Wuhan 430064, China
3 College of Bioengineering and Food, Hubei University of Technology, Wuhan 430068, China
* Correspondence: zuxiaoyan@hbaas.com; Tel.: +86-27-80839098; Fax: +86-27-87380171
† These authors contributed equally to this work.

Abstract: Electron-beam irradiation (EBI) is an efficient, safe, and nonthermal sterilization technique that is extensively used in food preservation research. Here we report the effects of different EBI doses (0, 4, 8 kGy) and preservation temperatures (room temperature [RT], 4 °C) on the muscle water distribution and muscle quality indices of silver carp chunks (SCCs). The highest entrapped water content was found in the 4-kGy-irradiated/4-°C-stored samples. The expressible moisture content (EMC) of the SCCs increased with increasing irradiation dose and was significantly lower in the RT group than in the 4 °C group. The irradiation dose and preservation temperature had no significant effect on the moisture content, whiteness value and protein content of SCCs ($p > 0.05$). When the irradiation dose reached 8 kGy, AV value, POV value and TVB value were significantly increased ($p < 0.05$). The myofibrillar protein content and actomyosin content of the SCCs in the 4 °C group was higher than that of the specimens in the RT group by 0.29–0.98 mg/mL ($p < 0.05$) and 36.21–296.58 μg/mL ($p < 0.05$), respectively. Overall, EBI treatment (4 kGy) and low-temperature preservation (4 °C) helped retain the muscle water content of the SCCs and preserve their quality, thereby endorsing the EBI treatment of silver carp products.

Keywords: electron-beam irradiation; silver carp chunks; water distribution; muscle quality

Citation: Li, H.-L.; Li, M.-J.; Zhao, Q.; Huang, J.-J.; Zu, X.-Y. Analysis of Water Distribution and Muscle Quality of Silver Carp (*Hypophthalmichthys molitrix*) Chunks Based on Electron-Beam Irradiation. *Foods* **2022**, *11*, 2963. https://doi.org/10.3390/foods11192963

Academic Editor: Thierry Astruc

Received: 11 August 2022
Accepted: 19 September 2022
Published: 22 September 2022

1. Introduction

Silver carp or white carp (*Hypophthalmichthys molitrix*), which is a fish species that belongs to the Cyprinidae family in the order Clupeiformes is a globally abundant freshwater resource [1]. Silver carp is popular among consumers for being nutritious, tender, and inexpensive [2]. However, the flesh of silver carp is susceptible to microbial and endogenous enzymatic activities as well as biochemical reactions, which can lead to spoilage during storage as well as the production of volatile basic nitrogen and free amino acids [3]. In addition to traditional processing and preservation methods such as curing, refrigeration, freezing, and heat treatment, advanced techniques such as modified atmosphere packaging [4], ultrahigh-pressure sterilization [5], snap freezing [6], and irradiation [7] have been adopted to maintain the nutritional quality and edible value of silver carp flesh. Among these methods, ultrahigh-pressure sterilization and snap freezing are cost ineffective owing to their requirement of expensive equipment and high costs, which hinder their widespread adoption. Moreover, the utilization of the modified atmosphere technology is limited by the variation in the gas components in the modified atmosphere system from product to product and the stringent requirements for packaging materials.

Irradiation is a safe and effective cold-sterilization technique, with electron-beam irradiation (EBI) being a particularly noteworthy food-decontamination technology that is cost effective, easy to manipulate, and contamination free [8,9]. Electron accelerator produced high energy electron beams can sterilize food products and extend their shelf life while maintaining their original quality and flavor [10]. As an advanced technique used in food preservation, EBI has been found to effectively reduce or eliminate nitrates and nitrites in cured meat [11]. H. Yu et al. found that EBI treatment at 10 kGy significantly reduced the total volatile basic nitrogen content in cod and increased its water and ash contents [12]. H. Guo et al. studied the effects of EBI on the volatile flavor substances of salmon fillets and showed that the best sensory flavor of fish flesh was achieved with 1 kGy irradiation treatment [13]. Q. Yu et al. found that EBI had no effect on the pH of shrimp and that higher irradiation doses resulted in fish with more odorous volatiles [14]. Recent studies on the EBI of aquatic products have primarily focused on changes in their physicochemical properties and sensory flavor; however, investigations related to the influence of EBI on the water distribution in silver carp flesh have been rarely reported.

Low-field nuclear magnetic resonance (LF-NMR) is a rapid, accurate, and nondestructive analytical detection technique. Using this method, the flow and distribution of hydrogen protons in tissues of food products can be determined by assessing the relaxation properties of hydrogen protons in a constant magnetic field, which reflect the water content and migration processes in samples [15]. LF-NMR has been applied to various sea products, including oysters [16], hake [17], and sea cucumbers [18]. However, studies focusing on the LF-NMR analysis of freshwater products are relatively scarce.

Therefore, this study took SSCs as the research object to explore the effects of different preservation temperatures (room temperature [RT], 4 °C) and different EBI doses (0, 4, 8 kGy) on the water distribution of SSCs, as well as analyzing the correlation between EBI, preservation temperature and muscle quality of SSCs. The results reported herein are anticipated to enable improvements in the processing and quality control of silver carp products through EBI treatment.

2. Materials and Methods

2.1. Materials

Analytical-grade chemicals including hydrochloric acid, boric acid, petroleum ether, isopropyl alcohol, trichloromethane, sodium hydroxide, anhydrous sodium sulfate, and TBA were purchased from Sinopharm Group Chemical Reagent Co., Ltd. (Shanghai, China). A Bradford Protein Assay Kit (P0006) was purchased from Shanghai Beyotime Biotechnology Co., Ltd. (Shanghai, China), and a total protein assay kit (A045-4-2) was procured from Nanjing Jiancheng Bioengineering Institute.

2.2. Sample Preparation and Processing

Fresh silver carp (n = 20) with weights and body lengths of 2.0 ± 0.2 kg and 54 ± 2.3 cm, respectively, were purchased from Wushang Supermarket (Agricultural Science City Store), Hongshan District, Wuhan, China. The silver carp was washed, and its head, tail, scales, internal organs, and bones were removed. Subsequently, the flesh was cut along the backbone into chunks weighing 20 ± 2.0 g. These portions were placed individually in low-density-polyethylene bags, sealed with a sealing machine, and then heated in boiling water for 2 min to cook the flesh to medium-rare. The fish samples were cooled, skinned, and randomly divided into two groups with equal sample sizes based on temperature—room temperature (RT; 25 °C) and 4 °C—with the 4 °C group samples being stored in iceboxes. The samples from both groups were separated into three subgroups, which were then subjected to EBI treatment at doses of 0, 4, and 8 kGy, respectively, at Wuhan Aibang High Energy Technology Co., Ltd. (Wuhan, China). After the EBI treatment, the samples were placed in iceboxes and then stored at RT or 4 °C. Subsequently, the samples were experimentally investigated to determine relevant indices. The actual doses received by the samples were calibrated with low- and high-dose-range silver dichromate dosimeters containing 0.35 and 2.5 mmol/L silver dichromate, respectively [19]. The dosimeters monitored the irradiation of the samples and were sent to the Irradiation Engineering Center of Hubei Province; the actual absorbed doses of the three subgroups were determined to be 0, 4.25, and 7.84 kGy, respectively.

2.3. Determination of Water Distribution

LF-NMR and magnetic resonance imaging (MRI) analyses were performed using a slightly modified version of the method described by L. Wang et al. [20]. The SCCs were cut into $1.8 \times 1.8 \times 1.8$ cm^3 pieces, padded dry, and placed in MRI-compatible test tubes. The distributions of bound, entrapped, and free water were monitored using an NMR analyzer (NMI20-025V-I, Suzhou Niumag Analytical Instrument Co., Ltd., Suzhou, China). The water in the SCCs could be classified into the following three categories based on how tight the water was bound to the tissues: bound water that is tightly tethered to muscle macromolecules (T_{2b}), entrapped water retained by myofibrils (T_{21}), and free water held in interfascicular space and by sarcoplasmic proteins (T_{22}) [21]. The following parameters were used for the T_2 measurements: resonant frequency (SF), 20 MHz; 90° pulse width (P1), 8 µs; 180° pulse width (P_2), 16.48 µs; spectral width (SW), 100 kHz; waiting time (TW), 1000 ms; econ time (TE), 1 ms; echo number (NECH), 700; and repetitive scans number (NS), 4. The sample signals were acquired using the CPMG sequence and NMR analysis software. The inverse transformation was performed using SIRT100000 to obtain the relaxation time and peak area for each water type. Additionally, the pseudo-color maps of the water molecules were constructed based on the analysis of water proton densities.

2.4. Determination of Moisture Content and Expressible Moisture Content (EMC)

The moisture content was determined using the method described by Shi et al. with slight modifications [22]. Clean aluminum weighing bottles were dried in an oven at 55 °C for 1 h (DHG-9203A, Shanghai Yi Heng Co., Ltd., Shanghai, China), removed, covered, and then transferred to a desiccator to cool for 20 min before being weighed. The drying and cooling processes were repeated until a constant weight was achieved, which was based on the difference between two successive weights not exceeding 2 mg. The SCCs were minced, weighed, and placed in the prepared weighing bottles. The aforementioned drying, cooling, and weighing processes were repeated, and the final constant weights were recorded. The water content was calculated using the following equation:

$$\text{Moisture content } (\text{g}/100\text{ g}) = \frac{m_1 - m_2}{\text{m}_1} \times 100, \tag{1}$$

where m_1 is the weight of the SCCs immediately after mincing (g), m_2 is the constant weight of the minced SCCs (g), and 100 is the conversion factor.

The EMC was determined using a slightly modified version of the method described by Jiao et al. [23]. Briefly, the SCCs were weighed, wrapped with filter paper, and then centrifuged (TGL-24MC, Changsha Pingfan Instrument Co., Ltd., Changsha, China) at 2000 rpm for 15 min. The centrifuged samples were weighed, and the EMC was calculated using the following equation:

$$\text{EMC } (\%) = \frac{m_1 - m_2}{m_1} \times 100\%, \tag{2}$$

where m_1 and m_2 (g) are the masses of the SCCs before and after the centrifugation, respectively.

2.5. Determination of Whiteness and pH Values

As an important appearance related indicator of aquatic products, whiteness value directly affects the initial assessment of these products by consumers [24]. Based on the method described by Gulcan et al. [25], the brightness (L^*), redness (a^*), and yellowness

(b^*) values of the cut surface of SCCs were measured using a colorimeter (CR-400, Konica Minolta, Japan), and the whiteness values were calculated thereafter as follows:

$$Whiteness = 100 - \sqrt{(100 - L*)^2 + a*^2 + b*^2}, \quad (3)$$

pH is an important indicator of the quality and freshness of aquatic products [26]. For the pH analysis, 4 g of an SCC sample was mashed and mixed with 40 mL of distilled water. The mixture was thoroughly agitated and left to stand for 20 min before being filtered. The pH value of the collected filtrate was then determined using a pH meter (Mettler-Toledo Instruments Co., Ltd., Shanghai, China).

2.6. Determination of AVs, Peroxide Values (POVs), and TBA Values

AVs, POVs, and TBA values are important indicators for evaluating the extent of lipid hydrolysis as well as primary and secondary lipid oxidation [27]. AVs were determined according to the GB5009.229-2016 standard and the method reported by Wei et al. [28]. The AVs of the SCCs were calculated as follows:

$$\text{AV (mg/g)} = \frac{(V - V_0) \times 0.01 \times 56.1}{\text{m}}, \quad (4)$$

where V and V_0 (mL) are the volumes of the standard titration solution consumed for determination of the samples and the corresponding blanks, respectively; 0.01 is the molar concentration of a potassium-hydroxide standard solution (mol/L); 56.1 is the molar mass of potassium hydroxide (g/mol); and m is the weight of the SCC sample (g).

POVs were determined according to the GB5009.227-2016 standard and the method reported by Wang et al. [29]. The POVs of the SCCs were calculated as follows:

$$\text{POV (g/100 g)} = \frac{(V - V_0) \times 0.01 \times 0.1269}{\text{m}} \times 100, \quad (5)$$

where V and V_0 (mL) are the volumes of the sodium thiosulfate standard solution consumed by the samples and blanks, respectively; 0.01 is the molar concentration of the sodium-thiosulfate standard solution (mol/L); 0.1269 is the mass of elementary iodine equivalent to 1 mL of the sodium-thiosulfate standard titration solution [$c(Na_2S_2O_3) = 1.000$ mol/L]; m is the weight of the SCC sample (g); and 100 is the conversion factor.

The TBA values were determined using a slightly modified version of the method described by Salih et al. [30]. Five grams of an SCC sample was placed in a centrifuge tube and mixed with 25 mL of a trichloroacetic acid solution (20% volume fraction). The mixture was uniformly stirred and left to stand for 1 h before being centrifuged at 2000 rpm for 10 min and then filtered. Distilled water was added to the collected filtrate to achieve a final volume of 50 mL. A portion of this filtrate (5 mL) was mixed with 5 mL of TBA solution (0.02 mol/L) and reacted in a boiling water bath for 20 min. After cooling, the mixture was subjected to absorbance (A) analysis at 532 nm using a spectrophotometer (UH5300, Hitachi Co., Ltd., Tokyo, Japan), and the TBA values were calculated as follows:

$$\text{TBA (mg/100 g)} = A \times 7.8, \quad (6)$$

where A is the absorbance of the solution measured at 532 nm, and 7.8 is a constant.

2.7. Determination of Total Protein Content, Myofibrillar Protein Content, and Actomyosin Content

The total protein contents were determined according to the GB5009.5-2016 standard and the method reported by Yang et al. [31]; the corresponding values of the SCCs were obtained as follows:

$$\text{Protein content (g/100 g)} = \frac{(V_1 - V_2) \times 0.05 \times 0.0140}{\text{m} \times 10/100} \times F \times 100, \quad (7)$$

where V_1 and V_2 are the volumes of the hydrochloric-acid standard titrant consumed by the specimens and blanks, respectively; 0.05 is the concentration of the HCl standard titrant (mol/L); 0.0140 is the mass of elementary nitrogen equivalent to 1 mL of the HCl standard titrant [c(HCL) = 1.000 mol/L], m is the weight of the SCC sample (g); 10 is the volume of digested sample used for titration (mL); F is the nitrogen-to-protein conversion factor; and 100 is the general conversion factor.

The content of myofibrillar protein (W/V) were determined using the method described by Benjakul et al. (1997) with slight modifications [32]. One gram of an SCC sample was mixed with 10 mL of precooled 0.1 mol/L KCl solution, followed by homogenization at 10,000 rpm for 1 min. The dispersion was then centrifuged at 10,000 rpm for 20 min at 4 °C, and the supernatant was discarded. The resulting precipitate was resuspended in a 0.6 mol/L KCl solution with a volume 8× its original value and homogenized for 1 min (XHF-D, Ningbo Xinzhi Co., Ltd., Ningbo, China). The mixture was then left to stand at 4 °C for 1 h and centrifuged thereafter at 12,000 rpm for 30 min at 4 °C. The supernatant was collected, and the myofibrillar protein content of the samples was determined using a quantitative protein assay kit.

The actomyosin contents were determined using the method described by Zhou et al. with slight modifications [33]. Two grams of an SCC sample was minced and mixed with 10 mL of precooled 0.6 mol/L KCl solution. The mixture was homogenized at 10,000 rpm for 30 s and then centrifuged at 5000 rpm for 30 min at 4 °C. The supernatant was collected and diluted in precooled distilled water with a volume 3× the original value. The diluted supernatant was further centrifuged at 5000 rpm for 20 min at 4 °C, and the precipitate was collected and resuspended in a precooled 1.2 mol/L KCL solution with equal volume. The resulting mixture was blended using a magnetic stirrer (DF-101S, Wuhan Ke'er Instrument Co., Ltd., Wuhan, China) for 30 min and then centrifuged at 5000 rpm for 20 min at 4 °C. The supernatant was collected, and the actomyosin content was determined using a quantitative protein assay kit.

2.8. Data Analysis

Origin 2019 software was used for graphically visualizing the results, whereas SPSS 26.0 and Microsoft Office Excel 2016 were used for statistical data analysis. The results are presented as mean ± standard deviation (Mean ± SD). Analysis of variance (ANOVA) was used for significance analysis, with a P-value less than 0.05 ($p \leq 0.05$) considered to be statistically significant.

3. Results and Discussion

3.1. Effects of Different EBI Doses on the Water Distribution in SCCs

The peak areas in T_2 spectrums reflect the relative water content of the SCCs in the corresponding states. As shown in Figures 1a and 2b, the peak area of the water content T_{2b} in samples that received the same dose of irradiation treatment was lower in the 4 °C group than in the RT group ($p < 0.05$), whereas T_{21} showed a rightward shift. These results indicate that low temperatures could promote the transformation of bound water to entrapped and free water, which may lead to a weakened water-binding capability of silver carp flesh and, consequently, an increase in EMC (Figure 4b). Compared with that in the unirradiated group, T_{22} of samples in both the irradiated groups (RT and 4 °C) showed a leftward shift and an increase in the peak areas, indicating that the irradiation reduced the loss of free water from the SCCs and transformed it into entrapped water. This was possibly caused by the change in the state of free water through the EBI-induced weakening of hydrogen bonds [34]. For the unirradiated samples, the sum of the T_{21} and T_{22} peak areas of samples in the 4 °C group was greater than that of specimens in the RT group, suggesting that low temperatures could also moderately reduce the drip loss of silver carp flesh. Figures 1b and 2 show that the signal from the bound water disappeared in the 4-kGy-irradiated/4-°C-stored group. This disappearance was probably due to the large peak area and high peak intensity of the signal from the entrapped water as well as the conversion of bound and free water to entrapped water, which led to a further decrease in the originally low bound-water proportion. The peak area of the signal representing entrapped water increased with increasing irradiation dose at RT. However, at 4 °C, the entrapped-water content was the highest in the 4-kGy-irradiated samples whereas it was reduced in the 8-kGy-irradiated specimens. In conclusion, low temperatures and irradiation could both reduce the drip loss of silver carp flesh, and the high content of entrapped water in the 4-kGy-irradiated/4-°C-stored group suggested that irradiation at 4 kGy and preservation at 4 °C improved the tenderness and quality of the silver carp flesh.

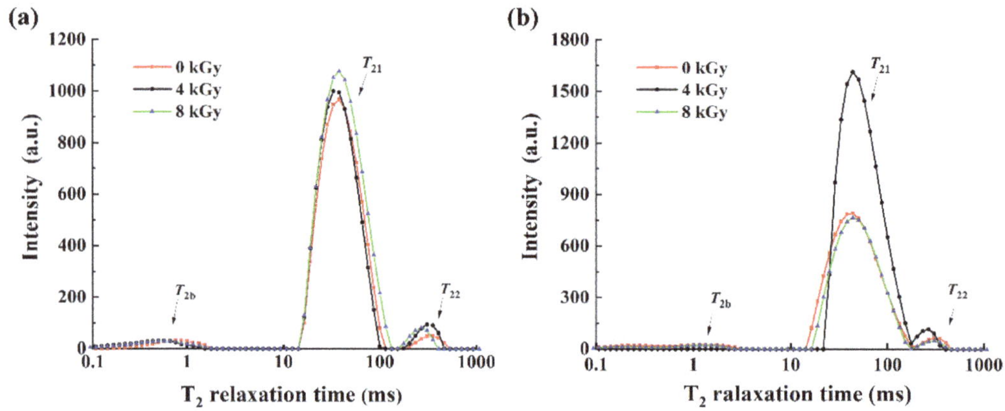

Figure 1. Effect of irradiation dose on moisture distribution of silver carp chunks stored at room temperature (**a**) and 4 °C (**b**).

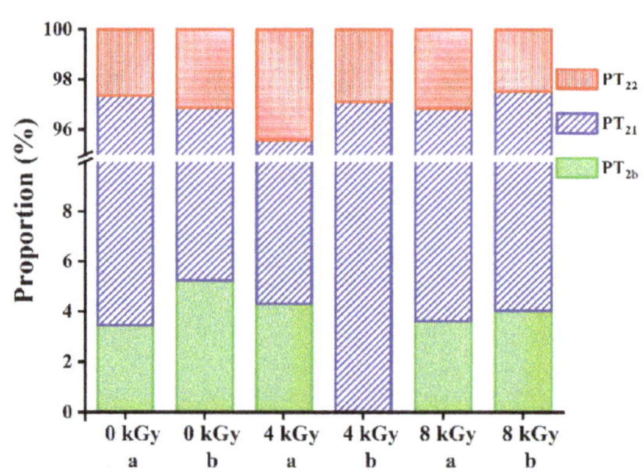

Figure 2. Diagram of relative moisture content of silver carp chunks at different irradiation doses stored at room temperature (**a**) and 4 °C (**b**).

The pseudo-color maps of water proton density constructed under different treatment conditions are shown in Figure 3. These maps reflect the distribution of water, with the areas featuring strong signals presented in red and the signal-free zones indicated in blue. The brightness of an image increases with increasing proton density, indicating a higher water content in the sample and less drip loss [35]. As shown in Figure 3, the brightness of the proton density pseudo-color maps decreased in the following manner: (d) > (e) > (b) > (c) > (a) > (f); this is consistent with the results shown in Figure 1a,b. Among the SCCs stored at RT, the samples of the irradiated group were darker than those in the unirradiated batch, and the red signals in the maps intensified with increasing EBI dose.

The signal intensity of the 4-kGy-irradiated/4-°C-stored samples was the strongest and higher than that of the RT stored equivalents. This may be because T_{22} showed a leftward shift most obviously under this condition, which reduced the loss of free water to the greatest extent. These results indicate that EBI could effectively reduce the drip loss of silver carp flesh and, in essence, inhibit the loss and diffusion of water in SCCs. Moreover, the preservation temperature of 4 °C and EBI dose of 4 kGy were effective in maintaining the quality of the silver carp product.

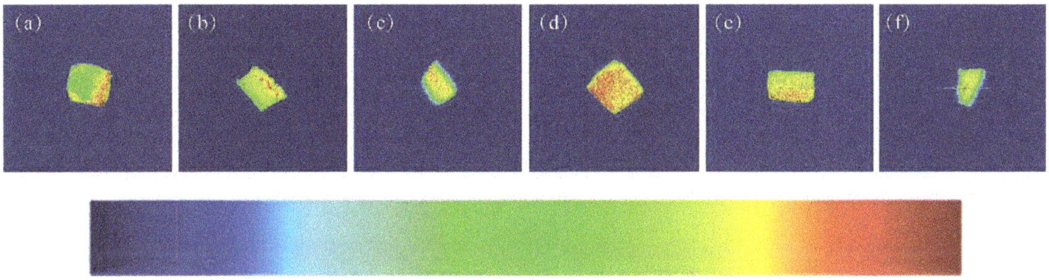

Figure 3. Pseudo-color maps of water molecular proton density of silver carp chunks with different treatment methods. (**a**) Room temperature, 0 kGy; (**b**) 4 °C, 0 kGy; (**c**) Room temperature, 4 kGy; (**d**) 4 °C, 4 kGy; (**e**) Room temperature, 8 kGy; (**f**) 4 °C, 8 kGy.

3.2. Effects of Different EBI Doses on the Moisture Content and EMC Loss of SCCs

All As shown in Figure 4a, the moisture content of SCC samples in the 4 °C group was not significantly different from that of the RT group ($p > 0.05$). Moreover, the moisture contents of the SCC samples treated at the same temperature but irradiated with different doses were not significantly different from those of the unirradiated samples ($p > 0.05$). Yang et al. showed that the moisture content of EBI-treated vacuum-packed Atlantic salmon fillets was not significantly correlated with the irradiation dose, which is consistent with the findings reported herein [36].

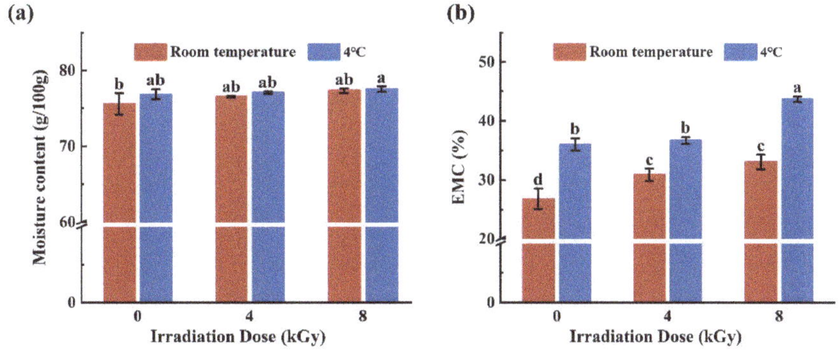

Figure 4. Effect of different doses of electron beam irradiation on moisture content (**a**) and EMC (**b**) of silver carp chunks. Different lowercase letters indicate significant difference among groups ($p < 0.05$).

Figure 4b shows that the rate of EMC loss of the centrifuged SCC samples was significantly lower in the RT group than in the 4 °C group ($p < 0.05$). Under ambient conditions, the EMC loss rates of the samples in the irradiated groups were significantly higher than those of the unirradiated samples ($p < 0.05$), whereas no significant difference was observed between the irradiated groups ($p > 0.05$). For the samples stored at 4 °C, the rates of EMC loss were not significantly different between the 4-kGy-irradiated and unirradiated samples

($p > 0.05$), whereas the 8-kGy-irradiated samples showed significantly higher values than those of the 4-kGy-irradiated counterparts ($p < 0.05$). The EMC was used to determine the water-holding capacity of the irradiated SCCs by exploiting the inverse proportionality between these parameters [37]. These results suggest that low temperatures and high irradiation doses may lead to a reduced water holding capacity and an increased EMC loss rate of SCCs.

3.3. Effects of Different EBI Doses on the Whiteness and pH Values of SCCs

As shown in Figure 5a, the effect of EBI on the whiteness value of the SCC samples was not significant at both RT and 4 °C ($p > 0.05$), indicating that EBI did not cause significant color related changes in the fish flesh. Zhang et al. conducted EBI treatment of vacuum-packed grass carp surimi and found that the whiteness values did not differ significantly among irradiated groups after EBI on the zeroth day of storage ($p > 0.05$), which is similar to the results reported herein [38].

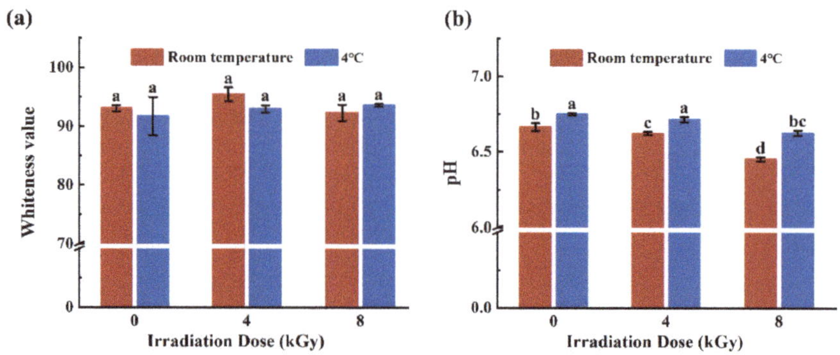

Figure 5. Effect of different doses of electron beam irradiation on whiteness value (**a**) and pH (**b**) of silver carp chunks. Different lowercase letters indicate significant difference among groups ($p < 0.05$).

The pH values of the SCCs stored at RT decreased significantly ($p < 0.05$) with increasing irradiation dose (Figure 5b). This may be because irradiation can moderately facilitate the breakdown of muscle glycogen, which produces acids such as ATP, lactate, and phosphocreatine, thereby leading to the decrease in sample pH and, consequently, accelerated food spoilage [39]. Ham. Y.-K et al. also found that the pH value of cooked pork sausages irradiated with electron-beam decreased with increasing absorbed dose level ($p < 0.05$) [40]. The pH values of the SCC samples were significantly higher in the 4 °C group than in the RT group ($p < 0.05$). Among the samples in the 4 °C group, the pH values of 4-kGy-irradiated samples were not significantly different from the unirradiated counterparts ($p > 0.05$). These results indicate that storage at 4 °C and EBI treatment at 4 kGy assisted in maintaining the original pH of the SCCs, and the low temperature could reduce the EBI-induced quality loss of SCCs.

3.4. Effects of Different EBI Doses on AVs, POVs, and TBA Values of SCCs

Temperature had minor effects on the AVs and POVs of the samples (Figure 6a,b). No significant difference ($p > 0.05$) was found in the AVs and POVs between the samples treated with the same irradiation dose but at different temperatures. Overall, the AVs and POVs of samples in both temperature groups increased with increasing irradiation dose. However, the AVs and POVs of the 4-kGy-irradiated samples were not significantly different from those of the unirradiated specimens ($p > 0.05$), but were significantly lower than those of the 8-kGy-irradiated counterparts ($p < 0.05$). Oxidation of fats and oils yields unsaturated fatty acids, which can be further oxidized upon exposure to light and heat, resulting in the production of organic acids and, consequently, food rancidity; moreover, higher AVs indicate a greater degree of fat and oil oxidation [41]. The 4-kGy-irradiation treatment minimally affected the AVs of the SCCs, whereas the 8-kGy-irradiation treatment accelerated their rancidification, resulting in higher AVs. The elevated POVs of the 8-kGy-irradiated SCCs may be due to the expedited lipid oxidation induced by the free radicals in the tissues that are generated by high-dose irradiation [42,43].

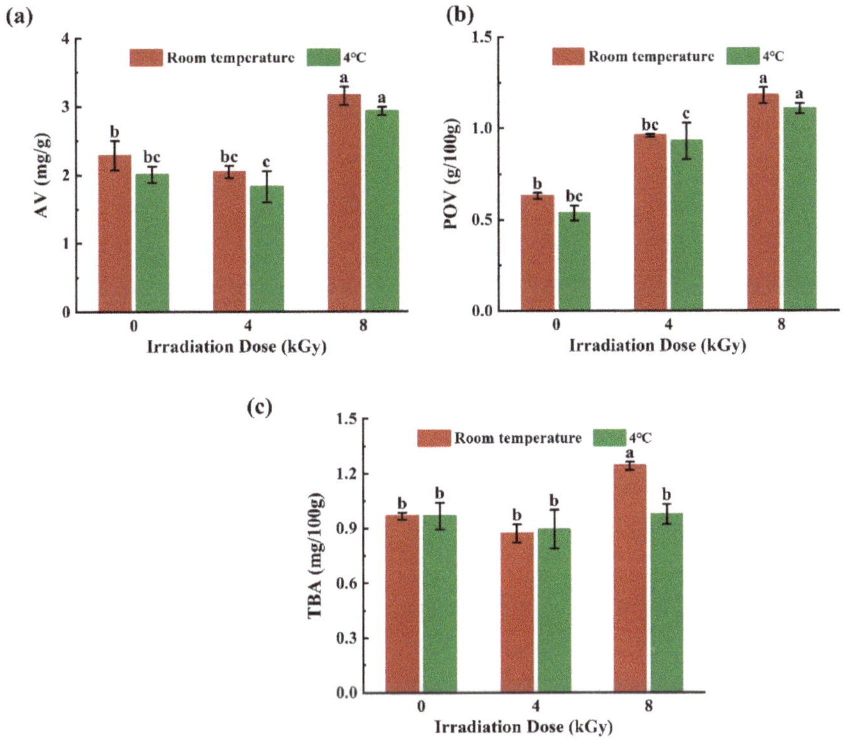

Figure 6. Effects of different doses of electron beam irradiation on TBA value (**a**), AV value (**b**) and POV value (**c**) of silver carp chunks. Different lowercase letters indicate significant difference among groups ($p < 0.05$).

The TBA values of the 8-kGy-irradiated samples in the RT group were higher than those of the 4-kGy-irradiated and unirradiated counterparts (Figure 6c; $p < 0.05$). However, in the 4 °C group, the TBA values of the 8-kGy-irradiated samples were not significantly different from those of the 4-kGy-irradiated equivalents ($p > 0.05$). This indicates that high dose EBI under ambient conditions could promote the decomposition of unsaturated fatty acids and accelerate lipid oxidation, whereas low-temperature preservation could delay lipid oxidation and thus counteract the adverse effects of irradiation. The TBA values and POVs of EBI treated pork jerky have been found to increase in an EBI-dose-dependent manner [44]. EBI can catalyze the production of free radicals in fish products and accelerate lipid oxidation, triggering a free-radical chain reaction that increases the TBA values [45], which is similar to the mechanism by which EBI increases the POVs.

3.5. *Effects of Different EBI Doses on Total Protein Content, Myofibrillar Protein Content, and Actomyosin Content*

Fish meat is a major source of animal protein for consumers owing to its high protein content [46]. The total protein contents of samples in all groups ranged from 16 g/100 g to 17 g/100 g, and the variation between all samples was not significant ($p > 0.05$; Figure 7a).

The aforementioned findings are consistent with those of Fallah et al. who found that irradiation minimally affected the crude protein content in camel meat [47].

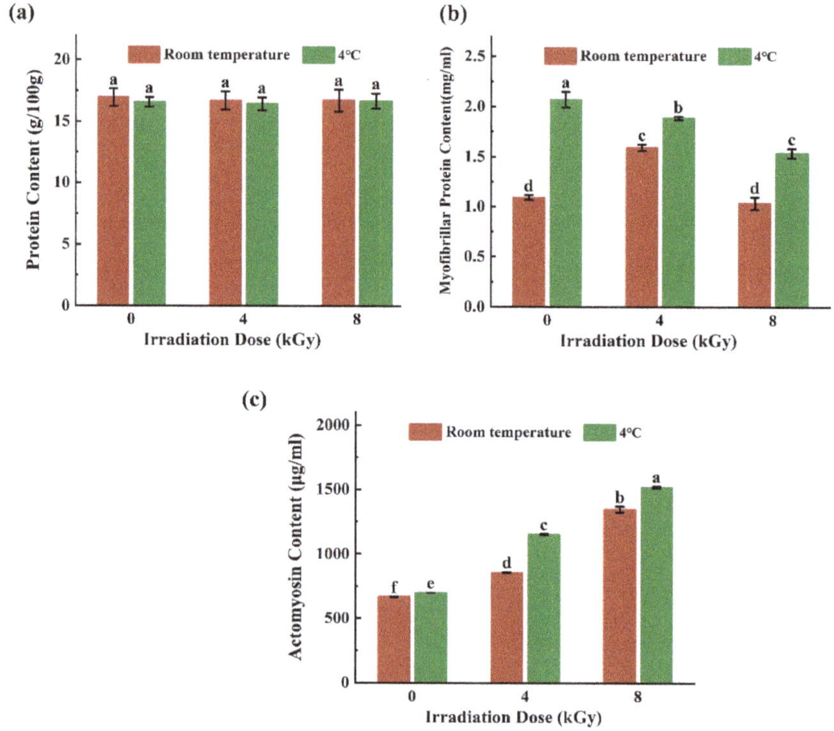

Figure 7. Effects of different doses of electron beam irradiation on protein content (**a**), myofibrillar protein content (**b**) and actomyosin content (**c**) of silver carp chunks. Different lowercase letters indicate significant difference among groups ($p < 0.05$).

Myofibrillar protein constitutes the myofibrils in muscles and directly affects the juiciness, texture, and elasticity of meat products. The myofibrillar protein content in the 4-kGy-irradiated/RT-stored samples was 1.59 mg/mL (Figure 7b), which was a 31.64% increase over that of the unirradiated samples ($p < 0.05$); moreover, the difference between the 8-kGy-irradiated and unirradiated samples was not significant ($p > 0.05$). This indicated that the 4 kGy EBI treatment and RT storage helped improve the elasticity and texture of the SCCs. Additionally, the myofibrillar protein content of the SCC samples in the 4 °C group was higher than that of the specimens in the RT group by 0.29–0.98 mg/mL ($p < 0.05$), possibly due to the low-temperature-induced inhibition of myofibrillar protein oxidation [48].

The actomyosin contents of the SCC samples in both temperature groups increased in an EMI-dose-dependent manner (Figure 7c), with the actomyosin contents of the irradiated samples being significantly higher than those of the unirradiated specimens ($p < 0.05$). Additionally, the actomyosin content of the samples in the 4 °C group was significantly higher than that of the samples in the RT group and ranged from 36.21 to 296.58 μg/mL ($p < 0.05$). Actomyosin is the main component of myofibrillar protein, and the denaturation and loss of myosin are considered valid indicators of quality loss of fish meat [49]. These results suggest that low-temperature preservation (4 °C) and EBI treatment could inhibit the denaturation and decomposition of actomyosin and effectively improve the quality of fish protein.

4. Conclusions

In this study, the effects of different preservation temperatures and EBI doses on the quality of SCCs were investigated. The results showed that low preservation temperatures and 4 kGy EBI treatment could prevent the partial free water loss in the meat of silver carp and consequently improve the tenderness and quality of the meat. The centrifugation induced EMC loss of the SCCs was significantly and positively correlated with the irradiation dose, and the water holding capacity of the samples in the 4 °C group was lower than that of the samples in the RT group. Low preservation temperature prevented the loss of myofibrillar protein and actomyosin, and EBI effectively inhibited actomyosin degeneration. Moreover, the myofibrillar protein content tended to decrease with increasing irradiation dose. When the irradiation dose reached 8 kGy, the oxidation of silver carp was accelerated and the pH value was increased. Overall, preservation at 4 °C and EBI treatment at 4 kGy were determined to be the optimal EBI processing conditions for SCCs. The findings reported herein are in support of the irradiation-based preservation of silver carp products.

Author Contributions: Conceptualization, Investigation, Validation, Writing-original draft, H.-L.L.; Methodology, Software, Data curation, Writing—original draft, M.-J.L.; Investigation, Data curation, Software, Q.Z.; Methodology, Supervision, Investigation, J.-J.H.; Supervision, Validation, Funding acquisition, Project administration, Writing—review & editing., X.-Y.Z. All authors have read and agreed to the published version of the manuscript.

Funding: This work was financially supported by National Key Research and Development Program of China (2019YFD0902000).

Institutional Review Board Statement: Not applicable.

Informed Consent Statement: Not applicable.

Data Availability Statement: The data showed in this study are contained within the article.

Acknowledgments: We appreciate Jun Cai for the provision of master's students.

Conflicts of Interest: The authors declare no competing financial or other interest.

References

1. Abdollahi, M.; Rezaei, M.; Jafarpour, A.; Undeland, I. Sequential extraction of gel-forming proteins, collagen and collagen hydrolysate from gutted silver carp (*Hypophthalmichthys molitrix*), a biorefinery approach. *Food Chem.* **2018**, *242*, 568–578. [CrossRef] [PubMed]
2. Majdoubi, F.-Z.; Ouizgane, A.; Farid, S.; Mossetti, L.; Droussi, M.; Guerriero, G.; Hasnaoui, M. Fry Survival Rate as a Predictive Marker of Optimal Production of Silver Carp (*Hypophthalmichthys molitrix, Valenciennes 1844*): A Biostatistical Study in Deroua Fish Farm, Morocco. *Proc. Zool. Soc.* **2021**, *75*, 152–160. [CrossRef]
3. Li, D.P.; Li, Q.; Zhang, Y.M.; Liu, X.C.; Hong, H.; Luo, Y.K. Quality changes and microbiological spoilage analysis of air-packed and vacuum-packed silver carp (*Hypophthalmichthys molitrix*) fillets during chilled storage. *J. Food Process. Preserv.* **2018**, *42*, e13389. [CrossRef]
4. DeWitt, C.A.; Oliveira, A.C. Modified Atmosphere Systems and Shelf Life Extension of Fish and Fishery Products. *Foods* **2016**, *5*, 48. [CrossRef] [PubMed]
5. Qiu, C.; Xia, W.; Jiang, Q. Pressure-induced changes of silver carp (*Hypophthalmichthys molitrix*) myofibrillar protein structure. *Eur. Food Res. Technol.* **2014**, *238*, 753–761. [CrossRef]
6. Yu, L.; Jiang, Q.; Yu, D.; Xu, Y.; Gao, P.; Xia, W. Quality of giant freshwater prawn (*Macrobrachium rosenbergii*) during the storage at −18 °C as affected by different methods of freezing. *Int. J. Food Prop.* **2018**, *21*, 2100–2109. [CrossRef]
7. Annamalai, J.; Sivam, V.; Unnikrishnan, P.; Kuppa Sivasankara, S.; Kaushlesh Pansingh, R.; Shaik Abdul, K.; Lakshmi, N.M.; Chandragiri Nagarajarao, R. Effect of electron beam irradiation on the biochemical, microbiological and sensory quality of Litopenaeus vannamei during chilled storage. *J. Food Sci. Technol.* **2020**, *57*, 2150–2158. [CrossRef]
8. Fernandes, A.; Antonio, A.L.; Oliveira, M.B.; Martins, A.; Ferreira, I.C. Effect of gamma and electron beam irradiation on the physico-chemical and nutritional properties of mushrooms: A review. *Food Chem.* **2012**, *135*, 641–650. [CrossRef]
9. Dong, S.; Guo, J.; Yu, J.; Bai, J.; Xu, H.; Li, M. Effects of electron-beam generated X-ray irradiation on the postharvest storage quality of Agaricus bisporus. *Innov. Food Sci. Emerg. Technol.* **2022**, *80*, 103079. [CrossRef]
10. Aguirre, J.; Rodríguez, M.R.; González, R.; García de Fernando, G. E-beam irradiation affects the maximum specific growth rate ofBacillus cereus. *Int. J. Food Sci. Technol.* **2013**, *48*, 382–386. [CrossRef]

11. Ravindran, R.; Jaiswal, A.K. Wholesomeness and safety aspects of irradiated foods. *Food Chem.* **2019**, *285*, 363–368. [CrossRef]
12. Yu, H.; Zhang, J.; Li, H.; Zhao, Y.; Xia, S.; Qiu, Y.; Zhu, J. Effects of E-beam irradiation on the physicochemical properties of Atlantic cod (*Gadus morhua*). *Food Biosci.* **2022**, 101803. [CrossRef]
13. Guo, H.; Feng, T.; Qi, W.; Kong, Q.; Yue, L.; Wang, H. Effects of electron-beam irradiation on volatile flavor compounds of salmon fillets by the molecular sensory science technique. *J. Food Sci.* **2021**, *86*, 184–193. [CrossRef]
14. Yu, Q.; Pan, H.; Qian, C.; Shao, H.; Han, J.; Li, Y.; Lou, Y. Determination of the optimal electron beam irradiation dose for treating shrimp (*Solenocera melantho*) by means of physical and chemical properties and bacterial communities. *Lwt* **2022**, *153*, 112539. [CrossRef]
15. Luo, J.; Li, M.; Zhang, Y.; Zheng, M.; Ming Ling, C. The low-field NMR studies the change in cellular water in tilapia fillet tissue during different drying conditions. *Food Sci. Nutr.* **2021**, *9*, 2644–2657. [CrossRef]
16. Cheng, S.; Zhang, T.; Yao, L.; Wang, X.; Song, Y.; Wang, H.; Wang, H.; Tan, M. Use of low-field-NMR and MRI to characterize water mobility and distribution in pacific oyster (*Crassostrea gigas*) during drying process. *Dry. Technol.* **2017**, *36*, 630–636. [CrossRef]
17. Sanchez-Alonso, I.; Martinez, I.; Sanchez-Valencia, J.; Careche, M. Estimation of freezing storage time and quality changes in hake (*Merluccius merluccius*, L.) by low field NMR. *Food Chem.* **2012**, *135*, 1626–1634. [CrossRef]
18. Tan, M.; Lin, Z.; Zu, Y.; Zhu, B.; Cheng, S. Effect of multiple freeze-thaw cycles on the quality of instant sea cucumber: Emphatically on water status of by LF-NMR and MRI. *Food Res. Int.* **2018**, *109*, 65–71. [CrossRef]
19. Wang, G.; Wang, D.; Qing, C.; Chen, L.; Gao, P.; Huang, M. Impacts of electron-beam-irradiation on microstructure and physical properties of yam (*Dioscorea opposita Thunb.*) flour. *LWT* **2022**, *163*, 113531. [CrossRef]
20. Wang, L.; Wang, X.; Ma, J.; Yang, K.; Feng, X.; You, X.; Wang, S.; Zhang, Y.; Xiong, G.; Wang, L.; et al. Effects of radio frequency heating on water distribution and structural properties of grass carp myofibrillar protein gel. *Food Chem.* **2021**, *343*, 128557. [CrossRef]
21. Al-Habsi, N.A.; Al-Hadhrami, S.; Al-Kasbi, H.; Rahman, M.S. Molecular mobility of fish flesh measured by low-field nuclear magnetic resonance (LF-NMR) relaxation: Effects of freeze–thaw cycles. *Fish. Sci.* **2017**, *83*, 845–851. [CrossRef]
22. Shi, H.; Zhang, M.; Yang, C. Effect of low-temperature vacuum frying assisted by microwave on the property of fish fillets (*Aristichthys nobilis*). *J. Food Process Eng.* **2019**, *42*, e13050. [CrossRef]
23. Jiao, X.; Cao, H.; Fan, D.; Huang, J.; Zhao, J.; Yan, B.; Zhou, W.; Zhang, W.; Ye, W.; Zhang, H. Effects of fish oil incorporation on the gelling properties of silver carp surimi gel subjected to microwave heating combined with conduction heating treatment. *Food Hydrocoll.* **2019**, *94*, 164–173. [CrossRef]
24. Peng, J.; Zheng, F.; Wei, L.; Lin, H.; Jiang, J.; Hui, G. Jumbo squid (*Dosidicus gigas*) quality enhancement using complex bio-preservative during cold storage. *J. Food Meas. Charact.* **2017**, *12*, 78–86. [CrossRef]
25. Gulcan, U.; Candal Uslu, C.; Mutlu, C.; Arslan-Tontul, S.; Erbas, M. Impact of inert and inhibitor baking atmosphere on HMF and acrylamide formation in bread. *Food Chem.* **2020**, *332*, 127434. [CrossRef]
26. Raeisi, S.; Ojagh, S.M.; Pourashouri, P.; Salaun, F.; Quek, S.Y. Shelf-life and quality of chicken nuggets fortified with encapsulated fish oil and garlic essential oil during refrigerated storage. *J. Food Sci. Technol.* **2021**, *58*, 121–128. [CrossRef] [PubMed]
27. Mohdaly, A.A.A.; Mahmoud, A.A.; Ramadan, M.F.; Roby, M.H.H. Biochemical and microbiological characteristics of some Mediterranean salted fish products. *Rend. Lincei Sci. Fis. E Nat.* **2021**, *32*, 343–355. [CrossRef]
28. Wei, D.; Li, L.; He, S.; Yu, J.; Tian, X.; Wu, Z. Improving the lipid oxidation in pork fat processing for Chi-aroma Baijiu through pretreatments and segmented soaking with liquor. *LWT-Food Sci. Technol.* **2020**, *130*, 109624. [CrossRef]
29. Wang, D.; Dong, Y.; Chen, X.; Liu, Y.; Wang, J.; Wang, X.; Wang, C.; Song, H. Incorporation of apricot (*Prunus armeniaca*) kernel essential oil into chitosan films displaying antimicrobial effect against Listeria monocytogenes and improving quality indices of spiced beef. *Int. J. Biol. Macromol.* **2020**, *162*, 838–844. [CrossRef]
30. Salih, A.M.; Smith, D.M.; Price, J.F.; Dawson, L.E. Modified extraction 2-thiobarbituric acid method for measuring lipid oxidation in poultry. *Poult. Sci.* **1987**, *66*, 1483–1488. [CrossRef] [PubMed]
31. Yang, M.; Yang, L.; Xu, J.; Nie, Y.; Wu, W.; Zhang, T.; Wang, X.; Zhong, J. Comparison of silver carp fin gelatins extracted by three types of methods: Molecular characteristics, structure, function, and pickering emulsion stabilization. *Food Chem.* **2022**, *368*, 130818. [CrossRef]
32. Benjakul, S.; Seymour, T.A.; Morrissey, M.T.; An, H. Physicochemical changes in Pacific whiting muscle proteins during iced storage. *J. Food Sci.* **1997**, *62*, 729–733. [CrossRef]
33. Zhou, A.; Lin, L.; Liang, Y.; Benjakul, S.; Shi, X.; Liu, X. Physicochemical properties of natural actomyosin from threadfin bream (*Nemipterus* spp.) induced by high hydrostatic pressure. *Food Chem.* **2014**, *156*, 402–407. [CrossRef]
34. Zhai, Y.; Pan, L.; Luo, X.; Zhang, Y.; Wang, R.; Chen, Z. Effect of electron beam irradiation on storage, moisture and eating properties of high-moisture rice during storage. *J. Cereal Sci.* **2022**, *103*, 103407. [CrossRef]
35. Cheng, S.; Wang, X.; Li, R.; Yang, H.; Wang, H.; Wang, H.; Tan, M. Influence of multiple freeze-thaw cycles on quality characteristics of beef semimembranous muscle: With emphasis on water status and distribution by LF-NMR and MRI. *Meat Sci.* **2019**, *147*, 44–52. [CrossRef]
36. Yang, Z.; Wang, H.; Wang, W.; Qi, W.; Yue, L.; Ye, Q. Effect of 10 MeV E-beam irradiation combined with vacuum-packaging on the shelf life of Atlantic salmon fillets during storage at 4 degrees C. *Food Chem.* **2014**, *145*, 535–541. [CrossRef]
37. Zheng, M.J.; Liu, X.; Chuai, P.J.; Jiang, Z.D.; Zhu, Y.B.; Zhang, B.; Ni, H.; Li, Q.B. Effects of crude fucoidan on physicochemical properties, antioxidation and bacteriostasis of surimi products. *Food Control* **2021**, *122*, 107806. [CrossRef]

38. Zhang, H.F.; Wang, W.; Zhang, S.F.; Wang, H.Y.; Ye, Q.F. Influence of 10-MeV E-Beam Irradiation and Vacuum Packaging on the Shelf-Life of Grass Carp Surimi. *Food Bioprocess Technol.* **2016**, *9*, 830–838. [CrossRef]
39. Ucar, Y.; Özogul, Y.; Özogul, F.; Durmuş, M.; Köşker, A.R. Effect of nisin on the shelf life of sea bass (*Dicentrarchus labrax* L.) fillets stored at chilled temperature (4 ± 2 °C). *Aquac. Int.* **2020**, *28*, 851–863. [CrossRef]
40. Ham, Y.-K.; Kim, H.-W.; Hwang, K.-E.; Song, D.-H.; Kim, Y.-J.; Choi, Y.-S.; Song, B.-S.; Park, J.-H.; Kim, C.-J. Effects of irradiation source and dose level on quality characteristics of processed meat products. *Radiat. Phys. Chem.* **2017**, *130*, 259–264. [CrossRef]
41. Wang, X.Y.; Xie, J.; Chen, X.J. Differences in lipid composition of Bigeye tuna (Thunnus obesus) during storage at 0 °C and 4 °C. *Food Res. Int.* **2021**, *143*, 110233. [CrossRef]
42. An, K.A.; Arshad, M.S.; Jo, Y.; Chung, N.; Kwon, J.H. E-Beam Irradiation for Improving the Microbiological Quality of Smoked Duck Meat with Minimum Effects on Physicochemical Properties During Storage. *J. Food Sci.* **2017**, *82*, 865–872. [CrossRef]
43. Hassanzadeh, P.; Tajik, H.; Rohani, S.M.R.; Moradi, M.; Hashemi, M.; Aliakbarlu, J. Effect of functional chitosan coating and gamma irradiation on the shelf-life of chicken meat during refrigerated storage. *Radiat. Phys. Chem.* **2017**, *141*, 103–109. [CrossRef]
44. Kim, H.J.; Kang, M.; Yong, H.I.; Bae, Y.S.; Jung, S.; Jo, C. Synergistic Effects of Electron-beam Irradiation and Leek Extract on the Quality of Pork Jerky during Ambient Storage. *Asian-Australas. J. Anim. Sci.* **2013**, *26*, 596–602. [CrossRef]
45. Arshad, M.S.; Kwon, J.H.; Ahmad, R.S.; Ameer, K.; Ahmad, S.; Jo, Y. Influence of E-beam irradiation on microbiological and physicochemical properties and fatty acid profile of frozen duck meat. *Food Sci. Nutr.* **2020**, *8*, 1020–1029. [CrossRef]
46. Oyekunle, J.A.O.; Ore, O.T.; Durodola, S.S.; Oyinloye, J.A.; Oyebode, B.A.; Ajanaku, O.L. Heavy metal levels and changes in trimethylamine content of smoked fish and meat under different storage conditions. *Sn Appl. Sci.* **2020**, *2*, 1–8. [CrossRef]
47. Fallah, A.A.; Tajik, H.; Farshid, A.A. Chemical Quality, Sensory Attributes and Ultrastructural Changes of Gamma-Irradiated Camel Meat. *J. Muscle Foods* **2010**, *21*, 597–613. [CrossRef]
48. Zhao, N.N.; Yang, X.Q.; Li, Y.J.; Wu, H.H.; Chen, Y.P.; Gao, R.C.; Xiao, F.; Bai, F.; Wang, J.L.; Liu, Z.Y.; et al. Effects of protein oxidation, cathepsins, and various freezing temperatures on the quality of superchilled sturgeon fillets. *Mar. Life Sci. Technol.* **2022**, *4*, 117–126. [CrossRef]
49. Fan, X.; Konno, K.; Lin, X.; Yu, X.; Liu, Y.; Dong, X. The effect of fish freshness on myosin denaturation in flounder Paralichthys olivaceus muscle during frozen storage. *Fish. Sci.* **2020**, *86*, 1111–1120. [CrossRef]

foods

MDPI

Article

Effects of Modified Atmosphere Packaging with Different Gas Ratios on the Quality Changes of Golden Pompano (*Trachinotus ovatus*) Fillets during Superchilling Storage

Xiaofan Zhang [1,2,3], Chuang Pan [3,*], Shengjun Chen [2,4,*], Yong Xue [1], Yueqi Wang [2,3] and Yanyan Wu [3]

1 College of Food Science and Engineering, Ocean University of China, Qingdao 266003, China; zhangxiaofan19@sina.com (X.Z.); xueyong@ouc.edu.cn (Y.X.)
2 Sanya Tropical Fisheries Research Institute, Sanya 572000, China; wangyueqi@scsfri.ac.cn
3 Key Laboratory of Aquatic Product Processing, Ministry of Agriculture and Rural Affairs, National R&D Centre for Aquatic Product Processing Technology, South China Sea Fisheries Research Institute, Chinese Academy of Fishery Sciences, Guangzhou 510300, China; wuyygd@163.com
4 Collaborative Innovation Center of Seafood Deep Processing, Dalian Polytechnic University, Dalian 116034, China
* Correspondence: silverpfoxc@hotmail.com (C.P.); chenshengjun@scsfri.ac.cn (S.C.); Tel.: +86-020-89020911 (C.P.); +86-020-89108310 (S.C.)

Abstract: The quality changes of golden pompano fillets in air packaging (AP) and modified atmosphere packaging (MAP) with 30% CO_2/70% N_2, 50% CO_2/50% N_2, and 70% CO_2/30% N_2 were evaluated under superchilling (−3 °C). The results showed that the whiteness of fillets decreased during storage. The rate of pH increase of MAP was significantly slower than in AP groups, in which MAP with 70% CO_2/30% N_2 effectively suppressed the PH. Interestingly, the hardness decreased on day five following the treatments, followed by a relatively stationary trend. MAP could greatly suppress the increase of total volatile basic nitrogen (TVB-N) contents of fillets compared to fillets packed in AP. All MAP groups of fillets maintained first-grade freshness throughout storage, while the AP samples decreased to second-grade freshness on about the 25th day. MAP with 70% CO_2/30% N_2 and MAP with 50% CO_2/50% N_2 had the best results in inhibiting protein degeneration and explanation. Unexpectedly, drip loss of fillets in MAP far exceeded the AP group during storage, which causes sensory discomfort. Anaerobic plate count (APC) of fillets in AP exceeded the consumption limit of 6.7 log CFU/g on day 26 (6.75 log CFU/g on the 26th day), whereas the MAP was still microbiologically acceptable after 30 days of storage (6.43, 6.41, 6.22 log CFU/g, respectively). Considering physicochemical and microbiological parameters, the shelf life of fillets packed in AP was 25 days. MAP treatments could prolong the shelf life of fillets by ~4–5 days compared to AP. Overall, MAP with 70% CO_2/30% N_2 gas ratio was best for inhibiting the quality deterioration of fillets. Furthermore, principal component analysis (PCA) was performed to evaluate the critical indicators of quality deterioration of the fillets. Two principal components were determined by dimensionality reduction, in which the contribution of the first principal component was centrifugal loss > hardness > TVB-N > APC > CO_2 solubility > TBARs > drip loss > pH, which mainly reflected the degree of microbial proliferation, protein hydrolysis, and oxidation. The contribution of the second principal component was pH > TBRAs > drip loss > APC > CO_2 solubility > TVB-N > hardness > centrifugal loss, indicating a high correlation between lipid oxidation and microbial proliferation index.

Keywords: *Trachinotus ovatus*; golden pompano; modified atmosphere packaging; superchilling storage; physicochemical properties; principal component analysis

Citation: Zhang, X.; Pan, C.; Chen, S.; Xue, Y.; Wang, Y.; Wu, Y. Effects of Modified Atmosphere Packaging with Different Gas Ratios on the Quality Changes of Golden Pompano (*Trachinotus ovatus*) Fillets during Superchilling Storage. *Foods* **2022**, *11*, 1943. https://doi.org/10.3390/foods11131943

Academic Editor: Fatih Öz

Received: 9 June 2022
Accepted: 28 June 2022
Published: 29 June 2022

1. Introduction

Golden pompano (*Trachinotus ovatus*) is a marine aquaculture economic fish of the genus *Trachinotus* in the family *Carangidae*, it has white, delicate flesh a delicious taste and is rich in protein and polyunsaturated fatty acids [1]. At present, golden pompano is mainly sold fresh, frozen or refrigerated, which is susceptible to protein, lipid oxidation and other spoilage phenomena during storage, resulting in short shelf life and reducing the economic value of golden pompano products [2].

Most aquatic products use low-temperature preservation, but traditional refrigeration (4 °C) shelf life is short. Moreover, frozen (−18~−20 °C) processing can easily cause protein denaturation, lipid oxidation and drip loss in the thawing process of aquatic products. However, superchilling is a temperature between refrigeration and freezing storage methods; the food temperature is usually controlled at 1–2 °C below freezing point [3]. Most microbial activities, protein deterioration and lipid oxidation are inhibited, moreover, the drip loss of aquatic products was at a lower level until the end of superchilling storage [4]. Compared with refrigeration, superchilling storage can effectively extend the shelf life of fillets by 1.5 to 4 times [5].

A previous study showed that superchilling (−3 °C) storage can effectively reduce microbial growth, lipid oxidation and proteolytic degradation in common carp surimi [6]. Mi et al. [4] reported that lower drip loss and higher water holding capacity under superchilling (−2 °C) storage were observed compared to frozen (−18 °C) storage and the shelf life of superchilled grass carp was 21 days. Liu et al. [7] also found that superchilling storage can inhibit the growth of microorganisms and retard the hydrolysis of proteins more effectively than at 0 °C in grass carp fillets. Superchilling preservation technology has many advantages in preserving food, but the temperature fluctuation, repeated freezing and thawing caused by the instrumentation or external environment can have adverse effects on the protein quality of the product [8,9]. Therefore, it is necessary to combine other preservation technologies to meet the market requirements for the freshness of aquatic products.

Modified atmosphere packaging (MAP) preservation technology has been widely used in aquatic preservation [10]. MAP refers to the use of high barrier performance packaging material to package food in which O_2, CO_2 and N_2 gas are mixed into the packaging material to inhibit microbial growth [11], reduce the enzymatic reaction and slow down the rate of lipid oxidation [12]. It was illuminated that MAP significantly reduced the growth rate of pH, TVB-N and TBARs values and prolonged the shelf life of bream blocks [10]. Similar results were reported in a study of rainbow trout fillets [13]. The storage temperature can directly affect the deterioration rate of aquatic products and the combination of MAP and superchilling can significantly improve product quality [14]. Some researchers studied the preservation effect of MAP (60% CO_2/40% N_2) on fresh Atlantic salmon fillets at −2 °C and 4 °C. It was revealed that the quality of MAP salmon fillets deteriorated slowly at superchilling storage, which could effectively maintain the good quality of salmon fillets for 24 days [15]. It was also claimed that the APC and thermophilic bacteria were lower during storage at −0.7 °C compared to 4 °C in catfish, meanwhile, the TVB-N and TMA proved that the quality of MAP catfish could be better maintained in the −0.7 °C [16]. The packaging cost of MAP products is mainly determined by the pre-cooling conditions, sales environment and packaging materials. The first two belong to the one-time investment of the factory, and the cost is determined according to scale. Packaging materials mainly have gas and packaging bags, etc.; according to a rough calculation, a product packaging cost price is about 0.071~0.092 EURO, more than ordinary air packaging and vacuum packaging cost price is slightly higher, but gas-packaged products' shelf life and fresh quality is better, so the sales price is higher than ordinary packaging but greatly enhances the market competitiveness of the product [17].

MAP combined with superchilling technology has been applied to many aquatic products. However, the exact gas ratio that is best for golden pompano storage has not been explored and our research will help enrich the market for packaged fillet products and contribute to improving the economic value of golden pompano or other fish crops. Therefore, the main objectives of this study were: (1) to investigate the effect of MAP with different gas ratios on physicochemical and microbial changes, such as pH, texture, water retention, CO_2 solubility, whiteness, TBARs, TVB-N and APC of golden pompano fillets by using MAP combined with superchilling; (2) to evaluate the critical indicators of quality deterioration of the packaging fillets during storage by PCA.

2. Materials and Methods

2.1. Sample Preparation

Fresh golden pompano (each fish weighing 500~600 g) (Figure 1A) were purchased from China Resources Vanguard supermarket in Guangzhou, China, accompanied by ice and returned to the laboratory within 5 min. Mucus was quickly washed off the surface with running water. Intact fillets (Figure 1B) of fish were removed from guts, black film, blood clots, etc. The surface water of the fillets was gently dried with kitchen paper and they were put into bags (Figure 1C) with high gas barrier performance (300 × 200) mm (Sinopharm Chemical Reagent Co., Ltd., Shanghai, China). Three different ratios of gas mixtures (30% CO_2/70% N_2; 50% CO_2/50% N_2; 70% CO_2/30% N_2) were selected using a gas conditioning preservation packaging machine (MAP-D400; Suzhou Senruy Co., Ltd., Suzhou, China) to create aerated packages named MAP1, MAP2 and MAP3 respectively. The control group used air packages named AP. Initial gas/fish volume ratio was 3:1 for all packages. 96 pieces of fillets were randomly divided into 4 groups for storage and 4 pieces were randomly taken from each group for parallel analysis during each experiment. All samples were stored together at -3 ± 0.5 °C in a precision low-temperature incubator (IN612C; Yamato Science Chongqing Co., Ltd., Chongqing, China).

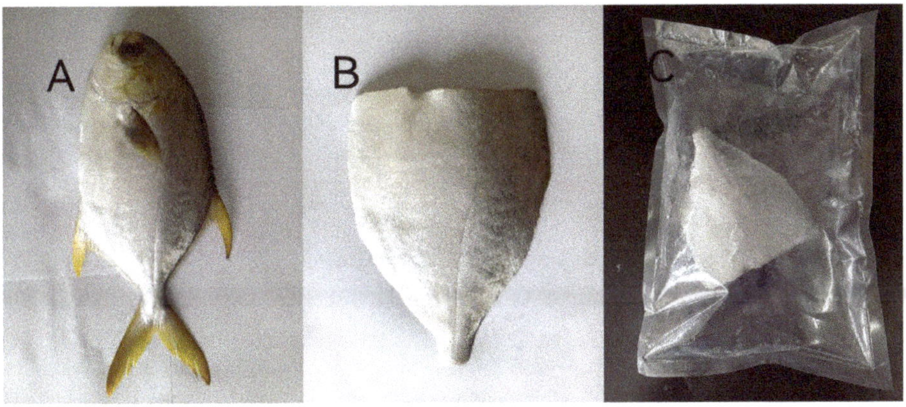

Figure 1. Images of golden pompano (**A**), fillet sample (**B**) and packaging method (**C**).

The samples were taken at 0, 5, 10, 15, 20, 25 and 30 days after storage and analyzed for changes in muscle hardness, elasticity, CO_2 solubility, whiteness, APC, pH, TVB-N, TBARs and water retention capacity.

2.2. Water Retention Analysis of Golden Pompano Fillets

2.2.1. Determination of Drip Loss

Drip loss of fillets was estimated according to the method of Liu and Liang [7]. The weight of the fillets was taken before they were packed and recorded as W_1. After storage, the juice on the surface of the fillets was absorbed using kitchen paper. The fillets were then weighed and recorded as W_2. Triplicate samples were used to calculate drip loss according to Equation (1).

$$drip\ loss(\%) = \frac{(W_1 - W_2)}{W_1} \times 100\% \quad (1)$$

2.2.2. Determination of Cooking Loss

Cooking loss of fish fillets was estimated using the method of Delles and Xiong [18] with light modification. The fillets were steamed for three minutes in a pot drawer containing boiling water at 100 °C. They were then weighed after gently absorbing the surface juice of the fillets with kitchen paper and recorded as W_3. Triplicate samples were used to calculate drip loss according to Equation (2).

$$cooking\ loss(\%) = \frac{(W_2 - W_3)}{W_2} \times 100\% \quad (2)$$

2.2.3. Determination of Centrifugal Loss

Determination of centrifugal loss according used the method of Delles and Xiong [18] with modification. A suitable size of filter paper was weighed on a balance and recorded as W_4. Fish samples of 2 g were then accurately weighed, wrapped with filter paper and placed in a centrifuge tube. The samples were then put it in a freezing centrifuge at 1500 r/min for 15 min. After centrifugation, the residual fillets were removed from the filter paper. The mass of the filter paper after centrifugation was recorded as W_5. The calculation (3) of centrifugal loss is as follows:

$$centrifugal\ loss(\%) = \frac{(W_5 - W_4)}{W_4} \times 100\% \quad (3)$$

2.3. Physical Analysis of Golden Pompano Fillets

2.3.1. Texture Profile Analysis

Texture profile analysis (TPA) was carried out by a texture analyzer (CT3; Brookfield, IL, USA). Fish blocks of 4 × 2 × 1.5 cm in length, width and height were taken for TPA texture analysis. The samples were placed on the platform and two cycles of TPA experiments were performed with a 4 mm diameter TA44 cylindrical probe with a trigger point load of 5 g; the test target was a distance of 5.0 mm; the test speed was 0.5 mm/s. The TPA parameters, such as hardness (kg) and elasticity (mm), were obtained. Six parallel samples were measured for each group.

2.3.2. Determination of pH and Whiteness

2.0 g of each fillet was homogenized in 20 mL of distilled water. The pH was measured using a digital pH meter (IS128; Yimai Shanghai, Shanghai, China) and this was repeated three times.

The color of fillets was measured using a Chroma Meter (CR-400; Konica Minolta, Tokyo, Japan) with the CIE color system. The results were expressed as brightness value L^*, red-green value a^* and yellow-blue value b^*, where L^* refers to lightness (0 is black and 100 is white), a^* indicates greenness ($a < 0$) or redness ($a > 0$) and b^* measures blueness ($b < 0$) or yellowness ($b > 0$) of samples. The colorimeter was calibrated using an instrument-configured whiteboard according to the manufacturer's instructions. The used illuminant for colour measurement was D65 combined with a standard 2° observer, port/viewing area of ⌀11 mm. Each sample was repeated six times. The calculation of whiteness was referred to as the method of Pathare, Opara and Al-Said [19], and the calculation (4) of whiteness is as follows:

$$\text{whiteness} = \sqrt{(100 - L^*)^2 + a^{*2} + b^{*2}} \quad (4)$$

2.3.3. Determination of Solubility of CO_2

As previously described, the solubility of CO_2 in fillets was determined using the method of Jakobsen and Bertelsen [20] with slight modifications. Two Brinell flasks were connected with latex tubes. A mixture of 25 mL of 12% perchloric acid and 25 mL of ethanol was added to one flask. 30 mL of 0.05 mol/L $Ba(OH)_2$ standard solution was added to the other flask. Then 20 g of the churned fish was transferred to the flask containing the mixture of perchloric acid and ethanol. CO_2 was released from the sample and dissolved into $Ba(OH)_2$ solution, and the flask system was kept connected for 18 h. The $Ba(OH)_2$ solution was rinsed from the flask with 20 mL of distilled water, filtered and the filter was rinsed with 10 mL of distilled water. The collected filtrate was titrated with 0.1 mol/L of HCl standard solution and three drops of phenolphthalein indicator. The solubility of CO_2 was calculated as mL CO_2/g sample. Samples were performed in triplicate.

2.4. Microbiological and Chemical Analysis of Golden Pompano Fillets

2.4.1. Determination of the Aerobic Plate Count (APC)

As per Yin et al. [13], 5 g of fillets was mixed with 45 mL of sterile saline. The content was diluted sequentially and 1 mL of each homogenate was added to a sterile culture dish. Plate count agar medium was added and the plate rotated to mix the content. After 72 h of incubation at 37 °C, the APC number was counted. The data were recorded as colony-forming units (CFU), expressed as log CFU/g. The samples were performed in triplicate.

2.4.2. Determination of Total Volatile Basic Nitrogen (TVB-N) Value

The TVB-N content of fillets was determined using a modified assay from Zhu et al. [16]. The 10 g samples were stirred and then 75 mL of distilled water was added to a reaction flask and mixed. The flask was then left to stand for 30 min before 10 mL of oxidase solution was added and the mixture was distilled using a Kjeldahl nitrogen analyzer (KDN-19A, Shanghai Qianjian Instrument Co., Ltd., Shanghai, China). The distillate was collected in a receiving flask to which 30 mL of 20 mg/mL boric acid solution and 3 drops of the mixed indicator were added, which was generated by dissolving 0.1 g methyl red and 0.1 g methylene blue into 100 mL of ethanol. Finally, the fractions were titrated with 0.1 mol/L hydrochloric acid standard solution. TVB-N values were expressed as mg/100 g of samples. Three replicate experiments were performed.

2.4.3. Determination of Thiobarbituric Acid-Reactive Substances (TBARs)

This was carried out according to the method of Cheng, Sun, Pu, Wang and Chen, [21] with slight modifications. Briefly, 5 g of fish samples were weighed and 25 mL of 7.5% trichloroacetic acid (containing 0.1% EDTA) was added, homogenized, shaken and extracted for 30 min. The mixture was then centrifuged for 5 min, 5 mL of supernatant was taken, and 5 mL of 0.02 mol/LTBA solution was added and heated in a boiling water bath for 20 min. The solution was left to turn pink and then cooled under running water. 5 mL of chloroform was then added and the mixture was shaken. This was then centrifuged for 5 min and the absorbance of the supernatant at 532 nm was measured. Three replicate experiments were conducted.

2.5. Statistical Analysis

The experiment was repeated three or six times and plotted using Origin 2021. The data were analyzed by one-way ANOVA and analysis of the significance of differences ($p < 0.05$) was carried out using SPSS 26.0. All the data were fitted to a normal distribution and correlation and dimensionality reduction analysis of indicators were performed using principal component analysis.

3. Results and Discussion

3.1. Effect of Different Gas Ratios on Water Retention of Golden Pompano Fillets

Drip loss will greatly affect consumers' willingness to buy in that it might be seen to affect the juiciness, flavor, appearance and texture of fillets [7]. Figure 2A shows that drip loss of fillets in all groups increased during storage. AP had lower drip loss than all MAP groups, indicating that CO_2 greatly affected the drip loss of fillets during superchilling storage. Generally, muscle water loss is mainly due to mechanical damage caused by ice crystallization and protein denaturation [22]. Meanwhile, CO_2 is readily dissolved in aqueous and lipid phases, to produce carbonic acid that reduces the pH and contributes to protein denaturation [23]. Kimbuathong et al. [22] showed that the ambient group had lower drip loss with white shrimp than the MAP group. Several studies have reported that high CO_2 gave higher drip loss and severe shrinkage of aquatic muscles due to reduced water holding capacity through protein decomposition [24,25]. Therefore, CO_2 in the MAP significantly affected the degradation and oxidation of protein and broke the muscle tissue structure. Figure 2B shows that centrifugal loss of golden fillets increased continuously in all groups during storage. The centrifugal loss of AP increased faster than in the MAP groups. The maximum centrifugal loss of AP was 22.56% on the 30th day. However, the centrifugal loss of MAP2 was the lowest on the 30th day with 18.58%, which was the opposite of the results obtained from the drip loss of AP and MAP2. This was probably due to the excess water already precipitated from the fillets during superchilling storage, resulting in less water content in fillets. As shown in Figure 2C, the cooking loss of fillets in all groups showed a downward trend, among which the cooking loss of fillets in MAP2 was generally lower than in other groups.

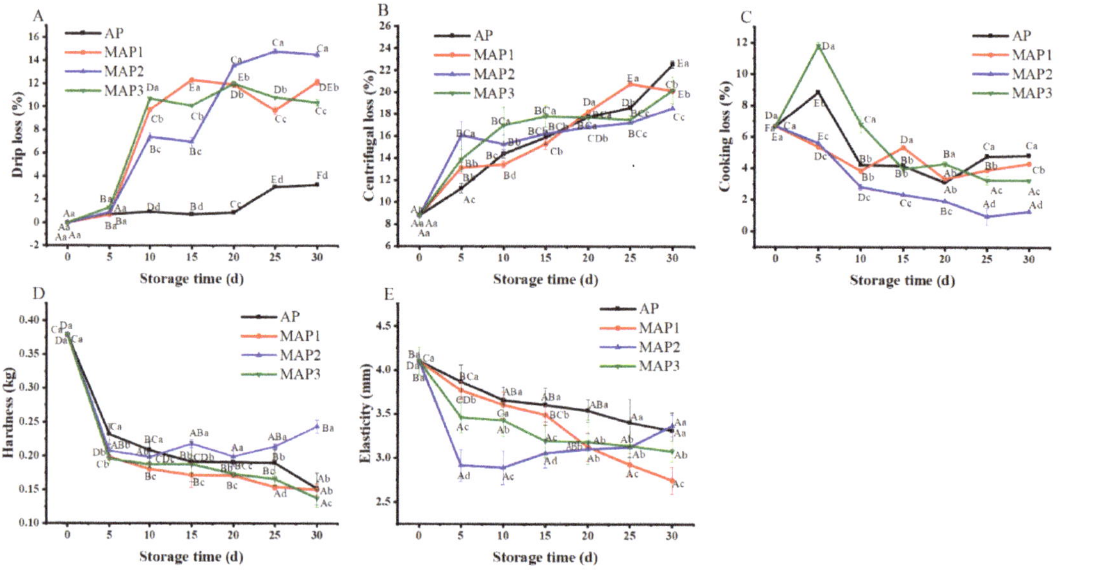

Figure 2. Effect of different gas components packaging on drip loss (**A**), centrifugal loss (**B**), cooking loss (**C**), hardness (**D**) and elasticity (**E**) of golden pompano fillets during storage (−3 °C). Packaging system: AP (control group), MAP1 (30% CO_2/70% N_2), MAP2 (50% CO_2/50% N_2), MAP3 (70% CO_2/30% N_2). Bars indicate the standard error. Different uppercase letters (A–F) indicate significant difference ($p < 0.05$) in means (n = 3) between display times within the same packaging system; different lowercase letters (a–d) indicate significant difference ($p < 0.05$) in means (n = 3) between packaging systems on the same day.

3.2. *Effect of Different Gas Ratios on Physicochemical Changes of Golden Pompano Fillets*

3.2.1. Changes in Texture

Autolysis occurs immediately after the death of aquatic products due to the redistribution of water in the muscle and protein oxidation denaturation [26]. The changes in muscle hardness values of golden pompano fillets during storage are shown in Figure 2D. The initial hardness value of fresh fillets was 0.379 ± 0.003 kg. A sharp decrease in all groups was found in the first 5 days ($p < 0.05$), then the hardness value of fillets with AP decreased continuously with the extension of storage time and reached its lowest value of 0.152 ± 0.024 kg at the end of the 30th day. Decreased hardness is attributed to protease activity released from microorganisms and changes that take place within proteins [26]. Nevertheless, the hardness value of the MAP groups had no significant difference during storage ($p < 0.05$). This indicates that the MAP had an excellent preservation effect on the hardness value of fillets. In Figure 2E, the overall trend of the elasticity value of fillets was similar to hardness. The initial value of the elasticity of fillets was 4.10 ± 0.16 mm. When stored to the 30th day, the elastic value of fillets in the AP was high at 3.31 ± 0.20 mm, and the low elastic value in the MAP1 was 2.75 ± 0.15 mm. The decrease of texture in all groups was insignificant ($p > 0.05$) due to the better resistance to freezing of golden pompano [27], and the good antibacterial effect of MAP with CO_2 during storage, which slowed down the growth of microorganisms in muscle. Meanwhile, the low temperature reduced the endogenous protease activities, thus inhibiting the oxidative denaturation of muscle.

3.2.2. Changes in Whiteness

During the storage process, fillets will produce a lot of metabolic accumulation substances or oxidation of fish protein and lipids due to microorganisms and endogenous enzymes, resulting in unsatisfying colors such as yellow, red and dark [28]. As shown in Figure 3A, the whiteness of fillets in all groups showed a trend of rising and then falling, and the whiteness of fillets in MAP3 rose to the highest on the 5th day of storage. The reasons for whiteness variation may be due to the changes in muscle structure and water-binding state in the fillets caused by the superchilling temperature, which led to the strengthening of scattering intensity and the increase of whiteness. The water floated on the surface of the fillets after thawing due to the decrease in the water holding power of the fillets, which enhanced the reflection of light [29]. At the end of storage on the 30th day, all the groups differed significantly ($p < 0.05$). The MAP2 dropped to the lowest whiteness value, contrary to that, MAP2 had the highest drip loss during storage. It is demonstrated that the strength of muscle water-holding capacity affects the changes in color and brightness of fillets. Overall, the MAP1 with 30% CO_2/70% N_2 gas ratio had the slightest changes in whiteness value during storage, which maintained the best whiteness.

3.2.3. Changes in CO_2 Solubility

The pH and acid-base equilibrium were closed related to the solubility of CO_2 in fillets. As shown in Figure 3B, the solubility of CO_2 in all groups showed an increasing trend in the early stage (first 10 days) of storage, among which the AP, MAP1 and MAP2 showed a fast rate of increase. In contrast, the increase rate of CO_2 solubility in the MAP1 group was slow and much lower than in the other three groups. In the middle and late storage stages (10–30 days), the increase of CO_2 solubility in all groups tended to level off with little difference ($p > 0.05$), indicating that saturation was reached in the presence of CO_2. Moreover, the solubility of CO_2 was basically in the range of 2.82 ± 0.03~4.33 ± 0.34 mL/g, which was 87.22~90.05% higher than the initial CO_2 solubility value.

3.2.4. Changes in pH Value

Generally speaking, the pH value of aquatic animals shows a decreasing trend and then increases after death. As shown in Figure 3C, the pH value of AP dropped to its lowest level on the 5th day of storage, then the pH value increased with storage time. The rate of pH increase of AP was significantly greater than MAP groups ($p < 0.05$). The initial decreases in pH might be related to the production of lactic acid and inorganic phosphate, while the increases during the later stages of storage may be due to the accumulation of ammonia and trimethylamine resulting from autolytic and microbial reactions [7,21]. The MAP groups had different degrees of decline in the first 10 days of storage, of which the MAP2 had the most significant decline ($p < 0.05$). This is probably due to the dissolution of some CO_2 gas into the fillets or the difference in its material. The pH of the MAP groups started to rise slowly after 10 days, suggesting that MAP can better prevent microbial spoilage, which may lead to alkaline components, such as ammonia and trimethylamine [7]. Overall, the MAP3 with 70% CO_2/30% N_2 was more suitable for storage under superchilling (−3 °C).

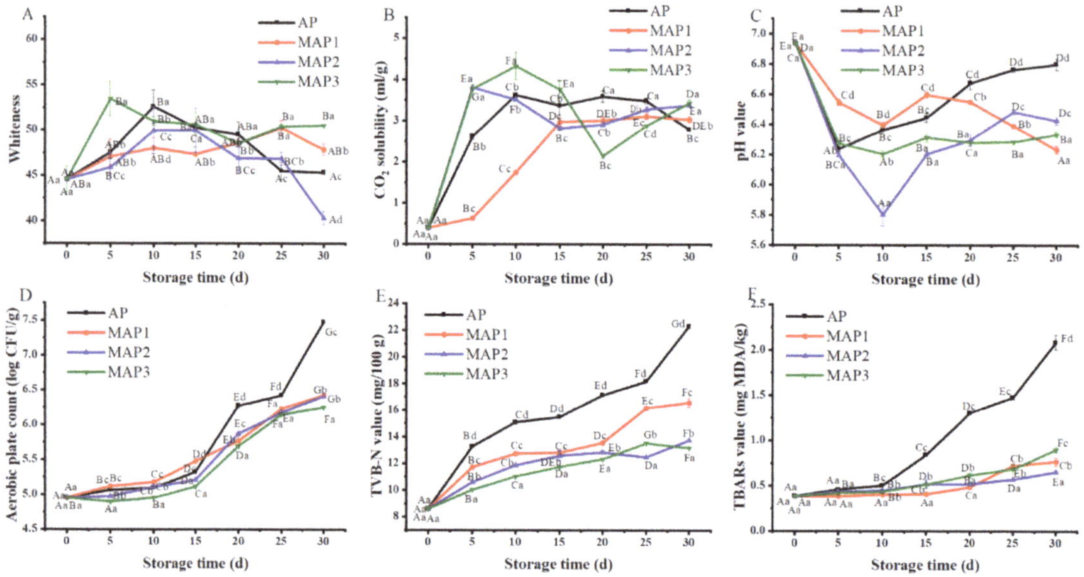

Figure 3. Effect of different gas components packaging on whiteness (**A**), CO_2 solubility (**B**) and pH value (**C**), APC (**D**), TVB-N value (**E**) and TBARs value (**F**) of golden pompano fillets during storage (−3 °C). Packaging system: AP (control group), MAP1 (30% CO_2/70% N_2), MAP2 (50% CO_2/50% N_2), MAP3 (70% CO_2/30% N_2). Bars indicate the standard error. Different uppercase letters (A–G) indicate significant difference ($p < 0.05$) in means (n = 3) between display times within the same packaging system; different lowercase letters (a–d) indicate significant difference ($p < 0.05$) of means (n = 3) between packaging systems on the same day.

3.2.5. Changes in TVB-N Value

TVB-N is the collective name of the protein decomposition of aquatic products in the storage process generated by ammonia, amines and other volatile alkaline nitrogenous substances [30]. As shown in Figure 3E, the TVB-N values of all groups increased and did not exceed 30 mg/100 g until the storage end. The fastest increase of TVB-N values could be seen in AP, followed by the MAP. The TVB-N values of the AP reached their highest level of 22.33 ± 0.16 mg/100 g on the 30th day of storage. This phenomenon might be due to formatted lactic acid dissolved in the filletswhich would result in a more remarkable pH change and cause a rise in TVB-N value [31]. However, the slowest increase was that of the MAP groups, and the highest TVB-N value in MAP1 was only 16.59 ± 0.34 mg/100 g, far below the AP. This is mainly because the higher CO_2 content in MAP could suppress bacterial activity and delay chemical reactions [15]. According to the National Aquatic Industry Standard-Fresh and Frozen Pomfret (SC/T 3103-2010 standard), the content of the TVB-N does not exceed 18 mg/100 g for first-grade freshness and does not exceed 30 mg/100 g for second-grade freshness. All MAP groups of fillets maintained first-grade freshness throughout the storage, while the AP samples decreased to second-grade freshness on about the 25th day. In terms of TVB-N content, compared to the AP, MAP with superchilling storage could extend the shelf life of fillets by 5 days. The MAP3 with 70% CO_2/30% N_2 gas ratio had the best results, which were similar to the results of Lan, Xie, Zhu and Zhu [32], who found that the MAP could extend the shelf life of refrigerated pomfret by about 2 ~ 8 days and that the gas-conditioning ratio of 80% CO_2/20% O_2 was the best.

3.2.6. Changes in TBARs Value

TBARs value is commonly used to determine the degree of fatty acid oxidation of aquatic products [29]. TBARs values indicate lipid oxidation in fillets (Figure 3F). The MAP can effectively inhibit the lipid oxidation of fillets and the trend of TBARs values in AP was similar to that of TVB-N during storage. It shows a slow increase followed by a rapid increase, while the TBARs values in the MAP increased slowly throughout the storage process. There were almost no significant changes in the TBARs value of all groups in the first 10 days ($p > 0.05$), indicating that the initial period of storage (first 10 days) was the induction period of fatty acid oxidative decay [26]. After 10 days of storage, TBARs values in AP increased rapidly due to the contact area of fatty acids with O_2 promoting the lipid oxidation in fish through the destruction of muscle cells [33,34]. The MAP groups maintained a low growth rate in the late storage period (20~30 days), indicating that CO_2 and N_2 could effectively inhibit lipid oxidation in fillets. The best effect of inhibition of lipid oxidation was found in MAP3. An increase in CO_2 tended to give lower TBARs values, possibly due to limited microbial growth and the release of a lipolytic enzyme [22].

3.3. Effect of Different Gas Ratios on APC of Golden Pompano Fillets

Bacterial growth was one of the main factors causing the spoilage of aquatic products [35,36]. Bacteria can use the rich protein, glycogen and other nutrients in aquatic products to reproduce during storage, increasing APC [35]. As shown in Figure 3D, the APC of all groups showed slow growth in the first 15 days and rapid growth in later storage. In many countries, the microbial guideline for fish and fish products indicates that 6.7 log units represent the consumption limit. However, the APC number alone cannot be used as an absolute inspection standard [6]. According to this guideline, the shelf life of AP was less than about 26 days (6.75 log CFU/g at the 26 day), whereas the MAP was still microbiologically acceptable after 30 days of storage (6.43, 6.41, 6.22 log CFU/g, respectively). Esteves, Guerra and Anibal [37] noted that the APC number of the MAP was always lower than 7.0 log CFU/g in the fresh samples edible line. López-Caballero, Gonçalves and Nunes [38] found that MAP with 80% CO_2/5% O_2/15% N_2 and 60% CO_2/5% O_2/35% N_2 effectively inhibited microbial growth in the first 3 days. The MAP groups can extend shelf life by about 4 days in terms of the APC indicator. The MAP3 with 70% CO_2/30% N_2 had the lowest APC number. These results showed that higher CO_2 tended to limit microbial growth. CO_2 extended the lag phase, reduced the growth rate of microorganisms in their logarithmic phase and inhibited the activity of succinate dehydrogenase and malate dehydrogenase enzymes in the Krebs Cycle of microorganisms [39,40].

3.4. Principal Component Analysis of Quality Indicators of Golden Pompano Fillets

3.4.1. Pearson Correlation Analysis of Quality Indicators

As shown in Table 1, correlation analysis was used to investigate the correlation between CO_2 solubility, pH, drip loss, centrifugal loss, hardness, APC, TVB-N and TBARs. The strength of the correlations was expressed using Pearson correlation coefficients. Specifically, the correlation coefficients between CO_2 solubility, pH, drip loss, centrifugal loss, and hardness were −0.611, 0.369, 0.649 and −0.663. It showed significance at the 0.01 level, thus indicating a highly significant correlation with these physical indicators. As the CO_2 solubility in fillets increased, the pH and water retention decreased, resulting in higher drip loss and centrifugal loss, lower hardness and deterioration of the quality of fillets. The correlation coefficient values in APC, TVB-N and TBARs were all greater than 0.800 and showed a significance at the level of 0.01. This proved that there was an extremely significant positive correlation with these indicators. With the enhancement of microbial and enzymatic decomposition, the content of TVB-N and TBARs of fillets showed a significant increasing trend. The growth of microorganisms during the storage of fillets had a significant effect on the chemical quality deterioration of fillets. The correlation coefficient of TVB-N and TBARs was 0.858, proving that protein decomposition and lipid oxidation of fillets affected each other to a great extent. The protein decomposition promoted lipid oxidation and vice versa. The correlation coefficients in centrifugal loss, APC, TVB-N and TBARs were all greater than 0.500, with a significance level of 0.01. It suggested that the decreased water retention of fillets is related to microorganisms, endogenous enzymes and oxidation actions.

Table 1. Pearson correlation analysis of quality indicators in golden pompano fillets during storage.

Quality Indicator	CO_2 Solubility	pH	Drip Loss	Centrifugal Loss	Hardness	APC	TVB-N	TBARs
CO_2 solubility	1.000							
pH	−0.611 **	1.000						
Drip loss [1]	0.369 **	−0.435 **	1.000					
Centrifugal loss [2]	0.649 **	−0.370 **	0.555 **	1.000				
Hardness [3]	−0.663 **	0.617 **	−0.513 **	−0.795 **	1.000			
APC [4]	0.266 *	0.082	0.407 **	0.745 **	−0.495 **	1.000		
TVB-N [5]	0.404 **	−0.063	0.150	0.720 **	−0.644 **	0.817 **	1.000	
TBARs [6]	0.251 *	0.227 *	−0.068	0.579 **	−0.373 **	0.813 **	0.858 **	1.000

* indicates highly significant correlation at the $p < 0.05$ level. ** indicates highly significant correlation at the $p < 0.01$ level. [1] Defreeze lost water after storage. [2] Water holding of muscles. [3] Force necessary to obtain a given deformation.[4] Anaerobic plate count. [5] Total volatile basic nitrogen. [6] Thiobarbituric acid-reactive substances.

3.4.2. Principal Component Analysis of Quality Indicators

The principal component analysis was conducted using SPSS 26.0 software. The characteristic roots, variance contribution rates and cumulative variance contribution rates of each principal component were obtained. As shown in Table 2, there were two principal components with characteristic roots greater than 1.000 and the sum of the contribution of the first and second principal components was 80.231%. The two principal components could reflect most of the information of the original variables. From Table 3, it can be seen that the eight indicators have a significant correlation and information overlap. Therefore, the indicators during storage were reduced from the initial eight to two principal components to achieve the purpose of dimensionality reduction. CO_2 solubility, pH, drip loss, centrifugal loss, hardness, APC, TVB-N and TBARs all have high loadings on the first and second principal components, indicating that the two principal components can mainly reflect the information of these indicators. The score coefficients indicate the degree of influence of each indicator on the principal components, and the score coefficients can be used to linearly combine each variable to establish the functional relationships between the first principal component (Y_1) and the second principal component (Y_2). The indicators are CO_2 solubility (X_1), pH (X_2), drip loss (X_3), centrifugal loss (X_4), hardness (X_5), APC (X_6), TVB-N (X_7) and TBARs (X_8). The score coefficient model for the eight variables was obtained from Table 3. Thus, the first principal component was extracted as Equation (5) and the second principal component score coefficient model was Equation (6).

$$Y_1 = 0.448X_1 - 0.412X_2 + 0.403X_3 + 0.386X_4 + 0.332X_5 + 0.225X_6 - 0.193X_7 + 0.323X_8 \quad (5)$$

$$Y_2 = -0.022X_1 + 0.214X_2 + 0.296X_3 + 0.332X_4 - 0.301X_5 - 0.335X_6 + 0.573X_7 + 0.480X_8 \quad (6)$$

In the principal component loading matrix of Table 3, the absolute values of detection reflect the contribution to the principal components, and the magnitude of contribution in the first principal component was centrifugal loss > hardness > TVB-N > APC > CO_2 solubility > TBARs > drip loss > pH. The absolute values of component coefficients of centrifugal loss rate, hardness, TVB-N and APC indicators were all greater than 0.800 ($p < 0.01$), and the absolute values of CO_2 solubility and TBARs indicator were all greater than 0.600 ($p < 0.01$). This indicated that they were highly correlated with the first principal component. The first principal component mainly reflected the degree of microbial proliferation, protein hydrolysis and oxidation of fillets. The contribution of the second principal component was pH > TBARs > drip loss > APC > CO_2 solubility > TVB-N > hardness >

centrifugal loss. The absolute values of the component coefficients of pH, TVB-N, and TBARs were all greater than 0.600 ($p < 0.01$). This certainly reflected the high correlation between protein denaturation and lipid oxidation.

Table 2. Principal component analysis of quality indicator in golden pompano fillets during storage.

Component [1]	Initial Eigenvalueinitial Eigenvalue			Extract Square and Load		
	Latent Root	Variance Contribution/%	Accumulative Variance Contribution/%	Latent Root	Variance Contribution/%	Accumulative Variance Contribution/%
1	4.388	54.847	54.847	4.388	54.847	54.847
2	2.031	25.384	80.231	2.031	25.384	80.231
3	0.808	10.101	90.332			
4	0.350	4.381	94.713			
5	0.168	2.102	96.815			
6	0.132	1.652	98.467			
7	0.076	0.950	99.417			
8	0.047	0.583	100.00			

[1] Comprehensive indicators formed by data conversion and dimensionality reduction for eight quality indicators.

Table 3. Loading matrix for principal component analysis of quality indicator in golden pompano fillets during storage.

Quality Indicator	Component 1 Coefficient	Coefficient of Component 1 Score	Component 2 Coefficient	Coefficient of Component 2 Score
CO_2 solubility	0.941	0.448	−0.046	−0.022
pH	−0.864	−0.412	0.305	0.214
Drip loss [1]	0.845	0.403	0.422	0.296
Centrifugal loss [2]	0.808	0.386	0.459	0.322
Hardness [3]	0.695	0.332	−0.429	−0.301
APC [4]	0.534	0.255	−0.478	−0.335
TVB-N [5]	−0.405	−0.193	0.816	0.573
TBARs [6]	0.677	0.323	0.684	0.480

[1] Defreeze lost water after storage. [2] Water holding of muscles. [3] Force necessary to obtain a given deformation. [4] Anaerobic plate count. [5] Total volatile basic nitrogen. [6] Thiobarbituric acid-reactive substances.

4. Conclusions

The quality changes of golden pompano fillets in AP and MAP with different gas ratios were evaluated at superchilling (−3 °C) storage. The results showed that MAP could effectively slow down the increase of pH, TVB-N value, TBARs value, hardness, elasticity, whiteness and APC of golden pompano fillets compared to AP. However, the water retention (drip loss, centrifugal loss, cooking loss) was poor in MAP. This phenomenon is not conducive to the sensory experience of packing fillets, making consumers lose interest. Pearson analysis showed that CO_2 solubility was significant, positively correlated with drip loss and centrifugal loss and negatively correlated with pH and hardness ($p < 0.01$). Moreover, the coefficient values of the interrelationships among APC, TVB-N and TBARs were all greater than 0.800. This showed a significantly positive correlation ($p < 0.01$), which reflected the close relationship between microorganisms and deterioration of chemical quality. These quality indicators could be simplified into two principal components by PCA with the cumulative variance contributions of 54.847% and 80.231%, respectively, reflecting the original information better. A score coefficient model of the principal components was established, providing a reference for comprehensive quality evaluation of fillets. In conclusion, superchilling (−3 °C) combined with MAP could effectively maintain the quality of golden pompano fillets, and the gas ratio of 70% CO_2/30% N_2 has the best comprehensive freshness preservation effect.

Author Contributions: Conceptualization, X.Z.; Data curation, X.Z.; Formal analysis, X.Z. and C.P.; Funding acquisition, S.C.; Investigation, X.Z. and C.P.; Methodology, X.Z. and C.P.; Project administration, S.C. and Y.W. (Yanyan Wu); Resources, S.C. and Y.X.; Supervision, S.C. and Y.W. (Yueqi Wang); Validation, C.P.; Visualization, Y.X.; Writing—original draft, X.Z.; Writing—review & editing, X.Z. and C.P. All authors have read and agreed to the published version of the manuscript.

Funding: Chinese Academy of Fishery Sciences (2020TD69), China Agriculture Research System (CARS-47), 2020 Research Program of Sanya Yazhou Bay Science and Technology City (SKJC-2020-02-013). Special Funds for Promoting Economic Development in Guangdong Province (For Modern Fishery) (YueNong 2019B14).

Institutional Review Board Statement: Not applicable.

Informed Consent Statement: Informed consent was obtained from all subjects involved in the study.

Data Availability Statement: The data presented in this study are available on request from the corresponding author. The data are not publicly available due to important national-level projects.

Acknowledgments: Special Scientific Research Funds partially supported this work for Central Non-profit Institutes.

Conflicts of Interest: The authors declare no conflict of interest.

References

1. Li, P.; Zhou, L.; Ni, S.; Xu, M.; Yu, Y.; Cai, J.; Wei, S.; Qin, Q. Establishment and characterization of a novel cell line from the brain of golden pompano (*Trachinotus ovatus*). *In Vitro Cell. Dev. Biol. Anim.* **2016**, *52*, 410–418. [CrossRef]
2. Qiu, X.; Chen, S.; Liu, G.; Lin, H. Characterization of Farmed Ovate Pompano (*Trachinotus ovatus* Linnaeus) Freshness during Ice Storage by Monitoring the Changes of Volatile Profile. *FSTR* **2014**, *20*, 79–84. [CrossRef]
3. Magnussen, O.M.; Haugland, A.; Torstveit Hemmingsen, A.K.; Johansen, S.; Nordtvedt, T.S. Advances in superchilling of food—Process characteristics and product quality. *Trends Food Sci. Technol.* **2008**, *19*, 418–424. [CrossRef]
4. Mi, H.; Qian, C.; Zhao, Y.; Liu, C.; Mao, L. Comparison of Superchilling and Freezing on the Microstructure, Muscle Quality and Protein Denaturation of Grass Carp (*Ctenopharyngodon idellus*): Superchilling and Freezing of Grass Carp. *J. Food Process. Preserv.* **2013**, *37*, 546–554. [CrossRef]
5. Banerjee, R.; Maheswarappa, N. Superchilling of muscle foods: Potential alternative for chilling and freezing. *Crit. Rev. Food Sci. Nutr.* **2017**, *59*, 1256–1263. [CrossRef]
6. Liu, Q.; Kong, B.; Han, J.; Chen, Q.; He, X. Effects of superchilling and cryoprotectants on the quality of common carp (*Cyprinus carpio*) surimi: Microbial growth, oxidation, and physiochemical properties. *LWT Food Sci. Technol.* **2014**, *57*, 165–171. [CrossRef]
7. Liu, D.; Liang, L.; Xia, W.; Regenstein, J.M.; Zhou, P. Biochemical and physical changes of grass carp (*Ctenopharyngodon idella*) fillets stored at −3 and 0 °C. *Food Chem.* **2013**, *140*, 105–114. [CrossRef]
8. Allende, A.; Oñez, A.; Bolton, D.; Chemaly, M.; Davies, R.; Cesare, A.; Bover-Cid, S. The use of the so-called 'superchilling' technique for the transport of filletsery products. *EFSA J.* **2021**, *19*, 6378. [CrossRef]
9. Wu, C.; Yuan, C.; Ye, X.; Hu, Y.; Chen, S.; Liu, D. A Critical Review on Superchilling Preservation Technology in Aquatic Product. *J. Integr. Agric.* **2014**, *13*, 2788–2806. [CrossRef]
10. Garrido, M.D.; Hernández, M.D.; Espinosa, M.C.; López, M.B. Enhanced Quality Characteristics of Refrigerated Seabream (*Sparus aurata*) Fillets Packed Under Different Systems (Modified Atmosphere vs. Vacuum). *J. Aquat. Food Prod. Technol.* **2016**, *25*, 156–168. [CrossRef]
11. Qiu, L.; Zhang, M.; Tang, J.; Adhikari, B.; Cao, P. Innovative technologies for producing and preserving intermediate moisture foods: A review. *Food Res. Int.* **2019**, *116*, 90–102. [CrossRef]
12. Bassey, A.P.; Chen, Y.; Zhu, Z.; Odeyemi, O.A.; Gao, T.; Olusola, O.O.; Ye, K.; Li, C.; Zhou, G. Evaluation of spoilage indexes and bacterial community dynamics of modified atmosphere packaged super-chilled pork loins. *Food Control* **2021**, *130*, 108383. [CrossRef]
13. Yin, C.; Wang, J.; Qian, J.; Xiong, K.; Zhang, M. Quality changes of rainbow trout stored under different packaging conditions and mathematical modeling for predicting the shelf life. *Food Packag. Shelf Life* **2022**, *32*, 100824. [CrossRef]
14. Mei, J.; Liu, F.; Fang, S.; Lan, W.; Xie, J. High-CO_2 Modified Atmosphere Packaging with Superchilling (−1.3 °C) Inhibit Biochemical and Flavor Changes in Turbot (*Scophthalmus maximus*) during Storage. *Molecules* **2020**, *25*, 2826. [CrossRef]
15. Sivertsvik, M.; Rosnes, J.T.; Kleiberg, G.H. Effect of Modified Atmosphere Packaging and Superchilled Storage on the Microbial and Sensory Quality of Atlantic Salmon (*Salmo salar*) Fillets. *J. Food Sci.* **2003**, *68*, 1467–1472. [CrossRef]
16. Zhu, Y.; Ma, L.; Yang, H.; Xiao, Y.; Xiong, Y.L. Super-chilling (−0.7 °C) with high-CO_2 packaging inhibits biochemical changes of microbial origin in catfish (*Clarias gariepinus*) muscle during storage. *Food Chem.* **2016**, *206*, 182–190. [CrossRef]
17. *China Fishery Statistical Yearbook*; China Statistics Press: Beijing, China, 2021; pp. 87–98.
18. Delles, R.M.; Xiong, Y.L. The effect of protein oxidation on hydration and water-binding in pork packaged in an oxygen-enriched atmosphere. *Meat Sci.* **2014**, *97*, 181–188. [CrossRef]

19. Pathare, P.B.; Opara, U.L.; Al-Said, F.A.-J. Colour Measurement and Analysis in Fresh and Processed Foods: A Review. *Food Bioprocess Technol.* **2013**, *6*, 36–60. [CrossRef]
20. Jakobsen, M.; Bertelsen, G. Predicting the amount of carbon dioxide absorbed in meat. *Meat Sci.* **2004**, *68*, 603–610. [CrossRef]
21. Cheng, J.-H.; Sun, D.-W.; Pu, H.-B.; Wang, Q.-J.; Chen, Y.-N. Suitability of hyperspectral imaging for rapid evaluation of thiobarbituric acid (TBA) value in grass carp (*Ctenopharyngodon idella*) fillet. *Food Chem.* **2015**, *171*, 258–265. [CrossRef]
22. Kimbuathong, N.; Leelaphiwat, P.; Harnkarnsujarit, N. Inhibition of melanosis and microbial growth in Pacific white shrimp (*Litopenaeus vannamei*) using high CO_2 modified atmosphere packaging. *Food Chem.* **2020**, *312*, 126114. [CrossRef] [PubMed]
23. Zhang, M.; Meng, X.; Bhandari, B.; Fang, Z. Recent Developments in Film and Gas Research in Modified Atmosphere Packaging of Fresh Foods. *Crit. Rev. Food Sci. Nutr.* **2016**, *56*, 2174–2182. [CrossRef] [PubMed]
24. Mastromatteo, M.; Danza, A.; Conte, A.; Muratore, G.; Del Nobile, M.A. Shelf life of ready to use peeled shrimps as affected by thymol essential oil and modified atmosphere packaging. *Int. J. Food Microbiol.* **2010**, *144*, 250–256. [CrossRef] [PubMed]
25. Qian, Y.-F.; Xie, J.; Yang, S.-P.; Wu, W.-H.; Xiong, Q.; Gao, Z.-L.; Shi, J.-B. Effect of CO_2 on Chemical and Microbial Changes of Pacific White Shrimp during Modified Atmosphere Packaging. *J. Aquat. Food Prod. Technol.* **2016**, *25*, 644–655. [CrossRef]
26. Thepnuan, R.; Benjakul, S.; Visessanguan, W. Effect of Pyrophosphate and 4-Hexylresorcinol Pretreatment on Quality of Refrigerated White Shrimp (*Litopenaeus vannamei*) Kept under Modified Atmosphere Packaging. *J. Food Sci.* **2008**, *73*, S124–S133. [CrossRef]
27. Yang, Z.; Liu, S.; Sun, Q.; Zheng, O.; Wei, S.; Xia, Q.; Ji, H.; Deng, C.; Hao, J.; Xu, J. Insight into muscle quality of golden pompano (*Trachinotus ovatus*) frozen with liquid nitrogen at different temperatures. *Food Chem.* **2022**, *374*, 131737. [CrossRef]
28. Erdağ, M.; Ayvaz, Z. The Use of Color to Determine Fish Freshness: European Seabass (*Dicentrarchus labrax*). *J. Aquat. Food Prod. Technol.* **2021**, *30*, 847–867. [CrossRef]
29. Messina, C.M.; Bono, G.; Renda, G.; La Barbera, L.; Santulli, A. Effect of natural antioxidants and modified atmosphere packaging in preventing lipid oxidation and increasing the shelf-life of common dolphinfish (*Coryphaena hippurus*) fillets. *LWT Food Sci. Technol.* **2015**, *62*, 271–277. [CrossRef]
30. Fu, L.; Zhang, S.; Zhou, J.; Liu, C.; Lin, J.; Wang, Y. Alterations of Protein Expression in the Muscle of Pacific White Shrimp (*Litopenaeus vannamei*) Contribute to Postmortem Changes. *J. Shellfish Res.* **2014**, *33*, 815–823. [CrossRef]
31. Genç, İ.Y.; Esteves, E.; Aníbal, J.; Diler, A. Effects of chilled storage on quality of vacuum packaged meagre fillets. *J. Food Eng.* **2013**, *115*, 486–494. [CrossRef]
32. Lan, W.Q.; Xie, J.; Zhu, F.; Zhu, R.Q. The Effects of Quality Changes in Pomfret (*Pampus argenteus*) Fillets from Different Treatments during Chilled Storage. *AMR* **2014**, *1051*, 378–382. [CrossRef]
33. Arashisar, Ş.; Hisar, O.; Kaya, M.; Yanik, T. Effects of modified atmosphere and vacuum packaging on microbiological and chemical properties of rainbow trout (*Oncorynchus mykiss*) fillets. *Int. J. Food Microbiol.* **2004**, *97*, 209–214. [CrossRef] [PubMed]
34. Zakrys, P.I.; Hogan, S.A.; O'Sullivan, M.G.; Allen, P.; Kerry, J.P. Effects of oxygen concentration on the sensory evaluation and quality indicators of beef muscle packed under modified atmosphere. *Meat Sci.* **2008**, *79*, 648–655. [CrossRef] [PubMed]
35. Daskalov, H. The importance of Aeromonas hydrophila in food safety. *Food Control* **2006**, *17*, 474–483. [CrossRef]
36. Fernández-Saiz, P.; Sánchez, G.; Soler, C.; Lagaron, J.M.; Ocio, M.J. Chitosan films for the microbiological preservation of refrigerated sole and hake fillets. *Food Control* **2013**, *34*, 61–68. [CrossRef]
37. Esteves, E.; Guerra, L.; Anibal, J. Effects of Vacuum and Modified Atmosphere Packaging on the Quality and Shelf-Life of Gray Triggerfish (*Balistes capriscus*) Fillets. *Foods* **2021**, *10*, 250. [CrossRef]
38. López-Caballero, M.; Gonçalves, A.; Nunes, M. Effect of CO_2/O_2-containing modified atmospheres on packed deepwater pink shrimp (Parapenaeus longirostris). *Eur. Food Res. Technol.* **2002**, *214*, 192–197. [CrossRef]
39. Singh, R.; Giri, S.K.; Kotwaliwale, N. Shelf-life enhancement of green bell pepper (*Capsicum annuum* L.) under active modified atmosphere storage. *Food Packag. Shelf Life* **2014**, *1*, 101–112. [CrossRef]
40. Zhang, J.; Song, S.; Li, D.; Luo, Y. Microbial communities and biogenic amines of crucian carp (*Carassius auratus*) fillets during partial freezing and chilled storage. *Int. J. Food Prop.* **2017**, *20*, S1053–S1064. [CrossRef]

MDPI
St. Alban-Anlage 66
4052 Basel
Switzerland
Tel. +41 61 683 77 34
Fax +41 61 302 89 18
www.mdpi.com

Foods Editorial Office
E-mail: foods@mdpi.com
www.mdpi.com/journal/foods

www.ingramcontent.com/pod-product-compliance
Lightning Source LLC
Chambersburg PA
CBHW061610220926
51850CB00005B/468